Chemistry
and Control
of Enzyme Reactions

Chemistry and Control of Enzyme Reactions

K. G. SCRIMGEOUR

Department of Biochemistry,
University of Toronto,
Canada

1977

ACADEMIC PRESS
London   New York   San Francisco

A Subsidiary of Harcourt Brace Jovanovich, Publishers

ACADEMIC PRESS INC. (LONDON) LTD.
24/28 Oval Road,
London NW1

United States Edition published by
ACADEMIC PRESS INC.
111 Fifth Avenue
New York, New York 10003

Library of Congress Catalog Card Number: 77 71802
ISBN: 0 12 634150 8

38386

Printed in Great Britain by C. F. Hodgson & Son Ltd., London

Preface

This book is a description of the properties of enzyme molecules. It is arranged as a series of general principles and representative examples, progressing from simple to complex properties. For clarity, I have used a descriptive approach and straightforward English. Readers will find this book useful as a guide to the original literature of enzymology.

Enzymology is an experimental science, and newer data may alter interpretations (sometimes only temporarily). Scientific evidence is not generated by apparatus, but it is the result of human planning and execution of experiments. Scientists take a personal experimental approach, and use slightly differing techniques. Because of this, controversies often arise when investigators try to relate their results to those of other researchers. Readers will encounter several unresolved controversies in this book and many projects that have not been completed. I have presented the pertinent facts available at present. In time, these problems will be resolved.

In the preparation of this book, I have been aided greatly by access to the thoughts of other enzymologists through their books and articles, and through seminars and symposia. I would first like to thank Frank Huennekens who taught me so much about science during my time as a student, and as a member of his department. I next wish to thank all my colleagues for helpful discussions. In particular, I thank John Mangum, Dieter Vonderschmitt, Peter Candido, Bruce Dunlap, Nicholai Gnuchaev, Theo Hofmann, Peter Lewis, Bob Murray, Gordon Tener and Jeffrey Wong. I would also like to thank A. J. Cornish-Bowden for his helpful and stimulating review of the manuscript. I have been revitalized each year by participation in the Gordon Research Conference on enzymes and I therefore acknowledge both the organizers of and contributors to this meeting. My thanks go to authors and journals for their permission to reproduce figures in this text. Preparation of this book would not

v

have been possible without the support of the Medical Research Council of Canada. The valuable collaborations I have had with my graduate students have been of benefit to me, and will always be pleasant to remember. Jim Charlton assisted in the preparation of the molecular models. A reference not found in the later part of the book has been my weekly guide: "Current Contents, Life Sciences." (Perhaps it could be described best as "uncited but not uncitable".) I am indebted to Lynn Hanley and Anne Tirpak for their tireless help in the preparation of this manuscript. I now realize why all authors thank their families in the Preface. I appreciate their patience over the past 3 years, and their preparation of the author index. I have not dedicated this book to any specific person. Instead, I would like to give credit to all the scientists and students who have contributed and will continue to contribute to our knowledge of enzyme chemistry.

April, 1976   *Gray Scrimgeour*

Contents

Abbreviations

Amino acids:

Ala	alanine	Leu	leucine
Arg	arginine	Lys	lysine
Asn	asparagine	Met	methionine
Asp	aspartate	Phe	phenylalanine
Cys	cysteine	Pro	proline
Gln	glutamine	Ser	serine
Glu	glutamate	Thr	threonine
Gly	glycine	Trp	tryptophan
His	histidine	Tyr	tyrosine
Ile	isoleucine	Val	valine

Å	ångström (10^{-10} m $= 10^{-1}$ nm)
ACP	acyl carrier protein
ADP	adenosine-5′-diphosphate
AMP	adenosine-5′-monophosphate, or adenylate
ATase	adenylyltransferase
ATCase	aspartate transcarbamylase
ATP	adenosine-5′-triphosphate
ATPase	adenosine triphosphatase
B1 and B2	subunits of ribonucleotide reductase
BCCP	biotin carboxyl carrier protein
cAMP	adenosine-3′,5-monophosphate, or cyclic AMP
C-chain	catalytic chain
CDP	cytidine-5′-diphosphate
CMC	critical micelle concentration
CM-cellulose	carboxymethyl-cellulose
CMP	cytidine-5′-monophosphate, or cytidylate
CoA or CoASH	coenzyme A
CTP	cytidine-5′-triphosphate

dATP	2′-deoxyadenosine-5′-triphosphate
dCTP	2′-deoxycytidine-5′-triphosphate
DEAE-cellulose	diethylaminoethyl-cellulose
DFP	diisopropyl phosphofluoridate (or fluorophosphonate)
dGTP	2′-deoxyguanosine-5′-triphosphate
DNP	2,4-dinitrophenyl
DPG	2,3-diphosphoglycerate
dTTP	thymidine-5′-triphosphate
dUMP	2′-deoxyuridine-5′-monophosphate, or deoxyuridylate
E	enzyme
E_0'	standard reduction potential at pH 7
EDTA	ethylenediaminetetraacetate
e.p.r. (or e.s.r.)	electron paramagnetic resonance (or electron spin resonance)
ES	enzyme–substrate complex
ETF	electron-transferring flavoprotein
FAD	flavin adenine dinucleotide
FdUMP	5-fluoro-2′-deoxyuridylate
FeS	iron–sulphur
FIGLU	formiminoglutamate
FMN	flavin mononucleotide
g	gram
g	unit of gravitational field (981 cm s^{-2})
G	free energy (Gibbs)
GDP	guanosine-5′-diphosphate
GlcNAc	N-acetylglucosamine
GMP	guanosine-5′-monophosphate, or guanylate
H	enthalpy
Hb	hemoglobin
his	histidine operon (genetic locus)
HPr	histidine-containing phosphocarrier protein
I	inhibitor
IMP	inosine-5′-monophosphate, or inosinate
k	velocity constant
K	equilibrium constant or binding constant
kcal	kilocalorie ($10^3 \times$ cal)
α-KG	α-ketoglutarate or 2-oxoglutarate

K_i	inhibitor constant (dissociation constant of EI complex)
K_m	Michaelis constant
lac	lactose operon (genetic locus)
LDH	lactate dehydrogenase
M	metal cofactor ion
M	molar, or moles per litre
mg	milligram (10^{-3} g)
ml	millilitre (10^{-3} l)
mM	millimolar (10^{-3} M)
μM	micromolar (10^{-6} M)
MurNAc	N-acetyl muramic acid
n	number of binding sites per oligomer, or number substituted
NAD$^+$	nicotinamide adenine dinucleotide
NADH	the reduced form of NAD$^+$
NADP$^+$	nicotinamide adenine dinucleotide phosphate
NADPH	the reduced form of NADP$^+$
nm	nanometre (10^{-9} m)
NMN	nicotinamide mononucleotide
n.m.r.	nuclear magnetic resonance
OAA	oxaloacetate
P	product; P, Q, R... used in multi-product reactions
P_i	orthophosphate
P_{II}	regulatory protein for glutamine synthetase modification
PEP	phosphoenol pyruvate
pK_a	negative logarithm of an acidic ionization constant
PLP	pyridoxal phosphate
PMP	pyridoxamine phosphate
PP_i	pyrophosphate
PRPP	5-phosphoribosyl-1-pyrophosphate
PRR	proton relaxation rate
PTH	3-phenyl-2-thiohydantoin
R-chain	regulatory polypeptide
R-state	high-affinity conformational state
RNase	ribonuclease
S	entropy
S	substrate; A, B, C... used in multi-substrate reactions

S, $S_{20,w}$	Svedberg unit (sedimentation coefficient), corrected to zero concentration at 20° C in water if specified $S_{20,w}$
SDS	sodium dodecyl sulphate
T	temperature
T-state	low-affinity conformational state
TMP	thymidine-5′-monophosphate, or thymidylate (same as dTMP)
TPP	thiamine pyrophosphate
Tris	tris(hydroxymethyl)aminomethane or 2-amino-2-hydroxymethylpropane-1,3-diol
UDP	uridine-5′-diphosphate
UDPG	uridine diphosphate glucose
UDPGal	uridine diphosphate galactose
UMP	uridine-5′-monophosphate, or uridylate
UTP	uridine-5′-triphosphate
v	initial velocity
V	maximal velocity
Xyl	xylose
Y	fraction of binding sites occupied

Part I

Fundamental Principles

1 Introduction

Enzymes, the protein molecules which catalyze the chemical reactions which support and reproduce life, are the centre of biochemistry and biology. Development of the enzyme theory at the end of the nineteenth century triggered the establishment of biochemistry. Now that enzymatic phenomena can be explained in chemical terms, this theory has become fact. In the past twenty years, knowledge of enzymes has progressed from a concept of proteins having specific catalytic activity to descriptions of macromolecular chemicals of defined structure, function and control. The rapid expansion of enzymology, the study of enzymes, has attracted scientists other than biochemists. Important research on the properties and uses of enzymes is now performed by many organic, inorganic, physical and industrial chemists, engineers, biologists, pharmacologists and health scientists in general.

My aim in this book is to transmit to students and research workers some of the fascination I have with enzymes. I am well aware of the great number of monographs and periodicals published on topics related to enzymology. There are also excellent books dealing with the chemistry of catalysis, but fewer devoted to either modern enzymology techniques or control of enzymic reactions. All three of these subjects are discussed here. The approach is an empirical one, using enzymes rather than theoretical models as examples of basic principles. I have tried to be as comprehensive as possible by discussing enzymes and coenzymes which illustrate a diversity of properties, mechanisms and experimental treatment. In view of the fact that all enzymology is based upon experiments, it would be most desirable to present experimentally derived data to support each major conclusion reached, but because of the limitations of space, this is not possible. Instead, some topics have been selected for fuller examination in preference to others. Study of these topics, including the original literature cited, should enable the reader to comprehend those topics which have been condensed. The references I have cited include many recent review articles, each containing an extensive list of references. Throughout the text, I have tried to refer to the most recent and most accessible papers and to fully documented accounts, rather than preliminary findings.

The sequence of chapters was planned to present gradually and logically the properties of enzymes, starting with the simpler principles and concluding with more complicated cases and problems. Within each topic, I have begun with a non-mathematical theoretical section, and followed this with appropriate examples of enzymatically catalyzed reactions. The text has three main divisions: an introductory section, outlining the properties of enzymes as proteins and the principles of catalysis, both non-enzymic and enzymic; a section dealing with examples of enzyme catalysis; and a section on control, describing biological regulations by control of enzyme activity.

Theoretical and practical material has been blended to assist all those studying enzymes, whether they are physical chemists, senior biochemistry students or clinical researchers. Those with a strong background in mechanistic theory should benefit from the examples and applications of theory. Others who have practical knowledge of enzymes may wish to gain insight into reaction mechanisms or to see how their own work relates to other enzyme chemistry. I have assumed that readers have completed an introductory course in biochemistry, and thus are familiar with the properties of proteins. Those not having studied biochemistry recently might profit from reading the section on proteins in a modern biochemistry textbook or by studying the monograph on protein chemistry by Haschemeyer and Haschemeyer (1973). The latter book fully explains many properties of enzymes in terms of protein structure. A sound background in organic and physical chemistry is desirable for complete appreciation of the descriptions of both the enzyme catalysts and the reactions they catalyze. When teaching enzyme chemistry, I have found it necessary to review basic chemical concepts; several chapters contain these reviews, which should aid readers who have relatively little chemical experience. I have not included lengthy descriptions of physical techniques, as these are available in general texts.

The remainder of this chapter is a résumé of the terminology used in enzyme chemistry, followed by a list of general references. I have furnished this material here so that the reader may be adequately prepared to tackle enzyme chemistry in the following chapters.

TERMINOLOGY

An *enzyme* is a protein molecule having both catalytic activity and specificity for its substrate or substrates. An enzyme does not perform a reaction, but it catalyzes the performance of the reaction. For example, lactate dehydrogenase does not reduce pyruvate but catalyzes the reduction of pyruvate by NADH. Although many enzymes are composed of just a protein unit, others require the presence of cofactors. A cofactor-requiring enzyme which has all its components present is called a *holoenzyme*. If the enzyme is lacking its cofactors, it is called an *apoenzyme*. The adjectives *enzymic* and *enzymatic* are short for "enzyme-catalyzed", and these three terms will be used interchangeably. Some enzymes are synthesized as inactive precursor molecules called *zymogens* or *proenzymes*. The conversion of a zymogen into an active enzyme is called *activation*. Another use of the word activation, related to an increase in the activity of a regulatory enzyme, should be easily recognized from the context.

A *substrate* is a reactant in an enzyme-catalyzed (or even non-enzymatic) reaction. Because most reactions are at least formally reversible, products of reactions can also be substrates. However, the term *product* refers to a compound formed in the normal forward direction of a reaction. A *ligand* is any compound which binds to an enzyme. The concentration of a ligand may or may not change during the reaction. An *inhibitor* is a compound that by binding to an enzyme slows the rate of the reaction which the enzyme is catalyzing.

The word *cofactor* is a general term for compounds other than the apoenzyme which are essential for full catalytic activity. Cofactors include both coenzymes and essential ions. *Coenzymes* are group-transferring reagents, carriers of mobile metabolic groups or oxidation-reduction equivalents. Coenzymes usually are low molecular weight organic compounds (Chapter 8), but some coenzymes are protein molecules (Chapter 9). Coenzymes can be divided into two types: cosubstrates and prosthetic groups. The word *cosubstrate* indicates that the coenzyme, carrying its metabolic group, leaves the catalytic site of one enzyme and is a substrate for a reaction catalyzed by another enzyme. A *prosthetic group* is a coenzyme bound to an enzyme throughout the cycle of catalytic action of the enzyme. In other words, it is a component of the active site. Some prosthetic groups are not coenzymes, but are possibly derived from amino acids of the polypeptide chains of enzymes. Metal ions which participate in the catalytic mechanisms of metalloenzymes are also prosthetic groups. The presence of specific cations or anions often is essential for the activity of an enzyme. These ions have often been called activators, but here will be referred to as ionic cofactors.

The name of an enzyme is usually derived from the name of its substrate and the type of reaction catalyzed, followed by the suffix "-ase". This style of nomenclature is also used for general descriptions of enzyme groups. For example, a *hydrolase* catalyzes a hydrolysis reaction. If the substrate of the hydrolase is the peptide bond of a protein, then the enzyme is called a *protease* or *peptidase*. The even more specific name *carboxypeptidase* indicates that the enzyme catalyzes the hydrolysis of the peptide bond at the carboxyl terminal of a protein. Both carboxypeptidases and aminopeptidases are called *exopeptidases* because they catalyze the hydrolysis of terminal peptide bonds. Proteases which specifically catalyze cleavage of interior peptide bonds are called *endopeptidases*. The International Union of Biochemistry (1961) has adopted a systematic classification of enzymes recommended by an International Commission on Enzymes. The Commission drew up a list of rules for naming and numbering enzymes, and classified over 800 enzymes. In the latest tabulation (Florkin and Stotz, 1973), the total has reached almost 1800 enzymes. The Commission also provided both systematic and trivial names for most enzymes. Trivial names, some differing slightly from those recommended by the Commission, are used in this book because of their common usage in the literature.

The Commission on Enzymes classified enzymes into six main divisions based on the types of reactions they catalyze. *Oxidoreductases* catalyze oxidation and reduction reactions. The name usually applied to an enzyme in this class is dehydrogenase. Most dehydrogenases utilize pyridine nucleotides, flavins or cytochromes as electron acceptors or donors. Other oxidoreductases are the oxidases and hydroxylases which have O_2 as an electron acceptor, and peroxidases which have H_2O_2 as the electron acceptor. *Transferases* are enzymes catalyzing group-transfer reactions. Most transferases either require a coenzyme or form covalent enzyme–substrate intermediates. *Hydrolases* catalyze hydrolytic reactions, and are classified by the bond hydrolyzed. Quite a few peptidases have retained trivial names that pre-date the knowledge of their substrate specificity. *Lyases* catalyze the non-hydrolytic removal of substituent groups. Lyases include decarboxylases and aldolases. *Isomerases* catalyze isomerization reactions, such as racemization, isomerization of double bonds or rearrangement of the carbon skeleton. *Ligases,* more often called synthetases, catalyze synthetic reactions in which the potential energy of ATP is used for bond formation.

The abbreviations used are explained when they are first introduced in the text, and are listed on pp. xiii–xvi at the beginning of the book.

GENERAL SUPPLEMENTARY READING

There are many excellent sources of review literature on the properties of enzymes. I have compiled a short list of some modern monographs and relevant review periodicals here. The references cited in this list are suitable either for clarification of simple points or for expanded discussion of the sub-topics in the list. Other references, both general and specific, are presented in each chapter.

Monographs

General biochemistry texts:

> "Biochemistry" A. L. Lehninger (1975). Second edition. Worth, New York.
> "Biological Chemistry" H. R. Mahler and E. H. Cordes (1971). Second edition. Harper and Row, New York.

General enzymology and enzyme mechanisms:

> "Enzymes" M. Dixon and E. C. Webb (1964). Second edition. Academic Press, New York and London.
> "Enzyme-catalysed Reactions" C. J. Gray (1971). Van Nostrand Reinhold, London.
> "Enzymic Catalysis" J. Westley (1969). Harper and Row, New York.
> "The Study of Enzyme Mechanisms" E. Zeffren and P. L. Hall (1973). Wiley-Interscience, New York.

Related chemical mechanisms:

"Bio-organic Mechanisms" T. C. Bruice and S. J. Benkovic (1966). Volumes I and II. W. A. Benjamin, New York.

"Catalysis in Chemistry and Enzymology" W. P. Jencks (1969). McGraw-Hill, New York.

"Mechanisms of Homogeneous Catalysis from Protons to Proteins" M. L. Bender (1971). Wiley-Interscience, New York.

Physical chemistry of enzymes:

"Enzymes: Physical Principles" H. Gutfreund (1972). Wiley-Interscience, London.

Periodicals

General:

"The Enzymes" P. D. Boyer, ed. Third edition. Academic Press, New York and London.

"Advances in Enzymology" A. Meister, ed. Interscience, New York.

"Annual Review of Biochemistry" E. E. Snell, ed. Annual Reviews, Palo Alto.

Techniques:

"Methods in Enzymology" S. P. Colowick and N. O. Kaplan, eds. Academic Press, New York and London.

Mechanism:

"Progress in Bio-organic Chemistry" E. T. Kaiser and F. J. Kézdy, eds. Wiley, New York.

Control:

"Current Topics in Cellular Regulation" B. L. Horecker and E. R. Stadtman, eds. Academic Press, New York and London.

"Advances in Enzyme Regulation" G. Weber, ed. Pergamon, New York.

REFERENCES

Florkin, M. and Stotz, E. H. (1973). "Comprehensive Biochemistry" Third edition, Vol. 13 "Enzyme Nomenclature". Elsevier, Amsterdam.

Haschemeyer, R. H. and Haschemeyer, A. E. V. (1973). "Proteins: A Guide to
    Study by Physical and Chemical Methods". John Wiley and Sons, New York.
Report of the Commission on Enzymes of the International Union of Biochem-
    istry (1961). Pergamon Press, Oxford.

2 Enzyme Structure

ENZYMES AS PROTEINS

The remarkable properties of enzymes have astounded biologists for over a century. A hundred years ago, a dispute which lasted for about 30 years started, with the vitalists stating that living yeast cells were required for the process of fermentation of sugars, and the mechanists saying that there were enzymes in the yeast cells which were chemicals catalyzing the reactions in the fermentation. In fact, the name "enzyme" is derived from the Greek for "in yeast" (*zymé*), and indicates that the catalysts were present intracellularly. After it had been shown that cell-free extracts could catalyze metabolic reactions, it was still many years before the chemical nature of enzymes was proved. In the 1920s, the first purifications of enzymes were achieved, and the first crystalline enzyme—urease from the jack bean—was obtained by Sumner (1926). Sumner showed that the crystalline globulin which he had prepared had extremely high enzymatic activity when it was dissolved.* In the following ten years, five more enzymes (pepsin, trypsin, chymotrypsin, carboxypeptidase and Old Yellow enzyme, a flavin-containing NADPH oxidase) were purified, crystallized, and found to be proteins also (Northrop *et al.*, 1948). Despite the scepticism of many eminent scientists then, and even later, all purified enzymes have been shown to be proteins. Some enzymes are simple polypeptides while others are complexes containing several polypeptide chains. Some require cofactors or are bound to cellular membranes, but the catalytic activity of all enzymes is a property of the major component—the protein.

---

* It is ironic that urease, used by Sumner as proof that a protein devoid of prosthetic groups can be an enzyme, has recently been shown to be a nickel metalloenzyme (Dixon *et al.*, 1975).

The almost magical property of yeast extracts to catalyze the conversion of sugar to ethanol or of body fluids to digest polymers in neutral solution and at low temperatures confounded early biochemists. Had they known about the enzymes' precise choices of substrates (specificity), the complex synthetic reactions catalyzed by enzymes, or the processes of regulation of enzyme activity by metabolites, their confusion would certainly have been greater. It is mainly by examination of the chemistry of the proteins—their shape, their functional groups, and their interaction with the coenzymes bound to them—that we approach explanations of these phenomena.

The ability to discuss enzymes or to use them in the laboratory requires an understanding of the structural properties of proteins. Enzymes are fragile globular proteins, not the tougher fibrous proteins. Purification or assay of an enzyme cannot be performed without taking precautions to prevent damage to the enzyme molecules.

Proteins are linear (i.e. unbranched) polymers of the 20 amino acids. Their properties are decided by the order of the amino acid monomers or *residues* in the polymer chain. In a globular protein, the long polypeptide chain is folded into a compact shape resembling a sphere. Each molecule of an enzyme assumes the same shape as all other molecules of the same enzyme, even though the number of configurations possible from bond rotations is astronomical. Experiments on the folding of enzymes indicate that this native shape, or *conformation,* is the most stable configuration of the molecules (Anfinsen, 1972). During biosynthesis of a protein, the folding into its native conformation is directed by the sequence of its amino acids. An excellent discussion of the protein structure of enzymes has been written by Gutfreund and Knowles (1967).

There are three levels of structural organization which hold enzyme molecules in their globular shape. These are the primary, secondary and tertiary structures of the protein. In addition, many large enzymes possess a quaternary structure, which is the aggregation of globulin subunits. The *primary structure* is the sequence of the amino acid residues, i.e. the covalent peptide bonds. Non-covalent bonds, much weaker than the covalent bonds, are responsible for both secondary and tertiary structure. *Secondary structure* is a result of formation of hydrogen bonds between carbonyl oxygens and amide-NH groups of the polypeptide backbone:

$$\overset{\displaystyle \diagdown}{\underset{\displaystyle \diagup}{C}}=O\cdots H-\overset{\displaystyle \diagup}{\underset{\displaystyle \diagdown}{N}}$$

The amino acid residues interacting to form these hydrogen bonds are not adjacent but are separated in the polypeptide chain. The two common secondary structures are the helix and the sheet. In the most stable helix, the α-helix, each hydrogen bond is formed between a carbonyl group and the amide group of the fourth amino acid residue toward the end of the protein having a free carboxyl group (i.e. there are 3.6 amino acid residues per turn of the helix). A minority of the residues of enzymes are in helical configurations (0–35% of the molecules, commonly). In sheets, the intrachain hydrogen bonds are formed between sequentially distinct regions of the protein which lie alongside each other. *Tertiary structure* is the folding of the protein, already coiled and partially stabilized, into the compact globular shape.* Interaction between the side chains (R-groups) of amino acid residues leads to tertiary structure. The hydrophobic bonds between side chains seeking an environment free of water are particularly important in forming tertiary structures. The amino acids valine, leucine, isoleucine, proline and phenylalanine have strongly hydrophobic side chains, and tryptophan, methionine and cysteine are relatively nonpolar. Other forces which assist in the maintenance of tertiary structure are ionic bonds, hydrogen bonds between side chains, and the covalent disulphide bonds. Although the non-covalent bonds are much weaker than covalent bonds, their great number and proper orientation give stability to the protein's three-dimensional shape. Presence of disulphide crosslinks makes the protein considerably more rigid. Many large proteins do not contain –S–S– bonds. Although proteins are basically rigid because of all the stabilizing bonds, some flexibility due to breaking of a few non-covalent bonds is possible, and probably occurs during the binding of substrates or in catalysis, as well as in regulation of activity. Structural analyses have shown that all the ionic side chains of enzymes

---

* Figure 1 illustrates the three levels of organization of the enzyme staphylococcal nuclease. This simple enzyme does not possess quaternary structure.

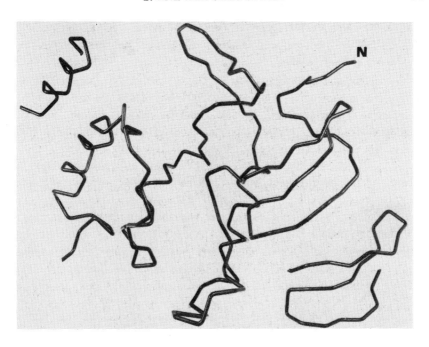

Fig. 1. The structure of staphylococcal nuclease (Cotton and Hazen, 1971). This nuclease has been isolated from the culture media of *Staphylococcus aureus*. It contains 149 amino acid residues, but lacks cysteine so there are no —S—S— bridges. The sequence or primary structure of the nuclease is known, extending from ^+H_3N—Ala^1— (indicated by N) to —Gln^{149}—COO$^-$. Because of a lack of clarity in the X-ray data, the positions of some of the C-terminal amino acids are not shown. Several elements of secondary structure are present in the enzyme. Three helical regions exist: residues 54–67, 99–107 and 122–134. The fragment of the structure in the upper left of the photograph is of the latter helix, placed adjacent to its position in the tertiary structure. Residues 13–36 form a three-stranded antiparallel pleated sheet. This sheet segment of the molecule is shown in the lower right corner of the photo, just below its position in the entire enzyme molecule. In the helices, 23% of the residues participate, and in the sheet, 16%. The remaining 61% of the protein's residues are in non-regular conformations (i.e. do not participate in any secondary structural stabilization). The tertiary structure of the nuclease is shown by the backbone of the polypeptide chain, with the bends in the wire of the model being the positions of the α-carbon atoms (p. 23). An atom of calcium and a molecule of thymidine-3′,5′-diphosphate (not shown) are present in the crystals of nuclease, and these ligands stabilize the tertiary structure.

are exposed to solvent, except for a few "buried" groups which often participate in the catalytic reaction. Although the interiors of enzymes are hydrophobic, not all apolar side chains are buried. Almost all the amino acids in an enzyme protein seem necessary for activity, with most acting to give the correct folding to the protein molecule. Some cuts in the polypeptide chain can be tolerated without loss of activity, but the combined fragments, held together by non-covalent bonds, have the same conformation as the native molecule.

In view of the fact that the folding of the polypeptide chain of an enzyme is so specific, alteration of this structure causes the loss of enzymatic activity. The disturbance of the natural three-dimensional structure of a protein is called *denaturation*. With some enzymes, spontaneous and rapid refolding (renaturation) into their original conformations can occur after complete denaturation. However, most enzymes are irreversibly inactivated upon denaturation either by heating or by chemical treatment. For some chemical experiments, it is desirable to effect complete denaturation of a protein (i.e. form a random coil, a flexible polypeptide chain free of any secondary or tertiary structure). The most effective denaturants are high concentrations of salts which disrupt hydrogen bonds and alter protein-solvent interactions (guanidinium chloride or urea); low concentrations of detergent molecules whose hydrophobic groups react with the hydrophobic groups of the protein; or oxidizing agents which break any $-S-S-$ bonds.

A fourth level of structural organization of enzymes, *quaternary structure*, is the formation of aggregates of protein molecules. Many globular molecules are bound by non-covalent bonds into multi-protein complexes of definite composition. Two types of enzymes show quaternary structure: oligomeric enzymes and multi-enzyme complexes. Each of these types of enzyme will be described in detail later. *Oligomeric enzymes* are also called multimeric; some may be allosteric, or regulatory, enzymes but I will leave their nomenclature to Chapter 12. The proteins which form a quaternary structure complex are described as being composed of *subunits*. To encompass all the types of enzymes having quaternary structure, several definitions of the word "subunit" are necessary. A subunit is a protein unit smaller than the whole protein complex. It can be either the simplest polypeptide unit of a protein complex or any separable functional portion of the complex. The usage of the latter two definitions for "subunit" are illustrated by the oligomeric enzymes lactate dehydrogenase, which exists in isoenzyme forms, and aspartate transcarbamylase, a regulatory enzyme. Lactate dehydrogenase is composed of four subunits, which are four protein molecules or chains of similar size and function, but not necessarily of identical primary structure. Aspartate transcarbamylase from *Escherichia coli* has 12 polypeptide chains, six of each two types, catalytic and regulatory. However, it contains five isolable subunits: two trimers of the catalytic proteins, and three dimers of the regulatory proteins. "Subunit" also is somewhat colloquially used as an adjective of broad meaning, so that a "subunit enzyme" is one that has some sort of quaternary structure. If all the subunits of an oligomeric enzyme are identical, the simplest unit is called a *protomer*. The separation of an enzyme into its subunits is called *dissociation*. Conversely, the aggregation of subunits is termed *association*. Association of subunits during the biosynthesis of an oligomeric enzyme is a reversible self-assembly, in which the state of minimum free energy is reached without external control. In this way association is analogous to the spontaneous conversion of a linear polypeptide into a globular protein.

DETERMINATION OF STRUCTURE

The complete determination of the structure of an enzyme is an Herculean task, but by modern methods it is possible to figure out the positions of individual atoms. X-ray crystallography can provide the three-dimensional structure but the crystallographic research must be preceded by much wet chemistry and by the design of a purification scheme which can supply large amounts of highly purified enzyme (Chapter 3). The resolution of the primary structure of a purified enzyme usually proceeds sequentially through the determination of:

1. the molecular weight;
2. the amino acid composition;
3. the terminal residues; and
4. the sequence of the amino acids.

Amino acid compositions and sequences are commonly reported using three-letter symbols (cf. Abbreviations Section). Then X-ray diffraction experiments, which are the only way to provide the exact secondary and tertiary structure, may be applied. The methods used for structural studies, proceeding from primary to quaternary structure, are outlined below. Mahler and Cordes (1971) give a fuller discussion of the structural organization of proteins and a bibliography of standard techniques, and volumes 11, 25, 26 and 27 of "Methods in Enzymology" are devoted to techniques for the study of enzyme structure.

Molecular Weight

In order for the data obtained from protein structural analysis to be quantitatively meaningful, the molecular weight of an enzyme must be established. Several methods should be used to give confidence to the resulting molecular weight value. Chemical or spectrophotometric analysis for the amount of a prosthetic group can give a minimal molecular weight from the assumption that one mole of the group is present per mole of enzyme or subunit. Several techniques relying on comparison of the protein under examination to proteins of known molecular weight are described in Chapter 3. These methods involve the position of elution from a molecular sieve column, the extent of migration in gel electrophoresis and the rate of sedimentation through a density gradient. Of these, gel electrophoresis is the most useful for both molecular weight determination and detection of heterogeneity of a protein preparation.

Direct or absolute determination of molecular weights can be obtained by analytical ultracentrifugation. Ultracentrifugal analysis depends on measurement of the concentration of protein at various points in the centrifuge cell during centrifugation. In high speed sedimentation velocity experiments, the rate at which the protein sediments in the rapidly rotating chamber (up to 70 000 rev/min) is observed either by a Schlieren optical system, requiring milligram quantities of protein, or by an absorption optical system, using as little as microgram amounts of protein (Schachman and Edelstein, 1973). Both methods rely on the measurement of the movement of the boundary between solvent at the top of the cell and the polymer sedimenting toward the bottom. The rate of sedimentation is expressed in Svedberg or S units. The determination of S by sedimentation velocity requires the knowledge of the diffusion coefficient of the protein and the assumption that the protein is spherical. The sedimentation equilibrium method does not have these two restrictions. For enzymes, the molecular weight can be roughly estimated from the sedimentation coefficient in water at $20°$ C ($S_{20,w}$) from the equation:

$$\text{Mol. wt.} = 6500 \ (S_{20,w})^{3/2}$$

or by reference to Table I. Observation of a single peak in a sedimentation velocity experiment indicates that the test material is homogeneous with respect to molecular weight, but minor contaminants may not be detected. Sedimentation equilibrium experiments are a more useful test for both molecular weight and purity. They are usually performed only after a sedimentation velocity analysis has indicated both the presence of only one protein and the approximate molecular weight. In sedimentation equilibrium experiments, the rotor speed is chosen so that an equilibrium between sedimentation of the protein and diffusion towards the centre of rotation is established. The concentration of protein as a function of the distance from the centre of rotation (x) is measured. If the plot of the log of concentration v. x^2 is linear, the enzyme is probably homogeneous. Polydisperse systems produce curves which are concave upward. To obtain the molecular weight of the protein, its partial specific volume

Table I. The approximate relationship between sedimentation coefficient and molecular weight of an enzyme.[a]

$S_{20,w}$	Molecular weight
2	20 000
3	35 000
4	50 000
6	100 000
8	150 000
10	200 000
15	400 000

[a] This table was constructed using empirical observations and the equation given in the text. The molecular weight values are correct within about ±25%. It is assumed that all enzymes, being globular proteins, have similar partial specific volumes (approximately 0.73).

must be measured or calculated from the amino acid composition. Usually, molecular weight values are calculated from the data obtained from sedimentation equilibrium experiments using computer programs. McCall and Potter (1973) have recently written a concise and useful introduction to the use of ultracentrifugation.

Amino Acid Analysis

Determination of the amino acid composition of a protein is necessary for its characterization. Also, quantitative amino acid analysis is used for end-group determination and in the sequencing of small peptides. An efficient method for automatic analysis of mixtures of amino acids has been devised, and the complete analysis can be performed with as little as a few milligrams of protein.

The first step in amino acid analysis is the hydrolysis of the protein to amino acids. Hydrolysis in strong acid is most satisfactory. About 0.5 mg of protein is dissolved in about 1 ml of constant-boiling (5.8 N) HCl, and heated in an evacuated and sealed glass tube to 110° C for about 24 h. After hydrolysis, the sample is dried and then dissolved in buffer for application to the chromatography column of the automatic analyzer. Standard acid hydrolysis of protein destroys tryptophan, so that a separate analysis for this acid must be performed using either the intact protein or an alkaline or proteolytic hydrolysate (Spies, 1967), or acid hydrolysis under conditions that do not destroy tryptophan must be used (Penke et al., 1974). Both serine and threonine are slowly degraded during the hydrolysis, and valine and isoleucine are not completely liberated until after 96 h. For accurate analysis, portions of the protein are hydrolyzed for 24, 48, 72 and 96 h and then analyzed. Serine and threonine are measured by extrapolation of their content to zero time, and valine and isoleucine are estimated from the 96-h values. Cystine and cysteine are estimated after conversion to cysteic acid, for example by inclusion of dimethyl sulphoxide in the HCl used for hydrolysis (Spencer and Wold, 1969).

The hydrolyzed samples are analyzed in an automatic amino acid analyzer by chromatography on an ion-exchange column from which the amino acids are differentially eluted under programmed conditions (Moore and Stein, 1963). The eluted solution is mixed with ninhydrin and heated for visualization of the amino acids.* The colour intensity is recorded on chart paper, and the amino acids are identified by their positions of elution. Each amino acid is quantitatively recovered and its amount is calculated from the area under its elution curve. The accuracy of the results is about ±1–2%. From the results of the analysis and knowledge of the molecular weight, the approximate number of residues of each amino acid in the protein can be calculated. Exact values of both the molecular weight and the number of amino acid residues are only known after the sequence has been determined.

Determination of N-Terminal Residues

The first step in the determination of the complete amino acid sequence of a protein is to find out which amino acids have the free α-amino and α-carboxyl groups. These amino acids are called the N-terminal and C-terminal residues, respectively, and their identification is called end-group analysis. The primary structure of a protein is numbered from the N-terminal position. Demonstration that an enzyme contains one mole of a certain amino acid as its N-terminal amino acid shows that the enzyme is pure and that its amino terminal is not substituted. Since the N-terminal amino acid is removed during its identification, it might be possible to repeat the end-group analysis on the remainder of the enzyme to identify the penultimate N-terminal residue, and so on. This type of chemical process can be used for sequence determination.

Several chemical methods are available for N-terminal analysis. Formation of 2,4-dinitrophenyl (DNP) derivatives by reaction of the free amino groups with 1-fluoro-2,4-dinitrobenzene was introduced by Sanger and used in the determination of the structure of insulin. The substituted protein or peptide is completely hydrolyzed with acid and then the yellow DNP-amino acid is extracted with ether and identified chromatographically. In the Edman degradation, the N-terminal residue is coupled with phenylisothiocyanate to form the phenylthiocarbamyl peptide. Next, the substituted peptide is cleaved

---

* For some uses, ninhydrin may be replaced by reagents which form fluorescent derivatives with primary amines, giving greater sensitivity in analyses (Udenfriend et al., 1972; Benson and Hare, 1975).

in anhydrous acid to the shortened peptide and a 2-anilino-5-thiazol-inone, which is converted to the more stable 3-phenyl-2-thiohydan-toin (PTH). The PTH-amino acid is identified by comparison to a series of standard PTHs in a chromatographic system. The same cycle

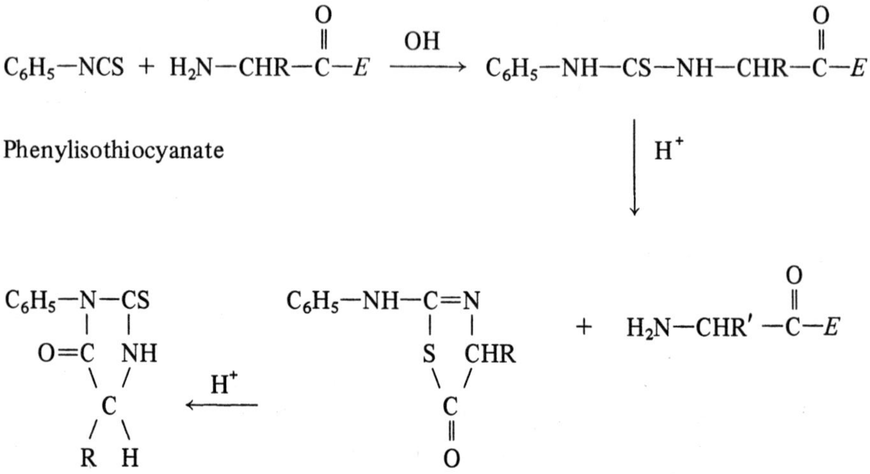

$$C_6H_5-NCS + H_2N-CHR-\overset{\displaystyle O}{\overset{\|}{C}}-E \xrightarrow{\text{OH}} C_6H_5-NH-CS-NH-CHR-\overset{\displaystyle O}{\overset{\|}{C}}-E$$

Phenylisothiocyanate

$$\Big\downarrow H^+$$

$$
\begin{array}{ccc}
C_6H_5-N-CS & C_6H_5-NH-C=N & \\
\ \ |\ \ \ \ | & \ \ \ \ \ |\ \ \ | & \\
O=C\ \ NH & S\ \ \ CHR & +\quad H_2N-CHR'-\overset{\displaystyle O}{\overset{\|}{C}}-E \\
\ \ \ \backslash\ / & \ \ \ \backslash\ / & \\
\ \ \ C & \ \ \ C & \\
\ \ / \backslash & \ \ \ \| & \\
\ R\ \ H & \ \ \ O &
\end{array}
$$

$$\xleftarrow{\ H^+\ }$$

Thiohydantoin (PTH)   Thiazolinone

of reactions can be performed until the order of the first 7–15 residues has been obtained. For quantitative estimation of N-terminal groups, carbamylation of denatured protein with potassium cyanate (KCNO) is used (Stark, 1972). After carbamylation at pH 8,

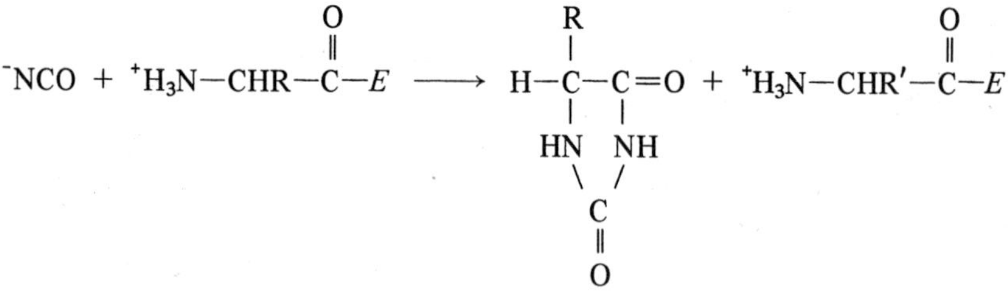

$$^-NCO + \ ^+H_3N-CHR-\overset{\displaystyle O}{\overset{\|}{C}}-E \longrightarrow H-\overset{\displaystyle R}{\overset{|}{C}}-C=O + \ ^+H_3N-CHR'-\overset{\displaystyle O}{\overset{\|}{C}}-E$$

$$
\begin{array}{c}
HN\ \ NH \\
\backslash\ / \\
C \\
\| \\
O
\end{array}
$$

Hydantoin

the peptide is heated in acid and the N-terminal residue is cleaved from the chain as a hydantoin. The hydantoin is purified by ion-exchange chromatography for identification or for hydrolysis to the free amino acid, which can be quantitatively analyzed using the amino acid analyzer. For identification of N-terminal groups in micro amounts of protein, a fluorescent sulphonylating reagent, dansyl chloride, is used (Gray, 1972). The labelled protein is hydrolyzed in HCl and the liberated dansyl amino acid can be identified by

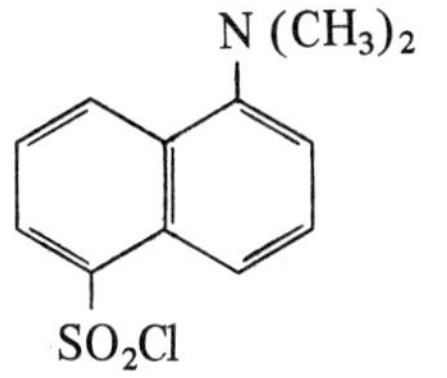

N $(CH_3)_2$

SO_2Cl

Dansyl chloride (1–dimethylaminonaphthalene–5–sulphonyl chloride)

the position of its fluorescence on a thin-layer chromatogram. Recently, there have been reports of the application of combined dansylation and Edman degradation for sequential analysis of nanomole amounts of protein (Percy and Buchwald, 1972; Weiner et al., 1972). The sequence of up to 15 residues from the N-terminal of a protein chain may be determined by these methods also.

Determination of C-Terminal Residues

The standard method for determination of C-terminal groups is to measure the rate of amino acid formation from a protein treated with a C-terminal specific exopeptidase. Pancreatic carboxypeptidase A, with higher specificity for amino acids with aromatic or large aliphatic R-groups, and carboxypeptidase B, specific for basic R-groups, have been used most often (Ambler, 1972). The protein to be analyzed ($0.02-2.00\,\mu$mol) is denatured with performic acid, then incubated with the carboxypeptidase. At convenient intervals, aliquots of the mixture are removed, the proteins precipitated with acetic acid and the amounts of amino acids released determined on the analyzer. From the rate of appearance of each amino acid residue, a short sequence may be determined. Jones and Hofmann (1972) have purified a carboxypeptidase from penicillium which has a wider specificity than the pancreatic enzymes and is thus more suitable for

C-terminal sequence analysis. However, because the automatic sequence analyzer (below) rarely allows analysis of the C-terminal residues, enzymatic analyses will probably be relied upon for completion of sequences.

Automated Sequence Analysis

The approach used for the determination of the primary structure of a protein has, until very recently, been based upon the pioneer work of Sanger (1959) with insulin. The principle was to fragment the protein into oligopeptides, purify these peptides, and find their amino acid composition and sequence. By using several different types of cleavage, overlapping peptides could be formed and from their overlapping sequences the total sequence could be worked out. For example, Sanger and Thompson (1953) identified the structure of 34 peptides obtained by acid hydrolysis and 16 peptides from enzymatic hydrolysis in their determination of the sequence of the A-chain (21 residues) of insulin. The determination of the sequence of each protein has typically taken many years of laboratory work.

Edman and Begg (1967) have designed an apparatus, called a sequenator, in which the steps (coupling, cleavage, extraction and drying) of the Edman degradation are carried out automatically and under strictly controlled conditions (reviewed by Niall, 1973). The thiazolinones are collected in a fraction collector and converted to the PTHs by heating in HCl. The PTH derivatives usually are identified by gas–liquid chromatography or thin layer chromatography. Because of the programmed control of the reaction conditions, the yields of products at each reaction step are better than when the degradation is performed manually. Yields of close to 100% per degradative cycle are now possible. When applied to intact proteins, analysis by the sequenator has been able to produce the sequence of from 20 to 60 amino acids at the N-terminal. Initial experiments can be performed with as little as 5–10 mg of pure protein.

A number of refinements in sequenator analysis methods have made it more feasible to determine larger portions of the sequence in this way. The major improvement is the ability to analyze large peptide fragments rather than only small ones. This greatly reduces the time for sequence analysis, so that only months rather than years may be needed to determine the primary structure of a protein. Hermodson *et al.* (1972) have discussed some of the recent refinements and applications of sequenator analysis, including alterations of some reaction conditions and methods for achieving the very limited and specific hydrolyses of the peptide in order to obtain suitable large polypeptide fragments. Chemical methods, such as cleavage at methionine residues by treatment with cyanogen bromide, or enzymatic methods, such as tryptic hydrolysis of proteins with either the lysine or arginine residues blocked, produce peptides of suitable size for automated analysis. A polypeptide with 30–200 residues appears best because the longer intact proteins are not sufficiently soluble for efficient analysis, and small peptides tend to be extracted from the analyzer. The length of sequence obtainable by sequenator analysis is limited by both the overall yield of the degradative reactions and mechanical losses of protein. The signal-to-noise ratio of the hydantoins becomes indecipherable after about 50–60 cycles, as background PTH-amino acids are formed.

Application of sequenator analysis to proteins can give information other than just sequences. Preliminary testing can show if a protein is pure, by indicating whether or not there is a single series of a few N-terminal residues. The number of subunits of an oligomeric protein can be assigned from the number of N-terminal residues or sequences found. Homologies between enzymes from different sources can be checked by comparison of their N-terminal sequences. Automated sequence analysis will prove to be a powerful tool for protein chemists. However, although the apparatus is commercially available, its purchase, installation, upkeep and operation are costly. This is also true of most other specialized tools (e.g. high resolution n.m.r., amino acid analyzers or X-ray diffraction units) and research workers often share equipment or collaborate with specialists.

X-Ray Crystallography

The application of crystallographic methods to the study of protein structure, an immense problem, has been so successful that it now is providing some of the most illuminating experimental results in enzymology. The first globular protein for which a detailed structure was obtained by X-ray crystallography was myoglobin (Kendrew, 1963). Since this initial achievement, the structures of several dozen enzymes—starting with lysozyme in 1965—have been described. Each year, the structural models of these enzymes are refined and the analyses of larger and more complicated enzymes are attempted successfully. Because data pertaining to enzyme structures are increasing so quickly, I suggest that the reader refers to the latest issues of the "Annual Review of Biochemistry" for the most up-to-date information. Dickerson (1964) and Eisenberg (1970) have written introductions to the techniques of X-ray crystallography, and the textbook by Dickerson and Geis (1969) includes photographs and drawings of the three-dimensional structures of many globular proteins.

Crystallography is the only known method by which one can measure the relative positions of almost all of the atoms of an enzyme. In the preliminary stages of analysis, the overall shape of a protein molecule, its molecular weight and the quaternary structure can be obtained. Suitable large crystals of enzymes or their derivatives are prepared. The intensities and positions of diffraction of X-irradiation from the crystals are then measured at various rotated positions of the crystals. These results are recorded either on photographic film or using a counting device (diffractometer). The smaller the wavelength used for the analysis (i.e. the outermost reflections), the finer the resolution obtained and the more detail seen. Because the number of reflections increases with the inverse cube of the resolution, high resolution experiments require much time and work. The enormous amount of data is accumulated and processed by electronic equipment and computers. It is converted into a three-dimensional map of the electron density of the enzyme crystal, with the electron-dense regions corresponding to atoms. The map is prepared showing the contours of electron density on stacked sheets of clear plastic. Each sheet represents a two-dimensional contour map of a slice of the crystal. From the maps, bond lengths and inter-bond angles can be calculated. Because electron density maps only have the definition necessary to distinguish groups, not individual atoms, they are interpreted by construction of molecular models which fit both the data on the maps and the other known structural properties (e.g. the sequence) of the enzyme. The maps can be converted to models and the quality of the models assessed using an optical comparator in which a large half-silvered mirror is used to superimpose the image of the model on that of the map (Richards, 1968). The models are usually skeletal structures in which rods of the correct lengths set at the correct angles represent the bonds formed between the centres of the atoms. North (1972) has summarized the techniques of drawing three-dimensional structures either for publication or for presentation at lectures. One of the most useful methods is a computer program which can convert the atomic co-ordinates of a structure into either a perspective diagram or a stereo pair diagram using a mechanical plotter. The program OR TEP (Johnson, 1970) has been used to draw some of the illustrations in this book. Diagrams cannot show the detail that can be seen in a correctly constructed model, though. Small backbone models of proteins, showing the positions of the α-carbons only, can be constructed rapidly and inexpensively from steel wire using a wire bender designed by Rubin and Richardson (1972). These models are useful in teaching enzymology.

The results obtained by X-ray diffraction are indisputable within the limits of experimental resolution and interpretation. Accurate protein structures can usually be obtained when the resolution achieved is 2–2.5 Å. The interpretation of the crystallographic data is much simpler if the primary structure of the enzyme is already known. In the absence of the sequence, though, the polypeptide backbone and the positions of easily recognizable side chains, such as aromatic rings or the sulphur of methionine, can be assigned. The present limits of interpretation are exemplified by work done with crystals of rubredoxin (p. 283). By examining only the electron density maps, refined to 1.5 Å, Herriott et al. (1973) have succeeded in identifying 50 of the 54 residues in the chemically determined sequence.

It can now be concluded safely that the structures obtained by X-ray crystallography correctly represent the molecular configurations in solution. The similarities of structures of molecules in different crystal forms and the retention of enzymatic activity in crystals strongly support this claim. As further support, the structure of the trypsin inhibitor from pancreas (Huber et al., 1971) matches the area of trypsin's structure with which it is in contact so exactly (Stroud et al., 1971) that the structures must represent those in solution.

Once the structure of an enzyme is known, examination of crystals of an enzyme–inhibitor or enzyme–substrate analogue complex can show the position of the ligand in the crystal and if any groups of the enzyme have moved as a result of binding of the ligand. Because structures of crystals represent static pictures of enzymes, crystallographic experiments must be combined with chemical results for a dynamic view of enzyme catalysis.

Sequences are known for many more proteins than are three-dimensional structures. Consequently, attempts are being made to predict secondary structures, and eventually tertiary structures, from the amino acid sequences of proteins (reviewed by Anfinsen and Scheraga, 1975). Because folding of a protein into its native conformation is a spontaneous process, predictions of protein shape from sequence are feasible. Methods for predicting helical regions, extended structures and bends in the chain exist. These schemes rely on the assignment of conformational preferences of each of the amino acid residues based on examination of their behaviour in synthetic polypeptides and in proteins whose three-dimensional structures have already been solved. A test of some of these prediction methods has been performed with adenyl kinase (Schulz *et al.*, 1974). Before the tertiary structure was published, Schulz asked several groups to predict the locations of helices, sheets and bends from the primary structure alone. Individual schemes had varying success when their predictions were compared to the determined structure, but taken together, the locations of most of the helices and bends and more than half of the sheet strands could be assigned. A similar exercise has been performed with T4 bacteriophage lysozyme (Matthews, 1975). With this enzyme, the predictive methods were less successful than with adenyl kinase. No single method was clearly superior to the others. Mathematical procedures for predicting the more distant interactions that result in globular shape with hydrophobic cores and polar shells will be much more complicated. However, work in this direction is important, as reliable prediction of native conformation from sequence data will be a great aid to both enzymologists and crystallographers.

Electron Microscopy

The quaternary structure of an oligomeric protein is difficult to evaluate. The composition in number of subunits can be estimated easily in several ways (e.g. the electrophoretic procedure described on p. 71); however, finding out the shape of the aggregated subunits can only be done using a high resolution technique such as X-ray crystallography or electron microscopy. Due to the high molecular weight and thus complexity of crystals, few oligomers have been examined by crystallography to date. Nevertheless, several protein complexes and oligomers have been successfully photographed by electron microscopy. These results and the structural interpretations of the electron microscopic images have been summarized recently by Haschemeyer and de Harven (1974). In those cases where the results are easily interpreted and in agreement with other structural data (e.g. glutamine synthetase from *E. coli*, p. 453), the results from electron micrographs have been spectacular. Superposition of images of different enzyme molecules has been used to artificially reduce the background noise and distortion in photographs. The results can be viewed with more confidence if the same protein shapes and symmetries are seen for both single molecules and crystals or other molecular arrays. There are limitations to electron microscopy, such as the desiccation of the protein molecules during slide preparation, and artefacts can occur. When cautiously applied, the method can yield much accurate information about enzyme substructure not yet available in any other way.

Chemical Synthesis of Proteins

In organic chemistry, synthesis has traditionally provided final proof of the structure of a molecule. Synthesis of proteins has now been achieved, and offers possibilities for many types of studies such as preparation of protein analogues for structure–function evaluations. Peptide synthesis requires not only the linkage of the amino group of one amino acid to the carboxyl group of another, but also protection of the amino and other side-chain groups that are not intended to react by selective substitution or blocking. The formation of the peptide bond is usually carried out either by activation of the carboxyl group, necessitating isolation of the carboxyl-activated intermediate, or by use of a dehydrating agent such as dicyclohexylcarbodiimide. Synthesis of small peptides has been possible for many years, but the enormous task of synthesizing an active protein from amino acids has awaited improved synthetic methods. One modification, called the solid-phase technique, has been introduced by Merrifield (1972) and his associates. In this procedure, an amino acid with a blocked amino group is covalently anchored to insoluble synthetic polymer beads through its carboxyl group. Then the blocking group is removed and a second blocked amino acid is condensed with the first residue, usually using carbodiimide as a dehydrating agent. A stepwise synthesis from the C-terminal residue is thus possible, with the completed protein being released from the beads and all protecting groups being removed as a last single step. At each synthetic step, the growing peptide remains bound to the solid particles and is easily separated from one set of reagents and prepared for suspension in the next set, for addition of the next blocked amino acid. This type of synthesis lends itself to automation. Despite several drawbacks, including side reactions, racemization and unpredictably low coupling yields in some cycles, the solid-phase technique—especially after methodological improvements are made—seems to be the only economical method to chemically synthesize large polypeptides.

Both reactions in solution using classical procedures and reactions on solid supports have been used for the synthesis of proteins (Merrifield, 1972). Over ten years ago, insulin was synthesized by the classical approach in three laboratories. The products were crystallized and shown to possess full hormonal activity. This success indicated that synthesis of larger proteins, though difficult, could be achieved. The first total synthesis of an enzyme was reported in 1969, when ribonuclease was synthesized by the solid-phase technique (Gutte and Merrifield, 1969) and by coupling of 19 fragments (Denkewalter et al., 1969). The solid-phase synthesis produced, after purification of the synthetic mixture, 85 mg of protein indistinguishable from ribonuclease A and having 80% the specific activity of the native enzyme. The classical approach produced small quantities of the ribonuclease S-protein (p. 189) which, when combined with the previously synthesized S-peptide, generated enzymatic activity. Ribonuclease, with 124 residues, was chosen for synthesis because of its stability and because it is one of the enzymes which can be refolded after denaturation. A number of experiments are now being performed to study the effect on enzyme function of the replacement of specific amino acids by using polypeptides synthesized by the solid-phase approach.

THE ACTIVE SITE

The reactions catalyzed by enzymes occur by interaction of the substrates with a portion of the surface of the enzyme called the *active site*. Enzyme chemists have defined the active site as the region of the enzyme in direct contact with the substrates during their conversion to products. It must also include any other groups having a direct function in the catalytic process. Not only substrates, but also transition states, any intermediates and the products are bound at the active site. The term *catalytic site* is used for the chemical groups or residues involved in the bond-making and bond-breaking reactions, and *binding site* for the residues attracting or positioning the substrates. Although division of the active site into these two parts is arbitrary and artificial, because the whole active site is needed for substrate binding, specificity and catalysis, I will none the less use catalytic site to refer to the groups that have been implicated mechanistically in an enzyme's function. The amino acids of the active site are widely separated in the primary structure but quite close in the tertiary structure of an enzyme. The active site is only a small part of the total surface area of an enzyme. Most enzymes react with substrates much smaller than themselves, so few amino acid residues are in contact with the substrates. Even those enzymes that catalyze reactions involving macromolecular substrates contact only small portions of the macromolecules. For example, lysozyme has a large cleft-shaped active site in which there is space for only a hexasaccharide unit of its polysaccharide substrate. X-ray studies have shown that most enzymes have pockets or clefts into which the substrates fit during the catalytic reaction. These active site areas are usually lined by hydrophobic amino acids, so that the active site is less polar than water.

Every enzyme forms a reversible enzyme–substrate (*ES*) complex, in which several non-covalent bonds hold the substrate to the enzyme surface. These bonds may include ionic interactions, hydrogen bonds and hydrophobic bonds. The reversible complex formed when the substrates are bound to the enzyme is referred to as the Michaelis complex. The structure of the enzyme provides the steric limitations necessary for substrate specificity. There is no covalent binding of intact substrates to enzymes, but some enzymes do covalently bind transferable groups formed from the substrate, as discussed in

Chapter 10. For those enzymes not forming covalent *ES* intermediates, X-ray studies provide the most direct evidence for the binding of coenzymes and of substrates. Coenzyme binding can also be measured by spectral techniques. Isolation and characterization of *ES* complexes is difficult because of the rapid formation and release of products. However, some intermediates can be observed spectrally, and Yagi and Ozawa (1964) have succeeded in crystallizing the Michaelis complex of D-alanine with D-amino acid oxidase under anaerobic conditions (i.e. in the absence of the ultimate electron acceptor, oxygen).

Specificity

An enzyme is specific both for the type of reaction it catalyzes and for the substrates it accepts. The degree of substrate specificity is variable, but most enzymes are limited to one or a few closely related substrates. Some enzymes show *absolute specificity* for reactants, with only one known substrate or pair of substrates being accommodated at the active site. Many enzymes originally described as possessing absolute specificity can accommodate synthetic substrate analogues as well as their single naturally occurring substrate. Obviously, the preparation of an artificial substrate does not make the

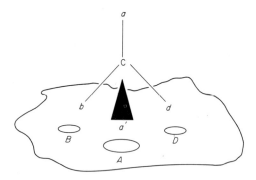

Fig. 2. Hypothetical interaction of the chemically symmetrical substrate (*Caa'bd*) with two binding sites (*B* and *D*) and one catalytic site (*A*) of an enzyme. Since group *b* can bind only to site *B* and group *d* only to site *D*, group *a'* but not group *a* can approach the catalytic site (after Ogston, 1948).

enzyme any less specific *in vivo*. Catalysis of the reactions of a series of structurally related substrates is termed *group specificity*. Group specificity can be quite limited or rather broad. Most enzymes are *stereospecific* because they accept only one isomer of an asymmetric pair.

Several interesting examples of the stereospecificity of enzymes involve the asymmetric reaction of symmetric compounds. These substrates have two identical substituents and two further but dissimilar groups at one carbon atom (*Caa'bd*). In 1948, Ogston proposed that if three points of the symmetric substrate were in contact with the asymmetric enzyme, then an asymmetric product could be formed. As shown in Fig. 2, if an enzyme binds the two dissimilar substituents at specific sites, then only one of the two similar groups can bind to the catalytic site. This proposal of three-point attachment was used to explain the occurrence of isotopic label from the 4-position of oxaloacetate in only the 1-position of α-ketoglutarate, despite the intermediate formation of the chemically symmetrical citric acid in the tricarboxylic acid cycle. This anomaly is explained by the stereospecificity of aconitase, which catalyzes the dehydration of citrate to *cis*-aconitate and also the hydration of the aconitate to isocitrate. Aconitase can distinguish between the two $-CH_2COOH$ groups of citrate. The enzyme removes only one specific hydrogen during the formation of aconitate and forms only one of four possible stereoisomers in the second step. When citrate and aconitase were incubated in D_2O until equilibrium was reached, only one atom of deuterium was incorporated into either citrate or isocitrate, showing that the reaction is stereospecific (Englard and Colowick, 1957). Another example of synthesis of an asymmetric product from a symmetric substrate is the formation of only L-glycerol-3-phosphate (L-α-glycerophosphate) from glycerol and ATP in the presence of

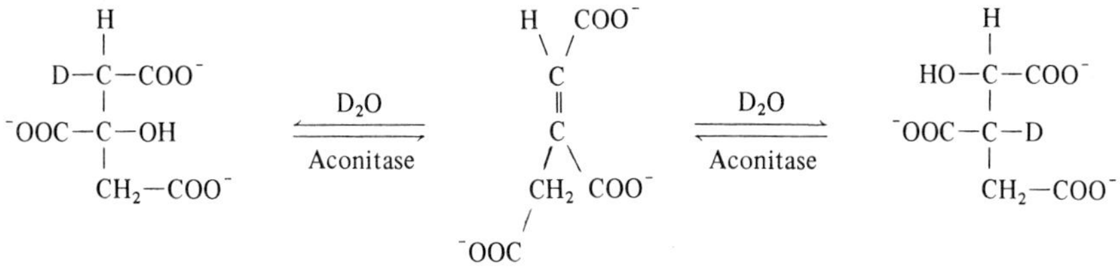

Citrate   *cis*-Aconitate   Isocitrate

glycerokinase (Bublitz and Kennedy, 1955). Because substrates such

$$
\begin{array}{ccc}
\begin{array}{c}
\text{CH}_2\text{OH} \\
| \\
\text{CHOH} \\
| \\
\text{CH}_2\text{OH}
\end{array}
&
\xrightarrow{\text{ATP}}
&
\begin{array}{c}
\text{CH}_2\text{OH} \\
| \\
\text{HO—C—H} \\
| \\
\text{CH}_2\text{OPO}_3^{2-}
\end{array}
\\[2em]
\text{Glycerol} & & \text{L–Glycerol–3–phosphate}
\end{array}
$$

as citrate or glycerol (both $Caa'bc$) can be converted to products possessing a chiral or asymmetric centre ($Cabcd$), they are said to possess *prochirality*. In these reactions, the enzyme molecules are chiral reagents catalyzing the formation of only one of the two possible stereoisomeric forms of the transition state or intermediate. These experiments and other examples of stereospecificity of enzymatic reactions are discussed in a review article by Popják (1970).

Some enzymes catalyze two (or more) reactions at the same active site. To prove that one enzyme is catalyzing both reactions, several criteria should be fulfilled. It should be impossible to separate the two enzyme activities, and the two activities should be purified at a constant ratio during the isolation of the enzyme. Also, both activities should be lost to similar extents during inactivation experiments such as heat denaturation. If the reaction rate when two substrates are mixed at saturating concentrations is the sum of the individual rates, then two separate enzymes or active sites are responsible for the reactions. As discussed above, aconitase catalyzes both the dehydration of citrate to *cis*-aconitate, and also the hydration of the *cis*-aconitate to isocitrate. Both reactions appear to occur at the same catalytic site (Glusker, 1971). Adenylosuccinase catalyzes two similar reactions in the purine biosynthetic pathway, the hydrolysis of adenylosuccinate to AMP and fumarate, and the hydrolysis of phosphoribosyl aminoimidazole succinocarboxamide to phosphoribosyl aminoimidazole carboxamide and fumarate (Ratner, 1972). Genetic studies have shown that mutants of *Neurospora crassa* which lack one adenylosuccinase activity are also lacking the other activity. Serine transhydroxymethylase from rabbit liver, which catalyzes the conversion of L-serine to glycine (p. 249), was found also to catalyze the cleavage of L-threonine and of allothreonine to glycine and acetaldehyde (Schirch and Gross, 1968). These latter cleavage reactions were previously believed to be catalyzed by threonine aldolase. Palekar *et al.* (1973) have found that the cytoplasmic serine transhydroxymethylase from rat liver catalyzes the cleavage of allothreonine (but not threonine) and also is responsible for the liver's aminomalonate decarboxylase activity. The mitochondrial serine transhydroxymethylase, though, has no activity with aminomalonate, allothreonine or threonine. Many enzymes are capable of catalyzing partial reactions, such as the steps in the overall reaction with which they are concerned (Chapter 10). Proteolytic enzymes often show esterase and amidase as well as peptidase activity, showing a lack of specificity for the leaving group.

Emil Fischer, in 1894, proposed a simple hypothesis to explain the substrate specificity of enzymes. He envisaged the enzyme as a rigid template of well-defined structure. A correctly shaped substrate would bind to the surface of the enzyme to form an *ES* complex, like a key fitting into a lock. The template theory explains the majority of catalytic phenomena but is too simple to explain some experimental observations of ligand binding. Several theories postulating flexibility of the enzyme and the substrate have been proposed to account for these anomalies. Since the binding of a compound is not sufficient for catalysis of its breakdown, Koshland has proposed the induced-fit theory of catalytic specificity (Koshland and Neet, 1968). This theory suggests that there is a change induced in the structure of the protein, in the presence only of good substrates, into a catalytically active conformation. Some compounds are thought to bind but are not of the correct structure to cause conformational changes suitable for catalysis. The binding of ligands to enzymes is often accompanied by conformational transitions (cf. Citri, 1973). For many reasons, we are sure that changes in protein conformation are important for enzymic function. The extent of change in shape may or may not correlate with reaction velocity. The major conformation change seen at the active site of crystalline carboxypeptidase A when the slowly hydrolyzed substrate glycyl-L-tyrosine is bound is often cited as one of the strongest pieces of evidence for the induced-fit theory (Hartsuck and Lipscomb, 1971). In this case, the substrate is assumed to induce an alignment of catalytic groups required for hydrolysis of good substrates, but the substrate is scarcely hydrolyzed by the crystalline enzyme. Factors other than induced-fit must therefore also be necessary for rapid catalysis. An alternative strain or distortion theory, in which the enzyme is relatively rigid and the substrate undergoes most of the distortion, is favoured by Jencks. Jencks and Page (1972) state that substrate specificity appears not only in binding but also in effects on the reaction rate. They point out that catalysis and specificity cannot be formally separated, since it is the binding of a substrate to the enzyme that is responsible for a great portion of the catalysis. Small conformational changes in the enzyme would not greatly decrease its catalytic power. The differences between these theories and their relative merits are assessed more fully later (pp. 163–165). In my opinion, induced-fit theory applies to the mode of binding of substrates and release of products, while the strain theory explains the central catalytic step, bond breakage and reformation. The theories need not be mutually inconsistent.

Catalytic Residues

Only amino acids which have ionizable side chains have the capability of acting as catalysts at the active site. The potential catalytic groups are: the β-carboxyl of aspartate, the γ-carboxyl of glutamate, the imidazole group of histidine, the thiol group of cysteine, the phenolic group of tyrosine, the ε-amino group of lysine, the guanidino group of arginine and the hydroxyl groups of serine and threonine. At pH 7, Asp and Glu will be ionized, Tyr will be uncharged, Lys and Arg will be protonated, and His will be in equilibrium between the charged and uncharged states. Because the catalytic groups are seldom freely accessible to solvent, their pK_a values (i.e. the pH at which the ionizable group is half in the protonated form and half in the ionized form) are often perturbed from what they would be in a simple peptide (Table II). Perturbation of the ionizations can be due to electrostatic or hydrophobic forces. In a hydrophobic environment, the ionization of carboxylic acids is depressed (the pK_a raised) but ammonium groups are affected very little. Anomalous pK_a values can also arise from stabilization of an ionizable group by formation of a hydrogen bond with a neighbouring group. The many effects of the local environment on a charged group make the measurement of the ionization constants of individual residues in proteins difficult. Hydrogen ion titrations usually measure only the ionizable groups on

Table II. The pK_a values of ionizable residues.

Group	Expected pK_a in proteins*	pK_a in free amino acids
Terminal α-carboxyl	3.1–3.8	1.7–2.6
Side-chain carboxyl	4.6	3.8–4.2
Imidazole	6.3	6.0
Terminal α-amino	7.5–7.8	8.8–10.4
Thiol	9.1–9.5	8.3–10.8
Phenol	9.6	10.1
ϵ-Amino	10.2–10.4	10.5
Guanidyl	>12	12.5

*The expected values in proteins are derived from the pK_a values of suitable model compounds in water at 25° C. Data from Tanford (1962), Steinhardt and Beychok (1964) and Mahler and Cordes (1971).

the surface of a protein, and thus show values close to normal. However, by other methods, perturbations of buried residues of up to 3 or 4 pH units have been noted. The determination of the pK_a values for specific side chains is an active area of research, with n.m.r. giving the most useful results.

The identification of the amino acids which participate in catalytic events of an enzyme can be attempted by both direct and indirect methods. Naturally, assignment of a catalytic role to a specific residue can only be made when the sequence is known. Interpretation of the complete three-dimensional structure of an enzyme and its enzyme–substrate complexes gives the most reliable assignment of catalytic residues. Examination of structural models of enzymes has implicated previously unsuspected amino acids in the mechanism. The assumption can be made that if there is a prosthetic group, either a coenzyme or a metal, it will be at the active site. The mechanisms of a number of coenzymes will be discussed in Chapter 8, and involvement of essential metal ions in Chapter 11. The most indirect, but often first method employed for assignment of catalytic roles, is to study the effect of pH on enzymatic activity (p. 120). This method can give only tentative assignments. Chemical modifications or destruction of reactive groups in the enzyme can be used to identify catalytic residues. Cohen (1970) has reviewed the principles and limitations of chemical modification of proteins for identification of groups essential for catalysis, and Means and Feeney (1971) have written a book summarizing the methods used for the chemical modification of active sites.

The reagents which give the most useful and direct data are compounds which, because of their similarity to substrates, can bind reversibly at the active site and then react covalently with a reactive amino acid at the catalytic site. These inhibitors have been called either *active site-directed irreversible inhibitors* (Baker, 1967) or *affinity-labels* (Singer, 1967). Rando (1974) has pointed out that most of the active-site directed inhibitors studied rely solely on the binding properties of the enzyme. He has suggested that compounds, which he calls k_{cat} inhibitors, selectively activated by the target enzyme's specific catalytic action, may be more useful still. Less direct modification experiments can still yield valuable information. There are many organic reagents which bind to reactive side chains of amino acids, with varying specificity. The reactions involved include acylation, alkylation, oxidation or reduction. For example, the reaction of p-hydroxymercuribenzoate with thiol groups is a common test for essential cysteine residues. The loss of enzymic activity upon chemical modification is often taken as involvement of the group modified in either the catalysis or binding process, or retention of active conformation of the enzyme. Data from titration with a modifying reagent can indicate the percentage activity as a function of the number of residues modified. If the substituted amino acid is a catalytically active group, then there should be complete loss of activity upon its substitution. Modification of the enzyme by a compound of wide specificity can be made more specific by reaction in the presence of substrate or inhibitor. The modified enzyme should retain activity since the active site would be protected by the ligand. Also, addition of a radioactively labelled modifying group to the ligand-protected enzyme should then allow identification of the reactive group at the active site. By using differential labelling and then hydrolyzing the radioactive enzyme, peptide fragments from the active site can be obtained for analysis. Since some groups are buried and some exposed, their reactivities may vary markedly. Some groups in enzymes—often but not always at the active site— have unusually high reactivity, and can be readily modified.

While assignment of specific roles to some residues is comparatively easy, it is more difficult to define the extent of the active site. Groups which affect the reactivity of the catalytic residues may not be in contact with the substrate, nor participate directly in the reaction, and thus not fit the standard definition of the active site. They play a more subtle but still essential role in catalysis. As I have mentioned, almost the entire protein molecule appears to be required for enzymic catalysis. Much of the structure is needed to give the proper spatial positioning to catalytic residues which are widely separated in the sequence but close in the tertiary structure. Other functions of the remainder of the protein may be to prevent the breakdown of reactive intermediates, or to bind regulatory ligands or bring about the conformation changes associated with regulation. As the research on enzyme proteins continues, the functions of more and more residues will be understood.

REFERENCES

Ambler, R. P. (1972). *In* "Methods in Enzymology" (Hirs, C. H. W. and Timasheff, S. N., eds), Vol. 25, pp. 143-154. Academic Press, New York.

Anfinsen, C. B. (1972). *Biochem. J.* **128**, 737-749.

Anfinsen, C. B. and Scheraga, H. A. (1975). *Adv. Protein Chem.* **29**, 205-300.

Baker, B. R. (1967). "Design of Active-Site-Directed Irreversible Enzyme Inhibitors". John Wiley and Sons, New York.

Benson, J. R. and Hare, P. E. (1975). *Proc. natn. Acad. Sci. U.S.A.* **72**, 619-622.

Bublitz, C. and Kennedy, E. P. (1954). *J. biol. Chem.* **211**, 963-967.

Citri, N. (1973). *Adv. Enzymol.* **37**, 397-648.

Cohen, L. A. (1970). *In* "The Enzymes" (P. D. Boyer, ed.), Third edition, Vol. 1, pp. 147-211. Academic Press, New York.

Cotton, F. A. and Hazen, E. E., Jr. (1971). *In* "The Enzymes" (P. D. Boyer, ed.), Third edition, Vol. 4, pp. 153-175. Academic Press, New York.

Denkewalter, R. G., Veber, D. F., Holly, F. W. and Hirschmann, R. (1969). *J. Am. chem. Soc.* **91**, 502.

Dickerson, R. E. (1964). *In* "The Proteins" (H. Neurath, ed.), Second edition, Vol. 2, pp. 603-778. Academic Press, New York.

Dickerson, R. E. and Geis, I. (1969). "The Structure and Action of Proteins". Harper and Row, New York.

Dixon, N. E., Gazzola, C., Blakeley, R. L. and Zerner, B. (1975). *J. Am. chem. Soc.* **97**, 4131-4133.

Edman, P. and Begg, G. (1967). *Eur. J. Biochem.* **1**, 80-91.

Eisenberg, D. (1970). *In* "The Enzymes" (P. D. Boyer, ed.), Third edition, Vol. 1, pp. 1-89. Academic Press, New York.

Englard, S. and Colowick, S. P. (1957). *J. biol. Chem.* **226**, 1047-1058.

Glusker, J. P. (1971). *In* "The Enzymes" (P. D. Boyer, ed.), Third edition, Vol. 5, p. 432. Academic Press, New York.

Gray, W. R. (1972). *In* "Methods in Enzymology" (Hirs, C. H. W. and Timasheff, S. N., eds), Vol. 25, pp. 121-138. Academic Press, New York.

Gutfreund, H. and Knowles, J. R. (1967). *In* "Essays in Biochemistry" (Campbell, P. N. and Greville, G. D., eds), Vol. 3, pp. 25-72. Academic Press, London and New York.

Gutte, B. and Merrifield, R. B. (1969). *J. Am. chem. Soc.* **91**, 501-502.

Hartsuck, J. A. and Lipscomb, W. N. (1971). *In* "The Enzymes" (P. D. Boyer, ed.), Third edition, Vol. 3, pp. 1-56. Academic Press, New York.

Haschemeyer, R. H. and de Harven, E. (1974). *A. Rev. Biochem.* **43**, 279-301.

Hermodson, M. A., Ericsson, L. H., Titani, K., Neurath, H. and Walsh, K. A. (1972). *Biochemistry* **11**, 4493-4502.

Herriott, J. R., Watenpaugh, K. D., Sieker, L. C. and Jensen, L. H. (1973). *J. molec. Biol.* **80**, 423-432.

Huber, R., Kukla, D., Rühlmann, A. and Steigemann, W. (1971). *Cold Spring Harb. Symp. quant. Biol.* **36**, 141-150.

Jencks, W. P. and Page, M. I. (1972). *In* "Enzymes: Structure and Function" (Drenth, J., Oosterbaan, R. A. and Veeger, C., eds), pp. 45-58. North-Holland, Amsterdam.

Johnson, C. K. (1970). "OR TEP: A Fortran Thermal-ellipsoid Plot Program for Crystal Structure Illustrations". Oak Ridge National Laboratory, Oak Ridge, Tennessee.

Jones, S. R. and Hofmann, T. (1972). *Can. J. Biochem.* **50**, 1297-1310.

Kendrew, J. C. (1963). *Science, N.Y.* **139**, 1259-1266.

Koshland, D. E., Jr. and Neet, K. E. (1968). *A. Rev. Biochem.* **37**, 380-389.

Mahler, H. R. and Cordes, E. H. (1971). "Biological Chemistry", Second edition, Chapters 3-4. Harper and Row, New York.

Matthews, B. W. (1975). *Biochim. biophys. Acta.* **405**, 442-451.

McCall, J. S. and Potter, B. J. (1973). "Ultracentrifugation". Baillière Tindall, London.

Means, G. E. and Feeney, R. E. (1971). "Chemical Modification of Proteins". Holden-Day, San Francisco.

Merrifield, R. B. (1972). *PAABS Revista* **1**, 105-195.

Moore, S. and Stein, W. H. (1963). *In* "Methods in Enzymology" (Colowick, S. P. and Kaplan, N. O., eds), Vol. 6, pp. 819-831. Academic Press, New York.

Niall, H. D. (1973). *In* "Methods in Enzymology" (Hirs, C. H. W. and Timasheff, S. N., eds), Vol. 27, pp. 942-1010. Academic Press, New York.

North, A. C. T. (1972). *In* "Enzymes: Structure and Function" (Drenth, J., Oosterbaan, R. A. and Veeger, C., eds), pp. 9-18. North-Holland, Amsterdam.

Northrop, J. H., Kunitz, M. and Herriott, R. M. (1948). "Crystalline Enzymes", Second edition. Columbia University Press, New York.

Ogston, A. G. (1948). *Nature, Lond.* **162**, 963.

Palekar, A. G., Tate, S. S. and Meister, A. (1973). *J. biol. Chem.* **248**, 1158-1167.

Penke, B., Ferenczi, R. and Kovács, K. (1974). *Analyt. Biochem.* **60**, 45-50.

Percy, M. E. and Buchwald, B. M. (1972). *Analyt. Biochem.* **45**, 60-67.

Popják, G. (1970). *In* "The Enzymes" (P. D. Boyer, ed.), Third edition, Vol. 2, pp. 115-215. Academic Press, New York.

Rando, R. R. (1974). *Science, N.Y.* **185**, 320-324.

Ratner, S. (1972). *In* "The Enzymes" (P. D. Boyer, ed.), Third edition, Vol. 7, pp. 182-197. Academic Press, New York.

Richards, F. M. (1968). *J. molec. Biol.* **37**, 225-230.

Rubin, B. and Richardson, J. S. (1972). *Biopolymers* **11**, 2381-2385.

Sanger, F. (1959). *Science, N.Y.* **129**, 1340-1344.

Sanger, F. and Thompson, E. O. P. (1953). *Biochem. J.* **53**, 353-374.

Schachman, H. K. and Edelstein, S. J. (1973). *In* "Methods in Enzymology" (Hirs, C. H. W. and Timasheff, S. N., eds), Vol. 27, pp. 3-59. Academic Press, New York.

Schirch, L. and Gross, T. (1968). *J. biol. Chem.* **243**, 5651-5655.

Schulz, G. E., Barry, C. D., Friedman, J., Chou, P. Y., Fasman, G. D., Finkelstein, A. V., Lim, V. I., Ptitsyn, O. B., Kabat, E. A., Wu, T. T., Levitt, M., Robson, B. and Nagano, K. (1974). *Nature, Lond.* **250**, 140-142.

Singer, S. J. (1967). *Adv. Protein Chem.* **22**, 1-54.

Spencer, R. L. and Wold, F. (1969). *Analyt. Biochem.* **32**, 185-190.

Spies, J. R. (1967). *Analyt. Chem.* **39**, 1412-1416.

Stark, G. R. (1972). *In* "Methods in Enzymology" (Hirs, C. H. W. and Timasheff, S. N., eds), Vol. 25, pp. 103-120. Academic Press, New York.

Steinhardt, J. and Beychok, S. (1964). *In* "The Proteins" (H. Neurath, ed.), Second edition, Vol. 2, pp. 139-304. Academic Press, New York.

Stroud, R. M., Kay, L. M. and Dickerson, R. E. (1971). *Cold Spring Harb. Symp. quant. Biol.* **36**, 125-140.

Sumner, J. B. (1926). *J. biol. Chem.* **69**, 435-441.

Tanford, C. (1962). *Adv. Protein Chem.* **17**, 69-165.

Udenfriend, S., Stein, S., Böhlen, P., Dairman, W., Leimgruber, W. and Weigele, M. (1972). *Science, N.Y.* **178**, 871-872.

Weiner, A. M., Platt, T. and Weber, K. (1972). *J. biol. Chem.* **247**, 3242-3251.

Yagi, K. and Ozawa, T. (1964). *Biochim. biophys. Acta.* **81**, 29-38.

3 Isolation of Enzymes

INTRODUCTION AND GENERAL PROCEDURES

Hundreds of different enzyme proteins occur in all living cells. For characterization as a chemical entity, each enzyme must be isolated from the other proteins and constituents of the cell of its origin. The isolation methods used for enzymes differ from most organic chemistry purification techniques because of the sensitive nature of the proteins. The methods which I shall describe allow fractionation of cellular proteins with minimal losses in catalytic activity. When properly used or chosen; the methods cause minimal denaturation of an enzyme. Because all enzymes are proteins, isolation of any one of them is a challenge; the investigator must discover and exploit the individual properties of that one enzyme.

The aim of a purification procedure is to increase the specific catalytic activity (activity per unit weight, mathematically defined in Chapter 5) of the enzyme during each purification step. Therefore, a rapid and adequate analysis for enzymic activity, i.e. an adequate *assay,* must be developed early in the research. The primary steps in a purification method usually are cell breakage, fractionation of the cellular homogenate by applying simple gross fractionation steps, and finally the application of more selective techniques such as column chromatography. In this chapter, I will describe the most common and useful methods for enzyme purification and also the supplementary techniques (e.g. removal of low molecular weight compounds and concentration of enzyme fractions) that must be used during enzyme isolation. The techniques are presented in the order in which they usually would be encountered or employed in a purification scheme.

Enzyme activity can be detected in unpurified cell extracts and sometimes even in whole cells, but for adequate characterization at

least partial purification is essential. The degree of purification necessary varies with the scientific objectives. For example, if you are using
an enzyme as a catalyst for an organic reaction—perhaps to prepare a
compound not available commercially—then the enzyme must be
purified sufficiently from other proteins to allow only the desired
reaction to occur when the substrates and enzyme solution are mixed.
If it is necessary to verify that the enzyme activities measured in a
crude extract are valid, or if an enzyme from a previously untested
source is being compared to a well-documented enzyme of similar
reaction specificity (e.g. by pH optimum and K_m values), then the
extent of purification might perhaps be 5- to 20-fold (i.e. an increase
in specific activity of 5–20 times when compared to the extract).
However, for a study of the enzyme itself and its reactions, the
major task of obtaining the enzyme in as pure a form as possible
must be attempted.

The purification of enzyme proteins is seldom easy, particularly
with enzymes composed of subunits. The establishment of a reproducible and efficient purification procedure can be both a challeng-
ing and time-saving experimental project in itself. It involves a
blending of both art and science. Usually, there are five to eight steps
in the isolation of an enzyme from its cell source, but the number of
steps and the recovery of activity at each step depend on the choices
which are made. Experience will tell whether or not to use a step
that gives a large increase in specific activity but a fairly low yield of
enzyme units. The source of the enzyme and the time and effort
expended in the purification must both be considered. If tissue
culture cells are to be used, then probably a purification technique
would be chosen that uses the smallest weight of cells to give the best
yield. When starting with easily available tissues such as beef liver
from a slaughterhouse, a technique that gives the fastest purification
in the least number of steps might be chosen, even if the recovery
was low. A purification procedure should involve a simple breakage
of the cells, followed by as few manipulations as possible to separate
the enzyme from all other cell material and proteins. An efficient
purification can be achieved, but only after testing many methods.
The amount of time spent on the purification procedure will vary
with the need for the enzyme. If many experiments using large
amounts of enzyme are planned, then one should try to devise as
efficient a scheme as possible. If only a few experiments are to be

performed, then spending months devising the best possible procedure would be unwarranted.

In view of the fact that enzymes are globulins, standard protein fractionation methods (Sober *et al.*, 1965) are usually applied during the purification. The choice of procedures usually follows a pattern which has proved over the years to be successful with many proteins. First, the cell extraction is performed. Then nucleic acid material is removed, the protein precipitated by addition of ammonium sulphate, and the partially purified solution which is obtained can be fractionated by a number of ways, including: adsorption, ion-exchange chromatography, gel filtration or electrophoresis. Affinity chromatography, relying on the specificity of binding of the enzyme, is a modern method often used in the later stages of purification. In addition, dialysis is usually used to remove low molecular weight contaminants or excess buffer. Dunnill and Lilly (1972) surveyed 62 isolation procedures for microbial enzymes, and found that ammonium sulphate precipitation, ion exchange on DEAE-cellulose or DEAE-Sephadex and removal of nucleic acids were all used in at least 80% of the procedures. The steps were applied in this order, and adsorption if used was applied last. The choice of order of steps usually involves consideration of the volume of material to be handled. For example, ammonium sulphate fractionation is much easier to perform on a large volume of extract than is chromatography. The techniques mentioned allow for ample variation in conditions to separate the enzyme or enzymes being purified from contaminating proteins. Each of these methods, plus other useful procedures, is discussed below.

The enzymologist has an advantage when purifying an enzyme rather than a non-enzymic protein in that the catalytic activity of an enzyme protein fraction is a measure of the amount of enzyme present in the fraction. During the purification, each fraction is analyzed for both enzyme activity and protein content. From these two determinations, the specific activity can be obtained and the course of purification followed. The assay method should be rapid (Dixon and Webb, 1964), so that activity can be determined immediately after each purification step, before proceeding to the next step. For the rapid determination of protein concentration, the absorbance at 280 nm is measured, and an absorbance of 1.0 nm in a 1-cm path cuvette is assumed to be equivalent to a protein concentration of 1 mg ml^{-1}. Solutions from early purification steps must be diluted with a neutral buffer solution before reading their absorbance. Naturally, before the final reporting of results, both the enzyme activity and the concentrations of protein at each step should be determined as accurately as possible. Layne (1957) has summarized the most suitable methods for determination of protein concentration. Often in early steps of the purification, there are inaccuracies in activity determinations. These can be due to side reactions using up either the substrate or product being measured, or to the presence of inhibitors of the enzyme. The results of the purification should always be summarized in a standardized table showing, for each step, the volume of solution treated, the concentration of protein in mg ml^{-1}, the activity in units ml^{-1}, the specific activity obtained from the previous two figures by division, and the yield or recovery of activity. The table should give enough data to enable others to reproduce the purification exactly. A sample table is presented in the final section of this chapter. An investigator soon learns to expect a two- to four-fold purification with a simple ammonium sulphate fractionation, a greater increase in specific activity with a well-applied DEAE-cellulose column, and so forth. Attainment of a constant high specific activity may indicate that the protein is homogeneous, but other criteria must also be applied, as discussed at the end of this chapter. The yield of activity should be as high as possible. If a high yield is obtained in each fractionation step and only a single protein species then obtained, the existence of isoenzymes can be ruled out for that tissue or cell particle. Separation of the activity into several fractions, for example in chromatographic fractions, may indicate the existence of isoenzymes.

The purified enzyme should have all the properties that the enzyme has *in vivo*. Unfortunately, this is not always possible. Mitochondrial enzymes involved in electron transport, for example, show different properties depending on the method of solubilization. Mild acid or mild base may denature the protein with a loss in activity, or possibly deamidation of glutamine or asparagine residues may occur. Deamidation may cause an increase in the number of electrophoretically observable species which have enzyme activity, thus giving spurious results by suggesting artefactual isoenzyme forms. During treatment with some commonly used extraction or purification procedures, IMP dehydrogenase from *Bacillus subtilis* can undergo a change in both its pH optimum and its ability to be inhibited by GMP (Wu and Scrimgeour, 1973). Common precautions for preserving the native conformation through all the steps of a purification scheme include careful buffering of all solutions to a pH at which the enzyme has been found to be stable, and the performance of all manipulations at or near $0°C$ to prevent denaturation, proteolysis and bacterial growth. Because the thiol groups of cysteine residues can oxidize, causing aggregation or inactivation, many enzymes (especially dehydrogenases) often require the presence of a reducing thiol (about 10^{-3} M) in all solutions. If possible, 2-mercaptoethanol is used, but the more effective and expensive dithiothreitol or dithioerythritol may be needed. Reagents used for purification should be as pure as can be obtained or prepared, to prevent enzyme inhibition by contaminants such as traces of heavy metals.

It is probable that some enzymes which have been purified have been partially degraded by proteolysis during the early steps in their purification, and that this degradation is as yet undetected. Biotin carboxyl carrier protein from *E. coli* was originally thought to have a molecular weight of either 9100 or 10 400. Further examination has shown that the smaller fragments were a result of proteolytic breakdown of a larger molecule, a dimer with a molecular weight of 22 500 for each subunit (Fall and Vagelos, 1972). Leucyl-tRNA synthetase is isolated from extracts of *E. coli* in two forms, E_{II} which is completely active and E_I which requires the addition of a polypeptide of molecular weight 3000 for full activity (Rouget and Chapeville, 1971). E_I and the peptide are probably formed by proteolysis of E_{II}. Phosphofructokinase from yeast has a molecular weight of 800 000 when purified rapidly, but it is hydrolyzed to an enzyme of 570 000 daltons plus some small peptides when purified by traditional methods (Diezel *et al.*, 1973). Several other yeast enzymes have been isolated in partially hydrolyzed forms. Trypsin has been used to solubilize some membrane-bound proteins, but because the resulting protein may be degraded it should only be used as a last resort. Cytochrome b_5 obtained from liver microsomes has a molecular weight of 12 000 when it is isolated using trypsin treatment, but when solubilized with detergents it has a higher molecular weight (Ito and Sato, 1968; Spatz and Strittmatter, 1971). The larger native protein is composed of two domains (or globular sections)—a heme-containing polar domain and an apolar domain joined by a proteolysis-susceptible peptide length. Thus, it can readily be converted into the lower molecular weight hemopeptide by treatment with trypsin (p. 492).

PURIFICATION TECHNIQUES

The most complete collection of articles on the purification of enzymes is to be found in the "Methods in Enzymology" series. As well as containing the purification procedures for isolation of hundreds of enzymes in these monographs, one volume (Volume 22) is devoted to purification techniques and should be consulted for details of some of the techniques which I will discuss.

Choice of Cell Source

In starting to purify an enzyme, the first consideration is the tissue or cell source to be used. If the objective is to examine the alcohol dehydrogenase from a newly isolated strain of yeast, or the lactate dehydrogenase from a tumour cell line, then the source is predetermined and the methods used normally would be adaptations of published procedures. If it is the reaction itself that is of interest, then it is wise to search for a tissue source that has an elevated level of enzyme activity. Bacteria can often be grown under conditions that force them to overproduce the enzyme to be studied. Regulation of enzyme synthesis in animals appears to take place by a different mechanism than that of bacteria, and the levels of enzyme activity in animals are not readily amenable to manipulation. Some tissues in an animal may, however, have higher activities of an enzyme than others. Diploid strains of cells may have twice the activity of normal cells. Judicious choice and growth of the cell source may save countless days of enzyme purification.

The rate of synthesis of bacterial enzymes often is affected by the presence of substrates or end-products. Some enzymes in catabolic pathways are not normally present but can be induced in the presence of substrate or substrate analogues. Enzymes in some anabolic pathways are not synthesized when the products of the pathway (*corepressors*) are supplied in the growth medium or produced at sufficient levels by the cells. This type of control is termed feedback repression. To obtain the maximal rate of synthesis of an inducible enzyme, a suitable *inducer compound* is added to the medium. For maximal synthesis of a repressible enzyme, the end-product must be removed or limited. It is thought that both induction and feedback repression occur by a similar control mechanism involving proteins called *repressors* (cf. p. 564). The active form of a repressor appears to prevent synthesis of specific messenger RNA (and thus enzyme) and would be the free repressor for an inducible enzyme or a repressor–corepressor complex for a repressible enzyme. The inactive form of a repressor (inducer–repressor complex, or repressor not bound to corepressor) has no control over enzyme synthesis. Research on microbial protein biosynthesis and its control is currently one of the most active (and difficult) biochemical endeavours.

Bacterial enzymes that are under repressor control can be increased in level by manipulation of the contents of the growth media. An enzyme which is inducible can be increased in extreme cases up to 1000-fold by addition of inducer. An enzyme that is controlled by feedback repression can be increased if the corepressor is eliminated from the medium or decreased in concentration. Auxotrophic mutants are mutant cells which require a metabolite for their growth. By growing auxotrophic mutants in the presence of the essential compound and then transferring them to a medium free of the metabolite, many of the enzymes in the pathway leading to the essential metabolite may be increased in level by derepression. Lowering the phosphate concentration on which *E. coli* is grown until the phosphate is the growth-limiting constituent derepresses alkaline phosphatase to the extent that this enzyme becomes 6% of the cellular protein (Garen and Levinthal, 1960). In fact, this enzyme activity is only manifested in *E. coli* under conditions of limiting phosphate. Gerhart and Holoubek (1967) have utilized a mutant strain of *E. coli* which was prepared specifically to produce high levels of aspartate transcarbamylase. A mutant which had limited ability to synthesize pyrimidines was made diploid for aspartate transcarbamylase. The organism was grown in the presence of uracil, but during the last few generations the uracil became exhausted, the organism derepressed and the aspartate transcarbamylase activity increased over 500 times. The genetic and growth manipulations resulted in the transcarbamylase being 8% of the cellular protein, a source of large quantities of enzyme for the experiments described in Chapter 13. By growing *E. coli* in a continuous culture system in the presence of sufficient lactose for growth but not enough for induction, Novick and Horiuchi (1961) selected mutants able to synthesize β-galactosidase without induction. These bacteria were able to produce β-galactosidase as 25% of their protein. In a review article, Demain (1971) discusses these and other types of variations in the growth of micro-organisms for enzyme production.

Some enzymes occur at elevated levels in cells because of the metabolic requirements or anomalies of the cells, and therefore these cells are excellent sources of particular enzymes. Glyceraldehyde-3-phosphate dehydrogenase is almost 10% of the extractable protein of rabbit muscle, so a ten-fold purification gives a pure enzyme. Formyl tetrahydrofolate synthetase in the purine-fermenting organism *Clostridium cylindrosporum* acts in a substrate-level oxidative phosphorylation which gives the only ATP formed in the degradation of purines. This enzyme constitutes about 3% of the dry weight of the *Clostridium* and requires only ten-fold purification from autolysates to obtain crystalline enzyme (Rabinowitz and Pricer, 1962). Dihydrofolate reductase levels increase about 25 times in leucocytes after treatment of a leukemic patient with anti-folate drugs. As discussed later, this increase is probably because of a slowed breakdown of the enzyme. Bacterial mutants resistant to the otherwise toxic anti-folates are now being used as sources of dihydrofolate reductase for mechanistic and structural studies.

Preparation of Cell Extracts

There are many ways to extract or disrupt tissues and cells. Any of the methods may suit the cell source which has been chosen, but some methods are better than others. With microbial cells sonication is used most often, and with mammalian cells homogenization is commonly used, but this does not mean that these are always the most suitable methods. On the contrary, some enzymes may be denatured by these techniques, or other procedures may release more enzyme. A variety of methods, with their advantages and disadvantages, is discussed below. Some tissues may require preliminary separation, such as removal of connective tissue or of leaf veins, before the cell extract is prepared.

Microbial cells may be disrupted by either mechanical or chemical means. Sonication is the standard mechanical technique. Cells are suspended in buffer solution, then exposed to short bursts of sonic oscillation, followed by cooling for several minutes in an ice bath. The number of 30-second bursts required to release the enzyme from the cells can be determined empirically by assaying the extract obtained after each sonication. Other mechanical methods include: extrusion of cell pastes through a French press or other pressure device; homogenization with a blender and glass beads; several sequences of freezing and thawing of a cell suspension; or grinding a cell paste with alumina using a mortar and pestle. Solvents such as cold acetone can be used to break the cell walls. Autolysis is often used with large quantities of cells such as yeast, but denaturation can occur easily during the long time needed. The mildest method of chemical breakage is to use egg white lysozyme to hydrolyze the cell wall polysaccharides. Lysozyme alone can be used with Gram-positive organisms such as *B. subtilis*. Treatment of some Gram-negative bacteria such as *E. coli* with lysozyme and EDTA either causes partial lysis (Repaske, 1958) or releases enzymes such as alkaline phosphatase or 5'-nucleotidase from the space between the cell wall and the cell membrane. By placing cells treated with lysozyme-EDTA in a hypotonic solution, the intracellular contents may be released. Ribonuclease and deoxyribonuclease can be added to the lysozyme solution so that nucleic acids are hydrolyzed during the incubation.

Mammalian cells are usually broken by homogenization. If subcellular particles are needed, the homogenization must be performed by as gentle a technique as possible, such as with a Potter–Elvehjem homogenizer (Potter, 1955). This homogenizer is a plastic pestle with a stainless steel rod and a thick-walled glass tube which fits closely to the pestle. The pestle is driven by a heavy duty stirring motor (600–1000 rev/min) mounted firmly in a vertical position, such as on the laboratory wall. A suspension of chopped tissue is placed into cold buffer or isotonic sucrose in the glass tube, and the tube is raised so that the pestle is just submerged. The stirring motor is turned on and the tube is raised carefully. After each complete passage of the pestle, the glass tube is cooled in an ice bath. Usually three or more passes of the pestle are required for cell breakage. Air bubbles should not be allowed into the homogenate during the treatment. A Waring blendor can be used for homogenization if large quantities of tissue are being used or if the enzyme is stable.

The crude homogenates from some mammalian tissues may contain fat, other interstitial material or cell debris that can be removed either by passage through a filter funnel covered with several layers of cheesecloth or glass wool, or by centrifugation followed by separation of the layers obtained. (A large syringe with a blunt-tipped, large bore needle is often useful in performing this separation.) Preparation of a cold acetone-dried powder from the homogenate also will remove lipid material.

Determination of Subcellular Distribution

Acquaintance with the subcellular location of the enzyme under study is essential for successful isolation. Some enzymes are found outside the cells of origin. These extracellular enzymes include hydrolytic enzymes secreted by bacteria, moulds or the exocrine glands of animals, and enzymes occurring in blood serum. Extracellular enzymes, therefore, have already been separated from much of the cellular material which accompanies intracellular enzymes, and their isolation is partly completed. With intracellular enzymes, it is helpful to know if the enzyme being examined occurs in the cytoplasm of the cell or if it is bound to one of the subcellular particles. If the enzyme is present in one of the organelle fractions, then separation of this enzyme-containing organelle from other particles and from the cell sap can be a major step in its purification.

In order to obtain the subcellular particles from a cell, the outer membrane must be broken and the cellular contents separated into fractions of different density by differential centrifugation. Standard methods exist for the preparation of given fractions such as nuclei, mitochondria, chloroplasts, lysosomes and microsomes. The method used to isolate the subcellular fractions of rat liver has been used as the basis of most other techniques. Tissue is homogenized by the Potter–Elvehjem method in 0.25 M sucrose [9–10 ml per g tissue], isotonic sucrose aiding in preventing destruction of particulates by osmotic rupture. Cell components are then separated by fractional centrifugation (Hogeboom, 1955). For example, nuclei are removed from a homogenate of liver by centrifugation at $700 \times g$ for 10 min.

C

Mitochondria are sedimented from the nuclear supernant solution by centrifugation for 10 min at 5000 X g, and the microsomal fraction (ribosomes and endoplasmic reticulum particles) by 60 min at 54 000 X g. Enzymes which are not sedimented by the latter treatment are considered to exist in the cytoplasm of the cell or cytosol fraction. $CaCl_2$ (0.002 M) can be added to the sucrose to prevent aggregation of nuclei, and EDTA (approximately 1 mM) is often added when preparing mitochondria to bind cationic impurities. Beef serum albumin, which binds a variety of chemicals, is used to stabilize plant mitochondria during their isolation. A specifically designed sedimentation procedure is available for isolation of almost any organelle. Density gradient centrifugation, especially using a zonal centrifuge, allows separation of many bands differing only slightly in density. If an enzyme occurs in only one subcellular particle, isolation of the particles may be chosen as a first step in its purification. Subcellular particles can be broken or the enzymes solubilized by several methods, as will be described in Chapter 14.

Enzyme activity has often been used as an identification characteristic and a criterion of purity for subcellular particles. Such marker enzymes should occur only in the fraction containing the particle, should have stable activity and should remain firmly bound to the particle during its isolation. Few enzymes fulfil these features, and caution must be exercised in the interpretation of experiments of this type (Mahler and Cordes, 1971). Enzyme activities commonly used as identifying markers are NMN-adenyl transferase for nuclei, succinate dehydrogenase and cytochrome oxidase for mitochondria, glucose-6-phosphatase for the microsomal fraction, and lactate dehydrogenase and glucose-6-phosphate dehydrogenase for the cytosol. Cell fractionation followed by enzyme assay should be combined with cytochemical studies for confirmation of localization of enzymic activity in subcellular particles. Microscopic examination of the isolated particles may suffice, but more complicated methods such as electron cytochemistry, which is electron microscopy combined with enzymic cytochemical methods, hold promise for future studies (Shnitka and Seligman, 1971).

Removal of Small Molecules

There are several points in a purification at which small molecules must be removed from the protein solutions. One common point is prior to testing a crude extract for activity. Endogenous substrates and cofactors should be removed so that blank assay readings are not high, thus masking the activity of the extract. Before chromatography, adsorption or concentration, it is sometimes advisable to remove excess buffer ions. Either charcoal or washed Dowex resins can be added as dry particles to remove small organic impurities such as nucleotides, but by far the most common method of removing small solutes is dialysis.

Dialysis is the removal of low molecular weight components by allowing them to pass from a small volume of solution through a semi-permeable membrane into a large volume of water or dilute buffer. (Dialysis membranes are cellophane tubings intended for wrapping sausage meat.) The tubing should be washed before use and kept moist. Washing procedures vary from soaking in 1 mM EDTA, distilled water or dilute acetic acid, to boiling in dilute buffer or washing by a series of steps designed to remove expected contaminants (McPhie, 1971). The washed tubing is cut to the length desired, with extra length for the knots. Two knots are tied at one end, close together, and the sack is filled with water to test for leaks. The sack is then emptied, and filled with enzyme solution. The open end is tied so that a bubble of air is trapped inside the tube. The external solvent is placed in a large glass jar or a multi-litre beaker, and either the large volume of liquid is stirred with a magnetic stirrer or the sack is stirred by attaching it to a glass rod mounted in a slowly rotating stirring motor. The external solvent should be changed every few hours during the dialysis. After dialysis is completed, the sack is emptied into a tube or beaker by puncturing it carefully with a pair of surgical scissors.

Chromatography using a molecular sieve (p. 55) may also be used to remove small molecules rapidly.

Removal of Nucleic Acids

Nucleic acids and nucleoproteins are usually removed from the crude extracts as a first stage in purification. The usual methods are either to destroy nucleic acids with nucleases or to add a basic material which will combine with and precipitate the negatively charged nucleate. Streptomycin sulphate is usually used with bacterial extracts, but Taylor and Utter (1974) have found high concentrations of lysozyme to be preferable. Protamine sulphate is normally added to animal tissue extracts, but to achieve optimal and reproducible results, the protamine should be a good grade of salmine, freshly suspended in buffer. Careful acidification of the crude extract to pH 5 with a weak acid such as acetic acid can also be used, but the addition of the basic compounds is preferable. The crude extract with the added streptomycin or protamine is stirred for about 5–10 min. As with other stirrings, a magnetic mixer and Teflon stirring bar are used. The bar is turned at a rate that does not cause any frothing and hence denaturation of protein. The container in which the suspension is mixed is kept cool by surrounding it with chipped ice. This purification step removes nucleic acids, cell debris and a few proteins. There are some cases where a desired acidic enzyme is precipitable with the protamine. For example, Mazumder *et al.* (1963) used precipitation by protamine to effect a 12-fold purification of methylmalonyl CoA mutase from an ammonium sulphate fraction of sheep liver. The enzyme was redissolved by extraction of the precipitate with dilute phosphate buffer. However, most enzyme proteins remain in solution at this stage.

Heat Treatment

In some cases, it is advantageous to carry out a selective heat denaturation as an early step in purification. This is useful only if the enzyme under study is more stable than the other proteins under the conditions of the heat treatment. The addition of a substrate, cofactor or reversible inhibitor at a saturating concentration will often *stabilize* an enzyme towards heat denaturation or proteolysis. After coagulation of much of the unwanted protein, the solution is rapidly cooled in an ice bath. Serine transhydroxymethylase from chicken liver can be heated to 70° C for 2 min in the presence of 0.017 M serine (which is later removed by dialysis), with a three-fold increase in specific activity but little loss in total activity (Scrimgeour and Huennekens, 1962). In the absence of added serine, the transhydroxymethylase is largely denatured when heated under the same conditions. Heat treatment may alter the kinetic or physical properties of the enzyme and so should be used with caution. Factors which may affect the yield or activity of the enzyme obtained by heat treatment are the pH, the ionic strength and the duration and rate of heating.

Fractionation by Precipitation

Several methods are used for the precipitation of proteins. Of these, by far the most common is the use of ammonium sulphate. Globular proteins demonstrate a decrease in solubility at high salt concentrations (i.e. they may be *salted-out*) and at a given temperature and pH value, it will be found that each enzyme has a range of salt concentration at which it is precipitated. However, because each protein precipitates over a range of salt concentration, separations are not clean. Ammonium sulphate is used for salting-out because of its high solubility and because it has little deleterious effect on proteins. Enzyme Grade (purified) ammonium sulphate should be used if at all possible. The salt is added either as dry crystals or as a saturated solution. The solution should be neutralized with NH_4OH and added slowly to the stirred solution of enzyme using a pipette with a widened tip held just under the surface of the enzyme solution. The concentration range at which optimal purification of enzyme is obtained is determined empirically. At a saturation level of 35%, little material will precipitate if the nucleic acids have been removed first. Cuts or fractions of about 15% span should be attempted during initial trials at the purification. A simple equation for calculating the amount of saturated solution to be added is:

$(x + $ vol.$)$ (desired % saturation)

$$= \text{(previous \% saturation) (vol.)} + 100x$$

where $x = $ volume of saturated ammonium sulphate to be added and vol. $= $ volume of the original solution.

Solid ammonium sulphate should be added in small increments, making sure that each addition is completely dissolved before the next is added. The use of solid salt has the advantage of keeping the volume of solutions lower, but the disadvantages of being slightly acidic and giving local high salt concentrations as the salt dissolves. A nomograph for calculating the amount of solid ammonium sulphate to be added to achieve any given concentration has been published

(Dixon and Webb, 1964). After the ammonium sulphate has been added and is dissolved, the solution is stirred gently for about 15 min. The precipitate is then collected by centrifugation in a refrigerated centrifuge.

Organic solvents are seldom used for precipitation of protein fractions because of possible denaturation of the proteins. Kaufman (1971) has summarized the technique for fractionation using ethanol. His article lists the advantages and precautions for solvent fractionation. The most difficult task is the control of the temperature. The enzyme solution should be kept below $0°$ C at all times, with a dry-ice–ethanol cooling bath. Other solvents which have been used for enzyme purification include acetone, chloroform–ethanol and chloroform. Precipitated proteins can usually be dissolved in dilute buffer solutions and the protein concentration can be kept high to stabilize the enzyme.

When describing the results and method of purification using salt or solvent precipitation, one must describe fully the conditions used so that the procedure can be repeated. The salt concentration which will precipitate an enzyme varies with the concentration of the enzyme, so the protein concentration should always be recorded.

Fractionation by Gel Adsorption

Two adsorbent gels have been found useful as purifying agents for enzymes, calcium phosphate and alumina C_γ. These gels can be used either to selectively adsorb the enzyme or to adsorb only contaminating proteins. Adsorption is carried out at a slightly acid pH (near 6) at low buffer concentration. Successive amounts of gel are added to the enzyme solution, with each being removed by centrifugation before the next addition. Rapid assay of eluted enzyme will tell when the enzyme has been adsorbed. Elution is performed by mixing the precipitated gel with small volumes of buffer using a large rounded glass stirring rod. The buffer used is usually phosphate at pH 7.5, with a little neutral salt added if needed. Some enzymes tend to be unstable after elution from gels, so it is wise to carry out a concentration step such as an ammonium sulphate precipitation after adsorption and elution. If ion-exchange chromatography is to be attempted next, the buffer used for elution from the gel must be removed in any case, so excess ammonium sulphate can be dialyzed away at the same time. Both the more frequently used calcium phosphate gel (Keilin and Hartree, 1938) and alumina C_γ gel (Willstätter and Kraut, 1923) are easily prepared, but the procedures call for several weeks of storage before use. They are both available commercially, ready for use.

Ion-Exchange Chromatography

In the mid 1950s, a major tool for enzyme purification, which constituted a series of substituted cellulose ion-exchange compounds, was prepared by Peterson and Sober (1956). Although some proteins had previously been purified on ion-exchange resins such as Dowex-50, the substituted celluloses allowed adsorption and elution at the neutral pH values needed for mild treatment of enzymes. The electrostatic interaction of protein with the binding groups on the surface of the cellulose particles is not strong enough to disrupt the conformation of most proteins (i.e. cause denaturation). There are two types of ion exchange adsorbents, *anion exchangers* which have a weak base bound to the cellulose and *cation exchangers* which have a weak acid bound. The anion exchangers are best exemplified by DEAE-cellulose, which has diethylaminoethyl [cellulose$-$O$-$CH$_2$CH$_2-$N$^+$H$-$(CH$_2$CH$_3$)$_2$] ionizable groups incorporated. DEAE-cellulose is protonated at pH values lower than about 9, and at neutral pH will weakly bind negatively charged proteins or anionic areas of protein molecules. CM-cellulose is the cation exchanger most often encountered in enzyme purification. It has carboxymethyl (cellulose$-$O$-$CH$_2-$COO$^-$) functions substituting some of the cellulose hydroxyl groups, and will bind positively charged species.

Excellent preparations of both DEAE- and CM-cellulose are available from commercial sources, but for some purposes it is necessary either to test the capacity of the commercial materials or even to synthesize the exchanger. Himmelhoch and Peterson (1966) have described a method that can be modified to test the properties of any potential adsorbent. Directions for washing and preparation of columns are supplied by the manufacturers and a detailed account appears in an article by Himmelhoch (1971). Briefly, the celluloses should be preconditioned by washing with mild (less than 0.1 N)

NaOH, then water, followed by mild HCl, then water, NaOH again and finally water, to remove excess base. The fine particles should be removed with a pipette connected to a suction pump. Then the required amount of cellulose is slurried into a glass chromatograph column, allowed to settle to form the column, and washed with several column volumes of the buffer to be used for both addition of the sample and the start of elution. The protein solution is carefully layered over and allowed to enter the column. Elution is usually accomplished by increasing the ionic strength (*gradient elution*) using a simple gradient mixer made from two flasks connected by a stopcock. The reservoir flask contains the more concentrated buffer (usually 0.1–0.2 M) and the mixing flask, stirred with a magnetic stirrer, contains the initial buffer (usually 0.005–0.01 M). If the enzyme is readily separated from other proteins on the exchanger, it sometimes saves time to elute by adding batches of several buffers of increasing concentration (*stepwise elution*). The gradient method gives the best separation because a gradual increase in the concentration of counter-ions competing with proteins for the ionic site on the cellulose produces sharper resolution of the proteins. The eluted material is collected in tubes in a fraction collector. A carefully maintained fraction collector having electronic circuits that are shielded from moisture can generally be trusted for overnight separations in a cold room, but when in doubt it is best to plan experiments so that a close watch can be kept on the elution and the spillage of any fractions can be avoided. If the eluting buffer is too dilute or if the fine particles have not been removed, then the column may run too slowly, resulting in loss of time and possibly of enzyme activity. A column of DEAE-cellulose can give a purification of 8- to 20-fold, which is considerably better than the combination of several of the steps that have previously been described.

Sometimes DEAE-cellulose can be added to a large volume of enzyme solution to either adsorb the enzyme or to adsorb contaminants, and the exchanger removed by centrifugation or filtering. It is fortunate if conditions are found which allow batch adsorption of the enzyme, for then purification time can be greatly reduced. With large-scale purifications, the adsorption can also be used as a concentration step.

Other exchangers that can be used for similar types of purification include the positively charged aminoethyl- and triethylaminoethylcelluloses and the negatively charged phospho- and sulphoethyl-celluloses. Both types of molecular sieves (Sephadex and Biogel) have been substituted with ion-exchange groups, and with some enzymes a better separation can be obtained using these column materials than with the more frequently used substituted celluloses.

Chromatography on Hydroxyapatite

Tiselius and his coworkers (1956) described the preparation of chromatography columns suitable for protein fractionation formed from the hydroxyapatite form of calcium phosphate. This material has the chemical structure $Ca_{10}(PO_4)_6(OH)_2$, and is formed by heating brushite, $CaHPO_4 . 2H_2O$. Columns of hydroxyapatite tend to run slowly compared with either DEAE-cellulose or Sephadex. The slow flow characteristics can be overcome by mixing the hydroxyapatite with filtering aids such as cellulose or even brushite itself. Brushite can sometimes be used instead of hydroxyapatite with results that are just as successful, and since it is prepared as an intermediate, its use may save effort. Bernardi (1971) has summarized the methods and some successful applications of hydroxyapatite chromatography.

Gel Filtration

The purification techniques described so far depend on either the solubility of the enzyme (or insolubility of contaminants) or its ionic charge. Gel filtration or molecular sieving utilizes an entirely different property of the protein for its fractionation—its molecular weight or effective radius. Although the column materials used do have some ion-exchange or adsorption properties, the main effect is one of exclusion of large protein molecules from the bead-like gel material. In other words, small proteins can be retarded by inclusion inside the cross-linked gel support, and the degree of retardation is a function of the size of the protein. The exclusion limit is the molecular weight of the smallest polymer which is not retarded by the gel.

Three polymers are used for production of gel filtration beads. Dextran is used for Sephadex, and the crosslinks are glyceryl chains. Polyacrylamide gels are the base for Bio-Gel P, and purified agarose, dissolved in water at $100°$ C and then cooled, forms a gel (called either Sepharose or Bio-Gel A) which is insoluble below 40° C. Each of these three types of gel is available with different exclusion limits, from Sephadex G-10 which excludes molecules of very low molecular weight (about 1000) to Bio-Gel A-150m which only excludes particles of 1.5×10^8 daltons or higher. Eight Sephadex gels, ten Bio-Gel P products and nine agarose gels of differing exclusion limits are currently available.

The preparation, packing and storage of molecular sieving gels are clearly described in the manuals supplied by the manufacturers and the physical characteristics and a troubleshooting chart are given in the article by Rieland (1971). In this article, the suitable fractionation ranges for globular proteins are listed for all the gels. Since charge effects are not being utilized in the elution, there does not have to be gradient elution, simplifying the purification and giving a protein dissolved in a buffer of low ionic strength. More care must be taken with molecular sieve columns both in packing and when applying samples than is required with the substituted celluloses. The dextran and polyacrylamide gels are supplied as beads which must be hydrated, whereas the agarose gels are in the form of an aqueous slurry which must not be dried, frozen or heated.

Gel filtration columns are useful not only for purification of a protein but also for desalting and for determination of molecular weights. As we shall see below, the dry gels can also be used for concentration of protein solutions. Gel filtration is a simple, rapid and gentle method for purification. It is easily scaled up for larger isolations, and generally the recoveries of enzyme activity are good. Most of the difficulties can be minimized or eliminated by exercising care and observing the experimental guidelines suggested by the manufacturers.

Preparative Electrophoresis

The most useful electrophoretic technique for enzyme isolation is preparative electrophoresis in polyacrylamide gels. Because the capacity of a gel is rather small (100 mg protein or less), this method is used as one of the last steps in a purification scheme. The approximate molecular weight of the enzyme is known by this time from gel filtration, and its charge is known from its behaviour on ion-exchange chromatography. With this information, the appropriate electrophoretic procedure can be established (Shuster, 1971). The protein band can be located by staining a small longitudinal section of the gel, or by using a fluorescent dye on the surface of the intact gel (Hartman and Udenfriend, 1969). Enzymes can usually be eluted from polyacrylamide gels by careful homogenization of the gel with about 10 volumes of buffer in a small hand-held glass homogenizer and soaking overnight. A more complete description of polyacrylamide gel electrophoresis is given later in the text, under analytical electrophoresis (p. 68). The methods used are the same for both analytical and preparative experiments, except for the sizes of the gel and sample.

Isoelectric focusing (Vesterberg, 1971) can sometimes be used for a late step in purification, too. In this technique, the protein undergoes electrophoresis in a column of water (stabilized against convection by a density gradient or gel) containing a series of Zwitterionic buffers (ampholytes). The ampholytes migrate to a position in the column at which they are neutralized, and the protein is concentrated at the point in the pH gradient of the column which corresponds to its isoelectric pH. This is a powerful separation technique, allowing separation of proteins having only small charge differences. There are several disadvantages, though, such as the many hours required, the possible precipitation of some proteins in the column (which then precipitate through the lower zones), and the possibility of appearance of artefactual bands during the run. Because of these and other limitations, isoelectric focusing has not found general use for large-scale enzyme purification. It is primarily an analytical technique. Leaback and Robinson (1975) have been able to separate two proteins having isoelectric points too close for separation by isoelectric focusing. They adsorbed the enzymes on an ion-exchange cellulose, then displaced them with an appropriate mixture of ampholyte buffers.

Affinity Chromatography

The most promising new technique, affinity chromatography, relies on the binding specificity of enzymes for the mechanism of separation. This is another type of separation in addition to fractionation by either solubility, charge or size. Either a substrate or inhibitor (ligand) of the enzyme to be purified is covalently linked to a solid matrix which is used to form the chromatography column. A solution of partially purified enzyme (subsaturating in the desired enzyme) is applied to the column and eluted using as slow a flow rate as is practical to allow complete equilibration. Protein which is not bound to the specific adsorbent ligand passes through the column. The desired enzyme may be eluted either by specific elution (e.g. with a solution of its substrate), or by non-specific elution, with an increase in salt concentration or a change in the pH of the eluting solvent. Cuatrecasas and Anfinsen (1971a) stated that the support matrix should:

1. not adsorb proteins strongly;
2. have good flow characteristics;
3. have chemical groups which can be activated for binding of ligands; and
4. be mechanically and chemically stable.

The most suitable support matrix found to date is agarose in the gel bead form, although polyacrylamide beads have also been used. The covalent binding of ligands to agarose is by attachment of an amine function to the carbohydrate by cyanogen bromide activation of the agarose (reviewed in Cuatrecasas and Anfinsen, 1971b). The pathway of coupling appears to involve formation of a cyclic imido-carbonate which reacts with the amino substituent (H_2NR, ligand) to form an isourea:

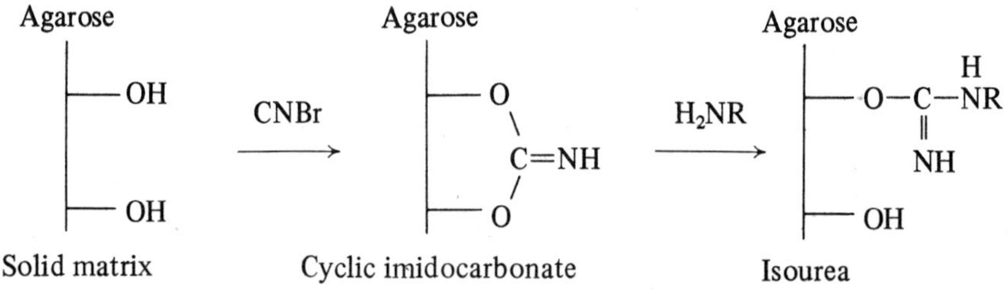

Solid matrix   Cyclic imidocarbonate   Isourea

Because the isourea bonds are not completely stable, a small but constant leakage of the ligand from the matrix occurs. Compounds having an amino group may be coupled directly to the agarose. However, it seems advantageous to react the agarose first with an amino compound which has an alkyl chain. The alkyl chain acts as a spacer arm connecting the ligand and positioning it away from the surface of the agarose. The use of the arm enhances the binding of either high molecular weight proteins or enzymes with relatively weak protein–ligand bonds. The correct length of spacer arm is needed so that the interaction of the bound ligand with the enzyme matches the ligand–enzyme interaction in free solution. Shaltiel and Er-el (1973) have suggested that unsubstituted aminoalkyl arms can bind accessible hydrophobic regions of proteins. For greatest specificity, they recommend that first the ligand be connected to the side chain, and then the elongated ligand be bound to the agarose. They have also used the alkyl agarose chains, with no specific ligands, for "hydrophobic chromatography". An excellent example of the use of hydrocarbon-coated agarose for enzyme separation and purification is resolution of the enzymes of the *E. coli* glutamine synthetase system on ω-aminoalkyl agarose beads (Shaltiel *et al.*, 1975; p. 457). Application of an ammonium sulphate fraction to a column of Sepharose$-NH-(CH_2)_5-NH_2$, followed by elution with a gradient of KCl, separates P_{II}, UTase, glutamine synthetase and ATase, and provides 10- to 20-fold purifications of each enzyme, with 60–90% recoveries. This work illustrates how systematic choice of adsorbent and of eluting conditions can be achieved, and how drastic elution conditions can be avoided. Columns of agarose, to which L-valine is attached by its amino group, adsorb a large proportion of proteins in a bacterial cell extract in the presence of 1 M potassium phosphate (Rimerman and Hatfield, 1973). The proteins can be eluted with a gradient of decreasing concentration of phosphate in an order related to their solubility in ammonium sulphate. Chromatography using this phosphate-induced binding (possibly hydrophobic) may have general application.

Two examples of agarose affinity column materials relying on *group specificity* of binding rather than *absolute substrate specificity* are that on which *p*-mercuribenzoate is bound and that on which NAD^+ or a derivative is bound. Group specific adsorbents may not be as selective as adsorbents containing more specific ligands, but judicious choice of loading and elution conditions can restore much of their selectivity. The organomercurial-agarose matrix has the structure:

Agarose

$$\text{---NH---CH}_2\text{CH}_2\text{---}\overset{\text{H}}{\text{N}}\text{---}\overset{\overset{\text{O}}{\|}}{\text{C}}\text{---}\langle\bigcirc\rangle\text{---Hg OH}$$

and has a high capacity for proteins having thiol groups (Cuatrecasas, 1970). These proteins can be eluted with either chelating agents or reducing agents. Few enzymes have been effectively purified on columns of this material. One example is creatine kinase, a protein abundant in skeletal muscle and one that had been purified from many animal species by classical fractionations. Because creatine kinase possesses an essential thiol group, it is a good candidate for purification on p-mercuribenzoate-agarose. Madelian and Warren (1975) used chromatography on mercuribenzoate-2-aminoethyl-Sepharose after partial purification as a final step in isolation of human muscle creatine kinase. Boegman (1975) utilized the organo-mercurial-Sepharose as an initial purification step, applying a crude supernatant fraction of fowl muscle. The affinity column retained the kinase but few of the contaminating proteins, most of which could be removed by subsequent chromatography on Sephadex. In both these purification procedures, creatine kinase was eluted with Tris buffer containing a thiol reductant. The extent of purification by affinity chromatography was not great (approximately three-fold), but the yields of activities were higher than those from classical purification methods.

General ligand affinity chromatography has been most extensively applied to the purification of NAD^+- and $NADP^+$-dependent dehydrogenases [reviewed by Mosbach (1974) and Harvey et al. (1974)]. The columns use agarose supports with NAD^+ (cf. p. 206 for the structure of NAD^+) or AMP derivatives covalently bound. These group-specific adsorbents can be used with a wide range of enzymes, and several are commercially available. Determination of which column material can best resolve one enzyme from a mixture is made by testing different ligands and elution conditions. Eluting solutions may contain NAD^+, NADH, a coenzyme plus a substrate or inhibitor, or maybe just salt or pH gradients (although need of one of these gradients suggests that the binding of the enzyme to the column is by a nonspecific process). Sometimes a pulsed amount of coenzyme may suffice for selective elution.

One column, ϵ-aminohexanoyl-NAD^+-Sepharose (Mosbach et al., 1972), has NAD^+ covalently linked, presumably by ester bonds of ribose hydroxyls, via a 6-aminohexanoic acid arm to the agarose. It can be used repeatedly with no loss of binding capacity, but may slowly hydrolyze even in the cold. Another Sepharose-based NAD^+

column in which the covalent binding is to the N^6-position of the AMP group of NAD^+ is available (Mosbach, 1974). It is more difficult to prepare, but its structure is known and it appears to be a more biospecific adsorbent (O'Carra et al., 1974). The NAD^+-agarose affinity columns have been supplemented with columns containing only the adenylate moiety of the coenzyme. For example, 8-(6-aminohexyl)-amino-AMP-Sepharose (Lee et al., 1974) and N^6-(6-aminohexyl)-AMP-Sepharose (Mosbach, 1974) have been synthesized and tested. Use of these AMP-Sepharose columns has proved advantageous in the purification of some enzymes. Luckily, the relative affinities of dehydrogenases to these and other available NAD^+- and AMP-agaroses vary markedly, so patient research can usually uncover a worthwhile purification step.

One recent discovery has been that blue dextran, a dye attached to dextran, can be used to purify some NAD^+- and AMP-requiring

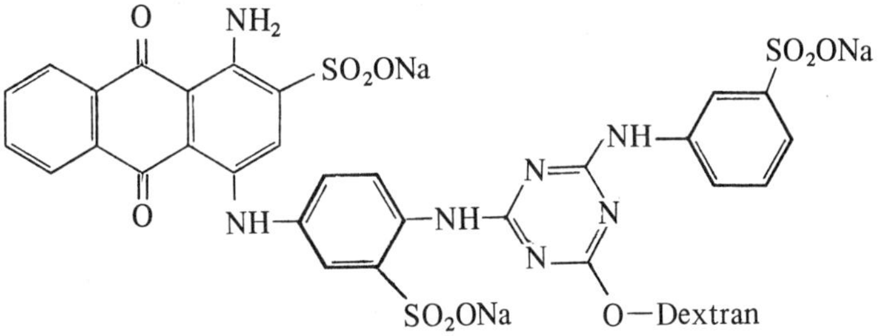

Blue dextran

enzymes. These enzymes bind to blue dextran in solutions of low ionic strength, but not in high ionic strength solutions. When blue dextran is coupled to Sepharose, some enzymes can bind specifically and then be eluted with ligands such as AMP or NAD^+, or with high concentrations of NaCl. Thompson et al. (1975) have suggested that blue dextran is a structural analogue of NAD^+ and specifically binds to those proteins containing the dinucleotide fold, a three-dimensional structure which occurs in some dehydrogenases and kinases.

Affinity chromatography has been used to great advantage by Stevenson and his colleagues for the simultaneous purification of proteases from pancreas glands (Lievaart and Stevenson, 1974). Moose pancreas tissue is extracted with dilute sulphuric acid at 4° C, which ruptures the cells and denatures some of the contaminating proteins. The extract is then clarified and the desired zymogens are obtained by an ammonium sulphate fractionation (40–65%), after which the extract is dialyzed and lyophilized. Portions of this crude pancreas extract are then dissolved and the zymogens converted to the active proteases by a short incubation. The mixed protease solution is applied to a column of 4-phenylbutylamine-Sepharose

$$\langle\bigcirc\rangle - (CH_2)_4 - NH - Sepharose$$

PBA–Sepharose

which selectively retains chymotrypsin. That the phenyl group of this matrix is an inhibitor of chymotrypsin, and hence a specific binding site, was shown by lack of binding of chymotrypsinogen to PBA-Sepharose or of chymotrypsin to unsubstituted Sepharose (Stevenson and Landman, 1971). The initial effluent from the first column is then applied to a column of lima-bean protease inhibitor(LBI)-

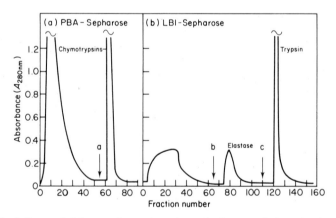

Fig. 1. Isolation of Moose chymotrypsins, elastase and trypsin by affinity column chromatography. The crude protease solution (in pH 8.0 buffer) was first applied to the PBA-Sepharose column. The initial fraction was collected and applied to the second column, LBI-Sepharose. Chymotrypsins were eluted from the PBA-Sepharose with 0.1 M acetic acid (point a). Elastase and trypsin were eluted from the LBI-Sepharose column with 50 mM sodium citrate, pH 3.0 (b) and 40 mM KCl, pH 2.0 (c), respectively. From Lievaart and Stevenson (1974), with permission.

Sepharose, which retains elastase and trypsin. The lima-bean inhibitor protein possesses two independent protease-binding sites. In these experiments, elastase binds to one site and trypsin to the other. The three protease activities are eluted—two chymotrypsins from the first column (these can be separated on CM-cellulose), and the separated and purified elastase and trypsin from the second column—as shown in Fig. 1. Both yields and purities of the proteases are high.

The ligands which have been used successfully for affinity columns include substrates, coenzymes, inhibitors and allosteric effectors. Perhaps this technique will be used to isolate the altered proteins, possessing substrate binding capability but little or no catalytic activity, in genetic disorders such as phenylketonuria. The application of affinity chromatography as a late—or even as an early—purification step may produce an enzyme solution close to homogeneity. A number of other examples of the use of affinity chromatography are discussed in later chapters. The theory and application of affinity chromatography is the subject of a recent book (Lowe and Dean, 1974), and of the entire Volume 34 of "Methods in Enzymology".

Concentration of Enzyme Solutions

An integral part of the purification of any enzyme is its concentration from dilute solution. Concentration is extremely important because most enzymes are labile in dilute protein solutions but are much more stable when present at a concentration of $10 \, \text{mg ml}^{-1}$ or greater. I have mentioned the use of ammonium sulphate precipitation as a method of concentration. Many enzymes are precipitated with ammonium sulphate after the purification is completed and stored frozen as pastes after centrifugation. It is desirable, though, to have the purified enzyme free of salt if possible. Two methods which are both simple and mild are ultrafiltration and drying in a dialysis sack.

Ultrafiltration under pressure, which relies on the passage of solvent but not high molecular weight solute through a selectively permeable membrane, is a rapid method for concentration of enzyme solutions. Several companies such as "Amicon" and "Millipore" produce both equipment and filters for different volumes of solution and different sizes of protein (Blatt, 1971). For enzyme purification, a simple magnetically stirred filtration unit connected to a tank of

N_2 gas can be kept set up in the cold room for use when needed. A somewhat similar procedure, using devices containing collodion membranes, can be used; this method is called either vacuum dialysis or pressure dialysis, depending on whether the solvent is forced or sucked through the membrane.

An inexpensive technique using membranes as selectively permeable agents is the drying of a solution that is inside a dialysis sack. Pervaporation (circulation of warm air around the sack) is a classical method of concentrating proteins, but not all enzymes can resist the length of time and heat required under pervaporation conditions. It is now much simpler to put the dialysis tube in a beaker in the refrigerator and cover it with dry Sephadex G-200. The Sephadex draws water from the dialysis tube, and leaves the protein inside the tube. The wet Sephadex should be scraped off every half hour or so until the desired volume is reached. The protein solution should not be overdried or it will be difficult to remove from the dialysis sack. The wet Sephadex can be dried and used several times. Sucrose crystals (granulated cooking sugar often is of high enough purity) can be used instead of the gel beads if it is not necessary to keep the concentrated enzyme solution free of sucrose. Sucrose in the enzyme solution can be an advantage, as sucrose or glycerol are sometimes used to prevent denaturation of proteins in solution. Polyethylene glycol (Carbowax) granules inside dialysis tubes have been used to adsorb water from protein solutions (Kohn, 1959), and also, dialysis sacks containing enzyme solution can be buried in Carbowax.

Other methods of concentration are described by Sober *et al.* (1965) and by Blatt (1971). These include lyophilization, adsorption, centrifugation and slow freezing. Crystallization (Jakoby, 1971) is an ideal method both for final purification and for concentration. Although formation of protein crystals does not assure homogeneity, crystallization can be used to remove traces of impurity and to stabilize the enzyme (*if* crystallization can be achieved).

Concluding Remarks on Enzyme Purification

Purification of each enzyme is a novel project. If the same purification could be used for all enzymes, then they could not be separated from each other! One makes use of the small differences between the enzyme and the contaminants, and exploits these differences. As I mentioned above, there is both an art and science to enzyme purification. The art involves correct performance of the techniques which have been discussed. The science of purification results from a rational choice and application of these techniques. Published methods for purification of the enzyme should be studied to avoid repetition of other workers' failures, or tested for possible modification or extension. Each step of a purification should be well understood. The conditions allowing maximum stability of the enzyme should be determined early in the work, and checked after both partial and final purification. The best method of cell breakage and the optimal conditions for each fractionation step should be determined. The methods described, with their variations, give an unlimited choice of experimental conditions which should allow for the purification of any enzyme. As with all experiments, the first attempt will not be as successful as the second and succeeding experiments. Proficiency costs both time and effort, but it results in a saving of both in the long run.

Large-scale isolations of enzymes present problems similar to the smaller purifications needed for laboratory use along with the additional difficulty of keeping the large amounts of material from denaturing. For example, the time needed to freeze and thaw a large volume of protein solution could lead to large losses in activity. Most techniques can be scaled up quite successfully; these include ion-exchange chromatography, gel filtration and ultrafiltration. The coarser sizes of gels and particles are needed for sufficient rates of flow. Solid ammonium sulphate is preferable to a saturated solution for large-scale work, so that the volume is as small as possible. Microbial cells, a favourite source of enzymes in smaller preparations, are even more often the enzyme source of choice in large-scale isolations because continuous growth and enzyme enrichment may be used. As the use of large amounts of enzymes in industrial and pharmaceutical applications increases, ingenuity applied to the purifications will also increase the number of modifications to standard purification techniques.

MOLECULAR WEIGHT AND HOMOGENEITY

Isolation of an enzyme should result in a protein fraction in which more than 99% of the protein is the enzyme desired. The presence of traces of other enzymes may go undetected by physical techniques, but their activity may be exerted by very small amounts. Some activities might obscure mechanistic studies and should be assayed. If present, these activities should be inhibited, or removed by further purification. If it can be shown that the major (or only) protein species is responsible for the activity which you set out to purify, then the physical properties of this protein obviously are those of the enzyme. The traditional methods used for study of the hydrodynamic properties of proteins, such as ultracentrifugation, can be applied to purified enzymes. More often, the new techniques using molecular sieves and electrophoresis on polyacrylamide gels are used to determine the molecular weight and subunit content of enzymes. These methods use small quantities of purified enzyme, so the experimental conditions can be varied to perform many experiments using but a few milligrams of the valuable enzyme.

The purity, or lack of purity, of an enzyme often becomes apparent from results of the latter steps in the purification. The presence of a skewed major peak or of several peaks of activity or of protein in a column chromatograph indicates that the enzyme is not yet pure or perhaps consists of subunits. If after meticulous assay the specific activity of each tube of a peak from a column is the same, this can be an indication of purity. Similarly, the achievement of a constant specific activity after further attempts at purification suggests either purity, a constant amount of denaturation during fractionation or poor choice of purification techniques. Physical techniques should therefore be used to confirm the purity and at the same time determine the molecular weight of the enzyme.

Sephadex and other gels can be used for the determination of molecular weight but are not as indicative of purity as is electrophoresis. The molecular weight results obtained by gel chromatography are only approximate (10% uncertainty), but the procedure is both simple and inexpensive, and can be applied to solutions of enzyme that are only partially purified. A comparison of the position of elution of a series of proteins of known molecular weights (purchased individually or as a molecular weight kit) is made, and a calibration curve of elution volume v. log of molecular weight is constructed. The elution position of the experimental sample (its protein and/or activity profile) is compared with those of the other

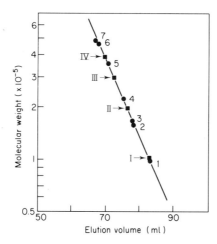

Fig. 2. Determination of the molecular weight of IMP dehydrogenase and its subunits by chromatography on Sepharose 4B. The Roman numerals I–IV show the positions of elution of enzyme-active fractions. The arabic numerals indicate the positions of elution of seven proteins of known molecular weights. The marker proteins were run in two separate batches. The elution volumes indicate that the four forms of IMP dehydrogenase have molecular weights of 105 000, 200 000, 300 000 and 390 000, respectively. Most of the activity was in the dimeric (peak II) and the tetrameric (peak IV) forms. From Wu and Scrimgeour (1973).

proteins. An example of the use of molecular exclusion chromatography for the determination of the molecular weight of an enzyme is presented in Fig. 2. The results described there were obtained using highly purified IMP dehydrogenase obtained from *B. subtilis*. This protein is an oligomeric enzyme, with molecular weights that are multiples of the 100 000-dalton protomer. Andrews (1965) has described the use of Sephadex G-100 and G-200 for estimations of molecular weights over the entire range of 10 000 to 800 000 daltons. He obtained anomalous results with glycoproteins, but found a smooth curve for carbohydrate-free globulins. For some presently unknown reason, a number of proteases give anomalously low molecular weights (Voordouw *et al.*, 1974).

Another method for determining the molecular weight of enzymes, either purified or in relatively crude preparations, is sucrose density centrifugation (Martin and Ames, 1961). The enzyme sample is layered on to the density gradient, centrifuged in the swinging bucket rotor of a preparative ultracentrifuge, then a hole is punched in the bottom of the centrifuge tube so that fractions can be collected. A standard protein of accurately known molecular weight is included. The sedimentation coefficients of the enzyme and the standard arc proportional to their mobilities, assuming that they have the same partial specific volumes. The approximate molecular weight of the enzyme can be obtained from the equation:

$$\frac{\text{migration of unknown}}{\text{migration of standard}} = \left(\frac{\text{mol. wt. of unknown}}{\text{mol. wt. of standard}}\right)^{2/3}.$$

Polyacrylamide Gel Electrophoresis

The technique which produces the greatest number of results with the least effort and least enzyme consumed is polyacrylamide gel electrophoresis. The original methods reported by Ornstein (1964) and Davis (1964) and the review by Chrambach and Rodbard (1971) should be referred to for the experimental details of gel formation and protein separation. The gel is formed from acrylamide and the crosslinking agent, methylenebisacrylamide, photopolymerized in small open-ended cylindrical glass tubes. Both these constituents are usually kept in a constant ratio (e.g. 30:1) for a set of experiments, and the amount of crosslinking is reported as the gram percentage of acrylamide monomer. A tracking dye (e.g. bromophenol blue) is added with the protein sample so that the rate of migration can be followed.

Apparatus required constitute a power supply (at least several hundred volts with an amperage of 80–100 mA), an electrophoresis chamber with connectors to the power supply and a rack for holding the gel tubes during polymerization. Caution should be taken not to touch the electrical apparatus during its operation, and not to touch dry acrylamide or pipette by mouth solutions of unpolymerized acrylamide. Protein may be located on the gel after electrophoresis by scanning in a special ultraviolet absorbance scanner or by staining the gel for protein. The most frequently used dye has been Coomassie Brilliant Blue R250. The gel is soaked in a solution of Coomassie Blue (dissolved in methanol and added to acetic acid) for 2–3 hours, then the gel is transferred to a solution of methanol–acetic acid for removal of excess dye (destaining) (Weber et al., 1972). About 12 hours are needed for removal of the dye by diffusion, but the process can be hastened by agitation and by changing the destaining liquid several times. Destaining can also be carried out by electrophoresis, but this method does not give as clear a background and some proteins may migrate during the second electrophoresis. When a new type of Coomassie Brilliant Blue, G250, is used, the background is not stained and therefore destaining is not required (Diezel et al., 1972). Optimal conditions for use of this rapid staining procedure (0.04% G250 dissolved in 3.5% perchloric acid) have been determined by Reisner et al. (1975). Before staining, the position of the tracking dye should be marked with a nick by a wire or by cutting the gel at the dye front. Photos are usually taken as records of electrophoretic runs, although the gels can be stored for several months. Enzyme activity can be determined either by incubation of the gel before staining with a specific staining assay mixture (Gabriel, 1971) or by elution of the enzyme from the gel slices and direct assay of the eluted fractions. A lengthwise slice of gel can be stained, and the protein stain compared with the enzyme assay for migration distance.

When testing a solution of enzyme for homogeneity, analytical electrophoresis should be run on several different amounts of protein. If only one band of protein is observed, even at concentrations which clearly overload the gel with the major peak, then it may be concluded that the protein is electrophoretically homogeneous under these experimental conditions. If possible, gels should be run at several pH values.

Polyacrylamide gel electrophoresis experiments can be used for the determination of molecular weight. Treatment of the enzyme with the denaturing agent sodium dodecyl sulphate (SDS, $C_{12}H_{25}$—SO_3Na), with inclusion of that agent in the electrophoretic system, allows the crosslinking property of the gel to be used for molecular sieving. SDS binds to almost all proteins at the constant ratio of 1.4 g SDS:1 g protein (Reynolds and Tanford, 1970). Treatment of protein molecules with SDS destroys all quaternary structure, and leads to the formation of partially helical rods of a constant diameter and of a length proportional to their molecular weight. With the hydrophobic groups of the SDS molecules bound to the proteins and the ionic sulphate groups exposed, the proteins all have a constant negative charge per unit mass. (Glycoproteins and extremely acidic or basic proteins may behave in an anomalous fashion when subjected to SDS gel electrophoresis.) Weber and Osborn (1969), extending the work of Maizel (Shapiro *et al.*, 1967) with SDS, outlined a simple and reproducible method. The proteins are reduced with 2-mercapto-ethanol and denatured with SDS to form protein monomer-SDS complexes that at pH 7.5 all migrate toward the anode. The original method called for 2-h incubation with 2-mercaptoethanol and SDS, but overnight incubation may be found necessary to completely disrupt some proteins. Weber *et al.* (1972) recommend heating of the protein sample for 2 min at 100° C to denature the proteases. With the charge-specificity effect removed by the SDS coating, the rate of migration becomes a function of protein molecular length alone. The mobility (distance migrated/distance tracking dye migrated) of the test protein is compared to the mobilities of a standard series of proteins, and its molecular weight is estimated from a graph of the log of the molecular weight v. mobility.

Since the electrophoretic mobility of a protein depends upon both charge and size, its molecular weight can be determined by studying the effect of gel crosslinking on electrophoresis at constant pH in Tris (Hedrick and Smith, 1968). A plot of the log of the mobility of a protein v. the gel concentration (weight of monomer per 100 ml of solvent) produces a straight line. The slope of this line is a function of the molecular weight of the protein. Above 50 000 daltons, the slopes vary directly with molecular weight. For estimations of lower

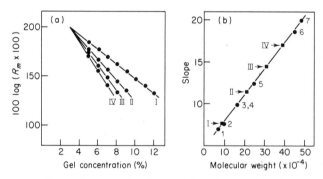

Fig. 3. Determination of the molecular weight of IMP dehydrogenase and its subunits by varying the gel pore size upon disc electrophoresis. Graph (a) shows the family of lines of varying slope obtained with the purified IMP dehydrogenase. Graph (b) compares the slopes of these four forms in graph (a) with the slopes obtained in similar experiments with seven proteins of known molecular weight. As in Fig. 2, the Roman numerals I–IV represent the monomeric through to the tetrameric forms of the dehydrogenase.

molecular weights, a slope–molecular weight standard curve must be constructed (Gonenne and Lebowitz, 1975). Variation of the acrylamide content over a 4% range is sufficient to determine the effect of size alone. Plots of relative mobility (R_m) v. log of molecular weight have given molecular weight estimates within 5% of those obtained from ultracentrifuge measurements. Figure 3 shows how this method was applied to the determination of the molecular weights of IMP dehydrogenase and its component protomers. The molecular weights of the four bands (95 000, 210 000, 310 000 and 390 000) agree with the values obtained by gel filtration (Fig. 2). A variation of this method, using electrophoresis in acetate at pH 4.0 containing 9 M urea, has been used successfully for molecular weight determination (Parish and Marchalonis, 1970).

To establish the quaternary structure of a protein (i.e. the number of subunits in an oligomer), an electrophoretic method was devised by Davies and Stark (1970). A solution of the enzyme is treated with the bifunctional amidinating agent dimethylsuberimidate:

$$\overset{\overset{+}{N}H_2}{\underset{\parallel}{}} \qquad \overset{\overset{+}{N}H_2}{\underset{\parallel}{}}$$
$$(MeO-C-(CH_2)_6-C-OMe),$$

an imidoester, in order to crosslink free ϵ-amino groups of lysine residues of adjacent subunits of the oligomer:

$$\overset{\overset{+}{N}H_2}{\underset{\parallel}{}} \qquad \overset{\overset{+}{N}H_2}{\underset{\parallel}{}}$$
$$(E-Lys-NH-C-(CH_2)_6-C-NH-Lys-E).$$

Under proper concentration conditions, the subunit, multiples of it, and the intact oligomer can be observed when treated by SDS electrophoresis. With an enzyme having identical subunits, the number of bands seen equals the number of protomers in the oligomer. Crosslinking within an oligomer can be distinguished from crosslinking between oligomers by lowering the protein concentration during amidination. Results of the crosslinking of quinonoid dihydropterin reductase (p. 262) are shown in Fig. 4. The mobilities of the two bands verify that the reductase is a dimer with a molecular weight of about 50 000. The chain length of the bifunctional modifier can be varied for best results. Davies and Kaplan (1972) have used dimethylpimelimidate to show that rabbit muscle pyruvate kinase is not a

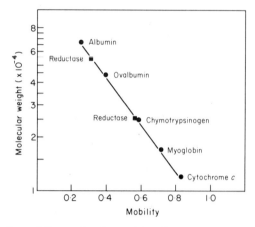

Fig. 4. Estimation of the molecular weight of sheep liver quinonoid dihydropterin reductase using SDS gel electrophoresis of enzyme treated with dimethylsuber-imidate. The mobilities of the five standard proteins are shown by circles, and the mobilities of the reductase dimer and monomer by squares. The crosslinked dimer migrated at a position suggesting a molecular weight of approximately 55 000. The other band, the monomer, had a migration rate corresponding to a molecular weight of 27 000 daltons. Untreated reductase (control) also had the same value, 27 000, as would be expected for a protein monomer-SDS complex. From Cheema *et al.* (1973).

symmetrical tetramer but has the four subunits arranged as two dimers. This conclusion was based on the observation that dimers and tetramers were the predominating species after crosslinking. Hucho *et al.* (1975) have extended this type of study of symmetry of oligomeric enzymes by using diimidates of chain lengths C_3 to C_{12}. If subunits are found in an enzyme, they can be shown to be identical in size by SDS gels and identical in charge if they show only one band upon regular electrophoresis at several pH values.

Polyacrylamide gel electrophoresis is the most sensitive method for detecting the presence of protein impurities in an enzyme preparation. As can be seen from the experiments described in the previous paragraphs, it can also be used for determination of the molecular weight and oligomeric structure. With the wide potential for application and the modest investment, every enzymologist should be familiar with these methods.

If the electrophoresis tests for homogeneity are positive, then the next steps in characterization of the enzyme are those applied to purified proteins (cf. Chapter 2), i.e. the determination of the N-terminal residues and the C-terminal residues, amino acid analysis and, if possible, determination of sequence and of three-dimensional structure by X-ray crystallography. The purification of an enzyme to homogeneity can be considered one hurdle, and so can the achievement of information on the physical properties of the enzyme. Sequence determination and the crystallographic examination are major projects, each of which may take several years or more.

AN EXAMPLE OF ENZYME PURIFICATION

As a final section in this chapter on purification of enzymes, I shall describe the purification of quinonoid dihydropterin reductase from beef liver (K. Korri and K. G. Scrimgeour, unpublished results). I am including this description to reinforce the previous discussion on methodology and to indicate the type of data that must be recorded when performing enzyme isolation.

The purification scheme used for the isolation of the beef liver reductase is a slight modification of methods used previously with sheep liver, sheep brain and beef adrenal (cf. Cheema *et al.*, 1973, and references cited therein). Because the object of the purification from beef liver was to obtain fairly large amounts of purified enzyme, procedures which give a final yield of about 10% were scaled up, and as it can be seen the cost was to drop the recovery to only 2.4% of the original activity (Table I). All purification steps were performed at 0-4° C, and 10^{-3} M 2-mercaptoethanol was included in all buffers to stabilize the enzyme. The procedures used are summarized in the flow sheet (Fig. 5), and the results obtained in Table I. Note that the presentation of experimental results conforms to the recommendations made on p. 41. This procedure has been performed a number of times, with reproducible results. One notable point is that the final specific activity is about double the highest previously reported, probably because of increased experience with handling the enzyme solutions and immediate assay of the final fractions.

Fresh or frozen and thawed liver is cut into one-inch pieces, and then homogenized in the presence of 0.03 M acetic acid. The reductase is stable at slightly acid pH values, so nucleic acid material can be removed in the initial step. Crude fractionation with ammonium sulphate more than doubles the specific activity of the enzyme and

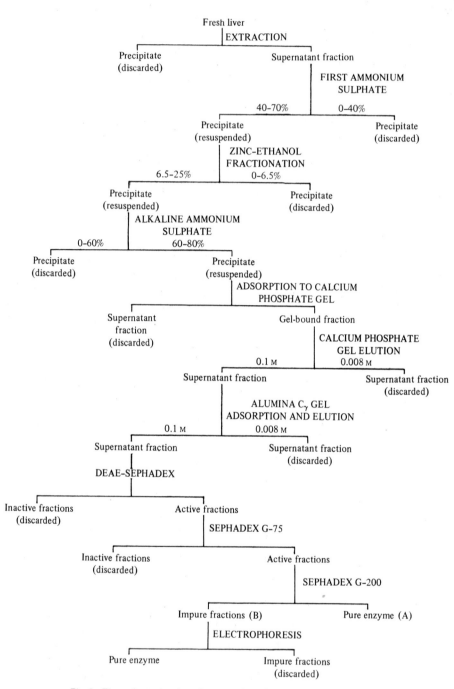

Fig. 5. Flow sheet showing the procedure for the isolation of quinonoid dihydropterin reductase from beef liver.

Table I. Summary of the purification of quinonoid dihydropterin reductase from beef liver (2660g). From K. Korri and K. G. Scrimgeour (unpublished data).

Step	Total Volume (ml)	Total Protein (mg)	Protein (mg ml^{-1})	Number of units ml^{-1}	Specific Activity (units (mg protein)$^{-1}$)	Total Activity (units)	Yield (%)
Extract	6500	186 000	28.6	8.1	0.3	52 700	100
First ammonium sulphate	1540	52 000	33.8	26.4	0.8	40 700	77
Zinc ethanol fractionation	710	17 600	24.8	39.4	1.6	28 000	53
Alkaline ammonium sulphate	380	8600	22.7	58.2	2.6	22 100	42
Calcium phosphate gel eluate	850	2200	2.6	11.8	4.5	10 000	19
Alumina C$_\gamma$ gel eluate	490	735	1.5	14.7	9.8	7200	14
DEAE-Sephadex A-50	150	75	0.5	34.7	69.4	5200	10
Sephadex G-75	23	39	1.7	136	80	3100	6
Sephadex G-200 (A)	4.3	6.5	1.5	192	126	830	1.6
(B)	5.1	11.7	2.3	206	90	1050	
Preparative disc electrophoresis	2.2	3.3	1.5	203	135	440	0.8
Total pure enzyme		9.8			130	1270	2.4

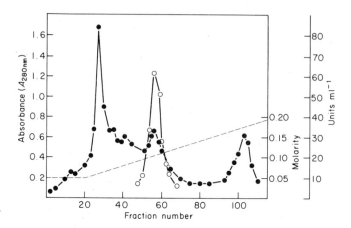

Fig. 6. Profile for the elution of quinonoid dihydropterin reductase from a column of DEAE-Sephadex A-50. A Tris–HCl buffer, pH 7.5, containing 10^{-3} M mercaptoethanol was used for elution. A gradient of 0.1–0.25 M Tris was applied, as shown by the dashed line. Protein concentration was measured by its absorbance at 280 nm. The enzyme activity was located between fractions 44 and 64, as noted by the open circles. Activity is reported as units ml^{-1} (μmol NADH oxidized min^{-1} ml^{-1} of eluted material). From K. Korri and K. G. Scrimgeour (unpublished data).

cuts the volume of the solution by 75%. The zinc–ethanol fractionation removes more contaminating proteins and further decreases the working volume. Ammonium sulphate fractionation at pH 8.2 allows the removal of some other contaminants. Two adsorption steps in which the reductase is eluted from the gels only by 0.1 M phosphate buffer give modest purifications. In the next step, ion-exchange chromatography on DEAE-Sephadex (Fig. 6), a nine-fold purification of the reductase is achieved, so that the enzyme is about 50% pure. One of the major contaminants removed in this step is beef serum albumin, which is not retained by the ion exchanger. The final steps have been chosen to remove the remaining impurities, primarily on the basis of differences in molecular radius. Pure enzyme, as judged by gel electrophoresis, is obtained from some fractions of the Sephadex G-200 column chromatography. More pure enzyme can be salvaged from the remaining active fractions of this column by preparative gel electrophoresis.

While allowing the isolation of about 10 mg of pure dihydropterin reductase, this purification method is time-consuming (it takes about 3 weeks of steady work) and the recovery of activity is lower than is desired. Therefore, several colleagues are attempting to apply affinity chromatography as a major step in the purification of this enzyme. Group specific affinity chromatography must be used because the specific substrate, quinonoid dihydropterin, is extremely unstable (p. 261).

REFERENCES

Andrews, P. (1965). *Biochem. J.* **96**, 595-606.
Bernardi, G. (1971). *In* "Methods in Enzymology" (W. B. Jakoby, ed.), Vol. 22, pp. 325-339. Academic Press, New York.
Blatt, W. F. (1971). *In* "Methods in Enzymology" (W. B. Jakoby, ed.), Vol. 22, pp. 39-49. Academic Press, New York.
Boegman, R. J. (1975). *FEBS Letters* **53**, 99-101.
Cheema, S., Soldin, S. J., Knapp, A., Hofmann, T. and Scrimgeour, K. G. (1973). *Can. J. Biochem.* **51**, 1229-1239.
Chrambach, A. and Rodbard, D. (1971). *Science, N.Y.* **172**, 440-451.
Cuatrecasas, P. (1970). *J. biol. Chem.* **245**, 3059-3065.
Cuatrecasas, P. and Anfinsen, C. B. (1971a). *In* "Methods in Enzymology" (W. B. Jakoby, ed.), Vol. 22, pp. 345-378. Academic Press, New York.
Cuatrecasas, P. and Anfinsen, C. B. (1971b). *A. Rev. Biochem.* **40**, 259-278.
Davies, G. E. and Kaplan, J. G. (1972). *Can. J. Biochem.* **50**, 416-422.
Davies, G. E. and Stark, G. R. (1970). *Proc. natn. Acad. Sci. U.S.A.* **66**, 651-656.
Davis, B. J. (1964). *Ann. N.Y. Acad. Sci.* **121**, 404-427.
Demain, A. L. (1971). *In* "Methods in Enzymology" (W. B. Jakoby, ed.), Vol. 22, pp. 86-95. Academic Press, New York.
Diezel, W., Kopperschläger, G. and Hofmann, E. (1972). *Analyt. Biochem.* **48**, 617-620.
Diezel, W., Böhme, H.-J., Nissler, K., Freyer, R., Heilmann, W., Kopperschläger, G. and Hofmann, E. (1973). *Eur. J. Biochem.* **38**, 479-488.
Dixon, M. and Webb, E. C. (1964). "Enzymes", pp. 27-53. Academic Press, New York.
Dunnill, P. and Lilly, M. D. (1972). *In* "Enzyme Engineering" (L. B. Wingard, Jr., ed.), pp. 103-105. Wiley-Interscience, New York.
Fall, R. R. and Vagelos, P. R. (1972). *J. biol. Chem.* **247**, 8005-8015.
Gabriel, O. (1971). *In* "Methods in Enzymology" (W. B. Jakoby, ed.), Vol. 22, pp. 578-604. Academic Press, New York.
Garen, A. and Levinthal, C. (1960). *Biochim. biophys. Acta.* **38**, 470-483.
Gerhart, J. C. and Holoubek, H. (1967). *J. biol. Chem.* **242**, 2886-2892.
Gonenne, A. and Lebowitz, J. (1971). *Analyt. Biochem.* **64**, 414-424.
Hartman, B. K. and Udenfriend, S. (1969). *Analyt. Biochem.* **30**, 391-394.
Harvey, M. J., Craven, D. B., Lowe, C. R. and Dean, P. D. G. (1974). *In* "Methods in Enzymology" (Jakoby, W. B. and Wilchek, M., eds), Vol. 34, pp. 242-253. Academic Press, New York.

Hedrick, J. L. and Smith, A. J. (1968). *Archs Biochem. Biophys.* **126**, 155-164.
Himmelhoch, S. R. (1971). *In* "Methods in Enzymology" (W. B. Jakoby, ed.), Vol. 22, pp. 273-286. Academic Press, New York.
Himmelhoch, S. R. and Peterson, E. A. (1966). *Analyt. Biochem.* **17**, 383-389.
Hogeboom, G. H. (1955). *In* "Methods in Enzymology" (Colowick, S. P. and Kaplan, N. O., eds), Vol. 1, pp. 16-19. Academic Press, New York.
Hucho, F., Müllner, H. and Sund, H. (1975). *Eur. J. Biochem.* **59**, 79-87.
Ito, A. and Sato, R. (1968). *J. biol. Chem.* **243**, 4922-4923.
Jakoby, W. B. (1971). *In* "Methods in Enzymology" (W. B. Jakoby, ed.), Vol. 22, pp. 248-252. Academic Press, New York.
Kaufman, S. (1971). *In* "Methods in Enzymology" (W. B. Jakoby, ed.), Vol. 22, pp. 233-238. Academic Press, New York.
Keilin, D. and Hartree, E. F. (1938). *Proc. R. Soc. B.* **124**, 397.
Kohn, J. (1959), *Nature, Lond.* **183**, 1055.
Layne, E. (1957). *In* "Methods in Enzymology" (Colowick, S. P. and Kaplan, N. O., eds), Vol. 3, pp. 447-454. Academic Press, New York.
Leaback, D. H. and Robinson, H. K. (1975). *Biochim. biophys. Res. Commun.* **67**, 248-254.
Lee, C.-Y., Lappi, D. A., Wermuth, B., Everse, J. and Kaplan, N. O. (1974). *Archs Biochem. Biophys.* **163**, 561-569.
Lievaart, P. A. and Stevenson, K. J. (1974). *Can. J. Biochem.* **52**, 637-644.
Lowe, C. R. and Dean, P. D. G. (1974). "Affinity Chromatography". John Wiley and Sons, London.
Madelian, V. and Warren, W. A. (1975). *Analyt. Biochem.* **64**, 517-520.
Mahler, H. R. and Cordes, E. H. (1971). "Biological Chemistry", Second edition, pp. 431-464. Harper and Row, New York.
Martin, R. G. and Ames, B. N. (1961). *J. biol. Chem.* **236**, 1372-1379.
Mazumder, R., Sasakawa, T. and Ochoa, S. (1963). *J. biol. Chem.* **238**, 50-53.
McPhie, P. (1971). *In* "Methods in Enzymology" (W. B. Jakoby, ed.), Vol. 22, pp. 23-32. Academic Press, New York.
Mosbach, K. (1974). *In* "Methods in Enzymology" (Jakoby, W. B. and Wilchek, M., eds), Vol. 34, pp. 229-242. Academic Press, New York.
Mosbach, K., Guilford, H., Ohlsson, R. and Scott, M. (1972). *Biochem. J.* **127**, 625-631.
Novick, A. and Horiuchi, T. (1961). *Cold Spring Harb. Symp. quant. Biol.* **26**, 239-245.
O'Carra, P., Barry, S. and Griffin, T. (1974). *In* "Methods in Enzymology" (Jakoby, W. B. and Wilchek, M., eds), Vol. 34, pp. 108-126. Academic Press, New York.
Ornstein, L. (1964). *Ann. N.Y. Acad. Sci.* **121**, 321-349.
Parish, C. R. and Marchalonis, J. J. (1970). *Analyt. Biochem.* **34**, 436-450.
Peterson, E. A. and Sober, H. A. (1956). *J. Am. Chem. Soc.* **78**, 751-755.
Potter, V. R. (1955). *In* "Methods in Enzymology" (Colowick, S. P. and Kaplan, N. O., eds), Vol. 1, pp. 10-15. Academic Press, New York.
Rabinowitz, J. C. and Pricer, W. E., Jr. (1962). *J. biol. Chem.* **237**, 2898-2902.
Reisner, A. H., Nemes, P. and Bucholtz, C. (1975). *Analyt. Biochem.* **64**, 509-516.

Repaske, R. (1958). *Biochim. biophys. Acta.* **30**, 225-232.
Reynolds, J. A. and Tanford, C. (1970). *J. biol. Chem.* **245**, 5161-5165.
Rieland, J. (1971). *In* "Methods in Enzymology" (W. B. Jakoby, ed.), Vol. 22, pp. 287-321. Academic Press, New York.
Rimerman, R. A. and Hatfield, G. W. (1973). *Science, N.Y.* **182**, 1268-1270.
Rouget, P. and Chapeville, F. (1971). *Eur. J. Biochem.* **23**, 459-467.
Scrimgeour, K. G. and Huennekens, F. M. (1962). *In* "Methods in Enzymology" (Colowick, S. P. and Kaplan, N. O., eds), Vol. 5, pp. 838-843. Academic Press, New York.
Shaltiel, S., Adler, S. P., Purich, D., Caban, C., Senior, P. and Stadtman, E. R. (1975). *Proc. natn. Acad. Sci. U.S.A.* **72**, 3397-3401.
Shaltiel, S. and Er-el, Z. (1973). *Proc. natn. Acad. Sci. U.S.A.* **70**, 778-781.
Shapiro, A. L., Viñuela, E. and Maizel, J. V., Jr. (1967). *Biochem. Biophys. Res. Commun.* **28**, 815-820.
Shnitka, T. K. and Seligman, A. M. (1971). *A. Rev. Biochem.* **40**, 375-396.
Shuster, L. (1971). *In* "Methods in Enzymology" (W. B. Jakoby, ed.), Vol. 22, pp. 412-437. Academic Press, New York.
Sober, H. A., Hartley, R. W., Jr., Carroll, W. R. and Peterson, E. A. (1965). *In* "The Proteins" (H. Neurath, ed.), Second edition, Vol. 3, pp. 1-97. Academic Press, New York.
Spatz, L. and Strittmatter, P. (1971). *Proc. natn. Acad. Sci. U.S.A.* **68**, 1042-1046.
Stevenson, K. J. and Landman, A. (1971). *Can. J. Biochem.* **49**, 119-126.
Taylor, B. L. and Utter, M. F. (1974). *Analyt. Biochem.* **62**, 588-591.
Thompson, S. T., Cass, K. H. and Stellwagen, E. (1975). *Proc. natn. Acad. Sci. U.S.A.* **72**, 669-672.
Tiselius, A., Hjerten, S. and Levin, O. (1956). *Archs Biochem. Biophys.* **65**, 132-155.
Vesterberg, O. (1971). *In* "Methods in Enzymology" (W. B. Jakoby, ed.), Vol. 22, pp. 389-412. Academic Press, New York.
Voordouw, G., Gaucher, G. M. and Roche, R. S. (1974). *Biochem. Biophys. Res. Commun.* **58**, 8-12.
Weber, K. and Osborn, M. (1969). *J. biol. Chem.* **244**, 4406-4412.
Weber, K., Pringle, J. R. and Osborn, M. (1972). *In* "Methods in Enzymology" (Hirs, C. H. W. and Timasheff, S. N., eds), Vol. 26, pp. 1-27. Academic Press, New York.
Willstätter, R. and Kraut, H. (1923). *Chem. Ber.* **56**, 1117-1121.
Wu, T. W. and Scrimgeour, K. G. (1973). *Can. J. Biochem.* **51**, 1380-1390.

D

4 Reaction Mechanisms

Chapters 4–6 comprise a section summarizing the basic theories of catalysis and the experimental results which support these theories. First, an introduction to chemical reaction mechanisms is presented in this chapter. Then, in Chapter 5, the methods used for studying the properties of enzyme catalysis by kinetic measurements are examined from an operational viewpoint. Finally, Chapter 6 is devoted to the theories of enzyme catalysis which have been proposed to explain the extraordinary catalytic activity of enzymes. Subsequent chapters describe some of the mechanisms of enzyme catalysis as they are presently envisaged. The next three chapters, therefore, provide the fundamentals of catalytic theory and an explanation of the experiments used to determine the mechanisms of the enzymic reactions described later.

Studies of the mechanisms of chemical reactions which are related to biological systems fall into the division of chemistry called bio-organic chemistry. The development over the past 20 years of this sub-science can be followed by reading the monographs of Ingraham (1962), Kosower (1962), Bruice and Benkovic (1966), Jencks (1969) and Bender (1971). Bio-organic chemistry relies mainly on physical organic chemistry for its interpretations and techniques, but as we shall see knowledge obtained from heterocyclic chemistry, co-ordination chemistry, physical chemistry and enzyme chemistry is also incorporated into the application of bio-organic chemistry to studies of enzyme mechanisms.

REACTION MECHANISMS

Elucidation of the mechanism of a chemical reaction involves learning as much pertinent detail of its chemistry as is possible. We must know the stoichiometry of the reaction and be able to detect or isolate any intermediates that occur in the reaction. A mechanism is the description of the exact pathways of all atoms and molecules during the reaction, including those of the substrates, products and solvent molecules. This delineation involves the changes in both chemical bonds and in chemical energies. The mechanism must be compatible with the effect of reaction conditions on the rate of the reaction or on individual steps of the process. Much of the bio-organic chemistry entails the study of this latter feature, the kinetics of reactions. In particular, the effects of catalysts on a reaction provide useful information about the mechanism of the reaction and about the functions of enzymes and coenzymes as catalytic agents. Studies of mechanisms—either non-enzymic or enzymic—have not yet reached the ultimate descriptions that are desirable. We must settle for whatever information can be obtained and seek further techniques to fill in the gaps. Mechanisms are often proposed which do not conflict with any known data, but later experiments may disprove or alter them. It is the ingenuity of chemists applied to scientific data which produces mechanistic theories. Likewise, it is the rapid rate of scientific discovery, not the naïvety of previous chemists, which will cause replacement of some of the mechanisms I will describe.

Types of Reactions

There are several ways of classifying chemical reactions. Although these will be reviewed briefly here, it would be wise for anyone not familiar with the principles or nomenclature to read "Organic Reaction Mechanisms" (Breslow, 1969) or any other modern organic chemistry text dealing with the dynamics of organic chemical reactions.

Since chemical bonds can be described as the sharing of pairs of electrons between two atoms, one description of a reaction can be the manner in which electrons react during bond cleavage (i.e. by the type of reactive intermediate which is formed). In ionic or *heterolytic* reactions, if bond cleavage results in retention of both electrons by a carbon atom, an ion known as a *carbanion* is produced:

$$R_3C:X \rightleftharpoons R_3C:^- + X^+.$$

In another type of organic ionic reaction, the reacting carbon atom may lose both electrons and form a cation, called a *carbonium ion* (described below). Amongst the methods able to detect carbanion or carbonium ion intermediates are spectroscopy, stereochemistry and kinetics. In *homolytic* reactions, the electron pair is broken so that one electron goes with each atom to form two compounds called *free radicals*, each with an unpaired electron:

$$RO:OR' \longrightarrow RO\cdot + \cdot OR'.$$

Transient free radical intermediates can be detected and characterized by the magnetic properties given to them by the unpaired electrons. Although most enzymatic reactions occur by ionic mechanisms, we shall encounter some examples of radical reactions. A type of reaction related to radical reactions is electron transfer, in which transition metal ions (notably ferrous or ferric) undergo oxidation or reduction.

Ionic organic addition reactions are classified, according to the types of chemicals reacting with the carbon atom in question, as either *nucleophilic* or *electrophilic*. Electron donors are compounds which are attracted to nuclei and therefore are called nucleophiles. Electron acceptors are termed electrophiles. An example of a nucleophilic reaction is the reversible addition of bisulphite to an aldehyde:

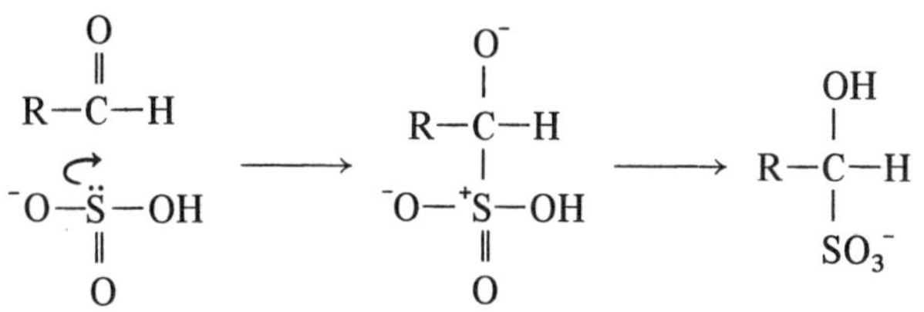

In the above mechanism, I have used the common notation of showing the motion of pairs of electrons in a reaction by a curved arrow. The nucleophilic reagent, bisulphite, adds to the carbonyl group as shown by the arrow. A very simple example of an electrophilic addition is the protonation of pyridine:

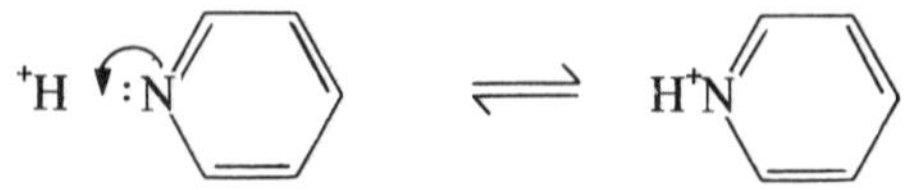

Here, the proton may be considered as the electrophile which adds to the negative centre of the larger molecule.

Ionic substitution reactions are separated into two types by their mechanisms. Substitutions proceed either by a preliminary cleavage or by a one-step attack by the nucleophile. The two alternatives are distinguished experimentally by kinetics and by the stereochemistry about the reacting carbon. If the substitution reaction proceeds by an ionization of the carbon atom, a positively charged ion—a carbonium ion—is formed:

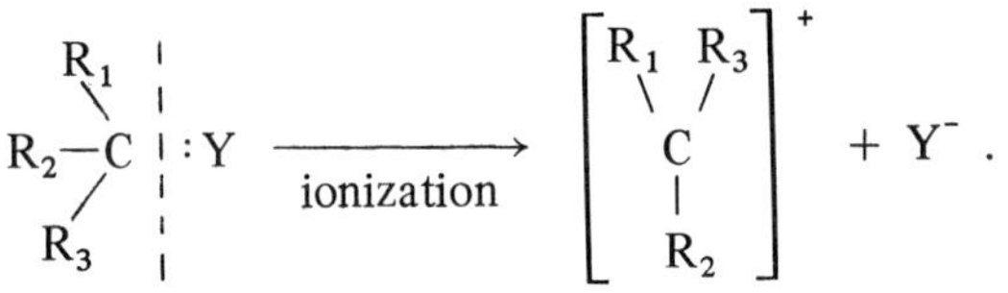

$$
\begin{array}{c}
R_1 \\
\diagdown \\
R_2-C \quad :Y \\
\diagup \\
R_3
\end{array}
\xrightarrow[\text{ionization}]{}
\left[
\begin{array}{c}
R_1 \quad R_3 \\
\diagdown \diagup \\
C \\
| \\
R_2
\end{array}
\right]^{+}
+ \; Y^{-} \; .
$$

Carbonium ion

This type of reaction is called unimolecular nucleophilic substitution or S_N1. The term unimolecular infers that only one molecule (that which ionizes) breaks a bond in the slowest step of the reaction. Carbonium ions normally have planar bonds and react extremely rapidly with a nucleophile (Z) or with solvent. The group Y which ionizes from the carbon is called the *leaving group*.

$$
R_3CY \rightleftharpoons R_3C^{+} + Y^{-} \xrightarrow{Z^{-}} R_3CZ
$$

The strength of the $C-Y$ bond depends on the basicity of Y. Weak bases (good leaving groups) have weak $C-Y$ bonds and ionize readily. Strong bases are efficient electron donors to the carbon nucleus and are therefore poor leaving groups. Protonation of Y will reduce its basicity and weaken the $C-Y$ bond. In this way, the proton acts as a catalyst for the ionization.

Another type of substitution reaction is bimolecular or S_N2. In an S_N2 reaction, the attacking group adds to the face of the carbon's tetrahedron, away from the corner bearing the leaving group. The entering group X displaces the leaving group Y in one rapid step, with no detectable intermediate. As a result, there is an inversion of configuration at the carbon. Many substitution reactions fall between the limits of the S_N1 and S_N2 processes in their behaviour. If a

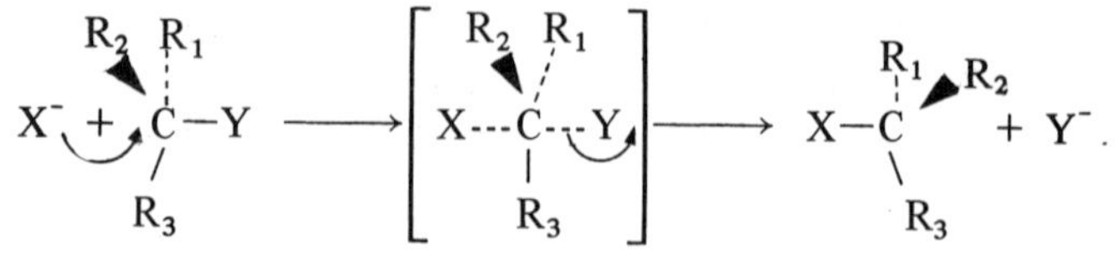

$$X^- + \overset{R_2}{\underset{R_3}{\overset{|}{C}}} - Y \longrightarrow \left[X \text{---} \overset{R_2}{\underset{R_3}{\overset{|}{C}}} \text{---} Y \right] \longrightarrow X - \overset{R_1}{\underset{R_3}{\overset{}{C}}} R_2 + Y^-.$$

Transition state

nucleophile is attached to the molecule undergoing substitution, intramolecular reaction may occur if this nucleophile is in a more suitable position for attack than if it were a nucleophilic group on a different molecule. Enhanced reaction of an internal nucleophile is described as the *neighbouring group effect*. Large increases in reaction rates can be observed by incorporating reactants into the same molecule, suggesting that enzymes could increase the rate of a bimolecular reaction by collecting the reactants at the active site (cf. proximity effects, p. 145).

Elimination reactions also may be categorized as either unimolecular (*E1*) or bimolecular (*E2*), depending on whether or not a base is required in the reaction. The *E1* mechanism involves initial formation of a carbonium ion, followed by loss of a proton from the adjacent carbon (rather than by addition of a nucleophile as in S_N1 reactions). Elimination reactions usually occur in strongly basic solutions, and if the base participates in the reaction, it is a bimolecular process. In an *E2* reaction, both the proton and the leaving group are lost at the same time. A third mechanism for elimination proceeds via loss of the proton to form a fairly stable carbanion, which loses the leaving group. This type of reaction is termed *E1cB* (unimolecular elimination with conjugate base). Elimination reactions of amino acids require the presence of the coenzyme pyridoxal phosphate, and they involve initial loss of a proton with stabilization of the carbanion by pyridoxal.

Transition State Theory

The rate at which a chemical reaction occurs depends upon the rate of effective collisions of the molecules involved. Two molecules must collide in the correct orientation, so that the reacting bonds are close, and they must possess sufficient energy when they collide so that they can approach the alignment of electrons in the bonds of the products. The minimum amount of energy (along the direction of their approach) needed for reaction to occur is called the activation energy. Only in a few of the many collisions do the reactants have sufficient energy to break the bond that is present and form the activated complex or *transition state*. If the progress of a reaction is represented graphically (Fig. 1), the transition state is the point of maximum energy. The reaction co-ordinate is the pathway of the reaction on the energy diagram. This two-dimensional diagram is actually based on a three-dimensional diagram showing the relationship between interatomic distances and energy levels (Jencks, 1969, p. 599 in his book). The transition state is not an intermediate, and cannot be isolated. It is an energized arrangement of the reacting atoms with sufficient energy to form products, not just to reform unchanged reactants. If a reaction is reversible, it passes through the same transition state in both directions.

By assuming that the difference in free energy ($\Delta G^{\ddagger}$) between the initial state and the transition state can be related to reaction rate measurements, thermodynamic equations have been derived to describe the transition state. The relative stabilities of the reactants and the transition state are given by the equilibrium constant for the formation of the transition state, $K^{\ddagger}$, which is related to both the

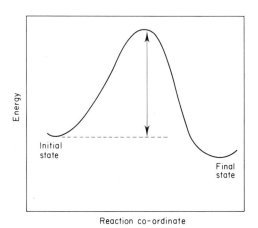

Fig. 1. Energy diagram for a single-step reaction. The magnitude of the energy of activation for the forward direction is indicated by the arrows.

entropy of activation ($S^\ddagger$) and the heat of activation ($H^\ddagger$):

$$\Delta G^\ddagger = - RT \ln K^\ddagger = \Delta H^\ddagger - T\Delta S^\ddagger.$$

These terms are analogous to the common thermodynamic parameters from which they are derived. $\Delta H^\ddagger$ may be considered to be the height of the activation barrier, i.e. the energy difference between the reactants and the transition state. $\Delta S^\ddagger$ reflects the constraint of the system, or the change in probability accompanying the formation of the transition state. For example, $\Delta S^\ddagger$ is negative for the association of two compounds, since the number of species has decreased, or there is more orderliness to the system (loss of translational and rotational freedom). Monomolecular reactions often have $\Delta S^\ddagger$ values near zero or positive.

The actual energy diagrams for most chemical reactions are more complex than that shown in Fig. 1. For example, in an endothermic or non-spontaneous reaction involving formation of a metastable but detectable intermediate such as a carbonium ion, the profile might be similar to Fig. 2. The slowest step in the forward direction is the formation of the intermediate. The energy diagrams which have been obtained for several enzymic reactions have at least three peaks.

If a reaction does occur every time there is a properly oriented encounter between the two reactants, it is called a diffusion-controlled

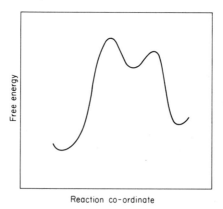

Reaction co-ordinate

Fig. 2. Reaction co-ordinate diagram for an endothermic reaction having a metastable intermediate (shown by the hollow). The reaction is endergonic because the difference in free energy between the initial and final states is positive.

reaction. The rates of these reactions give an upper limit for rates of reactions. Gutfreund (1972) has estimated that the limiting rate for diffusion-controlled formation of an enzyme–substrate complex is about $10^8 \, \text{M}^{-1} \, \text{s}^{-1}$.

A *catalyst* is a substance that speeds up the attainment of thermodynamic equilibrium by lowering the energy of activation necessary for reaction or by providing an alternative pathway of lower activation energy. This infers that the catalyst is present in the transition state and stabilizes it. After the products are formed, the catalyst dissociates from them. The catalyst may be changed during the reaction, but it is unchanged in the overall process. The activity of catalysts, including enzymes, is explained by their ability to decrease $\Delta G^{\ddagger}$ (or increase $K^{\ddagger}$) by either decreasing $\Delta H^{\ddagger}$ or increasing $\Delta S^{\ddagger}$. $\Delta H^{\ddagger}$ may be decreased by formation of a substrate–catalyst complex. The energies of activation for the formation and for the breakdown of the intermediate complex would both have to be lower than that of the uncatalyzed reaction (Fig. 3). While the decreasing $\Delta H^{\ddagger}$ is a major function, enzymes also alter the $\Delta S^{\ddagger}$ term by positioning reactants fairly rigidly before the transition state is reached, so that there is little or no decrease in entropy needed for conversion of the

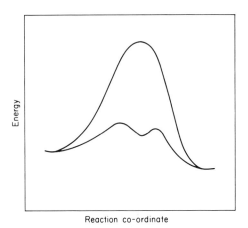

Reaction co-ordinate

Fig. 3. Comparison of the reaction co-ordinate curves for a catalyzed reaction (lower curve) with an uncatalyzed reaction (upper curve). In the catalyzed reaction, either the first step or the second step may have the higher activation barrier.

enzyme–substrate complex to the transition state. In Chapter 6, we shall examine both this entropy effect and the theory that the transition state binds more tightly to the enzyme than do the substrates. When bound to the active site, the reactants are placed further along the reaction co-ordinate towards the transition state.

Two variables which can influence the rate of reactions are temperature and concentration of salts. An increase in temperature causes more collisions of the reactants to have sufficient energy to form the transition state. The change of reaction rate with temperature is measured by a graph, called an Arrhenius plot, in which $\log k$ (the rate constant, defined below) is plotted against $1/T$. The energy of activation in kcal mol^{-1} for the reaction can be obtained by multiplying the slope of the plot by $-(2.303 \times 1.98)$. Because salts can markedly change reaction rates, a series of kinetic experiments is usually carried out at constant ionic strength. Salt effects are due to alterations in ion–water interactions or ion–ion interactions in water (Jencks, 1969, Chapter 7 in his book). Salt ions may have either general or specific effects on reaction rates. We will see cases of ions affecting the reactivity of proteins, as well as effects on simple nonenzymic reactions.

Chemical Kinetics

The mechanism for a chemical reaction is a description of that reaction in terms of space and energy. This requires both a stereochemical description of the reaction and the drawing of an activation energy diagram. Several tests are available for mechanism determination. The reaction products, including their stereochemistry, must be characterized. The locations of isotopic tracer atoms in the products are compared to their former positions in the reactants. Isotope effects—the effect of isotopic substituents on reaction rate—can also provide information on mechanisms and intermediates. However, the most powerful tool for mechanism determination is *kinetics,* the measurement of reaction rates. The rate of a reaction is proportional to the concentration of the transition state. This concentration depends on the concentration of the reactants and catalysts. In enzyme kinetics, the analogous statement is that the rate of reaction depends on the concentration of the appropriate enzyme–substrate complex. Kinetic studies measure the rate of formation of products or of disappearance of reactants as a function of concentration of reactants. Conclusions are reached after systematically studying these measurements under varying conditions of temperature, solvent composition and chemical substitution. Kinetic studies which are extremely relevant to our discussion are those which measure the effect of catalysts upon a reaction.

Kinetic studies give information only about the slowest step of a reaction—the rate-determining step. This step is the achievement of the transition state in the overall process, from either direction. Any mechanism derived from kinetics cannot distinguish between the supposed reactants and any other reactants of similar total composition and charge which might be in rapid equilibrium with them.

Kinetic measurements are expressed by equations which relate the experimentally determined rates to the concentrations of reactants. An outstanding explanation of the application of kinetics to mechanistic studies and the derivations of the equations with their graphical transformations is presented in Chapter 11 of "Catalysis in Chemistry and Enzymology" (Jencks, 1969). The experimental approach is to reduce the equations to their simplest form by adjustment of the concentrations of substances so that the rate measurements depend on variation of the concentration of only one reactant.

We have encountered the term molecularity already in this chapter. The *molecularity* of a reaction is a theoretical account of the number of species reacting in the rate-determining step. The *kinetic order* of a reaction is an empirical term which may or may not equal the molecularity. Consider a simple reaction in which the substrate A is converted to a product P, $A \rightarrow P$. The rate of the reaction (v) is dependent on the concentration of A (molar concentrations are shown by square brackets).

$$v = \frac{-\mathrm{d}A}{\mathrm{d}t} = \frac{+\mathrm{d}P}{\mathrm{d}t} = k[A]^1 .$$

The equation relating v to $[A]$ is called the rate equation or rate law, and k is the rate constant expressed in this case in reciprocal time units (s^{-1}). The order of the reaction is defined by the sum (n) of the exponents in the rate equation. Here, $n = 1$, so the reaction is first-order. Calculations of first-order reaction results are simple, so it is wise to adjust reaction conditions so that the measurements are made under first-order conditions.

In the reaction $A + B \rightarrow P + Q$,

$$v = k[A]^1[B]^1.$$

Since $n = 1 + 1$, the reaction is second-order, being first-order with respect to A and first-order with respect to B. The rate constant has the dimensions of reciprocal concentration and reciprocal time ($1 \, mol^{-1} \, s^{-1}$).

If we return for a moment to organic substitution reactions, we will see how S_N1 and S_N2 reactions can be differentiated by their kinetic dependences. The rate law for an S_N1 reaction is:

$$v = k[HY]$$

and that for an S_N2 reaction is:

$$v = k[HY][X^-].$$

If one (or more) of the reactants effectively remains in constant concentration during the reaction, its term can be eliminated in the rate expression. This is accomplished by keeping all other reactants in great excess over the one being studied. For example, in the hydrolysis of dilute sucrose in aqueous solution, water is present at a much higher concentration than the sugar, and the water concentration is virtually unchanged after the reaction. This type of reaction is called a *pseudo first-order reaction.* In kinetic studies with buffer catalysts, the catalyst is usually present in excess. This is done for

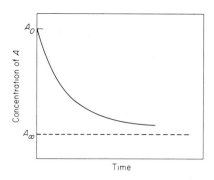

Fig. 4. Progress curve for a first-order reaction. A_0 is the initial concentration of the substrate and A_∞ is the concentration of substrate at equilibrium.

methodological reasons, often because the change in absorbance of the substrate is being measured (at a concentration of about 10^{-4} M) and the buffer is used (at about 0.05–0.2 M) both to catalyze the reaction and to maintain the pH. In enzyme studies, the substrate is present in much higher concentration than the catalyst, again for a technical reason.

Figure 4 is typical of the type of raw data which one would obtain for a first-order reaction. It shows the change in concentration of the reactant A with time. The end-point (A_∞) must be determined precisely.* By allowing a reaction to proceed for ten half-lives, the extent of reaction is 99.9%, which may be assumed to be completion. The rate of reaction, continually slower as the reaction progresses, is the negative of the slope of Fig. 4. The easiest way to determine the rate constant is by a graphical solution using an integrated form of the rate equation

$$\log (A_t - A_\infty) = -2.3 \, kt$$

plotted as a semi-log plot of $(A_\infty - A_t)$ against time (Fig. 5). By measuring the half-time of the reaction $(t_{1/2})$, k is obtained from the relationship

$$k = \frac{0.693}{t_{1/2}}.$$

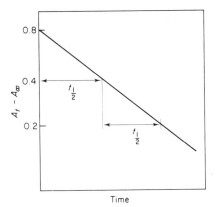

Fig. 5. Semi-logarithmic plot of the change of absorbance of a first-order reaction. A_t is the absorbance of reactant A at the time t of the reaction, and A_∞ its absorbance at the end-point.

* When it is not possible to determine A_∞, an alternative graphical method to that described here may be used, but not without caution (Jencks, 1969, pp. 562–564 in his book).

Because $t_{1/2}$ should be independent of the extent of a first-order reaction, results are obtained and checked easily.

CATALYSIS

Two types of catalysis which we shall encounter in enzyme mechanisms are covalent catalysis and acid–base catalysis (Jencks, 1969). Both types of catalysis are much better understood in non-enzymic reactions, but play important roles in the mechanisms of enzymes.

Covalent Catalysis

Examples of *covalent catalysis,* in which the substrate or part of it forms a covalent intermediate with the catalyst, are much easier to find in enzyme chemistry than in organic chemistry. Some of these enzymic reactions will be commented on in Chapter 10. The reactions which take place between most coenzymes and substrates can also be classified as covalent catalysis, with the coenzymes in turn being attached to the enzymes by either non-covalent or covalent bonds. Coenzymes function by binding covalently, in a transient fashion, groups involved in transfer reactions.

Jencks (1969, Chapter 2 in his book) gives examples of three types of covalent catalysis: nucleophilic, electrophilic and oxidative. Nucleophilic and electrophilic catalysis are closely related, since one type may be classified as the other when the reaction is carried out in the reverse direction. Jencks has therefore interpreted the reaction being catalyzed in terms of whether the reaction step deemed most important is facilitated by either a nucleophilic or electrophilic compound in the direction in which the reaction is usually written. Oxidative catalysis occurs when a catalyst reacts with a substrate in a manner allowing a reversible oxidation of the substrate to take place. An example of this type of catalysis is the epimerization of uridine diphosphate galactose using NAD^+ as the oxidative catalyst (p. 216). Covalent catalysis is often seen in the interaction of carbonyl groups with amino functions. A good illustration of this type of catalysis was discovered by Cordes and Jencks (1962). While studying the catalysis of the formation of the semicarbazone of p-chlorobenzaldehyde, they found that aniline has abnormally high catalytic ability. The aniline is reacting with the aldehyde in the rate-determining step to form a covalent intermediate, a Schiff base. The Schiff

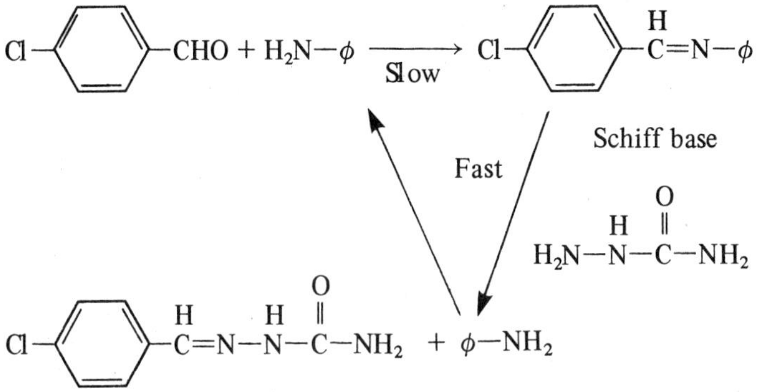

base rapidly reacts with semicarbazide to release the semicarbazone and the nucleophilic catalyst, aniline. Kinetic experiments proved that the Schiff base is a necessary intermediate in the aniline-catalyzed reaction. Aniline is a catalyst for this reaction because p-chlorobenzaldehyde reacts more readily with aniline than with semicarbazide and because the covalent intermediate is more susceptible to attack by semicarbazide than is the free aldehyde. The Schiff base is thermodynamically unstable, allowing product formation to occur. It is important to remember that any nucleophilic catalyst has attributes which are specialized to show both increased kinetic reactivity and thermodynamic instability of the substrate–catalyst intermediate.

Electrophilic catalysts withdraw electrons from the reaction centre, and therefore are termed electron sinks. The electron-withdrawing properties of metal ions in some metalloenzymes are discussed in Chapter 11.

Acid–Base Catalysis

Many reactions in both organic chemistry and enzymology are catalyzed by either *acids* (proton donors) or *bases* (proton acceptors). Although it is difficult to prove that an enzyme mechanism involves acid–base catalysis, it is reasonable to speculate on the basis of analogous chemistry that many enzymes involve proton transfer as at least one of the factors in their supercatalytic abilities. Therefore, we must have a sound understanding of simple cases of acid–base catalysis before attempting to deal with enzyme mechanisms.

Acid–base catalysis is divided into two categories—specific catalysis and general catalysis. Specific-acid catalysis of a reaction occurs when the rate is enhanced by solutions that are more acidic, but when the rate is independent of buffer concentration. Specific-base catalysis is similarly defined as catalysis by basic solution. Because the concentration of H^+ (or H_3O^+) or OH^- determines the rate of reaction, specific catalysis depends on the pH of the solution alone. General-acid and general-base catalysis refer to catalysis by proton donors or acceptors other than protons or hydroxyl ions. Because the concentrations of H^+ and OH^- are very low under physiological reaction conditions, it is general, not specific, acid–base catalysis that is important in enzyme chemistry. In bio-organic reactions, acid or base catalysts are usually used as buffer compounds as well, so in general catalysis the rate is dependent on the concentration of buffer.

Buffers can exist in two forms, either the protonated form BH^+ or the free base form B. For an amine, the protonated form would be $R-CH_2NH_3^+$ and the free base $R-CH_2NH_2$. The BH^+ form of acetic acid would be CH_3COOH and the free base form CH_3COO^-. (Note that BH^+ is not always a cation, nor B a neutral species. The abbreviations are intended to indicate the state of protonation, not the charge.) Catalysis by the BH^+ species is general-acid catalysis, and catalysis by B is general-base catalysis.

As I have mentioned, catalysis by transfer of protons is the most common mechanism for enhancement of rates in non-enzymic reactions, and so undoubtedly is an important mechanism in enzyme catalysis also. Two major types of reactions involve proton transfer, $C-H$ bond cleavage and $C-$(heavy atom) bond cleavage. In $C-H$ cleavage, a general base removes the hydrogen as a proton. In $C-$(heavy atom) cleavage, e.g. $C-N$, a general base removes H^+ from water to generate a supply of OH^- nucleophile at neutral pH values. Thus, the effective removal or formation of H^+ under what would normally be quite unfavourable conditions is responsible for the higher rates observed in the presence of acid–base catalysts. These are just the sort of conditions presented by physiological systems, where extremes of pH are fatal. An example of $C-H$ cleavage is the enolization of a ketone such as acetone:

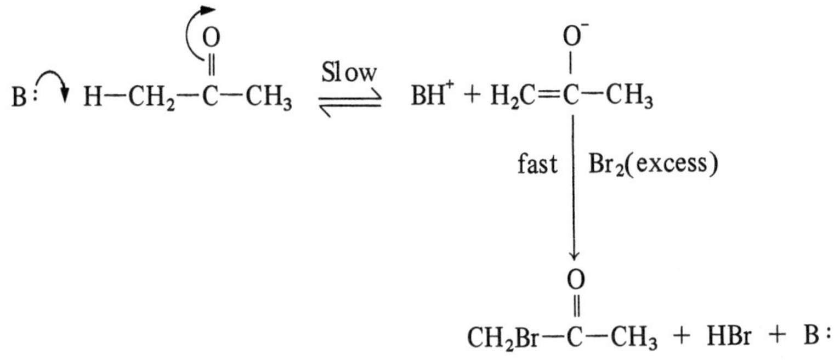

$$B: \quad H-CH_2-\overset{\overset{\displaystyle O}{\|}}{C}-CH_3 \quad \underset{\text{Slow}}{\rightleftharpoons} \quad BH^+ + H_2C=\overset{\overset{\displaystyle O^-}{|}}{C}-CH_3$$

$$\text{fast} \ \bigg| \ Br_2(\text{excess})$$

$$\overset{\overset{\displaystyle O}{\|}}{CH_2Br-C-CH_3} + HBr + B:$$

The base B removes a proton from the ketone, allowing formation of the enolate ion. This ion can be measured by trapping in a reaction with bromine. An example of general-base catalyzed cleavage of a C—N bond is the catalysis of acetyl imidazole hydrolysis by imidazole:

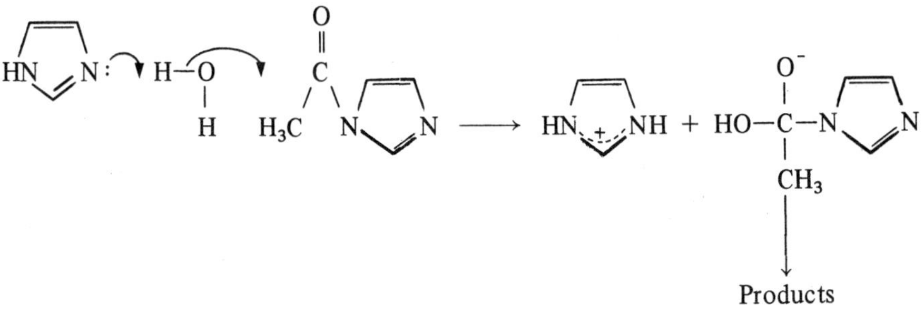

Products

By proton removal, the imidazole increases the nucleophilic strength of H_2O without actually forming free OH^- (Jencks, 1969, p. 168 in his book).

Establishing whether acid–base catalysis occurs in a chemical reaction is a straightforward experimental project. Kinetic measurements are recorded at varying pH values and buffer concentrations. The experiments needed to accurately plot the curves described below may seem tedious (as, indeed, may all research), but as with all the experiments I am describing, the reliable transfer of experimental data into mechanisms requires confidence in the results and as simple an interpretation as possible. The general rate equation for a reaction that can undergo acid–base catalysis has five terms:

$$k_{obs} = k_{H_2O} + k_{H^+}[H^+] + k_{OH^-}[OH^-] + k_{BH^+}[BH^+] + k_B[B].$$

The observed rate constant (k_{obs}) can be made up of one or more of these terms, and the experiments done are designed to eliminate the inoperative terms (and their corresponding mechanistic modes) as well as to verify the types of catalysis occurring. The first term will only be observed if the water solvent acts as a catalyst. The second or third terms apply only if specific acid or base catalysis occurs, and the last two terms represent the contributions of general acid or base catalysis. After examining several hypothetical examples, the method should be readily understood.

Figure 6 shows that the rate increases as buffer concentration is increased and also as the pH is increased. Because there is no detectable intercept, both specific acid–base catalysis and solvent catalysis may be ruled out. This is an example, then, of general-base catalysis. Figure 7 is slightly more complicated, since as well as general-base catalysis there is an observable intercept at both pH values. Clearly, then, there is general-base catalysis since the rate is dependent on the concentration of the free base form of the buffer. There is also specific-base catalysis since the intercept increases with increasing pH. Further experiments would have to be performed to either confirm or eliminate catalysis by solvent, by plotting the intercept values against pH. These experiments are all performed under pseudo first-order conditions, with the buffer being in excess over the reactant.

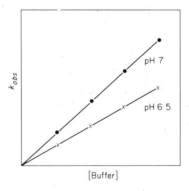

Fig. 6. Effect of pH change and buffer concentration on a general-base catalyzed reaction.

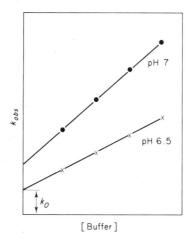

[Buffer]

Fig. 7. Effect of pH change and buffer concentration on a reaction that is both specific- and general-base catalyzed. k_0 represents the effect of solvent and OH$^-$ on the observed rate constant, k_{obs}.

To obtain a rate constant k_2' for the buffer-catalyzed reaction, the slopes of the lines may be calculated:

$$k_2' = \frac{(k_{obs} - k_0)}{[\text{Buffer}]_{\text{total}}}.$$

At each pH value at which a curve is obtained, the fraction of the buffer in the free base form is calculated from its pK_a using the Henderson–Hasselbalch equation:

$$pH = pK_a + \log \frac{[B]}{[BH^+]}.$$

The values of k_{BH^+} and k_B are determined by plotting the k_2' values against the fraction of the buffer in the free base form (Figs 8 and 9). Some reactions have either general-acid or general-base catalysis, and some show both types of catalysis. Kinetic experiments such as these show the types of catalysis that a reaction undergoes, and suggest other experiments that can be carried out to further document the mechanism.

The catalytic activity of bases (or acids) of similar structure but of varying strength can be related to their dissociation constants

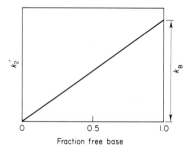

Fig. 8. Effect of buffer composition on the apparent second-order rate constant k_2' for a reaction in which general-base catalysis occurs. The catalytic rate constant (k_B) for general-base catalysis is obtained from the extrapolation of the curve to the right ordinate.

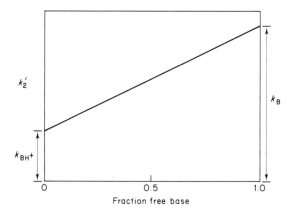

Fig. 9. Effect of buffer composition on the apparent catalytic constant k_2' for a reaction subject to both general-acid and general-base catalysis.

(pK_a values). This is called the Brønsted catalysis law. By plotting the log of k_B against pK_a, a straight line should be obtained for a reaction, with the largest deviations being observed for catalysts which are most dissimilar in their structure to the others tested. The slope of the Brønsted plot tells the sensitivity of the reaction to the basic (or acidic) strength of the catalysts.

As described above, the pathways of the two directions of a reversible reaction proceed through the same transition state, so that if the mechanism is delineated for one direction it is known for the other, also. If a catalyst acts as a general base (B) in the forward direction, then its conjugate acid (BH^+) acts as a general acid in the reverse reaction. Both these statements are examples of the principle of microscopic reversibility (Bender, 1971, p. 11 in his book).

Isotope Effects

Isotopes are used for three purposes in mechanistic experiments: to trace the pathway of atoms in reactions; to assay reactions; and to determine the location of bonds that are cleaved in multi-step reactions. The first two uses are much more familiar than the last, the kinetic isotope effect.

The effect of substitution of a heavier isotope for the normally predominant isotope is often to slow a reaction down. The replacement of hydrogen (^1H) by deuterium (^2H or D) gives a large change in the mass ratio, and so has been studied most. The ratio of the rate in the presence of the normal isotope to that with isotopic substitution (e.g. k_H/k_D) is called the *isotope effect*. If a hydrogen (protium) atom that participates in a reaction is replaced by a deuterium atom, then the rate of reaction normally is slowed down by a factor of 6–10. For example, in the enolization of acetone, if the acetone is completely substituted with deuterium, the rate of formation of the enolate is about ten-fold slower. This indicates that the rate-determining step is the cleavage of the C—H bond, because a C—D bond is stronger than a C—H bond. If tritium (^3H or T) is substituted for hydrogen, the isotope effect is further increased. Isotope effects which involve direct participation of the isotopic atom are called primary isotope effects.

A secondary isotope effect is observed if an isotopic substitution of an atom other than that which is transferred has an effect on the reaction rate. In S_N1 reactions, the formation of the carbonium ion is slowed by replacement of hydrogen by deuterium:

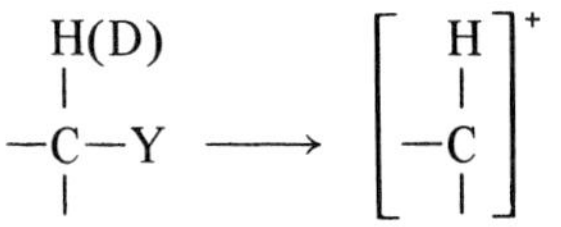

$$\begin{array}{c} H(D) \\ | \\ -C-Y \\ | \end{array} \longrightarrow \left[\begin{array}{c} H \\ | \\ -C \\ | \end{array} \right]^+ + \; Y^-$$

and the presence of a secondary isotope effect, k_H/k_D, of about 1.2 indicates that a carbonium ion is an intermediate in the reaction.

Replacement of H_2O by D_2O as the solvent may also have an effect on the rate of reaction. If transfer of a deuterium ion is involved in the reaction, then a deuterium isotope effect of 2–3 should be observed. This effect should be observed with all reactions in which general-acid or general-base catalysis is operative in the rate-determining step, because the hydrogen atoms of acids dissolved in D_2O rapidly exchange with the deuterium. Because the ionization of acids is less in D_2O, when the pD of a solution in D_2O is measured with a pH meter, the observed reading must be corrected by adding 0.4 units. Solvent isotope effects can distinguish between nucleophilic catalysis (no isotope effect) and acid–base catalysis (isotope effect).

The interpretation of the magnitude of isotope effects is complex (Jencks, 1969, Chapter 4 in his book; Gutfreund, 1972), but is important in mechanistic studies. Richards (1970) has described the use of kinetic isotope effects for experiments with enzymes. The use of isotopes other than those of hydrogen is more difficult, but these and other small isotope effects will probably be used for enzyme studies in the future.

Model Systems

There have been many studies of reactions which have been considered to be very similar to the catalytic reactions of enzyme processes. Some of these model reaction systems have been analogous, and others have been dissimilar in many ways (some almost totally). The study of good model systems is essential, though, since more often than not vital information pertaining to the reactants is learned. Before attempting to apply knowledge gained from model studies to enzyme mechanisms, we must first know what the similarities and differences are between the chemical model and the enzymic process. We must know if both reactions proceed by the same pathway, if the same functional groups are acting as catalysts, and when the properties of the model reaction are not analogous to the enzyme reaction. The study of model reactions is an important field, but one must be judicious in the application of the results.

Throughout the text, I shall discuss a wide variety of chemical reactions that appear to be reasonable models of enzyme action. Here I would like to describe research that has been performed on what was first hoped to be a model for enzymatic hydrolysis of esters and peptides, the imidazole-catalyzed hydrolysis of *p*-nitro-phenyl acetate. Although the mechanism of this reaction does not parallel the mechanism of peptidases, it has provided insight into catalysis by the imidazole group of histidine and into hydrolytic reactions in general.

In view of the fact that the imidazole group of histidine had been implicated in the activity of several hydrolases (chymotrypsin, trypsin and acetyl cholinesterase), the imidazole-catalyzed hydrolysis of *p*-nitrophenyl acetate was investigated by several groups (Bender and Turnquest, 1957; Bruice and Schmir, 1957) as a possible model for reactions catalyzed by the hydrolytic enzymes. Imidazole was used as a buffer catalyst, and the rate of the hydrolysis of the ester was assayed by measuring the formation of the *p*-nitrophenol (or *p*-nitrophenolate ion) spectrophotometrically. The reaction was found to be first-order with respect to both the imidazole and the substrate. Transient formation of an N-acetylimidazole intermediate showed that the imidazole was a covalent catalyst in the first step of the hydrolysis.

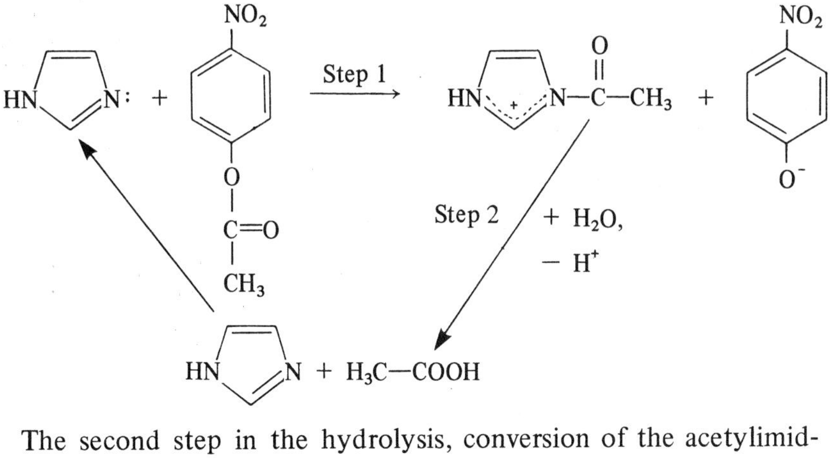

The second step in the hydrolysis, conversion of the acetylimid-azole to acetate and imidazole, is also catalyzed by imidazole. In this step, the free base form of imidazole is the catalyst. If it were a nucleophilic catalyst again, there would be no net reaction occurring, just acyl transfer between two molecules of imidazole, so the reaction must be one of general-base catalysis. Jencks and Carriuolo (1959) found that the hydrolysis of acetyl imidazole was catalyzed by acid, solvent and base. They showed that the rate of hydrolysis in D_2O

was 2.5 times slower than that in H_2O, consistent with a proton shift in the formation of the transition state. High concentrations of neutral salts strongly inhibited the reaction, indicating a decrease in availability of water for hydrolysis. These and other data suggest that in acid the hydrolysis of acetyl imidazole is a bimolecular (S_N2) reaction of water with the cationic form of acetyl imidazole:

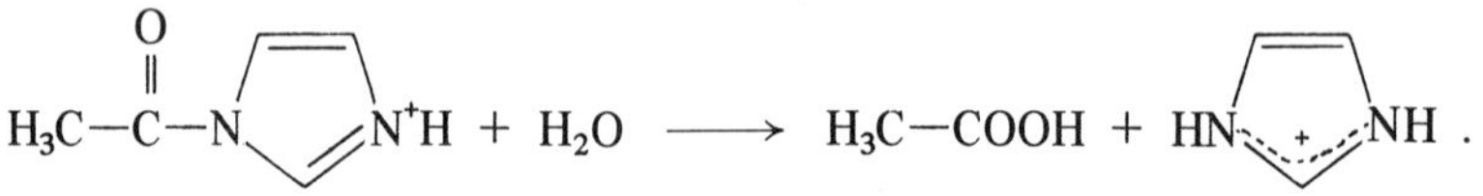

$$H_3C-\overset{\overset{\text{O}}{\|}}{C}-N\diagup N^+H + H_2O \longrightarrow H_3C-COOH + HN\overset{+}{\diagup}NH \ .$$

In basic solution, OH^- is the attacking species. The mechanism of catalysis by imidazole (p. 95) is to remove a proton from water in the transition state.

The role of imidazole as a basic catalyst for the second step of the overall hydrolysis of the phenyl acetate is indicative of the function of histidine in some hydrolases, such as ribonuclease. *O*-acyl serine rather than *N*-acyl histidine compounds are the first detectable intermediates formed by chymotrypsin, trypsin and acetyl cholinesterase. Thus, the initial discovery of the remarkable ability of imidazole to act as a covalent catalyst by forming a reactive *N*-acyl intermediate does not apply as a model for the enzyme reactions it was designed to mimic. The chemistry of the imidazole group, however, has been enriched by these experiments, and the system provides an excellent example of the use of the mechanistic techniques which I have described in this chapter. Like aniline in the catalysis of semicarbazone formation (p. 93), imidazole is effective as a covalent catalyst in the two-step hydrolysis reaction by having the correct increased reactivity with the ester and decreased stability of the intermediate. In addition, because the pK_a of imidazole is 7, it is the strongest free base that can exist in neutral solution.

REFERENCES

Bender, M. L. (1971). "Mechanisms of Homogeneous Catalysis from Protons to Proteins". Wiley-Interscience, New York.
Bender, M. L. and Turnquest, B. W. (1957). *J. Am. Chem. Soc.* **79**, 1652–1662.
Breslow, R. (1969). "Organic Reaction Mechanisms", Second edition. W. A. Benjamin, New York.

Bruice, T. C. and Benkovic, S. J. (1966). "Bioorganic Mechanisms", Vols 1 and 2. W. A. Benjamin, New York.

Bruice, T. C. and Schmir, G. L. (1957). *J. Am. Chem. Soc.* **79**, 1663-1667.

Cordes, E. H. and Jencks, W. P. (1962). *J. Am. Chem. Soc.* **84**, 826-831.

Gutfreund, H. (1972). "Enzymes: Physical Principles", pp. 157-175. Wiley-Interscience, London.

Ingraham, L. L. (1962). "Biochemical Mechanisms". John Wiley and Sons, New York.

Jencks, W. P. (1969). "Catalysis in Chemistry and Enzymology". McGraw-Hill, New York.

Jencks, W. P. and Carriuolo, J. (1959). *J. biol. Chem.* **234**, 1272-1285.

Kosower, E. M. (1962). "Molecular Biochemistry". McGraw-Hill, New York.

Richards, J. H. (1970). *In* "The Enzymes" (P. D. Boyer, ed.), Third edition, Vol. 2, pp. 321-333. Academic Press, New York.

5 Enzyme Kinetics

The study of the kinetics of enzyme reactions has supplied much of our present knowledge of enzyme chemistry. For many years, kinetic experiments provided the only evidence for the existence of the enzyme–substrate (*ES*) complex as an intermediate in enzyme-catalyzed reactions. Enzyme kinetics is a science in itself, relying heavily on advanced logic and algebra (Wong, 1975). My aim in this chapter is to show how a biochemist can use enzyme kinetics. I will not derive the kinetic equations, but will relate them to the rate equations presented in the previous chapter. For the history, the derivations and the involved mathematics of enzyme kinetics, readers should refer to review articles, texts and monographs on enzyme kinetics (e.g. Segal, 1959; Mahler and Cordes, 1971; Dixon and Webb, 1964; Westley, 1969; Cleland, 1970; Laidler and Bunting, 1973; Segel, 1975; Wong, 1975). The three basic uses of enzyme kinetics will be dealt with in the order in which they are used in the study of an enzyme: enzyme assay, effects of variables and applications to mechanistic studies. These three levels of use overlap. Some research calls only for assay methods, and some only for the first two methods. Complete characterization of an enzyme requires the use of all three procedures.

The principles we have learned with non-enzymic catalysts apply equally well to enzyme catalysts, as long as it is remembered that the enzyme is subject to denaturation. Simplifications of experimental conditions are often applied to enzyme kinetics, so it is necessary to be cautious of trying to relate kinetic assays to *in vivo* problems. In enzyme kinetics, it is common to apply pseudo first-order conditions also, with the distinction being that the limiting factor is the catalyst, not the substrate.

As a simplified example of enzyme action, let us examine the enzyme-catalyzed conversion of substrate S to product P. We have

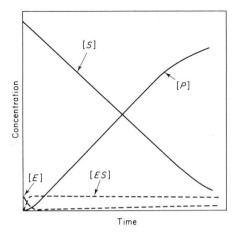

Fig. 1. Changes in the concentrations of reactants during a hypothetical reaction involving irreversible conversion of S to P via an ES complex.

seen that a catalyst is in contact with the transition state in a cata-lyzed reaction (p. 87) and will therefore assume that the substrate is complexed with the enzyme during the catalyzed reaction:

$$S + E \rightleftharpoons ES \rightarrow P + E.$$

Let us also assume that the equilibrium lies well towards formation of P, and that the slowest of the three steps shown in the equation is the conversion of ES into product and recycled enzyme. If S is present in a suitably large excess, soon after mixing of S with E the concentration of E approaches zero (Fig. 1). That is, the enzyme is almost all in the form of the ES complex and remains so until near the end of the reaction. Reaction conditions of this kind are called steady-state conditions.* Points to note from this hypothetical bi-molecular reaction are:

1. the substrate reacts with the enzyme for a finite time to form an ES complex;
2. the ES complex can either dissociate to form free $E + S$ or can react to form $E + P$;
3. S reacts continuously with E to replenish ES during the reaction; and

* Under steady-state conditions, it is not essential that the concentration of E be near zero. Figure 1 depicts a restricted case of the steady state, one in which substrate saturates the enzyme.

4. there is a lag in the formation of the product (greatly exaggerated in Fig. 1) before the steady state is reached.

This two-step reaction will be used as a simplified case to explain the principles of enzyme assays and the effect of substrate and inhibitor concentrations.

ENZYME ASSAYS

Because enzymes are characterized by the reactions which they catalyze, the concentration of an enzyme may be determined by measuring the rate at which it is able to catalyze its specific reaction. The analysis for enzyme activity is called an *assay*. The results obtained from enzyme assays are not in weight of enzyme, but units of catalytic activity. Enzyme assays must be carried out under conditions that are optimal for activity, with the pH, temperature and concentration of reagents used in a predetermined manner. The rate or velocity is measured either by the rate of disappearance of a substrate or, preferably, by the rate of formation of a product. (In practice, it is more precise to measure an increase of a small quantity than to measure a small decrease in a large quantity.) A graph of concentration of product formed against time, called a progress curve, is obtained (Fig. 2). To be valid, enzyme assays must be made under initial velocity conditions. By this, I mean that the rate of the

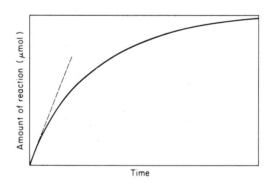

Fig. 2. Velocity–time or progress curve for a typical enzyme reaction. The initial velocity (v) is determined from the slope of the tangent (dashed line) at the origin of the curve. Either the formation of product or the utilization of substrate may be measured to obtain the initial velocity.

reaction must be taken in the initial linear-time period of the progress curve, because the curve flattens as equilibrium is approached. In the hypothetical conversion of S to P which has just been discussed, it was assumed that the equilibrium of the reaction greatly favoured formation of P. By measuring initial velocities, therefore, the reaction is examined under conditions which approach irreversibility. In addition to the possibility of back-reaction as an error in rate measurement, during the later stages of an enzyme reaction there might be denaturation of the enzyme, inhibition by a product or decrease of substrate concentration below the saturation levels needed to keep the enzyme as ES in the steady state.

Initial velocities (abbreviated as v) are measured as tangents at the origin of the progress curve (Fig. 2), and their units are μmol product formed min^{-1}. Analytical procedures for obtaining kinetic results may be either *continuous*, such as by direct recording of spectrophotometric changes or *discontinuous*, by removal of samples at various times and analysis for the product. Enzymologists prefer a continuous assay, but some reactions do not have properties that allow continuous monitoring. If a discontinuous assay procedure is used, single experimental points may be used if it is certain that the assay samples being taken are from the linear portion of the progress curve. This portion usually does not exceed the first 5–10% of the reaction. Therefore, there must be leeway in any assay to allow for fluctuations in the concentration of enzyme so that the reaction being measured is always first-order with respect to enzyme.

The assay should always be tested to show that the velocity v is proportional to the amount of enzyme added. Deviations from linearity with concentration of enzyme may be due to limitations in the assay conditions as described above or to the presence of inhibitors (Fig. 3). Inhibitors can be removed by enzyme purification. Assay limitations (Fig. 3, Curve B) are overcome by increasing the concentration of the limiting substrate so that v is again proportional to $[E]$. The concentration of each substrate and cofactor should be at least four-fold higher than its K_m value (p. 114) for best results. If the concentration of a substrate must be below its K_m value (perhaps because of solubility or high absorbance), the velocity of an assay may be reported in V (maximal velocity) units, obtained by substitution of experimental v values and the K_m in the Michaelis–Menten equation.

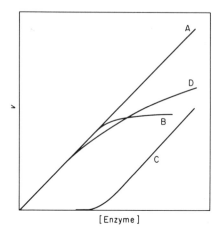

[Enzyme]

Fig. 3. Effect of concentration of enzyme on initial velocity. Curve A shows a reaction which behaves normally. Curve B is that of a reaction which has insufficient substrate to saturate the enzyme at high $[E]$. Curve C shows the type of curve obtained if an inhibitor for which the enzyme has high affinity is present in the substrate solution. Curve D shows the type of curve obtained if there is an inhibitory substance in the enzyme solution.

Standardized definitions and units are used to describe the results of enzyme kinetic experiments and the activities of enzymes. The initial velocity v is expressed in enzyme units. One unit of enzyme activity was defined by the Enzyme Commission in 1961 as the amount of enzyme which will catalyze the formation of $1\,\mu$mol product min^{-1} under specified conditions.* Therefore, all values of v should be expressed in μmol product formed (or substrate used) min^{-1}. The *specific activity* of an enzyme preparation is expressed in μmol product formed min^{-1} (mg protein)$^{-1}$. The extent of purification of an enzyme (cf. Chapter 3) is the ratio of the specific activity of the purified enzyme to the specific activity of the enzyme in the crude extract. If a pure enzyme sample has been obtained and tested, then the purity of other samples can be determined by dividing the specific activity of the sample by the specific activity of the pure enzyme and multiplying by 100%. The catalytic activity of a pure enzyme is often expressed as its turnover number or catalytic constant (k_{cat}, the rate constant for: $ES \rightarrow E + P$, p. 116). This

---

* The Commission on Biochemical Nomenclature (Hoffmann-Ostenhof, 1974) has recommended that enzyme activity be reported in katals (1 katal = amount of activity that converts 1 mol substrate $s^{-1} = 60 \times 10^6\,\mu$mol $min^{-1} = 6 \times 10^7\,\mu$mol min^{-1}). The katal has not yet been accepted by enzymologists.

value, having the dimension min^{-1}, is the number of catalytic events occurring per minute, i.e. number of μmol substrate converted to product $(\mu mol\ enzyme)^{-1}\ min^{-1}$. The reciprocal of k_{cat} is the time required for a single catalytic cycle with a single-subunit enzyme. All of the above measurements are to be made under defined assay conditions, with the temperature stated (preferably 25° or 30° C).

An assay method must fulfil several important criteria. The stoichiometry of the reaction should be found by analysis. This includes knowing the requirements for cofactors (coenzymes and activating ions). One of the first experiments is to show a requirement for each of the reagents added, by comparing samples with single omissions to a complete system. The validity of the procedure must be verified by demonstrating the coincidence of formation of product with utilization of substrate by independent analytical means. Dennis (1972), while testing several bacteria for the presence of GMP reductase, found that *B. subtilis* appeared to have high activity for the reductase when the rate of disappearance of NADPH was

$$GMP + NADPH + H^+ \xrightarrow{\text{GMP reductase}} IMP + NADP^+ + NH_3$$

measured in the presence of GMP. However, the rate of formation of IMP was low in comparison, and most of the NADPH was being oxidized by a GMP-stimulated NADPH dehydrogenase. Use of the spectrophotometric assay which was valid with other cells was useless with *B. subtilis* extracts. Finally, the optimal pH and the amount of each substrate needed for saturation of the enzyme must be empirically determined.

It is sometimes necessary to use coupled assay procedures, in which the formation of product from the reaction being tested is used as a substrate in a second enzymatic reaction which is more easily analyzed. Coupling may also be used to overcome an equilibrium unfavourable for assaying a reaction. Bertino *et al.* (1960) were unable to detect the presence of dihydrofolate reductase in extracts of leukemic leucocytes using the usual assay which measures the oxidation of NADPH. The presence in the extracts of an excess of a second enzyme, formyltetrahydrofolate synthetase, catalyzed formate- and ATP-dependent conversion of tetrahydrofolate, the product of the first reaction, to 10-formyltetrahydrofolate. This product was converted to the coloured methenyltetrahydrofolate when perchloric acid was added to deproteinize the reaction:

Dihydrofolate + NADPH + H⁺ ⟶⇌

                     Tetrahydrofolate + NADP⁺

Tetrahydrofolate + HCOOH + ATP ⟶⇌

           10-Formyltetrahydrofolate + ADP + P_i

10-Formyltetrahydrofolate $\xrightarrow{\text{HClO}_4}$

        5,10-Methenyltetrahydrofolate + H_2O.

In this case, the second enzyme was present endogenously, but puri-fied auxiliary or indicator enzymes also may be added for coupled assays. Segel (1975, p. 85 in his book) has summarized the require-ments for the validity of a coupled assay:

1. a first reaction that is zero-order in substrate and irreversible; and
2. a second reaction that is first-order in the product of the first reaction and irreversible.

If insufficient indicator enzyme is added, a lag will appear in the overall progress curve, so initial time-course experiments should be performed to ensure that enough of the second enzyme is present. The maximal velocity for the second reaction should be at least ten times that of the first step. A linked system requires that both enzymes function near their optimum at the conditions of the coupled assay, or that the assay be performed in two stages.

    After optimal assay conditions have been established, a protocol is written down for carrying out the assay reactions. This protocol lists the concentrations and volumes of all reagents used, and the order in which the reagents are mixed. Reactions are usually started by addition of either substrate or enzyme to assay mixtures that have reached the temperature of reaction. The protocol will tell which controls and blanks are essential to obtain the correct activity (e.g. omission of one substrate and omission of enzyme). It is often advis-able to prepare what are called "cocktail mixes" for enzyme assays. These mixtures contain all the reagents which are stable and can be mixed and stored together. For example, buffer, cofactors or sub-strates and water to make the correct volume may be premixed. As well as convenience, a cocktail mixture gives accuracy by using exactly the same concentration of all its reagents for all assays, thus avoiding some pipetting errors.

If an assay procedure is chosen that has been used previously, it is advisable to make sure that it is the most suitable assay for the purposes. It should be as simple, accurate and specific as possible. For analyzing quickly, such as in seeing which tubes from a column chromatograph contain enzyme activity, a less accurate estimate of activity may suffice, but all kinetic experiments should be performed using an accurate procedure. Some methods have been found to be inaccurate and non-specific, so a day or so in the library will often save months in the laboratory repeating inaccurate experiments. "Methods in Enzymology" contains articles describing assay methods for most enzymes.

Technical points to observe when performing assays include standard analytical procedures with the additional inclusion of caution to prevent undue denaturation of the enzyme. Glassware should be washed free of heavy metal ions (including dichromate from washing solutions). All the reagents should be as pure as possible and should either be prepared fresh or kept under conditions where they are stable. Distilled water is used for making all solutions. Pipetting of the small volumes of reagents must be done carefully using accurate pipettes. If piston-activated pipettes are used, they should be calibrated and used under conditions which give reproducible results (Ellis, 1973). Enzyme assay solutions are mixed gently (e.g. by inversion of cuvettes or tubes covered with small squares of Parafilm, or by adding the final reagent on a tiny plastic mixing spoon) so that the enzyme is not denatured by air bubbles. Because the initial velocity is dependent on temperature, the assays must be carried out under controlled temperature conditions. Constant temperature baths are used for test-tube assays, and circulating water heaters and thermospacers are needed for spectrophotometers. The weak acid or base used as a buffer in an enzyme assay must be tested to assure that it does not inhibit the enzyme. Gomori (1955) has described the preparations of 20 buffer solutions. In addition, Good et al. (1966) have prepared and tested 12 new buffers covering the pK_a range of 6.15–8.35. Henderson (1971) has described a rapid and precise method of measuring the initial velocity from a recorded tracing with a protractor and a calibration table.

E

For discontinuous assays, the reaction should be stopped by inactivating the enzyme with a precipitant such as trichloroacetic acid. In these assays, a zero-time analysis should also be performed as a control. Enzyme assays should always be performed in duplicate or triplicate. When studying the effect of a variable on initial velocity of a reaction, individual experimental points should show sufficient consistency with other points so that their accuracy is convincing.

Many enzyme assays are used for diagnostic testing of blood serum or other biological fluids. The assays may analyze either for a metabolite in the serum or more commonly for an enzyme activity. Standardized protocols have been devised for determination of several dozen enzymes ("Standard Methods of Clinical Chemistry"; Mattenheimer, 1971; Wolf et al., 1973). These procedures have been simplified and the technical details have been carefully described so that the results one obtains are readily reproducible. Cocktail mixes of incubation mixtures and enzyme standards are available as commercial kits, facilitating the use of enzymatic analyses in hospital and other clinical laboratories. Analyzer machines which can prepare samples, record results and/or calculate and present the data are now manufactured (Schwartz, 1971; White et al., 1972). These analyzers usually rely on assays which measure changes in either the absorbance or fluorescence of pyridine nucleotides. The same precautions used in the manual assays described above apply to automated assays. Both continuous (kinetic) and discontinuous (single-point) assays are performed, depending on the equipment available to the laboratory. With single-point assay methods, the electronic recorders of an automated analyzer may be adjusted to a sensitivity which ensures that only the initial velocity is being measured. Coupled assays often must be used, because neither substrate nor products of the test reaction absorb at a suitable wavelength. For example, in the determination of glutamate-oxaloacetate transaminase (GOT) activity, the formation of oxaloacetate is assayed by its oxidation to L-malate, allowing the formation of NAD^+ to be followed by the decrease in absorbance of NADH at 340 nm:

$$\alpha\text{-Ketoglutarate} + \text{L-aspartate} \xrightarrow{\text{GOT}} \text{L-glutamate} + \text{oxaloacetate}$$

$$\text{Oxaloacetate} + \text{NADH} \xrightarrow[\text{(excess)}]{\text{Malate dehydrogenase}} \text{L-malate} + NAD^+$$

Application of enzyme assays to clinical analysis requires standardization of the proven assay methods and the use of uniform reaction conditions (Bergmeyer, 1972) so that results from one laboratory can be compared with those from others.

EFFECTS OF VARIABLES

It is possible to learn a great deal about an enzyme by examining the effects of concentrations of substrates, temperature and pH on its activity. Each of these variables is usually tested while establishing optimal assay conditions, but in addition, data can be obtained to indicate the probable *in vivo* concentrations of substrates, the possible amino acid residues functioning at the catalytic site and the thermodynamic activation parameters for the enzyme-catalyzed reaction. The effect of these three variables on enzyme activity shall be studied next, and then some of the simpler cases of enzyme inhibition will be discussed. A thorough grasp of these effects is essential for an understanding of the mechanistic applications of enzyme kinetics.

Effect of Substrate Concentration

Early in this chapter, we saw how a simple two-step bimolecular reaction could be written for the conversion of S to P under the influence of the catalyst E:

$$E + S \underset{k_{-1}}{\overset{k_1}{\rightleftharpoons}} ES \overset{k_2}{\longrightarrow} E + P.$$

Let us now see what the effect of concentration of S is on the rate of this reaction and see how this effect applies in most enzymic reactions. As the concentration of S is increased, the initial velocity of the reaction produces a hyperbolic curve (Fig. 4). There are two regions of this hyperbola which are easily recognized mathematically. At low concentrations of S, v is directly proportional to $[S]$, or first-order with respect to S. The overall reaction is then second-order at these low concentrations, being first-order in both S and E. At high concentrations of S, when E is saturated with S by conversion of all the enzyme to the ES complex, v is independent of added S, or zero-order in S. These are the pseudo first-order reactions used for enzyme assays, where the rate depends only on $[E]$. Between the first-order and zero-order concentrations is a region of mixed-order kinetics.

The equation for a rectangular hyperbola such as Fig. 4 is

$$y = \frac{ax}{x + b}$$

where a and b are both constants. If we change the co-ordinates into the biochemical terms $y = v$ and $x = [S]$, with the constants being $a = V$ (maximal velocity) and $b = K_m$ (Michaelis constant), the rate law is

$$v = \frac{V[S]}{[S] + K_m}$$

which is called the Michaelis–Menten equation.

The Michaelis–Menten equation has been derived by two different approaches. Michaelis and Menten (1913) assumed that the rate-determining step in the conversion of S to P is the breakdown of ES to $E + P$, i.e. k_2 is much less than k_{-1}. Briggs and Haldane (1925) used a more generalized approximation, based on the steady-state assumption that $d(ES)/dt = 0$. The equation for all the possible reactions involving ES

$$d(ES)/dt = k_1[S][E] - k_{-1}[ES] - k_2[ES] = 0$$

and that for the formation of product P from ES

$$dP/dt = k_2[ES]$$

can be combined and rearranged to give the same Michaelis–Menten rate equation, but with a different significance to K_m.

By substituting $[S] = K_m$ into the Michaelis–Menten equation, the initial velocity is found to be $v = V/2$. The numerical value of the K_m, therefore, is the concentration of substrate at half-maximal velocity (Fig. 4). Any further physical significance to K_m depends on the mechanism of the reaction. K_m has concentration units (M) and is independent of the concentration of enzyme. From the Michaelis and Menten theory, $K_m = k_{-1}/k_1$ which is the dissociation constant (K_s) for the reaction $ES \rightleftharpoons E + S$. A low value of K_m indicates tight binding of substrate to enzyme. For some enzymes, the dissociation

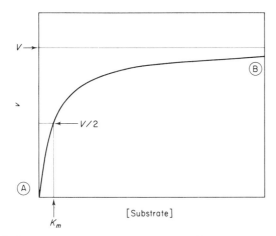

Fig. 4. Hyperbolic saturation curve showing the effect of the concentration of substrate S on the initial velocity of an enzyme-catalyzed reaction. In region A, v is first-order in S, and in region B where v is asymptotically approaching the maximal velocity V, it is zero-order in S. The concentration of S at which $v = V/2$ is the Michaelis constant for S.

constants obtained by physical, not kinetic, experiments are approximately equal to their K_m values for the substrates, but the assumption that they are always equivalent can be misleading. The more general equation for K_m obtained from the steady-state derivation is $K_m = (k_2 + k_{-1})/k_1$. This equation can apply to reactions in which any of the three rate constants is smallest.

It is logical that cells should utilize reactant levels at or near the proportional regions of an enzyme saturation curve. In doing so, the cell does not waste catalytic capability but has the capacity to deal with increases in substrate levels without accumulating large amounts of substrate. Therefore, determination of a K_m establishes the approximate intracellular concentration of the substrate (Cleland, 1967). *In vivo,* the levels of product should not be sufficiently high to cause direct product inhibition of enzyme. There have been reasonable suggestions that, to avoid product inhibition, the product is less tightly bound to an enzyme than is the substrate, thereby defining the natural direction of a reversible reaction. Evolution of suitable kinetic constants may exert crude controls in a few cases, but control of metabolic pathways at the enzyme level by products occurs by a more complex process (Chapter 13).

The maximal velocity V is the asymptotic value of v in Fig. 4,

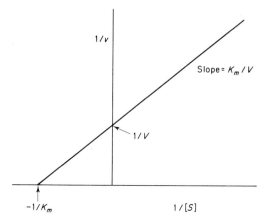

Fig. 5. Lineweaver–Burk double reciprocal plot used for the determination of both K_m and V from initial velocity measurements. This graph is preferable to a plot of v v. $[S]$ because its straight line can be extrapolated to obtain v at infinite $[S]$ (i.e. the value of V).

where the enzyme is completely saturated with substrate. From both Michaelis–Menten and steady-state considerations, when $[ES] = [E_0]$, the original concentration of E,

$$V = k_2[ES] = k_2[E_0].$$

Therefore, V varies with the concentration of enzyme. The constant k_2 can also be called k_{cat}, the catalytic constant or turnover number (cf. p. 108).

Simple v v. $[S]$ saturation curves are not used to determine kinetic constants because they do not show the true maximal velocity (the asymptote cannot be approached closely). Despite this limitation, always plot accumulated rate data by this method first, particularly in the case of subunit enzymes. In order to obtain K_m and V, three linear transformations of the Michaelis–Menten equation have been used. For all three methods, the concentration of substrate used should be well-spaced, and not cover just a small portion of the curve. The most commonly used plot is the Lineweaver–Burk or reciprocal plot (Fig. 5). K_m and V values can be obtained readily from its intercepts. Although it is the least effective linear plot on a statistical basis, virtually all enzymologists have reported their results using this graph (N.B. enzymologists, not enzyme kineticists). Because the data in the literature are reported in terms of the reciprocal plot, I will

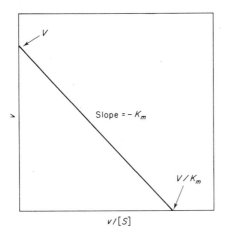

Fig. 6. Eadie–Hofstee single reciprocal plot used for the evaluation of K_m and V from experimental values of the initial velocity, v.

often refer to this graph, especially in enzyme-inhibition studies. The Eadie–Hofstee plot of v v. $v/[S]$ (Fig. 6) and the Hanes plot of $[S]/v$ v. $[S]$ both are more useful than the reciprocal plot because they show departures from linearity more obviously. The Eadie–Hofstee plot is the inverse of the Scatchard plot for ligand binding, to be described in Chapter 12 (p. 416). I do not advocate use of the reciprocal plot, but agree with Wong (1975, p. 30 in his book) in recommending that accurate kinetic experimental data should be examined by all three linear graphs for comparison. Use of these three plots and the original v v. $[S]$ plot should demonstrate the presence of any abnormal kinetic behaviour.

The most rapid and also most precise and accurate graphical method for determining K_m and V is a recently described direct linear plot (Eisenthal and Cornish-Bowden, 1974). For each observed result (v and $[S]$), the points are marked on the fourth quadrant of a v v. $[S]$ plot (Fig. 7). The points of intersection in the first quadrant give the co-ordinates of the best-fit values for V and K_m. When there are errors in the determinations, there is no unique intersection point; the median of each series of points gives the best estimate of the kinetic constants. Eisenthal and Cornish-Bowden have commented, quite correctly, that there is no justification for continued use of the Lineweaver–Burk plot for any purpose. Two series of observations (e.g. for inhibition or activation studies) can be displayed on one

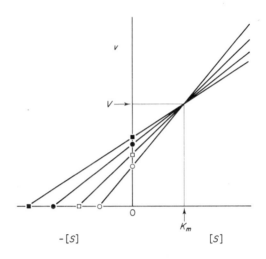

Fig. 7. The direct linear plot for determination of K_m and V. Each line represents the measured v at a given $[S]$ and is plotted with the intercepts of v and $-[S]$ on the graph. V and K_m are obtained from the point of intersection, as indicated. This method is extremely simple and rapid, and because no calculations are made, data can be plotted during the course of an experiment.

direct linear plot, but the authors suggest that the Hanes plot be used for presentation of multiple series of results.

Several practical advances may aid in determining K_m values. Gurr *et al.* (1973) have used a gradient-flow apparatus to automatically record saturation curves for the alcohol dehydrogenase reactions, an approach that would save time for variations and repetitions of kinetic experiments. Lee and Wilson (1971) have described a modification of the reciprocal plot which allows one to measure non-initial velocities for reactions that are not appreciably reversible. The use of computers and statistical treatments is desirable for advanced applications of kinetics.

With some enzymes, inhibition by substrate may be observed. In a v v. $[S]$ plot, inhibition produces a maximum in the curve, followed by a decrease in v. The reciprocal plot for a reaction in which there is a substrate inhibition shows a discontinuity which asymptotically approaches the ordinate (Fig. 8). An increase in concentration of one substrate may cause a simultaneous decrease in that of the second substrate. For example, in the hydrolysis of sucrose, a high molarity of the sucrose decreases the concentration of the second reactant, water (Dixon and Webb, 1964, p. 75 in their book). Another possible

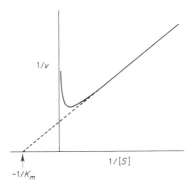

Fig. 8. Reciprocal plot for an enzyme reaction in which the substrate is an inhibitor.

type of substrate inhibition is exemplified by the inhibition of fumarase when the concentration of fumarate is much greater than its K_m concentration (Rajender and McColloch, 1967). This suggests that the active site has two positively charged groups, each one binding a carboxylate of one molecule of fumarate ($^-$OOC—CH= CH—COO$^-$). With excess fumarate, two molecules are bound per active site, one to each positive centre. Examples of activation of enzymes by increased concentrations of substrates will be discussed in Part III on the control of enzyme reactions.

Effects of Temperature

An increase in temperature can have several effects on an enzyme-catalyzed reaction. The two major effects are an increase in the velocity of the reaction catalyzed and an increase in the rate of denaturation of the enzyme. Since these are both competing reactions that increase with increasing temperature, no temperature optimum can be defined for an enzyme. Enzymes have probably evolved so that they have stable structures under the physiological conditions at which they function. Thermophilic bacteria, for example, contain enzymes which are stable at elevated temperatures. It has been noted (p. 42) that denaturation can be lessened during enzyme purification by working at low temperatures. Some enzymes are labile near 0° C, though, for reasons which will be discussed in Chapter 12 (p. 408).

The effect of temperature on the reaction rate can be interpreted for simple enzymic reactions if measurements are made over a temperature span in which the enzyme is stable. Maximal velocities, assumed to reflect the conversion of the ES complex to product, can be used to construct an Arrhenius plot and, if they are available for both directions of a reaction, a reaction co-ordinate diagram can be drawn for the enzymic reaction.

Effect of pH

Changes in the pH of an assay solution can cause a number of effects on initial-velocity measurements with an enzyme. The enzyme may be irreversibly denatured by acid or base. This effect can be tested by incubation of the enzyme at the pH in question, followed by assaying of the sample at the pH optimum. If a substrate ionizes (or is protonated) to an inactive form with a change in pH, then its total concentration must be raised to ensure a saturating concentration of the active form. The most important effects of pH are on the velocity of the reaction. These changes can be caused by alterations in the state of ionization of either binding or catalytic groups in the enzyme or groups involved in maintaining the conformation of the enzyme.

When v is plotted against pH, the result is a bell-shaped curve for many enzymic reactions (Fig. 9). The point of maximal activity is the *pH optimum*, which is a fixed value for the enzyme under the conditions of the assay. Routine assays for activity are performed at this pH. The pH–activity curve often is explained by assuming that the sigmoid curve on the left side is caused by the protonation of an ionizable group, B, and the sigmoid curve on the right is produced by ionization of another ionizable group, A. For activity, A must be protonated and B must be dissociated. The pH values of half-maximal activity on each side of the curve often correspond to the pK_a values of each of the ionizable groups. Knowledge of the pK_a values allows one to guess but not prove the identity of the groups.

The hydrolysis of several substrates catalyzed by chymotrypsin gives a bell-shaped pH–activity curve with pK_a values of about 7 and 8.5. The free base form of aspartate residue 102 is responsible for the pK_a of 7. This aspartate functions at the catalytic site of the enzyme by acting as part of an acid–base catalyst system (p. 321). The ionization at pH 8.5 has been attributed to the N-terminal isoleucine.

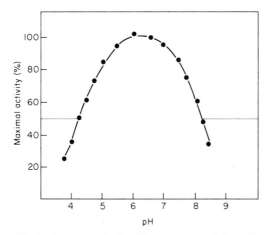

Fig. 9. A pH profile for the hydrolysis of benzoyl-L-arginine ethyl ester catalyzed by papain at $25°C$ (drawn from the data of Smith and Parker, 1958). The pK_a values of about 4.3 and 8.2 were assigned to a perturbed histidine imidazole which acts as a general base catalyst and to an unionized thiol which becomes acylated, respectively (Lucas and Williams, 1969). Polgár (1974) has presented evidence for an alternative mechanistic explanation, that the reactive form of papain is the mercaptide-imidazolium ion pair $[-S^--HB^+]$, in which the proton of Cys-25 is already on His-159. Presence of this ion pair is also represented by the bell-shaped curve.

The $-NH_3^+$ form of its amino group binds to the carboxylate of Asp-194. In basic solution, when this group becomes $-NH_2$, the chymotrypsin reversibly changes to an inactive structure in which the aspartate side chain seeks a different environment and blocks the active site. This shows us that the pH effect need not be on catalytic groups directly.

A knowledge of the effects of variables on an enzyme reaction may have practical as well as academic interest. Some enzymes in inborn metabolic diseases have K_m or V values which are altered from their normal counterpart enzymes. For industrial uses, mutant micro-organisms may be selected for the presence of enzymes with more desirable kinetic properties or stabilities. A familiar example was the inclusion of proteolytic enzymes in laundry detergents for several years. One of these proteases isolated from a mutant of *B. subtilis* had a pH optimum of 9–10 and was stable at $40°C$ for over 2 h (Chemical and Engineering News, 1969), making it useful under washing conditions. These enzymes, in conjunction with the deter-gents, were very effective in removing blood and gravy stains, but "the greatest advance in detergents in more than 20 years" also dissolved wool garments and soap-workers' lungs, and could cause allergies for frequent users. Enzymes of this type are now used more realistically in pre-wash preparations, for selective removal of protein stains.

Effects of Inhibitors

Inhibitors are compounds which lower the activity of enzymes by preventing either the formation or productive breakdown of enzyme-substrate complexes. In this chapter, I will deal with reversible inhibitors, compounds which bind to enzymes by non-covalent bonds. Other inhibitors such as the modifiers used in active-site titrations cause inactivation of enzymes by forming covalent bonds at or near the active site. Use of inhibitors in kinetic experiments can give information about the binding sites and specificity of an enzyme. Substrate and product inhibitors can aid in choosing between possible kinetic mechanisms, as we shall see below. Many drugs are enzyme inhibitors, and the pharmacological uses of some enzyme inhibitors as anti-metabolites form the subject of Chapter 15.

Most reversible inhibitors belong to one of three categories: competitive, noncompetitive or uncompetitive. These three classes are defined by their kinetic properties, and the differences between them are reflected by their effects on the slope or the intercept at the y-axis in Lineweaver–Burk plots with and without inhibitor. [For diagnosis of inhibitor action by other plots, see Wong (1975, p. 45 in his book) and Eisenthal and Cornish-Bowden (1974).]

Competitive inhibitors prevent the binding of substrate to the enzyme. In the presence of a competitive inhibitor (I), the enzyme can form an EI complex:

$$E + I \rightleftharpoons EI \text{ (inactive)},$$

rather than form an ES complex. When both I and S are present in a solution, the proportion of the enzyme which is able to form ES depends on the relative concentrations and affinities of S and I with E. A large increase in $[S]$ at constant $[I]$ can convert all the enzyme to the ES form. The effect of a competitive inhibitor is to raise the apparent K_m for S, as can be seen in the graphical representation of

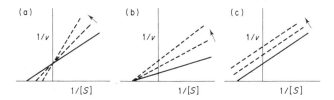

Fig. 10. Reciprocal plots for (a) competitive, (b) noncompetitive, and (c) uncompetitive inhibition. The solid lines are the reciprocal plots in the absence of inhibitor, and the dashed lines in the presence of two different concentrations of inhibitor. The arrows show the direction of increase in $[I]$.

competitive inhibition (Fig. 10a). Only the slopes of the reciprocal plot are affected by a competitive inhibitor. Although many competitive inhibitors have a structural resemblance to the substrate and bind to the same site on the enzyme (e.g. malonate and succinate with succinate dehydrogenase), this is not a necessary criterion for competitive inhibition. Negative feedback inhibitors of allosteric enzymes can be competitive with S even though they bind at different sites on the enzyme.

A *noncompetitive inhibitor* binds to a site on the enzyme that is not identical to the substrate-binding site, and does not affect the affinity of the enzyme for the substrate. Both EI and EIS complexes can form, but the EIS complex cannot be converted into product. A noncompetitive inhibitor, then, lowers the maximal velocity but does not change the K_m for S. The reciprocal plot for noncompetitive inhibition in Fig. 10b shows that the $1/v$ intercepts on the ordinate are increased (i.e. V decreases). A noncompetitive inhibitor can be envisaged as removing active enzyme from solution. Its inhibition cannot be reversed by an increase in the concentration of substrate. Noncompetitive inhibitors are rare.

Uncompetitive inhibitors bind only to the ES complex, not with free enzyme, to form an inactive EIS complex. The Lineweaver–Burk plot for an uncompetitive inhibitor (Fig. 10c) shows that the slopes are not altered, so the K_m is decreased and V is decreased. Uncompetitive inhibition usually occurs only with multi-substrate reactions.

If an EIS complex can be converted to products, then the inhibitor is termed a partial inhibitor. Other inhibitors show a behaviour which is a mixture of two actions. Because these and other reversible inhibitors do not fall into any of the three simple classes, further classifications of inhibitors are necessary (Wong, 1975).

The dissociation constant of an EI (or EIS) complex is called the inhibitor constant, K_i. The K_i for an inhibitor may be determined graphically, such as by plotting $1/v$ against $[I]$ at several fixed concentrations of substrate (Dixon and Webb, 1964, p. 327 in their book; Wong, 1975, p. 49 in his book). The value of K_i is obtained from the concentration on the abscissa at or below the point of intersection of the curves. Cornish-Bowden (1974) has introduced the use of $[S]/v$ v. $[I]$ plots for determining K_i values for mixed, uncompetitive and noncompetitive inhibitors.

Auld *et al.* (1972) have recently described a method which assesses the nature of a reversible inhibitor from experiments at only one concentration of substrate. They studied the hydrolysis of a series of N-dansylated substrates in the presence of carboxypeptidase A by a rapid-mixing method that allowed the concentration of ES to be determined directly. The carboxypeptidase reaction follows classical Michaelis–Menten kinetics:

$$E + S \underset{k_{-1}}{\overset{k_1}{\rightleftharpoons}} ES \overset{k_2}{\longrightarrow} E + P,$$

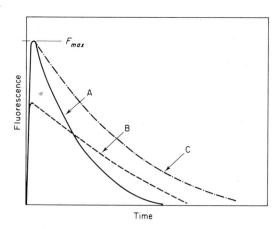

Time

Fig. 11. Diagram showing formation and decay of the ES complex measured by stopped-flow fluorescence (after Auld *et al.*, 1972). F_{max} shows the $[ES]$ for the uninhibited reaction (curve A). F_{max} is lowered by a competitive inhibitor (curve B), but is not affected by a noncompetitive inhibitor (curve C). The duration of the reaction was about 10 s.

where the second step is rate limiting and the K_s of the ES complex equals the K_m for S. While a dansylated peptide is bound to the enzyme, it fluoresces. Therefore, fluorescence is directly proportional to the concentration of the ES complex. Experimentally, the rapid increase in fluorescence when carboxypeptidase and the substrate were mixed and the slower decrease in fluorescence as the product was released from the enzyme were recorded on an oscilloscope (Fig. 11). The maximal fluorescence, F_{max}, indicates the maximum concentration of ES. The area under the oscilloscope curve is inversely related to k_2. A competitive inhibitor (curve B) would lower F_{max} but the area under the curve would be constant. A noncompetitive inhibitor (curve C) would decrease k_2 (i.e. increase the area) but not affect F_{max}. Mixed-action inhibitors would change both F_{max} and the area. In addition, values for K_m, K_i and k_2 can be calculated from the data, and the method is suitable for many applications.

MECHANISTIC APPLICATIONS OF ENZYME KINETICS

Steady-state kinetics can be used to suggest mechanisms for enzyme reactions. For complete discussion of the approach and techniques, it is possible to refer to recent reviews on the subject (Cleland, 1970; Plowman, 1972; Wong, 1975). In the space available here, I can only present sufficient material to enable one to understand the examples of the use of kinetics described later in the book. Some of the experiments can be performed with partially purified enzymes, but they should be verified when the enzyme is available in a pure form.

The simple two-step representation of an enzymatic reaction cannot be correct for a reversible reaction and should have at least one more step, the interconversion of ES and EP complexes. This should be obvious, because both S and P should be able to form complexes with E:

$$E + S \underset{k_{-1}}{\overset{k_1}{\rightleftharpoons}} ES \underset{k_{-2}}{\overset{k_2}{\rightleftharpoons}} EP \underset{k_{-3}}{\overset{k_3}{\rightleftharpoons}} E + P.$$

We now have six rate constants that must be considered, and any evaluation of all of these gives much more information about the mechanism of the reaction than does simpler Michaelis–Menten type data. Because most reactions involve more than one substrate, rate laws had to be derived for multi-substrate reactions. Fitting the kinetic properties of a reaction to the rate laws may impart the order of addition of substrates and release of products, and the site of the rate-limiting step. The steady-state kinetic techniques available to provide these data are initial-velocity studies, product-inhibition experiments and isotope exchange at equilibrium. All three of these are based on initial-velocity techniques. The use of fast-reaction methods can extend the initial-velocity measurements to give information about transient intermediates.

Initial Velocity Studies of Multi-substrate Reactions

In the last 20 years, attempts have been made to apply initial-velocity kinetics to the study of group transfer in multi-substrate reactions. Development of the principles, based on simpler kinetic work, and derivation of the basic rate laws were done primarily by Alberty, Dalziel, Wong and Hanes, and Cleland. A great aid is the nomenclature system of Cleland (1963), which is used in this book and by most enzymologists.

Cleland introduced a notation which abbreviated a kinetic mechanism to a horizontal line or group of lines. The sequence of steps proceeds from left to right, with the addition of substrates (lettered $A, B, C, \ldots$) to the enzyme (E) and the release of products $(P, Q, R, \ldots)$ from the enzyme indicated by vertical arrows. For specific examples, I will write names or abbreviations for the substrates and products in place of A and P. The forms of the enzyme (free enzyme or complexes) are written under the line. Enzyme complexes may be stable (i.e. covalent) or transitory complexes. Central complexes are transitory complexes in which the active site is filled (e.g. $EAB-EPQ$ for a bisubstrate reaction) and are written in parentheses. They correspond to the ES complex of the Michaelis theory. Enzyme-reactant complexes which can still bind another reactant are called non-central complexes. I have modified Cleland's notation so that the rate constants have positive numbers for the forward direction (written to the left of the arrow) and negative numbers for the reverse (right of the arrow). Reactions are divided into *sequential* and *ping-pong*. Sequential reactions are those which require combination of all substrates with the enzyme before any product is released. Sequential reactions can be either *ordered*, in which there is an obligatory order of addition of substrates and release of products, or *random*, in which there is no obligatory order of binding or release. Rapid-equilibrium reactions are sequential reactions in which the dissociation of products greatly exceeds the overall reaction rate. Ping-pong reactions occur when a product is released from the enzyme before all the substrates have been added. They are called ping-pong because of the existence of at least two stable forms of the enzyme (E and F) which oscillate during the reaction. Shown below are diagrams for the three basic bisubstrate–biproduct reactions. Rate constants have been written for the ordered reaction only.

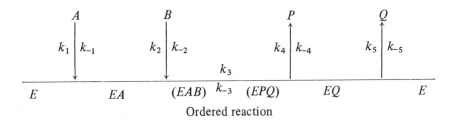

Ordered reaction

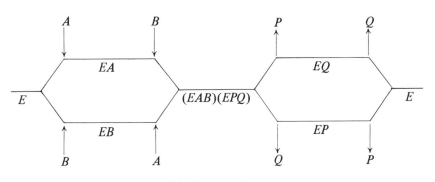

Random reaction

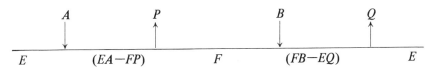

Ping–pong reaction

We shall see examples of NAD-dependent dehydrogenases where NAD⁺ is substrate A and NADH is product Q in the ordered mechanism. Creatine kinase is an example of an enzyme following the random bisubstrate mechanism. Transaminases obey ping-pong kinetics, with their pyridoxal forms being E and the pyridoxamine forms being F.

While Cleland's shorthand is suitable for showing kinetic mechanisms for many reactions, a more general approach such as that employed by Wong and Hanes (1962) is essential to delineate all the pathways found in some enzyme reactions. Considering bisubstrate reactions as a transfer of group G from the donor molecule GX to the acceptor Y, i.e.

$$GX + Y \rightleftharpoons X + GY,$$

Wong and Hanes have written the simplest general mechanism for all bisubstrate reactions as:

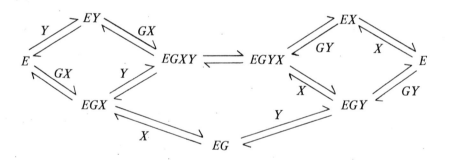

By eliminating certain steps, the ordered, random and ping-pong mechanisms (termed linear, branching and substituted enzyme mechanisms, respectively, by Wong and Hanes) can be obtained. For example, the ping-pong mechanism follows only the lower reactions via EGX, EG and EGY.

Initial-velocity patterns for multi-substrate reactions are obtained by varying the concentration of one substrate at different fixed levels of the other substrate(s) in the absence of product. If reciprocal plots are made for the variable substrate at several levels of the fixed substrate in a bisubstrate reaction, the lines will intersect for a sequential reaction and be parallel for a ping-pong reaction. Florini and Vestling (1957) and Dalziel (1957) reported a method for determining the

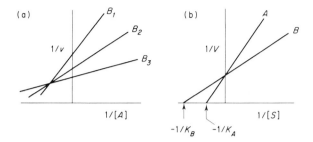

Fig. 12. Graphical method for determining the K_m values for a sequential bisub-strate reaction (Florini and Vestling, 1957). Graph (a) is a reciprocal plot of initial velocity for varying concentrations of substrate A at several fixed con-centrations of substrate B (B_1, B_2 and B_3) of the order of magnitude of the K_m for B. The intercept at the ordinate is the reciprocal of the maximal velocity at infinite $[A]$ for that concentration of B. These $1/V$ intercepts are plotted in another reciprocal graph, (b), against $1/[A]$ values, producing line A with the abscissa intercept being $-1/K_m$ for A (i.e. $-1/K_A$). A plot similar to (a) with varying $[B]$ at fixed levels of A is used to give line B in graph (b) to obtain $-1/K_B$.

K_m values for the substrates of a bisubstrate sequential system. The intercepts at the y-axis on the reciprocal graph are plotted against the reciprocals of the concentration of fixed substrate (Fig. 12). The ordinate intercepts of the replot are $-1/K_a$ and $-1/K_b$ and the point of intersection is the reciprocal of maximal velocity when both substrates are in high concentration. If the point of intersection of the original plots is on the abscissa, then it may be concluded that the binding of one substrate has no effect on the binding of the other (i.e. the mechanism is rapid equilibrium random).

Product Inhibition

For most reactions, initial-velocity patterns are not sufficient to indi-cate the type of kinetic mechanism, so product inhibition studies must also be done. Alberty (1958) suggested a technique, which Cleland (1963) has generalized, for using the effect of products as reversible inhibitors to determine from the reciprocal plots whether the effect of a product inhibitor is on the slope or the intercept at the ordinate. As we have already seen, competitive inhibitors affect only the slope, uncompetitive inhibitors affect only the intercept and noncompetitive inhibitors affect both slope and intercept. Mixed

Table I. Types of inhibition by products in bisubstrate-biproduct reactions
$(A + B \rightleftharpoons P + Q)$.

Kinetic mechanism	Product	A as variable substrate	B as variable substrate
Ordered	P	N	N
(linear)	Q	C	N
Random	P	C	C
(branching)	Q	C	C
Ping-pong	P	N	C
(substituted enzyme)	Q	C	N

N = mixed noncompetitive inhibition
C = competitive inhibition

noncompetitive inhibition,* in which the lines of a reciprocal plot cross to the left of the vertical axis but either above or below the horizontal axis, is often observed in product-inhibition studies. Because products are substrates for the reverse direction of a reaction, they act as inhibitors of the forward direction by binding enzyme in a non-productive form. If a compound affects the intercept of the reciprocal plot, then it binds to a different form of the enzyme than does the variable substrate. If a compound affects the slope of the reciprocal plot, then the inhibitor and the variable substrate react with the same form of the enzyme (or with different forms which are connected by a series of reversible steps).

Let us examine the three standard types of mechanisms for inhibition patterns. With the ordered reaction, product Q affects the slope with the leading substrate, A, since they both compete for free E, but Q is a mixed noncompetitive inhibitor with B. Product P can only bind to EQ, so is a mixed noncompetitive inhibitor of both A and B at non-saturating levels of these substrates. In the simplest random mechanism, Q competes with both substrates, and so does P. In the ping-pong mechanism, P would be expected to be competitive with B and noncompetitive with A, and Q competitive with A and noncompetitive with B. These behaviours are summarized in Table I.

* Cleland (1967) has used the term "noncompetitive" for this type of inhibition, but "mixed noncompetitive" is a preferable description.

Although many reactions approximate the three basic Cleland mechanisms, both theoretical and empirical exceptions occur. In addition, mechanisms which are much more complex than the three simple diagrams have been found. Further details on the distinctions between kinetic mechanisms are presented by Cleland (1970), Segel (1975) and Wong (1975).

Isotope Exchange at Equilibrium

Boyer (1959) was the first to suggest the use of isotopic exchange at equilibrium as a tool for enzyme kinetics. Alberty *et al.* (1962) and Wong and Hanes (1964) have further developed the theoretical basis of the technique. Chapter 10 in Segel (1975) describes the theory and practical application of isotope exchange. The principle is to add a tracer amount of a radioactively labelled reactant to an enzyme system in which the substrates and products have reached equilibrium (so no net chemical reaction occurs) and to determine the rate at which the label from the reactant appears in the appropriate product. It is assumed that there is no kinetic-isotope effect. The rate measured is the initial velocity of the appearance of label from substrate into product, and the method is really an extension of initial-velocity measurements to reaction steps that are inaccessible by chemical analysis. Isotopic-exchange experiments permit one to measure the relative magnitudes of exchange rates and to use these rates to distinguish between possible mechanisms which are not always separable by initial-velocity or product-inhibition studies. High purification of the enzyme is essential.

Ordered and random mechanisms can be distinguished easily by isotope exchange. We will first consider the exchange of isotopically labelled A with product Q for an ordered mechanism. Pairs of reactants, one substrate and one product, must be varied in constant ratio (to maintain chemical equilibrium). When A and Q are varied in

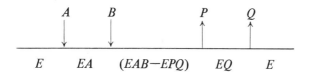

exchange rate against A–Q concentration is observed. If the concentrations of B and P are varied in constant ratio, there will be a substrate-inhibition type curve (like Fig. 8) for the effect of B–P concentration on A–Q exchange. This is because B is essential for exchange, but it removes all the EA complex from solution at saturation with B, leaving no complex for A to exchange with. Variation of B–P concentrations should have no inhibitory effect on B to P exchange. A clearcut example of the use of isotope exchange to demonstrate an ordered reaction is the experiment at pH 7.9 with lactate dehydrogenase (Silverstein and Boyer, 1964). The pyruvate–lactate pair inhibited NAD^+–$NADH$ isotope exchange, but had no effect on pyruvate–lactate exchange. Under these conditions, NAD^+ must bind to the enzyme before lactate, and pyruvate must be released before $NADH$.

In random mechanisms with no preferred pathway, no reactant can affect the release of any other reactant from the enzyme, so no substrate inhibition should be observed. Linear reciprocal plots would be obtained for all exchanges when substrate and product pairs are varied in concentration. In some cases, when a dead-end or non-reacting complex is formed, substrate inhibition is observed for exchange rates of all reactants.

Random-mechanism reactions can be tested by isotope exchange to see if they are rapid-equilibrium reactions (Westley, 1969; Cleland, 1970). If the slowest step is the interconversion of enzyme-bound substrates and enzyme-bound products (i.e. of the two central complexes), then all exchange rates are limited by the same step and the exchange rates for all the reactants will be the same. Large differences between exchange rates of the reactants would suggest limitation of the rate of the overall reaction by other steps.

For ping-pong reactions, isotopic-exchange experiments can be applied to each oscillation separately by omitting the second substrate from the reaction. For example, if A is varied at different fixed levels of P, parallel reciprocal plots should be obtained. If k_2 is

$$
\begin{array}{ccc}
A & & P \\
\uparrow & & \uparrow \\
k_1 \downarrow k_{-1} & & k_2 \uparrow k_{-2} \\
\hline
E & (EA-FP) & F
\end{array}
$$

measured by a chemical assay method, then k_1, k_{-1} and k_{-2} can be calculated from these data (pp. 125–127 in Westley, 1969). Independent exchange experiments with the second oscillation involving B, Q and F could give the other four rate constants for the reaction.

Rapid-reaction Techniques

Steady-state kinetic procedures are not able to measure the rates of very fast reactions such as the binding of a substrate to an enzyme. Rapid-reaction techniques have been designed to follow reactions occurring in less than a millisecond (initial-velocity assays require at least 15–20 s). Two basic experimental approaches, flow and relaxation, are used for studying fast reactions.

Flow procedures rely on rapid mixing of reactants and are limited in time resolution by the time needed for mixing. They are intermediate between conventional assays and relaxation methods. Stopped-flow systems have been used most often because they consume less enzyme than other flow procedures. A stopped-flow apparatus mixes two solutions within a few milliseconds inside a chamber that is monitored by a spectrophotometer or a fluorimeter connected to a storage oscilloscope. The experiments in Fig. 11 are an example of the use of a stopped-flow technique to measure the concentration of an ES complex. The concentrations of reactants can also be measured if they have suitable physical properties.

Relaxation methods detect the rate of attainment of a new equilibrium by a reversible-reaction system which has been perturbed from its original equilibrium. A change in equilibrium by rapid heating is the most usual perturbation. In a temperature-jump experiment, a test solution is heated about 5° C in 5 μs by an electric discharge through the solution. The changes in reactant concentrations can be measured for the next few hundred milliseconds. Results are also measured by spectrophotometry with an oscilloscope and a camera.

Gutfreund (1972) has described the methods, limitations and applications of rapid-reaction techniques. He clearly outlines the possibilities of measuring both changes of concentrations of reactants before the steady state is reached and transient intermediates. He states that data from rapid-reaction experiments have shown one of three processes to be the rate-determining step for different enzyme-catalyzed reactions:

1. a change in conformation of the initial *ES* complex to a reactive complex;
2. the chemical interconversion of substrates to products; or
3. the release of product from the enzyme, which may be controlled by a reversal of the substrate-induced conformation change.

At the outset, one would probably expect the chemical interconversion (i.e. the catalytic event in which covalent bonds are broken and reformed) always to be the slowest step. However, several types of kinetic evidence indicate that the catalytic mechanisms of many enzymes have developed so highly through evolution that further acceleration would not be useful, and the conformational changes associated with substrate binding or product release are rate limiting.

An isotopic-trapping kinetic method for comparing the rate of release of a reactant in a multi-substrate reaction with the rate of its conversion to product has been introduced by Rose *et al.* (1974). The technique involves incubation of enzyme and a labelled substrate (pulse) to form a binary *ES* complex, followed by a short incubation with a mixture of excess substrate (chase) and the remaining reactants. Rose and his colleagues used this method to show that, in the hexokinase-catalyzed reaction, glucose is much more rapidly converted to glucose-6-phosphate than it is dissociated from the enzyme–glucose–MgATP ternary complex. This method should prove useful in supplementing optical rapid kinetic techniques.

EXAMPLES OF MECHANISTIC USE OF KINETICS

I am concluding this chapter by examining the application of enzyme kinetics to two enzymic reactions. The two examples vary in complexity and in the kinetic devices used. The following descriptions of fumarase and alcohol dehydrogenase should point out the value of resourceful use of enzyme kinetics.

Fumarase

The enzyme fumarase (reviewed by Hill and Teipel, 1971) was one

$$^-OOC-CH{=}CH-COO^- + H_2O \rightleftharpoons {}^-OOC^4-C^3H_2-C^2HOH-C^1OO^-$$

Fumarate   Malate

of the first enzymes to be subjected to an intensive kinetic examination. Fumarase catalyzes the reversible hydration of fumarate to L-malate, one of the steps in the tricarboxylic acid cycle. Initial-velocity experiments are easy to perform because the appearance or disappearance of the double bond of fumarate can be measured spectrophotometrically; the enzyme has a high turnover number; and the equilibrium (malate : fumarate = about 4.4) allows the rate to be measured in both directions. The reaction is relatively simple, as can be seen from the stoichiometry. The effects of variables on the rate of the reaction catalyzed by crystalline pig heart fumarase were documented during the mid 1950s by Massey and Alberty. These measurements, along with several isotopic experiments, allow a mechanism to be written, but the large size of the enzyme (194 000 daltons, composed of four subunits) has delayed experiments which would indicate which amino acid residues are present at the catalytic site.

The kinetic experiments with fumarase show the value of enzyme kinetics in describing the physical properties of an enzyme and the reaction which it catalyzes. Except at high substrate concentrations, both the forward and reverse reactions follow Michaelis–Menten kinetics. The values of V at varying pH (p. 120) for both the hydration of fumarate and the dehydration of malate have typical bell-shaped curves (Frieden and Alberty, 1955). These results suggest that a general acid with a pK_a of 6.8 and a general base with a pK_a of 6.2 in the free enzyme (either two histidines or a histidine and a carboxyl) are essential for the enzyme's activity. Frieden and Alberty proposed a kinetic mechanism similar to that shown below, in which the active form of the enzyme has one of these two groups protonated:

$$EH_2^+ \text{ (inactive)} \underset{}{\overset{-H^+}{\rightleftharpoons}} EH \text{ (active)} \underset{}{\overset{-H^+}{\rightleftharpoons}} E^- \text{ (inactive)}$$

$$\text{Fum} + EH \underset{k_{-1}}{\overset{k_1}{\rightleftharpoons}} EH{-}\text{Fum} \underset{k_{-2}}{\overset{k_2}{\rightleftharpoons}} EH{-}\text{Mal} \underset{k_{-3}}{\overset{k_3}{\rightleftharpoons}} EH + \text{Mal.}$$

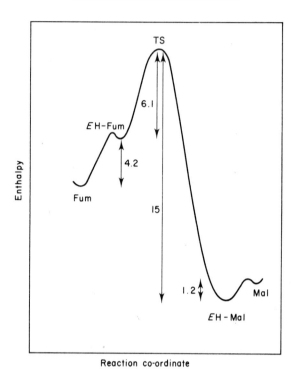

Fig. 13. Plot of enthalpy v. reaction co-ordinate for the fumarase reaction above 18° C. The energies of activation for each step (kcal mol⁻¹) are shown by the numerals in the figure. TS is the transition state. From the data of Massey (1953).

From the V values of the reaction in both directions and the ionization constants of the enzyme and the ES complexes, Alberty and Pierce (1957) calculated the rate constants or their limits for each of the six steps in the proposed mechanism. The rate constants for the binding of either fumarate (Fum) or malate (Mal) to EH, k_1 or k_{-3}, approached those of a diffusion-controlled reaction. The rate of release of products (k_{-1} and k_3) exceeded the rates of interconversion of EH—Fum and EH—Mal (k_2 and k_{-2}) by at least ten-fold. Therefore, the rate-limiting step is the interconversion of the ES and EP complexes, although the initial-velocity data cannot tell whether this is a single-step or a multi-step interconversion. Later experiments showed it to involve an intermediate. Massey (1953) examined the effect of variation of temperature from about 10–40° C on the rate of fumarase reactions. From the Arrhenius plots for both directions, the reaction co-ordinate curve shown in Fig. 13 was constructed. This type of mechanism, in which the slowest step is that in which bond breaking and reformation occurs, is intuitively pleasing and has been presented in texts as a generalized example of enzyme catalysis. However, as mentioned previously, other reactions will be encountered in which the release of products is slower than the catalytic events under the conditions of kinetic experiments.

A plausible mechanism which fits both the kinetic data and isotopic experiments was proposed by Alberty *et al.* (1957) and later amplified by others (cf. Hill and Teipel, 1971). The substrate is bound to the enzyme by two positively charged centres, as suggested by the substrate-inhibition experiments (p. 119). The first catalytic step in the dehydration of malate is the formation of a carbonium ion complex from $EH-Mal$:

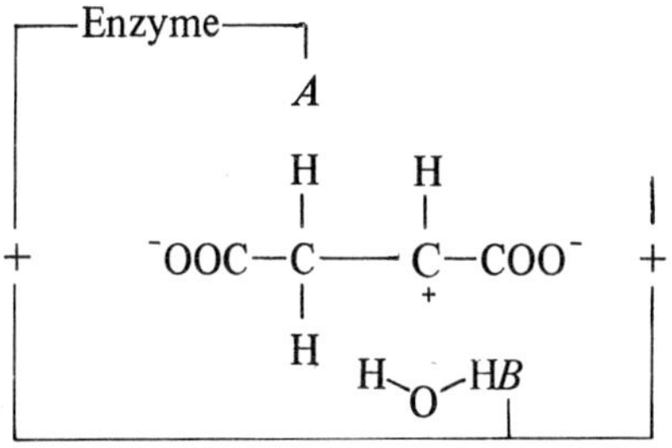

$$
\begin{array}{c}
\overline{\text{Enzyme}} \\
A \\
\text{H} \quad\text{H} \\
| \quad\quad | \\
+ \quad \text{-OOC}-\text{C}\!-\!\!-\!\!-\!\text{C}-\text{COO}^- \quad + \\
| \quad\quad\; {}_+ \\
\text{H} \quad \text{H}\!\diagdown_{\text{O}}\!\diagup\text{H}B
\end{array}
$$

The acid–base catalysts are designated by $-A$ and $-B$. The reaction is then an elimination process in which the carbonium ion intermediate can either lose a proton from C^3 to form $EH-Fum$ or bind OH^- at C^2 to form $EH-Mal$ again. Schmidt *et al.* (1969) have observed a secondary isotope effect k_H/k_T of 1.12 (equal to k_H/k_D of 1.09) for substitution of tritium for hydrogen at the C^3 position of malate. This isotope effect supports the existence of the postulated carbonium ion in the fumarase-catalyzed reaction. Verification of this mechanism awaits a more complete characterization of fumarase by protein chemistry and by transient-state kinetics.

Alcohol Dehydrogenase

No discussion on the use of kinetics would be complete without mention of alcohol dehydrogenase from horse liver. This alcohol dehydrogenase, isolated in crystalline form in 1948, catalyzes this reversible oxidation–reduction reaction:

$$CH_3CH_2OH + NAD^+ \rightleftharpoons CH_3CHO + NADH + H^+.$$

The enzyme has a broad specificity toward alcohol (Alc), aldehyde (Ald) and ketone substrates (reviewed by Sund and Theorell, 1963). The crystalline dehydrogenase can be separated by electrophoresis or column chromatography into several isoenzymes of differing substrate specificity. Alcohol dehydrogenase has a molecular weight of about 80 000 and consists of two identical subunits, each containing a coenzyme binding site. There are four molecules of zinc in the dimer, one of these at each catalytic site and two acting in a structural capacity. Each molecule of alcohol dehydrogenase can bind two molecules of NADH. The binding of NADH can be measured by the change in either absorbance or fluorescence of the coenzyme upon binding to the protein.

In 1951, Theorell and Chance performed kinetic studies with this enzyme using a rapid-flow method. The binding of NADH to the enzyme was determined by measuring both the concentration of NADH and of $E-NADH$ during the reaction, and by titrating the enzyme with NADH. These workers determined both a rate constant for the addition of NADH to the enzyme and a dissociation constant for the $E-NADH$ complex. From these two values, the rate of release of NADH was calculated. The binding of NADH was fast ($k = 3 \times 10^6 \ M^{-1} \ s^{-1}$) and tight ($K_s = 10^{-7} \ M$). The rate of release of NADH from $E-NADH$ was slow ($k = 1.1 \ s^{-1}$). The rate of the interconversion of the $E-NADH-Ald$ and $E-NAD^+-Alc$ complexes was rapid. The mechanism which was proposed, now called a Theorell–Chance mechanism, is shown in the diagram below:

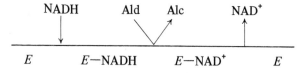

In this mechanism, the alcohol appears to be formed by direct reaction of acetaldehyde with the $E-NADH$ complex.

Several investigators reported results consistent with a simple ordered mechanism. Wratten and Cleland (1963) performed product-inhibition kinetics to confirm that this mechanism existed under the concentrations of reactants used. Their K_m values for NAD^+ and

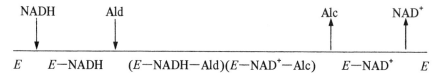

$$E \qquad E-NADH \qquad (E-NADH-Ald)(E-NAD^+-Alc) \qquad E-NAD^+ \qquad E$$

NADH were 1.7×10^{-5} M and 2.7×10^{-5} M, and for ethanol and acetaldehyde were 5.5×10^{-4} M and 2.4×10^{-4} M, respectively. They concluded that central complexes must exist (i.e. that $E-NADH$ was not converted directly to $E-NAD^+$ when it collided with Ald), and agreed with previous suggestions that the probable rate-limiting steps were the isomerization and/or dissociation of the enzyme-nicotinamide coenzyme complexes.

Hanes *et al.* (1972) have analyzed the kinetic properties at pH 8.6 and 27° C of the major isoenzyme of horse liver alcohol dehydrogenase. The basic mechanism is a random addition of ethanol or NAD^+ to the enzyme, with a compulsory release of acetaldehyde before NADH. In the presence of excess NADH or ethanol, an $E'-NADH-$

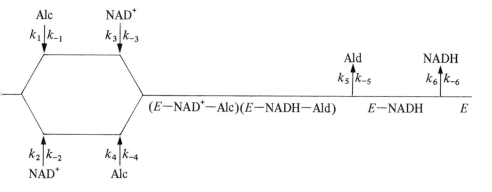

Alc complex is formed with the enzyme apparently having an altered conformation. From their data, these authors have calculated the rate constants for ten steps in the basic mechanism for the oxidation of ethanol (Table II). These rate constants indicate that the dissociation of $E-NADH$ (k_6) and $E-NAD^+$ (k_{-2}) are the rate-limiting steps, and that at low concentrations of ethanol (<4 mM) the ordered mechanism (i.e. steps 2, 4, 5 and 6) is followed. At medium concentrations of ethanol (7-8 mM), alcohol adds first to the enzyme, then NAD^+ (steps 1, 3, 5 and 6). Constant k_5 is a measure of the rate of conversion of the central complex to $E-NADH$ and acetaldehyde.

Table II. Rate constants for the oxidation of ethanol at $27°$ C catalyzed by horse liver alcohol dehydrogenase. Taken from the refined data of Wong and Hanes (1973) and expressed as activity per molecular weight of 80 000.

k_1	$1.04 \times 10^4\,\mathrm{M^{-1}s^{-1}}$	k_{-1}	$3.96\,\mathrm{s^{-1}}$
k_2	$7.34 \times 10^4\,\mathrm{M^{-1}s^{-1}}$	k_{-2}	$0.338\,\mathrm{s^{-1}}$
k_3	$7.16 \times 10^5\,\mathrm{M^{-1}s^{-1}}$	k_{-3}	$32.0\,\mathrm{s^{-1}}$
k_4	$2.86 \times 10^4\,\mathrm{M^{-1}s^{-1}}$	k_{-4}	$104\,\mathrm{s^{-1}}$
k_5	$176\,\mathrm{s^{-1}}$		
k_6	$9.5\,\mathrm{s^{-1}}$		

Since it greatly exceeds k_6, the C—H bond breakage during oxidation of ethanol must occur more rapidly than the release of NADH.

By correlating the kinetic data with that from many other investigations, we have a clearer picture of the mechanism of action of alcohol dehydrogenase. Under conditions in which the dissociation of NADH from the enzyme is accelerated by covalent modification of the enzyme (Plapp et al., 1973) or eliminated by direct oxidation of enzyme-bound NADH (Gershman and Abeles, 1973), the transfer of hydrogen from ethanol to NAD^+ becomes the rate-limiting step, with a primary isotope effect of about 4 with deuterioethanol. Bush et al. (1973) have reported that reduction of low concentrations of acetaldehyde by NADH labelled with deuterium in the transferable hydrogen causes a moderate isotope effect. Therefore, in the reverse direction, hydrogen transfer is slower than either addition of acetaldehyde (k_{-5}) or release of ethanol (k_{-4}).

The primary structure of the major isoenzyme of horse liver alcohol dehydrogenase has been determined, and the X-ray crystallography data to $2.9\,\text{Å}$ (Brändén et al., 1973) and to $2.4\,\text{Å}$ (Eklund et al., 1974) resolution is available for both the apoenzyme and an E—(ADP—Ribose) complex. Each subunit consists of two domains separated by the active site cleft. One domain binds the NAD^+ and the other provides the catalytic centre, including the catalytic zinc atom. Brändén has concluded that this zinc atom is close to C^4 of the nicotinamide ring of NAD^+ (p. 206) and that the metal ion binds ethanol in its alcoholate form. The catalytic step would then be the interconversion of these two species:

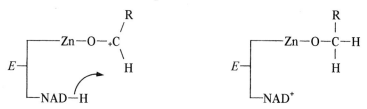

and would require a general base to accept the proton from ethanol. This mechanism is supported by other experiments but still has to be verified, and the nature of any rate-limiting conformation change upon release of NAD^+ or NADH is not yet known.

REFERENCES

Alberty, R. A. (1958). *J. Am. Chem. Soc.* **80**, 1777-1782.

Alberty, R. A. and Pierce, W. H. (1957). *J. Am. Chem. Soc.* **79**, 1526-1530.

Alberty, R. A., Miller, W. G. and Fisher, H. F. (1957). *J. Am. Chem. Soc.* **79**, 3973-3977.

Alberty, R. A., Bloomfield, V., Peller, L. and King, E. L. (1962). *J. Am. Chem. Soc.* **84**, 4381-4384.

Auld, D. S., Latt, S. A. and Vallee, B. L. (1972). *Biochemistry* **11**, 4994-4999.

Bergmeyer, H. U. (1972). *Clin. Chem.* **18**, 1305-1311.

Bertino, J. R., Gabrio, B. W. and Huennekens, F. M. (1960). *Biochem. biophys. Res. Commun.* **3**, 461-465.

Boyer, P. D. (1959). *Archs Biochem. Biophys.* **82**, 387-410.

Bränden, C.-I., Eklund, H., Nordstrom, B., Boiwe, T., Söderlund, G., Zeppezauer, E., Ohlsson, I. and Åkeson, Å. (1973). *Proc. natn. Acad. Sci. U.S.A.* **70**, 2439-2442.

Briggs, G. E. and Haldane, J. B. S. (1925). *Biochem. J.* **19**, 338-339.

Bush, K., Shiner, V. J., Jr. and Mahler, H. R. (1973). *Biochemistry* **12**, 4802-4805.

Chemical and Engineering News. (1969). February 3rd, pp. 16-17.

Cleland, W. W. (1963). *Biochim. biophys. Acta.* **67**, 104-137 and 173-196.

Cleland, W. W. (1967). *A. Rev. Biochem.* **36**, 77-112.

Cleland, W. W. (1970). *In* "The Enzymes" (P. D. Boyer, ed.), Third edition, Vol. 2, pp. 1-65. Academic Press, New York.

Cornish-Bowden, A. (1974). *Biochem. J.* **137**, 143-144.

Dalziel, K. (1957). *Acta chem. scand.* **11**, 1706-1723.

Dennis, A. W. (1972). Ph.D. thesis. University of Toronto, Canada.

Dixon, M. and Webb, E. C. (1964). "Enzymes", Second edition, Chapters 4 and 8. Academic Press, New York.

Eisenthal, R. and Cornish-Bowden, A. (1974). *Biochem. J.* **139**, 715-720.

Eklund, H., Nordstrom, B., Zeppezauer, E., Söderlund, G., Ohlsson, I., Boiwe, T. and Brändén, C.-I. (1974). *FEBS Letters* **44**, 200-204.
Ellis, K. J. (1973). *Analyt. Biochem.* **55**, 609-614.
Florini, J. R. and Vestling, C. S. (1957). *Biochim. biophys. Acta* **25**, 575-578.
Frieden, C. and Alberty, R. A. (1955). *J. biol. Chem.* **212**, 859-868.
Gershman, H. and Abeles, R. H. (1973). *Archs Biochem. Biophys.* **154**, 659-674.
Gomori, G. (1955). In "Methods in Enzymology" (Colowick, S. P. and Kaplan, N. O., eds), Vol. 1, pp. 138-146. Academic Press, New York.
Good, N. E., Winget, G. D., Winter, W., Connolly, T. N., Izawa, S. and Singh, R. M. M. (1966). *Biochemistry* **5**, 467-477.
Gurr, P. A., Wong, J. T. and Hanes, C. S. (1973). *Analyt. Biochem.* **51**, 584-588.
Gutfreund, H. (1972). "Enzymes: Physical Principles", pp. 176-232. Wiley-Interscience, London.
Hanes, C. S., Bronskill, P. M., Gurr, P. A. and Wong, J. T. (1972). *Can. J. Biochem.* **50**, 1385-1413.
Henderson, A. R. (1971). *Analyt. Biochem.* **42**, 143-148.
Hill, R. L. and Teipel, J. W. (1971). In "The Enzymes" (P. D. Boyer, ed.), Third edition, Vol. 5, pp. 539-571. Academic Press, New York.
Hoffmann-Ostenhof, O. (1974). *Eur. J. Biochem.* **45**, 1-2.
Laidler, K. J. and Bunting, P. S. (1973). "The Chemical Kinetics of Enzyme Action", Second edition. Clarendon Press, Oxford.
Lee, H.-J. and Wilson, I. B. (1971). *Biochim. biophys. Acta* **242**, 519-522.
Lucas, E. C. and Williams, A. (1969). *Biochemistry* **8**, 5125-5135.
Mahler, H. R. and Cordes, E. H. (1971). "Biological Chemistry", Second edition, Chapter 6. Harper and Row, New York.
Massey, V. (1953). *Biochem. J.* **53**, 72-79.
Mattenheimer, H. (1971). "Mattenheimer's Clinical Enzymology. Principles and Applications". Ann Arbor Science Publishers, Ann Arbor.
Michaelis, L. and Menten, M. L. (1913). *Biochem. Z.* **49**, 333-369.
Plapp, V. B., Brooks, R. L. and Shore, J. D. (1973). *J. biol. Chem.* **248**, 3470-3475.
Plowman, K. M. (1972). "Enzyme Kinetics". McGraw-Hill, New York.
Polgár, L. (1974). *FEBS Letters* **47**, 15-18.
Rajender, S. and McColloch, R. J. (1967). *Archs Biochem. Biophys.* **118**, 279-283.
Rose, I. A., O'Connell, E. L., Litwin, S. and Bar Tana, J. (1974). *J. biol. Chem.* **249**, 5163-5168.
Schmidt, D. E., Jr., Nigh, W. C., Tanzer, C. and Richards, J. H. (1969). *J. Am. Chem. Soc.* **91**, 5849-5854.
Schwartz, M. K. (1971). In "Methods in Enzymology" (W. B. Jakoby, ed.), Vol. 22, pp. 5-14. Academic Press, New York.
Segal, H. L. (1959). In "The Enzymes" (Boyer, P. D., Lardy, H. and Myrbäck, K., eds), Second edition, Vol. 1, pp. 1-48. Academic Press, New York.
Segel, I. H. (1975). "Enzyme Kinetics: Behavior and Analysis of Rapid Equilibrium and Steady-State Enzyme Systems". John Wiley and Sons, New York.
Silverstein, E. and Boyer, P. D. (1964). *J. biol. Chem.* **239**, 3901-3907.

Smith, E. L. and Parker, M. J. (1958). *J. biol. Chem.* **233**, 1387-1391.
"Standard Methods of Clinical Chemistry", Vol. 1f. Academic Press, New York.
Sund, H. and Theorell, H. (1963). *In* "The Enzymes" (Boyer, P. D., Lardy, H. and Myrbäck, K., eds), Second edition, Vol. 7, pp. 25-83. Academic Press, New York.
Theorell, H. and Chance, B. (1951). *Acta chem. scand.* **5**, 1127-1144.
Westley, J. (1969). "Enzyme Catalysis". Harper and Row, New York.
White, W. L., Erickson, M. M. and Stevens, S. C. (1972). "Practical Automation for the Clinical Laboratory", Second edition. C. V. Mosby, Saint Louis.
Wolf, P. L., Williams, D. and Von der Muehll, E. (1973). "Practical Clinical Enzymology". John Wiley and Sons, New York.
Wong, J. T. (1975). "Kinetics of Enzyme Mechanisms". Academic Press, London and New York.
Wong, J. T. and Hanes, C. S. (1962). *Can. J. Biochem. Physiol.* **40**, 763-804.
Wong, J. T. and Hanes, C. S. (1964). *Nature, Lond.* **203**, 492-494.
Wong, J. T. and Hanes, C. S. (1973). *Acta biol. Med. Germ.* **31**, 507-514.
Wratten, C. C. and Cleland, W. W. (1963). *Biochemistry* **2**, 935-941.

F

6 Theories of Enzyme Catalysis

The two previous chapters have presented the tools used for determining mechanisms—bio-organic techniques and enzyme kinetics. Physical organic experiments provide the theories of catalysis, and enzyme kinetic studies provide examples of large rate enhancements by enzymes. A *rate enhancement* or acceleration is a measure of how fast a reaction proceeds when catalyzed in comparison to the rate of the same reaction in the absence of catalyst. Enzymes are far more efficient catalysts than those used in non-enzymic reaction systems, by many thousands of times. An enzyme may speed up a reaction 10^{10}-fold or greater, while it may be more than a million times better at catalyzing a reaction than is a good synthetic catalyst.

The theories which account for the remarkable catalytic properties of enzymes can be related to the properties of protein molecules. Proteins are polyfunctional compounds, with amino-acid side chains and sometimes prosthetic groups, so several catalytic groups may be provided by one enzyme molecule. Covalent catalysis can be effected by nucleophilic groups in proteins, and general acid–base catalysis by the acidic and basic residues, including some metal ions. The hydrophilic and hydrophobic portions of an enzyme's surface might play roles which rely on their own properties of non-covalent bonding and altered dielectric constant. The flexibility of the tertiary structure of proteins in solution suggests that there might be small but decisive changes in conformation occurring during the action of enzymes.

Bio-organic chemists have proposed that a number of phenomena are responsible for the catalytic properties of enzymes. Because of their frequent occurrence in chemical catalysis, general acid–base and covalent catalysis would also occur with enzymes. The latter mode can be proved (not easily, but clearly) by isolation of the covalent compound formed from the enzyme and the substrate, and by demonstrating that this covalent compound is an intermediate in the reaction.

Proton transfer assisted by acid–base catalysis is probably involved in most (but not all) enzyme-catalyzed reactions. Direct evidence for general acid–base catalysis in enzyme reactions is difficult to obtain. The strongest evidence probably is the occurrence of acid–base catalysis in corresponding model reaction systems and the positions of potential acid–base catalyst groups in protein crystals. Jencks (1975) has commented that general-acid- and general-base-catalysis rate accelerations are usually 100-fold or less in non-enzymic reactions. This extent of rate acceleration alone cannot explain the efficient catalysis by enzymes. Therefore, in addition to these two modes of catalysis, other mechanisms must also be operating in enzyme reactions.

Three additional mechanisms are proximity effects, surface effects and bond distortion or destabilization effects. These properties of enzymes overlap with each other and also with acid–base or nucleophilic effects. They have been discussed under various names and given different weights in their relative contributions to rate accelerations by different investigators. These discussions have been summarized by Koshland and Neet (1968), Jencks (1969, 1973, 1975), Bruice (1970) and Mahler and Cordes (1971). Proximity consists of gathering the reactants and aligning them with the catalytic groups of the active site. Surface effects involve the entrance of substrates into a microenvironment on the surface of the protein which differs from free solution. Bond distortion is a strain or distortion of chemical bonds in the substrates and/or the enzyme. All of these effects are plausible and fit with the properties of proteins. Operation of these three effects and acid–base or covalent catalysis can account for the high catalytic power of enzymes, but quantitative assignments of each effect are not yet possible.

In this chapter, we will look at experiments on proximity, surface chemistry and bond distortion that can be applied to enzymic situations. These experiments are of gradually increasing complexity, with several effects combined in many cases. They will be extended in later chapters to examinations of enzyme reactions themselves, since ultimately all the mechanisms must be determined with enzymes.

PROXIMITY EFFECTS

Extensive discussions of proximity effects are given in articles by

Bruice (1971), Bruice and Benkovic (1966) and Jencks (1969, his Chapter 1; 1973). Bruice has termed these effects as "the propinquity effect" and Jencks (1969) as "approximation of the reactants" and later (1975) by the more general title, "probability and entropy". A dozen or more other names have been suggested for essentially the same phenomenon. The proximity effects are based on our concept of the enzyme as a site of concentration of reactants from dilute solution. Reaction rates are accelerated by a greater probability of collisions of sufficient energy and by specific geometric alignment or restriction of the substrates and catalytic groups at the enzyme surface.

Chemical model systems devised to test the effects of proximity of reactants usually compare the reactivity of a unimolecular reaction, in which both reactants are incorporated into the same molecule, with the rate of an intermolecular reaction involving the same reactive groups and proceeding by the same mechanism. Results are expressed in terms of the ratio of the unimolecular intramolecular rate constant (k_1) to the intermolecular bimolecular rate constant (k_2). Since k_1 has the units s^{-1} and k_2 the units $M^{-1} s^{-1}$, the ratio is expressed in molarity. Jencks (1973) has called this k_1/k_2 ratio the "effective molarity" of one group in the vicinity of the other reactive group. The ratio often exceeds attainable concentrations, so that intramolecular reactions must be faster for reasons other than just simple concentration effects. It is this extra rate acceleration, by providing optimum relative positioning of reacting groups, which will concern us now.

The catalysis of phenyl ester hydrolysis by the dimethylamino group provides a good example of an intramolecular reaction which is much faster than its bimolecular counterpart. Bruice and Benkovic (1963) compared the rate of hydrolysis of phenyl acetate by trimethylamine with the rate of hydrolysis of the phenyl ester of N,N-dimethylaminobutyrate. In both reactions, the mechanism is a nucleophilic displacement of the phenolate, with formation of an N-acyl intermediate. The rate of release of phenol (and hence the rate of formation of the intermediate) was measured by stopped-flow spectrophotometry. The ratio of the intramolecular rate constant to the bimolecular rate constant was 1260 M. The authors ascribed this large rate enhancement to the proximity of the substrate and the catalytic group in the intramolecular compound.

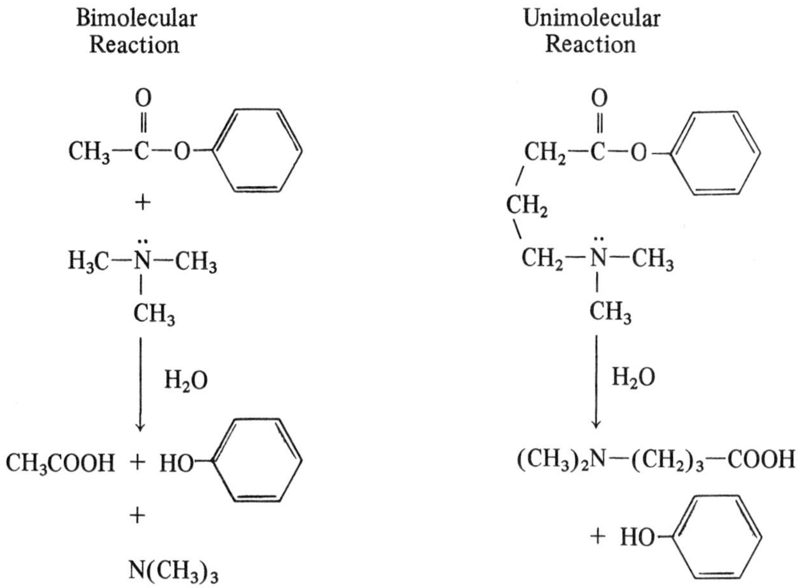

Bimolecular
Reaction

$$CH_3-\overset{\overset{\textstyle O}{\|}}{C}-O-\bigcirc$$

$+$

$$H_3C-\overset{..}{\underset{\underset{\textstyle CH_3}{|}}{N}}-CH_3$$

$\downarrow$ H_2O

$$CH_3COOH \ + \ HO-\bigcirc$$

$+$

$$N(CH_3)_3$$

Unimolecular
Reaction

$$\underset{\underset{\textstyle CH_2}{\diagup}}{CH_2}-\overset{\overset{\textstyle O}{\|}}{C}-O-\bigcirc$$

$$\underset{\underset{\textstyle CH_3}{|}}{\overset{\diagdown}{CH_2}}-\overset{..}{N}-CH_3$$

$\downarrow$ H_2O

$$(CH_3)_2N-(CH_2)_3-COOH$$

$+ \ HO-\bigcirc$

An instance of extremely large rate enhancement by proximity is the study of Bruice and Pandit (1960a, b; Bruice, 1970) on formation of anhydride intermediates in the hydrolysis of p-bromophenol monoesters of dicarboxylic acids. The rates of these reactions were compared with the rate for the nucleophilic attack of acetate on p-bromophenyl acetate. Rates for the release of the p-bromophenol were measured spectrophotometrically at constant pH values between 3 and 7 in dioxane–water solutions. The results, summarized in Table I, show that the "effective molarity" of the carboxylate ion (k_1/k_2) in the exo- 3,6-endoxo-Δ^4-tetrahydrophthalate derivative (IV) is more than 10^7 M. Bruice and Pandit noted that there was a 230-fold rate increase between compounds II and III, and a similar rate increase between III and IV. These rate enhancements were related to the number of single bonds allowing free rotation. The glutarate ester (II) has two bonds which can rotate, the succinate ester (III) has only one, and the bicyclic compound (IV) has none. Bruice and Pandit (1960a) suggested that the bicyclic compound exists in many fewer unreactive conformations, and that this increase in steric restriction causes acceleration of the reaction rate by positioning the reacting species in a configuration similar to that of the transition

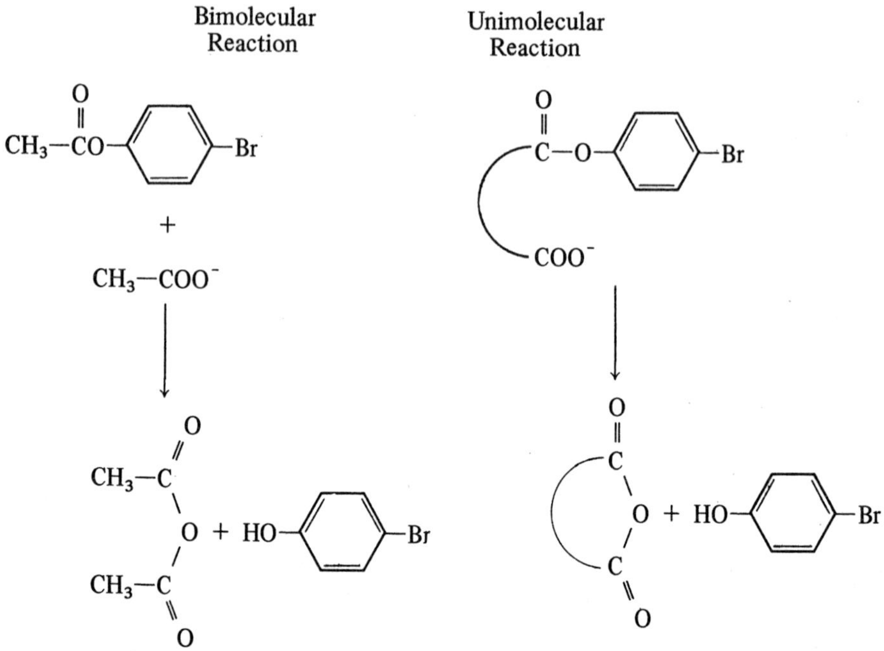

Bimolecular
Reaction

Unimolecular
Reaction

state. This is an excellent example of how proximity both increases effective molarity and provides geometric restrictions which produce a loss in entropy.

Another dramatic series of rate accelerations was obtained with intramolecular model systems by Cohen and his associates (Milstien and Cohen, 1972). These experiments also were based on the premise that there is a drastic loss of conformational freedom of a substrate when it is bound to an enzyme. In this case, freedom of rotation of bonds was controlled by van der Waals repulsion, by selective alkylation. For example, the rate of esterification of acetic acid by phenol (the bimolecular reaction) was compared with the rates of lactonization of o-hydroxycinnamic acid and a methylated analogue. Both the unsubstituted and substituted hydroxycinnamic acids have the same bond angles, and follow the same mechanism for internal esterification. There should be no steric hindrance at the reaction site caused by the additional methyl groups. The rate of lactonization of the methylated compound was about 10^{11} times faster than that of hydroxycinnamic acid. The ratio of k_1/k_2 for the methylated hydroxycinnamic acid compared to esterification of acetic acid was about 3×10^{15} M. These large rate accelerations were thought to be

Table I. Relative rates of formation of anhydrides. Prepared from the data of Bruice (1970).

	Ester[a]	k_1/k_2
I	$CH_3-\overset{\displaystyle O}{\overset{\displaystyle \|}{C}}-OR$ + CH_3-COO^-	
II	$\begin{array}{l} CH_2-\overset{\displaystyle O}{\overset{\displaystyle \|}{C}}-OR \\ / \\ CH_2 \\ \backslash \\ CH_2-COO^- \end{array}$	$9.5 \times 10^2\,M$
III	$\begin{array}{l} CH_2-\overset{\displaystyle O}{\overset{\displaystyle \|}{C}}-OR \\ \| \\ CH_2-COO^- \end{array}$	$2.2 \times 10^5\,M$
IV	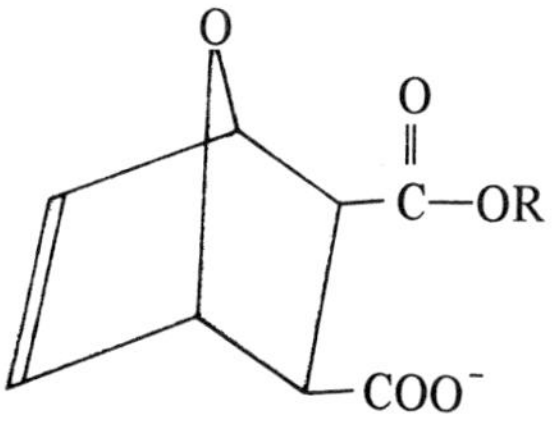	$5 \times 10^7\,M$

[a] $-OR$ = p-bromophenolate.

caused by production of a severely restricted geometry of the side chain that favoured the formation of the transition state. Cohen stressed that the rate enhancements observed were not only pro-o–Hydroxycinnamic acid   Lactone

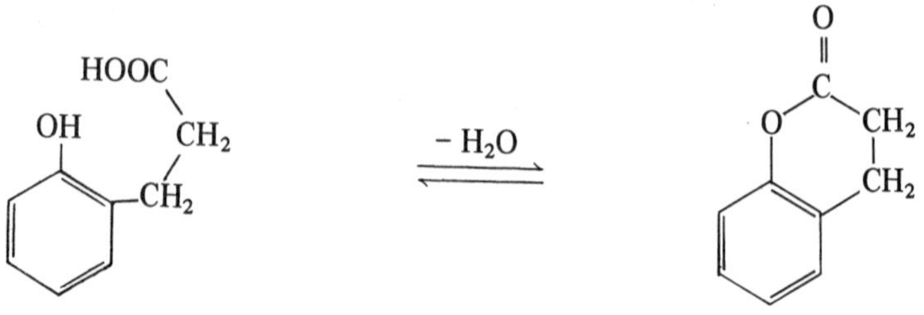

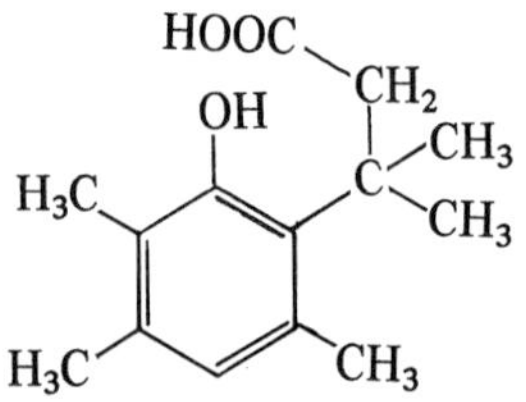

Hydroxycinnamic acid analogue

duced by the restriction of rotational freedom, but were also due to secondary effects which might have included improved orientation, reduction in solvation and interorbital distortion.

Recently, Storm and Koshland (1970, 1972) postulated that for a reaction to occur, molecular orbitals must be oriented much more precisely than had been estimated previously. They suggested that this "orbital steering" factor could account for the catalytic power of enzymes. Their postulate was based on rate measurements of the acid-catalyzed formation of five-membered lactones, similar in many respects to the experiments of Pandit and Bruice. Storm and Koshland suggested that precise orientation of each reactive group at the active site of an enzyme would produce a rate acceleration of 10^4, as calculated for their intramolecular model reactions. If four groups were steered in a fourth-order reaction (an unlikely event) the rate enhancement from the orientation factor alone might be as much as 10^{12} times. Such an enormous effect would not only answer the catalytic dilemma but would also explain how major changes in enzyme activity are caused by minor perturbations in enzyme structure by allosteric effectors (Chapter 12). Justifiably, introduction of the concept of orbital steering has been criticized (e.g. Bruice *et al.*, 1971; Page, 1972) as unnecessary, since the data of Storm and Koshland can be explained by traditional chemical principles (proximity). Misalignments of pairs of molecules would raise the activation energy over that for reaction of an aligned pair, but Bruice (1971) has concluded that orientation is not likely to be kinetically important if the angle of misalignment of orbitals is less than 10°. If orientations as precise as had been originally suggested* for orbital steering were necessary, one would be extremely lucky to synthesize such reactive intramolecular models!

Page and Jencks (1971) have interpreted the theory of catalysis by proximity in thermodynamic terms. The bonding process—either in intramolecular reactions or by formation of *ES* complexes—occurs with a loss of entropy. This loss is much greater than had been anticipated by many investigators. They have estimated that the entropy change for the formation of a moderately tight transition state from two reactants in solution is -35 entropy units. A loss of 35 entropy units at 27° C (300 K) would equal a decrease in $\Delta G^{\ddagger}$ of $T\Delta S^{\ddagger} = (300 \times 35) = 10.5$ kcal. This would correspond to a rate acceleration of 10^8, the approximate maximum. Jencks (1973) emphasizes that we therefore no longer may separate chemical catalysis from binding of substrates when considering enzymic catalysis. The loss of entropy caused by binding of reactants can readily result in high "effective molarities" for both intramolecular and enzyme-catalyzed reactions, as experimentally observed. Smaller entropy losses, from less rigid orientation, lead to smaller rate accelerations from proximity. As mentioned in Chapter 4 (p. 87), pre-positioning of reactants could cause the complete loss of freedom of rotation and translation of the reactants so that there is no further decrease in entropy in going from the *ES* complex to the transition state. Probably there must be a decrease in entropy in proceeding from the *ES* complex to the *E*-transition state, but the more firmly the substrates are bound, the closer to the minimum the $ES \rightarrow E-$TS entropy decrease will be.

---

* It has been estimated that the "orbital steering" orientation requirements were of the order of tenths of degrees of displacement.

Bifunctional Catalysis

Bifunctional catalysis is a special case of acid–base and proximity catalysis. The suggestion has been made that enzymes have two general acid–base groups which simultaneously interact with different parts of the substrate molecule. Such a mechanism would be of great catalytic advantage. Concerted removal and donation of protons would require correct alignment of the substrate and the catalytic groups. The classical chemical analogy is the mutarotation of tetra-methylglucose in benzene, catalyzed by 2-pyridone. Swain and Brown (1952) found that base catalysis of this reaction by pyridine or acid catalysis by phenol was slow, but a mixture of these two catalysts led to faster reaction. The rate law for the mutarotation was third-order:

$$v = k[\text{tetramethylglucose}] [\text{phenol}] [\text{pyridine}].$$

This indicates that both phenol and pyridine are present in the transition state. 2-Pyridone (the predominant tautomeric form of 2-hydroxypyridine) contains both catalytic groups and reacts with tetramethylglucose in a second-order reaction. It is a much better catalyst (7000 times better at 0.001 M) than a mixture of phenol and pyridine. Thus, combination of two catalytic groups on one molecule in an appropriate configuration leads to a large rate enhancement. A mechanism in which the basic carbonyl group accepts a proton from the 1-α-hydroxyl group, and the acidic N—H donates a proton to the ring oxygen to facilitate the formation of the free hydroxy-aldehyde, is now generally accepted. In ring closure, the proton transfers occur in the opposite direction, giving a mixture of the α- and β-anomers of the sugar.

It is important to note that the mutarotation was performed in a nonpolar solvent and that for catalysis to occur all reagents must be free of water. No catalysis of the mutarotation by 2-pyridone is found in aqueous solution, because water can donate and accept the protons rapidly itself. In dry benzene, the 2-pyridine is the only source or sink for protons. Several chemists have suggested that bifunctional catalysis of this kind could occur in the hydrophobic cleft at an enzyme's active site, but the cleft probably has access to some water molecules. Jencks (1972) has formulated a rule for the requirements of concerted acid–base catalysis in aqueous solutions, and says that bifunctional acid–base catalysis in water must be

α–D–glucose   Aldehyde form

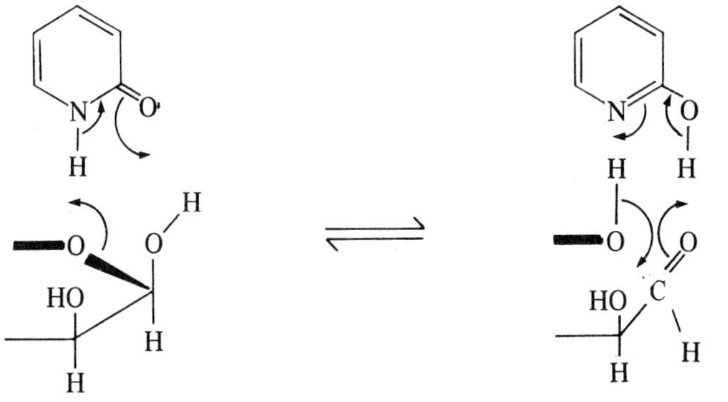

extremely rare. Although several catalytic groups will be found at the catalytic centres of enzymes, they could be used in stepwise not concerted reactions. Bifunctional catalysis of the Swain–Brown type by enzymes still remains a speculation, but a very attractive and reasonable speculation.

SURFACE EFFECTS

Of the possible causes of catalysis, the most difficult to assess are those associated with the properties of the surface of the protein. X-ray crystallographic experiments have shown that the hydrophobic residues of proteins tend to be directed away from the exterior, and that the active site of an enzyme occurs in a cleft usually lined with amino acids having hydrophobic side chains. Despite knowing these three-dimensional structures, we do not yet know what access water has to the active sites of enzymes. The solvent properties of the protein surface obviously are not identical to those of water. A decrease in dielectric constant would cause charge–charge interactions to become stronger. An increase in the electrostatic interaction between a transition state and the charged groups of the enzyme might stabilize the transition state sufficiently to assist product formation. The decrease in polarity of the solvent near the surface of the enzyme might also allow desolvation of substrate molecules and aid the approximation process by making juxtaposition of reacting orbitals easier. Let us now inspect several non-enzymatic systems in which hydrophobic interactions cause catalysis. These systems often demonstrate other features of enzyme reactions, too. Further elaboration of these projects may soon lead to a closer understanding of surface effects in enzyme catalysis.

Micelles

Reactions catalyzed by micelles show how hydrophobic interactions may influence reactions in aqueous solution. Micelles are molecular aggregates formed from compounds which have both hydrophilic and hydrophobic properties (reviewed by Tanford, 1973). These compounds are detergents or surfactants that have hydrocarbon side chains (tails) of from 8 to 18 carbons with cationic or anionic polar ends (heads). Micelle aggregates form because in water the hydrophobic chains of the detergent molecules associate, leaving the hydrophilic groups at the surface of the micelle exposed to solvent. From a prediction of micelle size, Tanford (1974) has concluded that micellar aggregates are not spherical but are rough-surfaced ellipsoids. Their optimal size and shape are determined by two opposing forces: hydrophobic attraction of the tails and repulsion between the heads. The interior of the micelle is much like a liquid hydrocarbon, but has some access to water. Podo *et al.* (1973) have used n.m.r. techniques on non-ionic detergent micelles to prove that no water penetrates to the hydrocarbon interior and that there is a gradual increase of contact with water at the polar end of the detergent molecules. In many ways, the properties of micelles are similar to those of globular proteins. Some micelles have catalytic properties, due either to the properties of their surfaces or to both surface- and specific-group effects. Comprehensive reviews of micellar catalysis have been written by Cordes and Dunlap (1969) and Cordes and Gitler (1973).

To understand how micelles catalyze reactions, we must first see how molecules in micelles differ from individual detergent molecules. As the concentration of a detergent is increased, there is an abrupt change in its physical properties at a narrow concentration range called the *critical micelle concentration* (CMC). At the CMC, the compound goes from a solute to a micelle because the level of hydrophobic groups is sufficient to associate. These groups stabilize themselves by seeking the least possible contact with water. Sodium

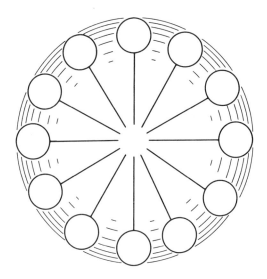

Fig. 1. Diagramatic cross-section of a micelle formed from individual detergent molecules. The circles represent the polar groups and the straight lines the hydrocarbon tails. The shading shows the approximate limits of the aqueous phase.

dodecyl sulphate (SDS, $C_{12}H_{25}OSO_3Na$) is typical of micelle-forming detergents. Above its CMC of 10^{-3} M, SDS forms micelles containing about 100 molecules (Fig. 1). With the hydrocarbon chains all pointing to the interior, the polar ends form a rough surface on the globular micelle. This surface is charged, but is not as polar as water. It is at this surface that catalysis takes place. Adjacent to the surface is a concentrated layer of counter-ions, such as Na^+ and H^+ (H_3O^+) for an SDS micelle, called the Stern layer. The fraction of charges on the surface neutralized by the counter-ions increases with increasing salt concentration.

Catalysis by micelles arises from two effects, electrostatic interaction of the transition state with the polar groups, and hydrophobic binding of portions of the substrate molecules. Rate accelerations are observed only if the transition state has a charge opposite to that of the surface of the micelle.

The hydrolysis of methyl orthobenzoate is catalyzed by micelles of SDS (Dunlap and Cordes, 1968). This reaction proceeds by proton transfer from H_3O^+ and release of a molecule of methanol. The carbonium ion formed then rapidly loses a second molecule of

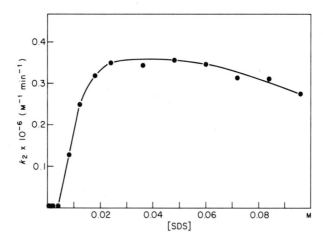

methanol. The rates of hydrolysis were assayed by measurement of the formation of the carboxylate ester at 228 nm. Methyl ortho-benzoate (5×10^{-5} M) was incubated at 25° C in 0.01 M acetate buffer, pH 5.4. First-order rate constants were obtained, and divided by the concentration of H_3O^+ to yield k_2. The effect of concentration of SDS is shown in Fig. 2. For the uncatalyzed reaction and at concentrations of SDS below the CMC, the rate is 4.5×10^3 M^{-1} min^{-1}.

Fig. 2. Effect of [SDS] on the rate of hydrolysis of methyl orthobenzoate. Second-order rate constants were determined as described in the text. From Dunlap and Cordes (1968) with permission.

Catalysis by SDS starts near the CMC, showing that catalysis is effected only by SDS micelles. The rate increases until a maximum rate of 3.6 $\times$ 10^5 M^{-1} min^{-1} (80-fold acceleration) is reached at a concentration of SDS where all the substrate molecules are adsorbed to the micelles. The reaction levels off, then there is an inhibition caused by Na$^+$ counter-ions neutralizing the surface charge. If the substrate were adsorbed into the core of the micelle, there would be no inhibition by Na$^+$. The reasons given for catalysis of this reaction by SDS micelles are

1. methyl orthobenzoate is adsorbed to the micelle (proximity effect);
2. the anionic micelle stabilizes the developing carbonium ion; and
3. reaction occurs at the Stern layer where cations, including protons, are concentrated (surface effects).

Gitler and Ochoa-Solano (1968) have prepared mixed micelles in which the micelle participates directly in the reaction by covalent catalysis. The micelles were formed from a mixture of N^{α}-myristoyl-L-histidine (MirHis) and an excess of the cationic detergent cetyl trimethylammonium bromide (CTABr, $CH_3(CH_2)_{15}N(CH_3)_3Br$). In the mixed micelles, the imidazole and carboxylate groups of MirHis and the trimethyl ammonium groups of CTABr would be at the surface.

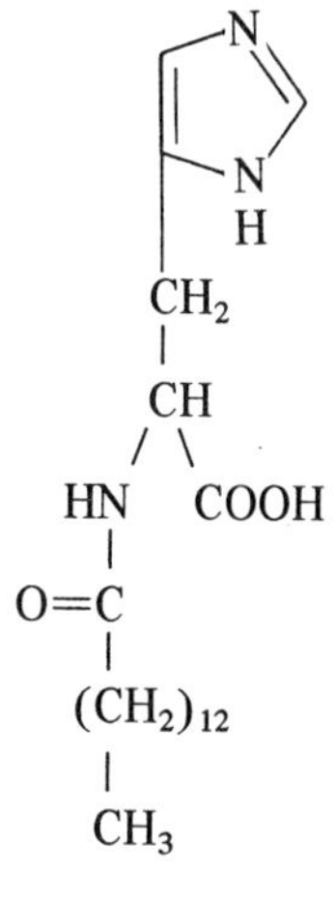

$$\begin{array}{c}
\text{N} \\
\text{imidazole ring} \\
\text{N} \\
\text{H} \\
| \\
CH_2 \\
| \\
CH \\
/ \quad \backslash \\
HN \quad COOH \\
| \\
O{=}C \\
| \\
(CH_2)_{12} \\
| \\
CH_3
\end{array}$$

MirHis

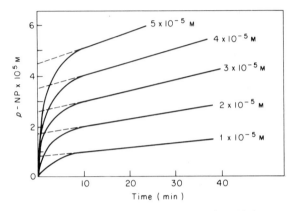

Fig. 3. Time-dependent liberation of p-nitrophenol (p-NP) from p-nitrophenyl-acetate in the presence of the concentrations of MirHis indicated at a ratio of CTABr : MirHis of 20 : 1. The dashed lines are extrapolations which show the amount of MirHis acylated during the burst. Adapted from Gitler and Ochoa-Solano (1968).

The reactions catalyzed are the hydrolysis of p-nitrophenol esters (with an increasing number of carbon atoms) in Tris buffer at pH 7.2. The rate of hydrolysis was measured by detection of the liberation of p-nitrophenol (cf. p. 101). Neither MirHis nor CTABr alone catalyzes the hydrolyses. Figure 3 shows the progress curve for hydrolysis of p-nitrophenyl acetate at five different concentrations of MirHis (all well below the ester concentration). The initial rapid formation of p-nitrophenol is followed by a slower linear phase. Approximately one mole of p-nitrophenol is released per mole of MirHis in the rapid phase. This type of "burst" is similar, but much slower, to that seen in catalysis of the same reaction by chymotrypsin. (It is as well to remember that with chymotrypsin, a serine residue not a histidine is acylated.) It is caused by rapid acylation of the imidazole group of MirHis, followed by a steady state of deacylation and reacylation. The intermediate acetyl MirHis is stable enough to be isolated by column chromatography.

When rate constants were measured with MirHis in excess over the substrate concentration, it was found that the rate of hydrolysis increased by increasing the chain length of the acyl group of the ester from C_2 to C_6 (Table II). These data show that the hexanoate ester is bound to the mixed micelle more tightly than is the acetate

Table II. Rate of hydrolysis of p-nitrophenyl esters by MirHis micelles. Pseudo first-order rate constants were determined and converted to second-order constants with corrections made for inclusion of only the basic form of the imidazole. The rate of these hydrolyses in the presence of N^{α}-acetyl-L-histidine was 9–16 M^{-1} min^{-1}, and in the presence of N^{α}-acetyl histidine and CTABr was 2–10 M^{-1} min^{-1}. From the data of Gitler and Ochoa-Solano (1968).

p-Nitrophenyl ester	k_2 (M^{-1} min^{-1})
Acetate	380
Propionate	870
Butyrate	1350
Valerate	3000
Hexanoate	7300

ester. The rates for micelle catalysis are much higher than those in the aqueous phase using acetyl histidine as the catalyst.

The MirHis–CTABr system has several features in common with enzyme catalysis besides the "burst" phenomenon. The rate of ester hydrolysis is high compared to that in free solution. There is formation of a micelle–substrate surface complex, in which the substrate must bind in the proper position relative to the imidazole. There are several groups in the micelle which may function during the catalysis —the imidazole group as an acyl acceptor and the cationic group of CTABr to stabilize the transition state. Finally, because there are several effects involved in catalysis of micelles, the assignment of quantitative roles to these effects is difficult.

Detergents can be used to dissolve water in nonpolar solvents. Solution is accomplished by formation of reverse micelles, in which the water is in the small core of the micelle, surrounded by the hydrophilic groups of the detergent. Studies of catalysis in reversed micelles may not produce any new concepts, but will aid in obtaining quantitative information about surface effects.

Inclusion Compounds

Inclusion compounds are another type of substance which form non-covalent complexes of a hydrophobic nature. Of these, the cyclodextrins have been used most in enzyme-model studies (Jencks, 1969, p. 408 in his book; Bruice, 1970). Cyclodextrins (or cyclo-amyloses) are α-1,4-polymers of D-glucose composed of six (α-), seven (β-) or eight (γ-cyclodextrin) glucose units and having a cyclic structure that is a cylinder with a central hole. The cavity can act as a hydrophobic binding site or as a site of lowered dielectric constant. Although rate enhancements in cyclodextrin reaction systems are not as large as in the intramolecular reactions we have discussed, other similarities to enzymatic reactions can be found. Benzoic acid binds in the cavity in a 1 : 1 stoichiometry, while p-nitrophenol co-ordinates with cyclodextrins at a rate close to that of a diffusion-controlled rate. Some of these reactions follow Michaelis–Menten type kinetics. Cyclohexanol is a competitive inhibitor of p-nitrophenyl acetate in one cyclodextrin system (Breslow and Overman, 1970). The chlori-nation of anisole by hypochlorous acid in the presence of α-cyclo-dextrin becomes a position-specific reaction, with chlorination occurring at only the para position, not in both ortho and para positions. Breslow and Campbell (1971) suggest that the ortho position of the anisole is blocked in the inclusion complex, and that a hydroxyl of the cyclodextrin forms a hypochlorite which then chlorinates the substrate (Fig. 4).

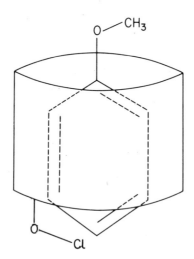

Fig. 4. Possible structure of the anisole–cyclodextrin inclusion complex.

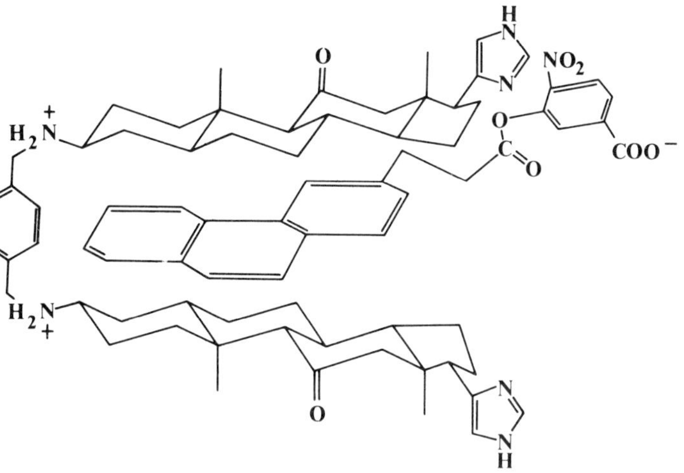

Fig. 5. Diagram showing the water soluble dimeric steroid catalyst complexed with a molecule of phenanthrene-bearing substrate. The adjacent imidazole nucleophilic nitrogen attacks the ester group of the substrate to form acylated catalyst and phenol. Reproduced from Guthrie *et al.* (1976) with permission.

Guthrie and his colleagues (1976) have synthesized and tested an extremely interesting water-soluble catalyst composed of two steroid molecules joined by a *p*-xylidene unit (Fig. 5). The catalyst is rendered soluble in water by the presence of four ionic substituents. Two of these substituents, the 17β-imidazolyl groups, are independent catalytic centres. The catalyst possesses two opposing hydrophobic binding surfaces, which give it specificity for inclusion of esters possessing aryl groups. Propionate esters with phenyl, naphthyl and phenanthrene hydrophobic groups are excellent substrates, with the latter being the best found. The rate enhancement for hydrolysis of the phenanthrene–propionate substrate catalyzed by the steroidal compound, when compared with hydrolysis catalyzed by free imidazole, is 143-fold. Pseudo first-order kinetics are observed when the steroid is in excess. When the substrate is in excess, "burst" kinetics are observed (rapid acylation of one catalytic centre of the catalyst, followed by slow deacylation). Guthrie has concluded that the conformation of the aryl propionate–imidazole transition state is strained by hydrophobic contact of the substrate with the steroid surfaces. This model system shows many similarities to enzymes, particularly the cleft with a hydrophobic lining.

Synzymes

A great many water-soluble linear polymers have been prepared, tested as catalysts for hydrolytic reactions, and found to exhibit modest rate enhancements. Recently, however, Klotz and his associates have synthesized a derivative of a branched polymer, polyethyleneimine, which gives remarkably high rate accelerations in several reaction systems. The structure of a segment of polyethyleneimine is:

$$H_2N-(CH_2-CH_2-NH)_x-(CH_2-CH_2-N)_y-(CH_2-CH_2-NH)_z-$$
$$\begin{array}{c} | \\ CH_2 \\ | \\ CH_2 \\ | \\ -N- \end{array}$$

The polyethyleneimine used as the matrix had a molecular weight of about 60 000. About 25% of the nitrogen atoms were primary amino groups, and to these primary amines, dodecyl ($C_{12}H_{25}$) groups were attached as substrate-binding sites and methyleneimidazole groups were attached as covalent catalytic functions (Klotz *et al.*, 1971). Klotz has termed the resulting catalytically active macromolecule a *synzyme*.

Arylsulphatases (Nicholls and Roy, 1971) are enzymes catalyzing the general reaction:

$$R-OSO_3^- + H_2O \longrightarrow ROH + H^+ + SO_4^{2-}$$

The function of these enzymes in vertebrates is unknown. Arylsulphatase type IIA has an essential histidine which may act directly in the hydrolytic reaction. This enzyme is commonly assayed by measuring the formation of 4-nitrocatechol from 2-hydroxy-5-nitrophenyl sulphate. Kiefer *et al.* (1972) have found that the substituted polyethyleneimine synzyme catalyzes the hydrolysis of this substrate at a rate about 100 times faster than the sulphatase from kangaroo liver. The relatively poorer catalysis by the sulphatase probably is due to the use of the nonspecific substrate. The synzyme gave a hyperbolic v v. $[S]$ curve, and showed a "burst" of nitro-catechol release. This indicated that a sulphated imidazole intermediate was formed. When the rate of catalysis by the synzyme was compared to that of free imidazole, it was found that there was a rate enhancement of 3.6×10^{12} (the highest of its kind reported so far).

This synthetic catalyst contains apolar and ionic regions, as well as a series of nucleophilic imidazole residues. The hydrophobic sites assist by proximity and surface effects and the ionic groups keep the polymer in solution. Thus, these workers have synthesized a very efficient catalytic polymer that has many of the properties of an enzyme. The main difference between the synzyme and the micellar systems described above is attachment of hydrophobic and catalytic groups to a polymer; the same kinds of catalytic mechanisms are proposed for the two types of catalysts.

BOND DISTORTION

One last factor which may play a major role in catalysis by enzymes is bond distortion. Strain placed upon the bonds of the substrate, the enzyme, or both when the enzyme–substrate complex is formed, may weaken the reacting bonds in the substrate sufficiently to aid in the formation of the transition state (Jencks, 1969, his Chapter 5). Jencks (1975) has proposed that the term "destabilization" be used because destabilizing mechanisms such as desolvation, electrostatic repulsion (both listed above under surface effects) and breaking of hydrogen bonds may be as important as geometric destabilization. Direct supporting evidence for a bond distortion effect (the best-known destabilization) is not yet available, but a number of indirect experimental methods suggest that the configuration of both the substrate and the enzyme are altered during catalysis. Because an enzyme must catalyze both directions of a reversible reaction, it seems unlikely that it would have optimal fit for either reactants or products. Instead, the enzyme probably has maximal affinity for the transition state. By distorting the bonds of the substrate towards the shape of the transition state, the enzyme would move the substrate along the activation barrier.

There are two main theories of enzyme catalysis involving distortion of bonds. These are called the *induced-fit theory* (Koshland and Neet, 1968) and the *strain* or *distortion theory* (Jencks, 1969, his Chapter 5). Let us first examine the induced-fit theory.

For many years, the "lock and key" hypothesis, suggested by Emil Fischer in 1894, was used to depict enzyme catalysis. In this simple theory, the enzyme is regarded as being a rigid surface which has a complementary slot in which the substrate fits and the reaction takes place. This suggests that the specificity of a reaction depends only on the K_m of a substrate, which is not necessarily true. In 1958, Koshland proposed a theory in which the presence of a substrate induces a change in conformation of the enzyme. The induced-fit theory has three basic postulates:

1. when small molecules bind to an enzyme, they may induce conformational changes (true);
2. precise orientation of catalytic groups with reactants is necessary for reaction to occur (false, cf. p. 151); and
3. only substrates induce the proper alignment of the catalytic residues.

Evidence which supports this type of theory was summarized by Koshland and Neet (1968) and Koshland (1970). Koshland feels that not all enzymes must exhibit flexibility, but that there may be a gradation of configurational changes. The effects of allosteric ligands are thought to be manifested through changes in conformation of enzymes, and undoubtedly conformational changes are induced by substrates. The induced-fit theory is based on the flexibility of the enzyme molecule but does not inherently consider the change in structure or energy of the substrate upon binding. The strain theory, in which both the substrate and the enzyme may have altered conformations, is a more general description of how bond distortion can be used to speed up a reaction.

The strain theory states that the binding forces between the substrate and the enzyme are used to cause bond distortion which facilitates the reaction (Jencks, 1969, pp. 249f in his book). Two extreme cases may be considered. If the enzyme is rigid, then it is the substrate which is distorted into a structure approaching the transition state. If the substrate is rigid relative to the enzyme, then the enzyme would be distorted to a form providing optimal fit for substrate, and the deformed enzyme in the *ES* complex would tend to return to its relaxed lower-energy state. In so doing it would force the substrate towards the transition state. In the latter example, the difference between the strain theory and the induced-fit theory is in the actual reason for bond distortion; in the strain theory its distortion is a driving force for the reaction, and in induced-fit it gives orientation for specificity. Jencks says that actual enzyme catalysis will be intermediate between the two extreme cases described above, and all available data support this statement. Induction of strain will cause at least some change in the conformations of both reactant and enzyme. If this strain is relieved in the transition state, there will be rate acceleration. If distortion causes rate acceleration, then it must be remembered that ultimately it is the substrate which is distorted. In simple terms, I find it difficult to imagine destruction of a stiff substrate by a very flexible enzyme. Indeed, enzyme proteins do not have great flexibility. One of the strongest confirmations of substrate bond distortion in enzymatic catalysis is the work done with transition-state analogues.

Transition-state Analogues

We have just seen that it would be advantageous for an enzyme to have greatest affinity for the transition state of the reaction that it catalyzes. We do not know the exact structures of transition states, but stable compounds having structures reasonably similar to transition states have been synthesized. The compounds, termed transition-state analogues, are potent enzyme inhibitors (Wolfenden, 1972; Lienhard, 1973). These transition-state analogues bind very tightly to the active sites (10^2–10^5 times better than substrates) and are specific in their binding properties. Kinetic studies using these analogues have confirmed postulated mechanisms (e.g. with lysozyme). Additional uses of transition-state analogues will probably include applications as highly specific chemotherapeutic compounds, as ligands for affinity chromatography, and as compounds to cocrystallize with enzyme to give a rough indication of the conformation of *ES* complexes. It is another feather in Emil Fischer's cap that his template theory of enzyme catalysis turns out to be correct, with the simple alteration that the key is the transition state, not the substrate.

SUMMATION OF EFFECTS

One of the first serious attempts to assign quantitative importance to catalytic effects was made by Bender and his colleagues in 1964. Their approach has been used by many other investigators. They compared the rates of OH^--catalyzed and chymotrypsin-catalyzed hydrolyses of N-acetyl-L-tryptophan amide. They transformed the second-order non-enzymic rate constant into a first-order rate of $4.8 \times 10^{-9} \, s^{-1}$, using several assumptions. The enzyme-catalyzed rate is $4.4 \times 10^{-2} \, s^{-1}$. They then rationalized this 10^7-fold difference in rates by assessing the factors by which the enzymic and non-enzymic catalytic processes differ. First, the non-enzymatic reaction is a hydrolysis and the enzymic reaction is an alcoholysis (the nucleophile is the $-CH_2OH$ of serine, which should react 100-fold faster with substrate than does water). This is a clear example of alteration by an enzyme of a reaction pathway to another of lower activation energy (cf. p. 87). A decrease in entropy of the substrate on binding to enzyme can account for a factor of 10^3, while a factor of 10^2 was assigned to a general-acid-catalyzed step in the enzymic process. The combination of these effects successfully accounted for the rate acceleration observed, with reasonable chemical analogies for all of them. Bender's instructive estimates could be criticized (e.g. for lack of similarity between the enzymic reaction and the reference reaction), but present estimates of catalytic effects with other enzymes are not much more advanced 12 years later.

We cannot yet assign exact quantitative roles to given effects for any enzyme, but the operation of particular catalytic modes are known for many enzymes. Chemical analogies can be found for all the basic effects. Each enzyme has evolved to a structure that catalyzes its specific reaction by making use of several, not just one, of the effects discussed: acid–base catalysis, covalent catalysis, proximity, multiple catalytic groups, surface effects and distortion. There can be considerable overlap between some of these fundamental catalytic contributions. Jencks (1975) has generalized that enzymes are better catalysts than simple chemicals because they use strong non-covalent binding interactions with substrates in addition to the well-known chemical mechanisms. In this manner, correct binding provides both specificity and catalysis.

REFERENCES

Bender, M. L., Kézdy, F. J. and Gunter, C. R. (1964). *J. Am. Chem. Soc.* **86**, 3714-3721.

Breslow, R. and Campbell, P. (1971). *Bioorg. Chem.* **1**, 140-156.

Breslow, R. and Overman, L. E. (1970). *J. Am. Chem. Soc.* **92**, 1075-1077.

Bruice, T. C. (1970). *In* "The Enzymes" (P. D. Boyer, ed.), Third edition, Vol. 2, pp. 217-279. Academic Press, New York.

Bruice, T. C. (1971). *Cold Spring Harb. Symp. quant. Biol.* **36**, 21-27.

Bruice, T. C. and Benkovic, S. J. (1963). *J. Am. Chem. Soc.* **85**, 1-8.

Bruice, T. C. and Benkovic, S. J. (1966). "Bioorganic Mechanisms", Vol. 1, p. 119. W. A. Benjamin, New York.

Bruice, T. C. and Pandit, U. K. (1960a). *Proc. natn. Acad. Sci. U.S.A.* **46**, 402-404.

Bruice, T. C. and Pandit, U. K. (1960b). *J. Am. Chem. Soc.* **82**, 5858-5865.

Bruice, T. C., Brown, A. and Harris, D. O. (1971). *Proc. natn. Acad. Sci. U.S.A.* **68**, 658-661.

Cordes, E. H. and Dunlap, R. B. (1969). *Accts Chem. Res.* **2**, 329-337.

Cordes, E. H. and Gitler, C. (1973). *Prog. bioorg. Chem.* **2**, 1-53.

Dunlap, R. B. and Cordes, E. H. (1968). *J. Am. Chem. Soc.* **90**, 4395-4404.

Gitler, C. and Ochoa-Solano, A. (1968). *J. Am. Chem. Soc.* **90**, 5004-5009.

Guthrie, J. P., McDonald, R. S. and O'Leary, S. (1976). In preparation.

Jencks, W. P. (1969). "Catalysis in Chemistry and Enzymology". McGraw-Hill, New York.

Jencks, W. P. (1972). *Chem. Rev.* **72**, 705-718.

Jencks, W. P. (1973). *PAABS Revista* **2**, 235-288.

Jencks, W. P. (1975). *Adv. Enzymol.* **43**, 219-410.

Kiefer, H. C., Congdon, W. I., Scarpa, I. S. and Klotz, I. M. (1972). *Proc. natn. Acad. Sci. U.S.A.* **69**, 2155-2159.

Klotz, I. M., Royer, G. P. and Scarpa, I. S. (1971). *Proc. natn. Acad. Sci. U.S.A.* **68**, 263-264.

Koshland, D. E., Jr. (1970). *In* "The Enzymes" (P. D. Boyer, ed.), Third edition, Vol. 1, pp. 341-396. Academic Press, New York.

Koshland, D. E., Jr. and Neet, K. E. (1968). *A. Rev. Biochem.* **37**, 359-410.

Lienhard, G. E. (1973). *Science, N.Y.* **180**, 149-154.

Mahler, H. R. and Cordes, E. H. (1971). "Biological Chemistry", Second edition, Chapter 7. Harper and Row, New York.

Milstien, S. and Cohen, L. A. (1972). *J. Am. Chem. Soc.* **94**, 9158-9165.

Nicholls, R. G. and Roy, A. B. (1971). *In* "The Enzymes" (P. D. Boyer, ed.), Third edition, Vol. 5, pp. 21-41. Academic Press, New York.

Page, M. I. (1972). *Biochem. biophys. Res. Commun.* **49**, 940-944.

Page, M. I. and Jencks, W. P. (1971). *Proc. natn. Acad. Sci. U.S.A.* **68**, 1678-1683.

Podo, F., Ray, A. and Némethy, G. (1973). *J. Am. Chem. Soc.* **95**, 6164-6171.

Storm, D. R. and Koshland, D. E., Jr. (1970). *Proc. natn. Acad. Sci. U.S.A.* **66**, 445-452.

Storm, D. R. and Koshland, D. E., Jr. (1972). *J. Am. Chem. Soc.* **94**, 5805-5825.
Swain, C. G. and Brown, J. F., Jr. (1952). *J. Am. Chem. Soc.* **74**, 2534-2537.
Tanford, C. (1973). "The Hydrophobic Effect: Formation of Micelles and Biological Membranes", Chapters 6-9. John Wiley and Sons, New York.
Tanford, C. (1974). *Proc. natn. Acad. Sci. U.S.A.* **71**, 1811-1815.
Wolfenden, R. (1972). *Accts Chem. Res.* **5**, 10-18.

Part II

Examples of Enzyme Catalysis

7 Enzymes without Prosthetic Groups

As the first examples of enzyme chemistry, I will describe three enzymes which require no cofactors. Each of these enzymes relies entirely on the protein molecule for the binding of substrate and for the catalytic site. These enzymes are lysozyme, ribonuclease and triosephosphate isomerase. A further example of an enzyme without a prosthetic group is chymotrypsin, discussed in the chapter on covalent catalysis (Chapter 10). The research performed with lysozyme can be used as a standard of achievement for other enzymes. Ribonuclease was one of the first enzymes for which multi-functional catalysis, i.e. the participation of several catalytic groups, was demonstrated. The puzzle of the mechanism of triosephosphate isomerase is in the process of being solved, and this enzyme, too, will probably be a milestone in research on enzyme mechanisms.

Enzymes without prosthetic groups have received a great deal of attention because of their relative simplicity. Some are quite small, and the determination of their structures has been correspondingly simpler than that for the majority of enzymes, which are more complex. We can also expect the mechanisms of the simpler enzymes to be less complicated and therefore more amenable to study. In theory, the kinetics of hydrolases and isomerases should be relatively simple, following Michaelis–Menten kinetics. With triosephosphate isomerase, the enzyme combines with only one substrate or one product. With lysozyme, water is an additional substrate, but is usually considered as being present in great excess. Ribonuclease catalyzes a two-step reaction. In the first of these steps, only one substrate participates, and only two substrates (one being water) are involved in the second step. For several reasons, the kinetics of both lysozyme and ribonuclease are not as simple as one would first expect.

X-ray crystallography has played a major role in the elucidation of mechanism in all three enzymes, and each one uses general-acid–base catalysis as one source of catalytic power. However, each of these three enzymes exemplifies different research approaches and different catalytic properties.

LYSOZYME

Lysozyme is perhaps the best understood enzyme in terms of mechanism of action. It was first crystallized in 1937, but studied for its protein properties rather than catalytic properties for many years. In 1965, the complete three-dimensional structure was reported, making lysozyme the first enzyme so characterized. In fact, most of our knowledge of its mechanism is a direct result of the X-ray crystallographic experiments. In part, this is due to the complexity of the natural substrate of lysozyme, an amino sugar–peptide polymer whose structure was only realized in the 1960s. A comprehensive review of the structure, properties, specificity and mechanism of lysozyme has been written by Imoto et al. (1972).

Lysozymes are characterized by their ability to lyse bacteria by catalyzing the hydrolysis of their cell walls. Lysozyme activity occurs in many tissues and secretions (e.g. tears, saliva and nasal mucus), but by far the best studied lysozyme is that isolated from hen egg-white, and the latter is the only lysozyme described here. Hen egg-white lysozyme is a basic protein (its isoelectric pH is near 11) which is heat stable. It is prepared by crystallization at pH 9.5 from egg-white by addition of NaCl to a concentration of 5% (Alderton and Fevold, 1946). The preparation and recrystallization are simple and the yield of enzyme activity high. Several forms of crystals can be obtained, depending on the pH at which crystallization is performed.

Lysozyme activity is usually assayed by using acetone-dried preparations of the Gram-positive bacterium *Micrococcus lysodeikticus* as substrate. The rate of hydrolysis of the dried cell walls is measured by the decrease in turbidity in the presence of added lysozyme. The compound giving shape and rigidity to bacterial cell walls is a polymer composed of alternating amino sugar residues, *N*-acetylglucosamine (GlcNAc) and *N*-acetyl muramic acid (MurNAc) connected by β-1,4-linkages (Fig. 1). A three-dimensional network,

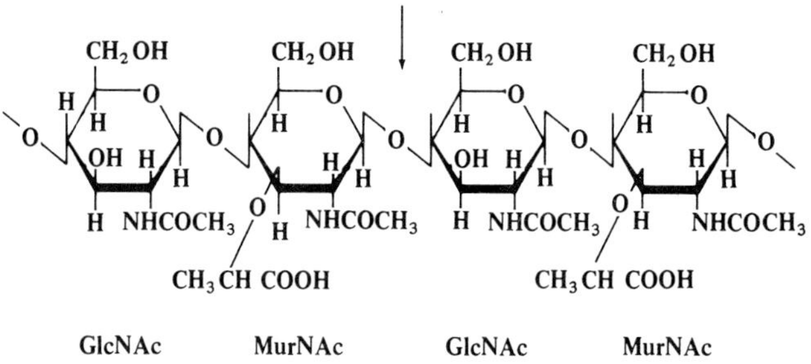

Fig. 1. Structure of the amino sugar polymer of bacterial cell walls. Both N-acetyl-D-glucosamine (GlcNAc) and N-acetyl muramic acid (3-O-lactyl-N-acetyl-D-glucosamine, MurNAc) are drawn in modified Haworth configurations. In the polymer, because of the β-1,4-bonds, the alternate residues would have their acetamido groups oriented in opposite directions. Also, the hexose rings would be in the chair configuration. The arrow shows the position of cleavage in the lysozyme-catalyzed reaction.

called a peptidoglycan, is formed by the crosslinking of some of the lactyl side chains of the MurNAc by oligopeptides, usually tetrapeptides. Gram-positive organisms have several layers of peptidoglycan in their cell walls, while in Gram-negative organisms a layer of peptidoglycan is sandwiched between two layers of membrane material, the outer layer being lipopolysaccharide. Lysozyme catalyzes the hydrolysis of the glycosidic linkage between the C^1 of MurNAc and the O^4 of GlcNAc. Chitin, the β-1,4 polymer of GlcNAc, can also serve as a substrate for lysozyme.

A brief look at the data upon which the structure of lysozyme is based is instructive. The increase in the precision of data over a long period shows the many experiments and the long time needed for the characterization of an enzyme. The molecular weight of lysozyme was measured between 1939 and the 1950s by many physical methods (reviewed by Jollès, 1960). Values estimated ranged from 13 000 to 18 000, with most indicating a molecular weight of 14 000–15 000. The amino-acid composition of lysozyme was first determined in 1951, and the initial analysis was remarkably close to the correct composition. In Table I, the data at three stages in the determination of the structure are shown. Column 1 gives the data available in 1951, column 2 the data in 1960, just before completion of the sequence studies, and column 3 the data based on the

Table I. Amino-acid composition of hen egg-white lysozyme. The first column of data shows results summarized by Fevold (1951), the second the preferred integers from data available in 1960 (Jollès) and the third the composition obtained after knowledge of the sequence (as given in Imoto et al., 1972).

	1951	1960	After sequence
Ala	10	12	12
Arg	11	10-11	11
Asx	20	22 ± 1	21
Asp	-	-	(7)
Asn	-	-	(14)
Cys	10	8	8
Glx	4	5	5
Glu	-	-	(2)
Gln	-	-	(3)
Gly	11	12	12
His	1	1	1
Ile	6	6	6
Leu	9	8	8
Lys	6	6	6
Met	2	2	2
Phe	3	3	3
Pro	2	2	2
Ser	10	12	10
Thr	7	7-8	7
Trp	8	5-6	6
Tyr	3	3	3
Val	6	6	6
Amides	(18)	(18)	(17)
Total	129	131-133	129

currently accepted primary structure of lysozyme. The composition showed that there was a high number of basic residues (eleven Arg and six Lys), as expected from the basic nature of lysozyme, and also a high proportion of aromatic amino acids. The presence of eight half-cystine residues indicated that there would be four dithiol bridges in the molecule. The complete sequence of lysozyme was reported independently by two laboratories in 1963 (Canfield, 1963;

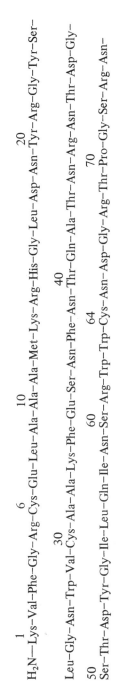

Fig. 2. Amino acid sequence of hen egg-white lysozyme. From Imoto et al. (1972).

1
6
10
20
H₂N—Lys–Val–Phe–Gly–Arg–Cys–Glu–Leu–Ala–Ala–Ala–Met–Lys–Arg–His–Gly–Leu–Asp–Asn–Tyr–Arg–Gly–Tyr–Ser–

30
40
Leu–Gly–Asn–Trp–Val–Cys–Ala–Ala–Lys–Phe–Glu–Ser–Asn–Phe–Asn–Thr–Gln–Ala–Thr–Asn–Arg–Asn–Thr–Asp–Gly–

50
60
64
70
Ser–Thr–Asp–Tyr–Gly–Ile–Leu–Gln–Ile–Asn–Ser–Arg–Trp–Trp–Cys–Asn–Asp–Gly–Arg–Thr–Pro–Gly–Ser–Arg–Asn–

76
80
90
94
100
Leu–Cys–Asn–Ile–Pro–Cys–Ser–Ala–Leu–Leu–Ser–Ser–Asp–Ile–Thr–Ala–Ser–Val–Asn–Cys–Ala–Lys–Lys–Ile–Val–Ser–

110
115
120
Asp–Gly–Asn–Gly–Met–Asn–Ala–Trp–Val–Ala–Trp–Arg–Asn–Arg–Cys–Lys–Gly–Thr–Asp–Val–Gln–Ala–Trp–Ile–Arg–

127
129
Gly–Cys–Arg–Leu—COOH

Jollès *et al.*, 1963). These sequences were obtained by the analysis of enzymically generated peptides. The sequences published were similar, but were not in agreement in several positions. The data from X-ray crystallography suggested that the environments of the questionable residues favoured the sequence proposed by Canfield, and this suggestion has been substantiated by a recent reinvestigation (Rees and Offord, 1972). Position 103 is Asn rather than Asp (J. Brown, quoted in Imoto *et al.*, 1972). The sequence of lysozyme is shown in Fig. 2. There are 129 amino-acid residues, with the molecule having a calculated molecular weight of 14 307. The disulphide bridges have been shown by three laboratories to be arranged in the order I–VIII (6–127), II–VII (30–115), III–V (64–80) and IV–VI (76–94).

The chemical synthesis of lysozyme has been attempted in several laboratories, but no homogeneous product of high activity has been obtained. Using the solid-phase technique, polypeptides having low activity have been prepared (Sharp *et al.*, 1973; p. 676 in Imoto *et al.*, 1972). Affinity chromatography can be used to purify these products, but the yield in the synthesis is low.

Crystallographic experiments designed to obtain the structure of lysozyme were started in about 1952. Ten years later, a model based on data of 6-Å resolution was published, and in 1965, Blake *et al.* reported the structure of a model based on resolution to 2 Å. Further work in Phillips's laboratory has extended these results, so that detailed structures for both lysozyme (Blake *et al.*, 1967a) and complexes of lysozyme with GlcNAc or a trisaccharide (Blake *et al.*, 1967b) were obtained. A skeletal model was built based on the electron-density map, bond angles and lengths of simple compounds, and the amino acid sequence. The structure of the model could be adjusted to the map density except for two regions on the surface of the enzyme where the side chains probably have some freedom of rotation. Lysozyme is ellipsoidal, with the approximate dimensions 45 × 30 × 30 Å. It has a small proportion of α-helix or other regular arrangements, and long stretches of rather irregular conformation. There is a large cleft dividing the molecule into two sections (Fig. 3); this cleft is the site of binding of the substrate. Most of the polar side chains are on or near the surface of lysozyme, while the side chains which form the interior of the molecule are primarily hydrophobic. The ionizable side chains in least-exposed environments are Asp-66, Asp-52, Tyr-53, His-15 and Glu-35.

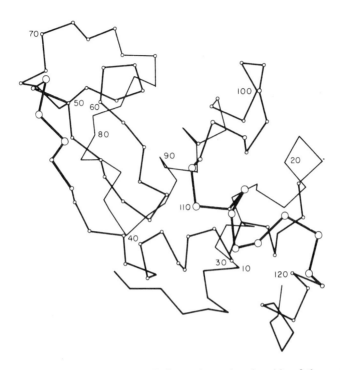

Fig. 3. View of the lysozyme molecule from the active-site side of the molecule. Only the positions of the α-carbon atoms are shown. Note the large cleft which is the site of binding of the substrate. The dithiol bonds (not shown) agree in position with the data from the chemical experiments. From p. 697 in Imoto *et al.* (1972).

X-ray studies of the saccharide–lysozyme complexes showed that the sugar molecules were bound in the cleft of the enzyme. The most useful information had been obtained with the trisaccharide tri-N-acetylchitotriose, $(GlcNAc)_3$, which could be slowly hydrolyzed in the presence of lysozyme but formed a stable complex with the crystalline enzyme. The structure of the $(GlcNAc)_3$–lysozyme complex showed that the free reducing group of the trisaccharide was bound towards the centre of the enzyme, and many nonpolar and hydrogen-bond contacts between the virtual substrate and the enzyme could be discerned. There were a few slight changes in the conformation of the enzyme upon binding of the trisaccharide. It was assumed that the complex observed was one in which the $(GlcNAc)_3$ was bound to the active site but not at the catalytic

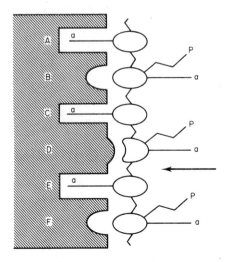

Fig. 4. Diagram showing the binding of a polysaccharide to the active site of lysozyme. The six subsites *A–F* bind the sugar residues. Alternate sites interact with acetamido groups (*a*). These sites cannot accommodate the lactyl side chains (*P*) of MurNAc. The sugar residue bound at subsite *D* is distorted. The arrow shows the glycosidic bond which is cleaved during lysozyme-catalyzed hydrolysis. From Imoto *et al.* (1972).

centre, and so represented only part of the Michaelis complex. Into the three-dimensional model of this complex, additional molecules of GlcNAc were fitted, extending from the reducing end of the (GlcNAc)₃. The fourth GlcNAc molecule could not be fitted into the model without being distorted into a half-chair configuration. Two additional molecules of GlcNAc, forming a hexasaccharide, could be fitted into the cleft and make contacts with appropriate binding groups without any difficulty. It was proposed that in the true *ES* complex, six saccharide residues are bound in the cleft, with three taking the places shown by the (GlcNAc)₃ at positions *A, B* and *C*, and the other three contacting the enzyme at positions *D, E* and *F* in the cleft (Fig. 4). The hydrolysis occurs at the reducing end of the distorted sugar, bound in position *D*. MurNAc residues, because of their large lactyl side chains, can only be fitted into positions *B, D* and *F*. This explains the specificity of the bond cleavage by lysozyme. The only two reactive amino-acid side chains located near the glycosidic bond between residues *D* and *E* are the carboxyl groups of Glu-35 and Asp-52. Glu-35 is in a nonpolar region of the cleft, and

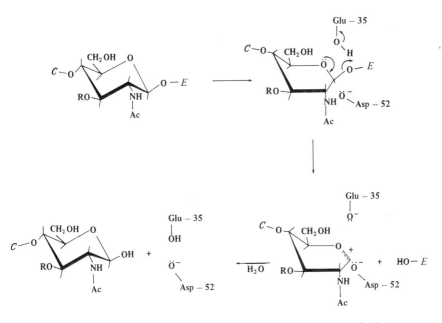

Fig. 5. Postulated mechanism for the hydrolysis of polysaccharides by lysozyme. The sugar structure shown is that of MurNAc residue *D* of the substrate. The first step is the binding of the substrate to the enzyme, with the strain of ring *D* into the half-chair conformation. Glu-35 then donates a proton to the glycosidic oxygen, so that the carbonium ion and the leaving group, HO—*E*, are formed. Asp-52 is in the correct position to form an ion pair with the carbonium ion. Finally, water reacts with the enzyme-bound intermediate to form product and regenerated enzyme.

Asp-52 is in a more polar environment. In the model of the proposed Michaelis complex, a carboxyl oxygen of Glu-35 is about 3 Å from the oxygen of the hydrolysis-susceptible glycosidic bond and is in a position to form a hydrogen bond with this oxygen. One of the oxygens of Asp-52 is close to the C^1 and O^5 atoms of sugar residue *D*. The X-ray data have been combined with chemical evidence for the proposal of a plausible mechanism (Vernon, 1967), often referred to as the Phillips–Vernon mechanism.

This mechanism for the catalysis of the hydrolysis of polysaccharides by lysozyme is based on three catalytic factors: distortion of the substrate; general-acid catalysis; and electrostatic interaction of the transition state with the enzyme (Fig. 5). Upon binding of the substrate to lysozyme, the ring of sugar residue *D* is distorted from the chair into the half-chair conformation. A half-chair conformation, in which C^2, C^1, O and C^5 are in a plane, would favour the formation of a carbonium ion (actually, an oxocarbonium ion, with the positive charge shared by C^1 and O^5), also believed to have the half-chair configuration. Glu-35, in a nonpolar environment and thus protonated at pH 5, acts as an acid catalyst by donating its proton to the glycosidic oxygen. Asp-52, being in a more polar environment, is negatively charged at pH 5 and promotes formation of the carbonium ion and stabilizes it after formation. To complete the hydrolysis, the disaccharide portion of the substrate bound in positions E and F diffuses away and is replaced by a molecule of water. The water can only approach the carbonium ion by the same pathway by which the disaccharide left, so that the β-configuration at C^1 is retained in the product. Transglycosylation by attack of nucleophiles other than water has been observed. Both the retention of configuration and the transglycosylation suggest formation of a covalent ES intermediate, but the X-ray data seem to rule out formation of a covalent bond between the carbonium ion and Asp-52. The only difference between a covalent catalytic step and the formation of the ion pair is the distance of approach of the carbonium ion to the oxygen of the carboxyl. Some enzymes (e.g. sucrose phosphorylase, p. 341) form a detectable covalent glycosyl intermediate by a very similar mechanism.

Since the proposal of this mechanism, many chemical experiments have been performed to test it, and almost all support it. Examination of the binding of oligosaccharides has verified the proposal of six sugar subsites. Both Asp-52 and Glu-35 have been found to be essential for activity by substitution experiments. Also, their pK_a values have been determined, and are in agreement with the mechanistic proposal. Secondary-isotope experiments support the intermediate formation of a carbonium ion, and a transition state analogue of the half-chair form is a potent inhibitor of lysozyme. Physicochemical experiments (summarized by Dunn and Bruice, 1973) suggest that general-acid catalysis by Glu-35 does not occur but that the neighbouring N-acetyl groups assist in the hydrolysis. However, some substrates lack the N-acetyl group, and the three-dimensional model suggests that participation of the N-acetyl groups is unlikely for steric reasons. Some pertinent experiments documenting the Phillips–Vernon mechanism are described below.

The free energies of association of sugars with lysozyme have been

Table II. Free energy of binding of substrates to lysozyme. Data from Imoto
et al. (1972).

Subsite	ΔG (kcal mol^{-1})
A	−2.3
B	−2.7
C	−4.6
D	+2.9 to +6.0
E	−4.0
F	−1.7

measured with many compounds and by many techniques (summarized by Imoto *et al.*, 1972). The *N*-acetimido group of GlcNAc must
be important for binding, because glucose itself is not bound. GlcNAc
oligosaccharides show an increase in affinity to lysozyme up to the
trimer, but not for longer oligomers. Table II shows the estimated
free energies for the binding of substrates to each of the six subsites
in the binding of substrates to each of the six subsites in the cleft.
These data support the theory that sugar residue *D* is distorted when
bound. The energy expended in filling subsite *D* is approximately
equal to that gained in filling sites *E* and *F*.

After the mechanism based on the X-ray structure was proposed,
several experiments were performed which clearly indicated the
essential nature of the carboxyl groups of Asp-52 and Glu-35. Lin
and Koshland (1969) reacted lysozyme with aminomethanesulphonic
acid ($H_2N-CH_2-SO_3H$) and a carbodiimide, so that the carboxyl
groups of the enzyme were substituted ($-C(=O)NH-CH_2-SO_3^-$).
Under denaturing conditions (guanidine > 3.5 M), even Glu-35 was
substituted. Modification in the presence of oligosaccharide substrates resulted in protection of Asp-52 and retention of 50% of the
activity of the enzyme. After removal of the substrate, Asp-52 could
be substituted with the resultant loss of all activity. These experiments eliminate all carboxyls other than Asp-52 and Glu-35 as
catalytically essential, and support the involvement of Asp-52 and
Glu-35 in catalysis. Parsons and Raftery (1969) prepared a derivative
of lysozyme in which only Asp-52 was esterified. This derivative has
no activity, but can still bind (GlcNAc)$_3$. Thus, substitution of

Asp-52 prevents catalysis but not binding of the substrate to sites *A–C*. Binding at site *D* would be blocked, though. Thomas *et al.* (1969) have used the 2′,3′-epoxypropyl-β-glycoside of (GlcNAc)$_2$ as an active site-directed inhibitor of lysozyme. The disaccharide (R) is

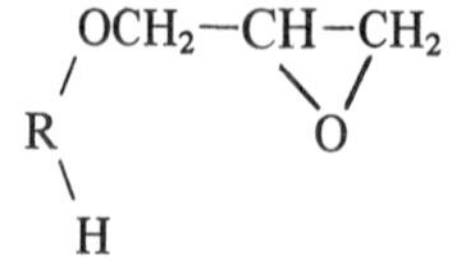

$$OCH_2-CH-CH_2$$

R\\O
\
H

bound to subsites *B* and *C* of the cleft, and the β-carboxylate of Asp-52 (at site *D*, and thus near the epoxide group) is alkylated. By first preparing this affinity labelled lysozyme, and then reducing the Asp-52 ester bond, Eshdat *et al.* (1974) have prepared lysozyme with homoserine at position 52. This modification causes loss of enzyme activity without introduction of a bulky group. The homoserine-52 lysozyme can bind the tetrasaccharide (GlcNAc—MurNAc)$_2$, which binds to sites *A–D*. These experiments prove the requirement of Asp-52 for catalysis. Selective oxidation of Trp-108 with iodine produces an oxindolyl ester of lysozyme, identified as the ester of Glu-35 (Imoto and Rupley, 1973). This modification does not

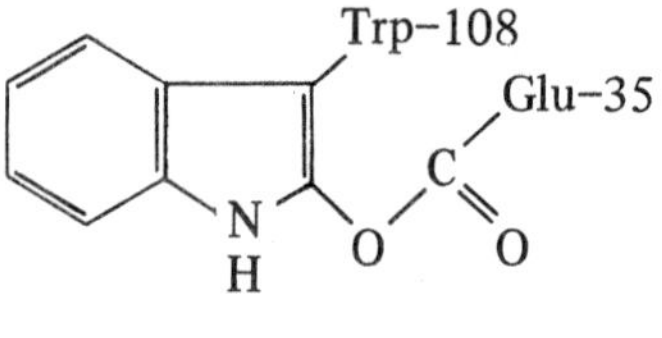

Trp–108
Glu–35
N
H O O

Oxindolyl ester

significantly lower the binding of saccharides, but the enzymatic activity is only 0.05–0.1% of native lysozyme. Crystallographic experiments have shown that the iodine oxidation alters only the interaction of Glu-35 at subsite *D*. Therefore, the drastic reduction in enzymatic activity must be due to the esterification of Glu-35, demonstrating unambiguously its catalytic function.

Acid–base titrations of lysozyme are consistent with the primary structure of the enzyme, if residue 103 is assumed to be Asn not Asp (cf. p. 176). The titration results for the carboxyl groups were in question until recently, when Roxby and Tanford (1971) obtained potentiometric evidence for ten carboxyls in a guanidine hydrochloride solution of lysozyme (2 Glu, 7 Asp and the C-terminal carboxyl). Timasheff and Rupley (1972) have titrated lysozyme in D_2O using infrared spectrophotometric analysis of both —COOH and —COO⁻ bands. They found pK_a values for carboxyl groups of 2.0 (1 group), 3.5 (3), 4.5 (3), 5.5 (2) and 6.5 (1). The two abnormal pK_a values, 2.0 and 6.5, have been assigned to Asp-66 (buried and hydrogen-bonded) and Glu-35, respectively. The effect of pH on initial velocity showed the requirement of two ionizable groups of pK_a 3–4 and 6–7. Titration of the monoesterified Asp-52 derivative suggested that Asp-52 had a pK_a of about 4.5 (Parsons and Raftery, 1972). Their evidence pointed to a pK_a of 5.8–6.5 for Glu-35, but on binding of the substrate glycol chitin, its pK_a rose to 8–8.5. Kuramitsu et al. (1974) have determined the dissociation constants of Glu-35 and Asp-52 by measuring the pH dependence of the circular dichroic band of Trp-108, which is near the two catalytic carboxyls. Their pK_a assignments of 3.4 for Asp-52 and 5.9 for Glu-35 were compatible with the values based on X-ray crystallography and with the activity–pH profile.

Raftery and his colleagues have measured the secondary isotope effects for the hydrolysis of C^1-substituted glycosides, both a 1-^2H-phenyl glucoside (Dahlquist et al., 1969) and $1,1',1''$-^3H-chitotriose (Smith et al., 1973). In these cases, k_H/k_D for lysozyme-catalyzed hydrolysis was 1.11 and 1.14 (corrected from $k_H/k_T = 1.19$), respectively. For acid-catalyzed hydrolysis, which proceeds via a carbonium ion (S_N1), k_H/k_D was 1.13, much closer to the isotope effect observed for lysozyme than that for base-catalyzed S_N2 hydrolysis (1.03). From these data, Raftery has concluded that there is considerable carbonium-ion character in the transition state. The data do not rule out formation of a covalent intermediate after the rate-determining step, though, allowing transglycosylation to occur.

Chemical evidence supporting the distortion of the pyranose ring of the sugar residue bound at the catalytic site is accumulating. Secemski et al. (1972) have compared the binding of the δ-lactone of tetra-N-acetylchitotetraose (below) to that of $(GlcNAc)_4$ itself. The δ-lactone should be in a slightly distorted half-chair, and so should be a transition-state analogue. It is a much more potent inhibitor of cell-wall lysis than is $(GlcNAc)_4$. Binding studies indicate that the half-chair transition-state analogue binds 6×10^3 times more tightly to subsite D than does the chair form, $(GlcNAc)_4$. This factor of 6×10^3 is considered to be an estimate of the contribution made

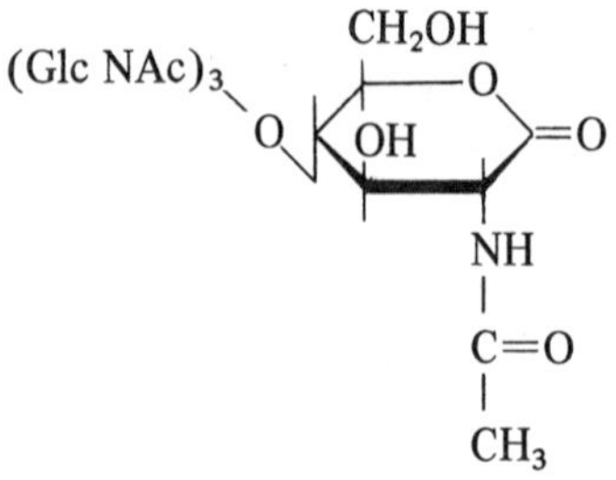

$$\text{(Glc NAc)}_3 \diagdown$$

CH$_2$OH

O — O

$O \diagdown$ OH $\diagup$ =O

NH

C=O

CH$_3$

to catalysis by tighter binding of the transition state than the substrate. The three-dimensional model of the (GlcNAc)$_6$-lysozyme complex indicates that steric hindrance to the C^6-hydroxymethyl groups of sugar D forces adoption of the half-chair configuration. An X-ray crystallographic analysis of the proposed transition-state analogue–lysozyme complex has recently been reported (Ford *et al.*, 1974). The lactone is, as expected, bound to site D. Its exact conformation has not been determined (it appears to be a sofa, rather than a half-chair), but the structural evidence supports the Phillips–Vernon mechanism. An N-acetylxylosamine derivative of chitotetraose, (GlcNAc)$_3$XylNAc, in which the —CH$_2$OH of the C^6 has been replaced by a proton, has been synthesized by van Eikeren and Chipman (1972). The Phillips–Vernon theory predicts that the XylNAc residue would bind to subsite D without distortion. The XylNAc residue is actually bound to subsite D with a free energy about 5 kcal mol^{-1} more favourable than either a MurNAc or GlcNAc residue. This evidence supports the crystallographic reasoning for distortion of MurNAc residue D. Sykes *et al.*(1971) have used n.m.r. spectra of GlcNAc—MurNAc—GlcNAc—MurNAc in the presence of lysozyme to measure conformational change of the reducing MurNAc ring. The observed spectral differences between free and bound substrate were consistent with a distortion of the terminal MurNAc towards a half-chair conformation.

Fast-reaction techniques, steady-state kinetics and kinetic isotope effects have been used by Banerjee *et al.* (1975) to indicate a pathway for the reaction of (GlcNAc)$_6$ with lysozyme. The kinetic studies are complicated by non-productive binding of the hexasaccharide at subsites A, B and C, with the reducing end-group at site C. The first step in the productive pathway is a rapid bimolecular reaction. This is followed by two isomerization steps, after which the productive ES complex has been formed. The authors concluded that even in the productive complex, interactions between the hexasaccharide and the D, E and F sites of lysozyme are not fully developed. The final rearrangement of the productive complex involves distortion of the substrate and/or enzyme so that there is complete bond cleavage. When the cleavage is performed in D_2O rather than H_2O, the reaction is slower ($k_H/k_D > 1$), consistent with general-acid catalysis in the rate-determining step. A complete free-energy reaction co-ordinate profile has been plotted from measurements of cleavage rates from 5–40° C. The $\Delta G^{\ddagger}$ values for the two isomerization reactions (15.9 and 17.9 kcal mol^{-1}) are similar but just slightly less than that for product formation from the productive complex (19.4 kcal mol^{-1}). By distributing the ES rearrangement over three steps, no single step makes a large destabilization contribution, providing an optimal catalytic effect.

It now seems certain that binding of the hexasaccharide substrate and distortion of the D sugar molecule towards the transition state by a relatively rigid enzyme is a major factor in catalysis by lysozyme. General-acid catalysis by the carboxyl side chain of Glu-35 and the formation of a carbonium ion during hydrolysis have been verified by chemical studies. The assignment of an electrostatic interaction function to Asp-52 relies only on X-ray data, but its essential nature has been demonstrated by modification experiments. Further definitive chemical experiments with lysozyme and greater refinement of the crystallography data will undoubtedly be performed in the next few years, and lysozyme will continue to be used as an outstanding example of the application of many techniques for the determination of detailed mechanisms of action.

RIBONUCLEASE

Development of the research on the ribonuclease (RNase) isolated from beef pancreas is another classical story in enzymology. The properties of RNase and polypeptides derived from it have proved many fundamental points in the relation of protein structure to enzyme activity. RNase was one of the first enzymes to be crystallized (in 1939), and it was the first enzyme to be completely sequenced. The extreme stability of this enzyme may be a disadvan-

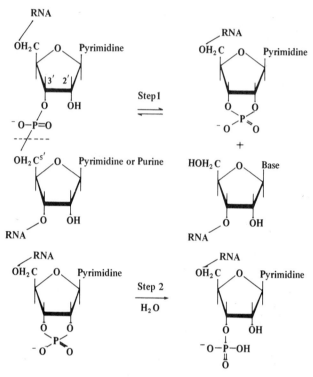

Fig. 6. The two steps in the ribonuclease-catalyzed hydrolysis of RNA, showing the 2′,3′-cyclic phosphate as an obligatory intermediate.

tage to nucleic acid chemists wanting to isolate undegraded RNA, but it is a distinct advantage to enzymologists. Because of this stability, many experiments have been performed on RNase that can be done with few other enzymes (e.g. easy reversible denaturation), and the chemical synthesis of RNase (p. 26) has thus been feasible. A recent review by Richards and Wyckoff (1971) documents many of the properties of this thoroughly studied enzyme.

Ribonuclease catalyzes the hydrolysis of phosphodiester bonds in RNA. The pancreatic RNase probably functions to hydrolyze dietary RNA in the intestine in most mammals. RNase-catalyzed hydrolysis takes place in two steps: a depolymerizing transesterification (trans-phosphorylation) in which the 2′-hydroxyl group replaces the leaving 5′-hydroxyl group to form a 2′,3′-cyclic phosphate; and the subsequent hydrolysis of the cyclic phosphate (Fig. 6). Step 1 is the more rapid and is reversible. Step 2 is slower but not reversible. RNase is specific for phosphodiester bonds in which cyclic $2',3'$-phosphates of pyrimidine nucleosides are formed. Activity usually is assayed either using yeast RNA as substrate and measuring the decrease in absorbance at 300 nm as the cyclic phosphate is formed (measuring step 1, primarily) or using cytidine-$2',3'$-cyclic phosphate and measuring the increased absorbance due to opening of the cyclic phosphate ring (a measure of step 2 only).

Ribonuclease is purified from pancreas by a method (Kunitz and McDonald, 1953) which would denature most enzymes. Minced pancreas tissue is extracted at $5°$ C with 0.25 N sulphuric acid for 18–24 h. The extracted protein is fractionated with ammonium sulphate, and the fraction containing RNase (65–80% saturation) is dissolved in water and adjusted to pH 3. The protein is then heated to 95–100° C for 5 min to remove proteolytic contaminants, then cooled and the denatured contaminants removed by filtering or centrifugation. RNase is precipitated by addition of ammonium sulphate, and then is crystallized—first with ammonium sulphate, and finally in dilute ethanol.

When examined by column chromatography, crystalline ribonuclease shows one major and several minor components. Some of the minor components occur naturally (they are glycoproteins of differing sugar content), but others are probably artefacts of purification or storage. The major fraction, called ribonuclease-A, was the second protein and first enzyme for which the primary structure was determined. Two laboratories (Anfinsen and his colleagues; Hirs, Stein and Moore and their coworkers) were involved in the determination of the sequence of the 124 amino acid residues (Fig. 7). RNase-A, like lysozyme, has four dithiol bridges. These are formed between residues 26–84, 40–95, 58–110 and 65–72. The molecule has ten acidic residues (5 Asp and 5 Glu) and 14 basic residues (10 Lys and 4 Arg). There are four histidines (His-12, His-48, His-105 and His-119), and two of these are present at the catalytic site. From the sequence, a molecular weight of 13 680 has been calculated for anhydrous RNase-A.

Ribonuclease-A has been modified by limited proteolytic digestion in the presence of subtilisin at pH 8 and $3°$ C (Richards and Vithayathil, 1959). In the product, RNase-S, only one peptide bond of RNase-A has been cleaved, usually the 20–21 bond, but also to a

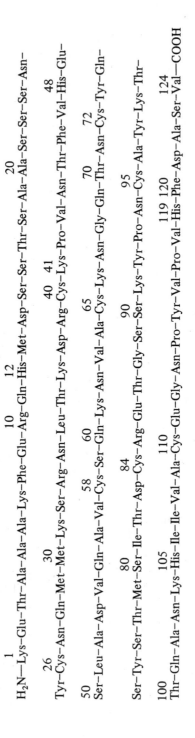

Fig. 7. Primary structure of bovine pancreatic ribonuclease-A. From Smyth *et al.* (1963).

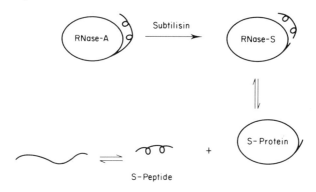

Fig. 8. Diagram showing the formation and reconstitution of ribonuclease-S. RNase-S is formed by cleavage of the peptide bonds between amino acid residues 20 and 21 of RNase-A. The two portions of RNase-A are held together by non-covalent bonds. RNase-S can be separated into S-peptide and S-protein. During reconstitution, the random-coil form of the S-peptide must be converted into the helical structure it possesses in RNase-S before being bound to the S-protein. Adapted from Richards and Wyckoff (1971).

limited extent the 21–22 bond is cleaved. RNase-S has the same catalytic activity as RNase-A but differs in a few kinetic properties. RNase-S can be separated into an S-peptide (residues 1–20) and S-protein (residues 21–124) by fractionation with trichloroacetic acid (Richards and Vithayathil, 1959) or by column chromatography on Sephadex G-25 (Gross and Witkop, 1966). Neither the S-peptide nor the S-protein has ribonuclease activity. When separated from the S-protein, the S-peptide loses its secondary structure. RNase-S can be reconstituted by mixing equimolar amounts of S-peptide and S-protein at neutral pH. In the reconstituted RNase-S' form, the S-peptide has regained its α-helical structure and essentially all catalytic activity is recovered. The interconversions of ribonuclease-S are summarized in Fig. 8.

These experiments with RNase-S demonstrate several principles of the structure of enzymes. First, the S-peptide can spontaneously regain its secondary structure when it forms the RNase-S complex with the S-protein. The importance of reversible denaturation is discussed below, with the description of denaturation studies with RNase-A. Second, the importance and the strength of non-covalent bonds in proteins is exhibited by the tight binding of the S-peptide to the S-protein. RNase-S and RNase-S , which seem identical, can be chromatographed on ion-exchange columns or crystallized without loss of the S-peptide. Finally, the catalytic site has been shown to involve amino acids present in distal parts of the primary structure. As we shall see below, His-12 (in the S-peptide) and His-119 (in the S-protein) are essential residues.

Ribonuclease-A can be denatured to what is probably a random coil by reduction of the four dithiol bonds with 2-mercaptoethanol in a solution of 8 M urea. The fully reduced RNase-A can be re-oxidized in air to slowly form a structure having the same enzyme activity and secondary and tertiary structure as RNase-A (White, 1961). An enzyme has been isolated from rat liver microsomes which accelerates the reactivation of the reduced RNase (DeLorenzo et al., 1966). This enzyme catalyzes the rearrangement of unnatural pairs of half-cystine residues to produce the correct dithiol bonds. The driving force for these rearrangements is the formation of the more stable conformation of the native tertiary structure. The ability of the polypeptide chain of RNase to spontaneously form the unique native conformation shows:

1. that the native conformation of the enzyme is the most stable form under physiological conditions (the lowest free energy of the kinetically accessible structures); and
2. that all the information needed for the biosynthesis of proteins is contained in the DNA's genetic code, which directs the formation of the primary structure.

The secondary and tertiary structures are determined by the primary structures of proteins, with enzymes folding into their active shape by co-operative formation of the many non-covalent bonds (and the correct $-S-S-$ bonds). In a review on the topic of protein folding, Wetlaufer and Ristow (1973) have commented that this type of self-assembly is a major function of any protein.

As discussed already (p. 26), the complete chemical synthesis of ribonuclease has been accomplished. Material having RNase-A activity (85 mg) was synthesized by the solid-phase method, and microgram quantities of RNase-S' were generated in the classical approach. The success of these syntheses proves that the completely denatured ribonuclease polypeptide can be renatured.

Crystallographic studies with both ribonuclease-A (Kartha et al., 1967; Carlisle et al., 1974) and ribonuclease-S (Wyckoff et al., 1967,

1970) have been reported. Structures have been deduced for the complexes of RNase-S with several substrate-related compounds. RNase-A and RNase-S are quite similar in conformation. RNase-S has been much more completely examined to date. However, there are still several regions in its structure where the position of the peptide backbone is not clearly defined. Ribonuclease has the same gross shape (globular or actually kidney shaped) and dimensions as lysozyme (approximately 38–45 Å on the shortest dimension). It has three lengths of α-helix and a long β-sheet region formed between residues 71–92 and 94–110. The active site is in a crevice formed by this sheet. Both the N- and C-terminal ends are folded close to the crevice. The active-site region includes the residues 11 and 12, 41–45 and 119–123. Note that these regions are close to each other in the tertiary structure, but distant in the primary structure. The chain containing the bond broken in the conversion of RNase-A to RNase-S is in a very exposed position, and is therefore susceptible to rapid proteolytic attack. The S-peptide is in an α-helical form in the crystals, and is bound to the S-protein by hydrophobic bonds.

Many chemical modifications have been performed on ribonuclease. Of the many modifications which inactivate the enzyme, those which seem of catalytic importance are substitution of Lys-41 and of His-12 and His-119. The ε-amino group of Lys-41 reacts rapidly with fluorodinitrobenzene. This reaction is prevented by the addition of competitive inhibitors of RNase. However, the function of this amino group is not yet known. Photo-oxidation of ribonuclease in the presence of methylene blue caused the oxidation of histidine residues and a complete loss of activity after three residues had been modified (Weil and Seibles, 1955). That about half of the activity was lost when 0.5 mol histidine was photo-oxidized suggested that destruction of only one histidine per molecule was sufficient to eliminate activity. With RNase-S, photo-oxidation of either of the isolated fragments caused a loss of activity (Richards, 1958). Since His-12 is the only histidine in the S-peptide, it must be involved in catalysis, along with a histidine residue in the S-protein. Later photo-oxidation studies with the S-protein showed that His-119 was essential (Kenkare and Richards, 1966). Barnard and Stein (1959) found that treatment with bromoacetate at pH 7 inactivates RNase by alkylating one of the four histidine residues. They identified the site of carboxymethylation as the histidine closest to the C-terminal

(His-119). Histidine in native ribonuclease can react with iodoacetate at pH 5.5–6, causing inactivation (Gundlach et al., 1959), but neither denatured RNase nor free histidine is alkylated by iodoacetate under similar conditions. A mixture of 1-carboxymethyl-His-119-RNase and 3-carboxymethyl-His-12-RNase is formed (Crestfield et al., 1963). Alkylation of either His-119 (the major site of alkylation) or His-12 prevents substitution of the other active-site histidine. The modification studies suggest that both His-12 and His-119 are cata-lytically essential and, as verified by the X-ray experiments, are close to each other in the active site.

A pK_a value for each of the histidine residues has been determined by n.m.r. spectroscopy in D_2O at 32° C (Meadows et al., 1968). Four peaks from the C^2 hydrogens of the four imidazole rings (numbered as peaks 1–4) are observed in the n.m.r. spectrum of RNase-A. Spectrophotometric titration of these four peaks showed that they have pK_a values of 6.7, 6.2, 5.8 and 6.4, respectively. Peak 4 showed an anomalous spectrum, and was assigned to the buried histidine, His-48. With carboxymethyl-His-12-RNase-A, both peaks 2 and 3 were altered, indicating they must be the active-site residues. This leaves peak 1 as His-105. The greatest difficulty was to identify which signal was due to His-12 and which to His-119. RNase-S was prepared and the C^2 hydrogen of the S-peptide was slowly and completely exchanged to a C—D bond by incubation for 5 days in D_2O. Then RNase-S' was formed by addition of S-protein, and only a peak at position 2 was absent when the n.m.r. spectrum was recorded. Therefore, peak 2 initially was assigned to His-12 and peak 3 to His-119. More recent experiments have verified the assignments of the pK_a values for His-48 and His-105, but have led to reversal of the values for His-12 and His-119 (Markley, 1975). (It appears that phosphate in the RNase-S' prepared by Meadows et al., 1968, altered the spectral position of peak 2.) Therefore, the active-site histidine residues are now assigned the pK_a values of His-12 = 5.8 and His-119 = 6.2.

Of several mechanisms proposed for the action of ribonuclease, that suggested in 1962 by Rabin (Findlay et al., 1962) is most con-sistent with chemical and physical experiments. Several elaborations or variations of this mechanism have been proposed (reviewed by Roberts et al., 1969; Richards and Wyckoff, 1971) but the details of the mechanism are not yet entirely clear. The Rabin mechanism

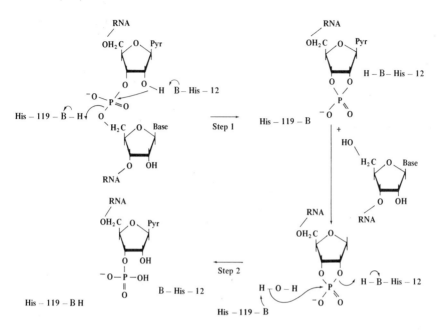

Fig. 9. Outline of the mechanism postulated for the hydrolysis of a phospho-diester bond in the presence of ribonuclease.

involves the two active-site histidines as acid–base catalysts. One, thought to be His-12, is a general base in the formation of the cyclic phosphate, and the other (His-119) is a general acid in this step (Fig. 9). The imidazole group of His-12 removes a proton from the 2′-oxygen of the ribose, allowing attack on the phosphorus atom by this nucleophilic oxygen and formation of a pentacovalent phosphorus intermediate. This intermediate then accepts a proton from the imidazole of His-119 to form a cyclic phosphate and the 5′-CH_2OH leaving group. These roles for His-12 and His-119 are compatible with the three-dimensional models of RNase-S : nucleotide complexes. Step 2 can be thought of as the reverse of step 1, but with water replacing the leaving group. Although it has been suggested that each step in the RNase-catalyzed hydrolysis occurs by concerted catalysis (cf. pp. 152–153), with proton donation by His-12 simultaneous with proton removal by His-119, there is no direct evidence to support this hypothesis. Neither is there any direct evidence yet in favour of stepwise proton transfer.

The thermodynamics of binding of mononucleotides to RNase-A have been measured by calorimetry (Flogel et al., 1975). Combining these results with other data on RNase, Biltonen has proposed that the protein provides a positively charged environment at the catalytic site that stabilizes a dianionic penta-co-ordinated phosphate intermediate. He has estimated that the electrostatic stabilization contributes a rate factor of between 5×10^2 and 10^6. Combined with proximity of the substrate to the catalytic residues, the electrostatic effect is sufficient to explain the rate of hydrolysis of $2',3'$-cyclic nucleotides catalyzed by RNase.

TRIOSEPHOSPHATE ISOMERASE

There are two stages in glycolysis: the formation of two molecules of glyceraldehyde-3-phosphate from one of glucose, and the energy-producing conversion of glyceraldehyde-3-phosphate to lactate. Triosephosphate isomerase catalyzes the last reaction in the first

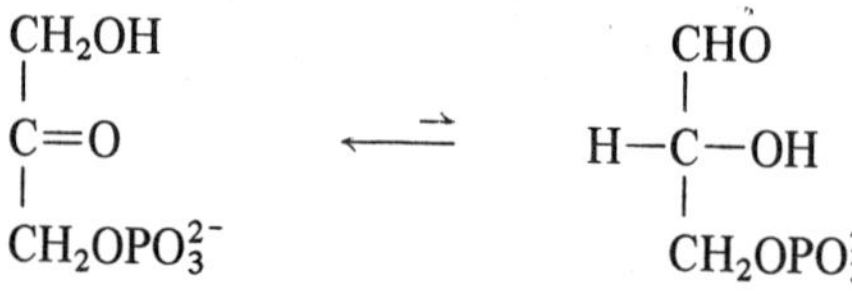

$$
\begin{array}{ccc}
\mathrm{CH_2OH} & & \mathrm{CHO} \\
| & & | \\
\mathrm{C{=}O} & \longleftarrow \overset{\longrightarrow}{} & \mathrm{H{-}C{-}OH} \\
| & & | \\
\mathrm{CH_2OPO_3^{2-}} & & \mathrm{CH_2OPO_3^{2-}}
\end{array}
$$

Dihydroxyacetone   Glyceraldehyde−3−phosphate
phosphate

stage, the interconversion of dihydroxyacetone phosphate and glyceraldehyde-3-phosphate. Triosephosphate isomerase is renowned for its extremely high catalytic activity. The turnover number in the forward direction is well in excess of $10^5 \mathrm{min}^{-1}$. Noltmann (1972) refers to the isomerase as "perhaps the 'fastest' enzyme in intermediary metabolism". The equilibrium of this aldo–keto isomerization favours the formation of dihydroxyacetone phosphate (>96%).

There has been a recent flurry of activity in studying the mechanism of triosephosphate isomerase, stimulated by the possibility of crystallographic examination of the isomerase–dihydroxyacetone phosphate Michaelis complex. This would be the first true ES complex studied by crystallography. The triosephosphate isomerase from chicken muscle has already been found to be suitable for high resolution X-ray experiments, and the electron density maps at 6 Å resolution (Banner *et al.*, 1971) and 2.5 Å (Banner *et al.*, 1975) have been calculated. Each subunit of the triosephosphate isomerase is roughly spherical. Amino acid side chains are distributed in a manner similar to that observed with other globular proteins, i.e. most non-polar residues are shielded from the solvent. The presumed active-site region of each subunit is a pocket which contains the Glu-165 residue discussed below. Crystals of the enzyme–dihydroxyacetone phosphate complex which are isomorphous with the native enzyme have been obtained and are being examined for the position and orientation of the substrate molecules.

At present, most research is being conducted using the triosephosphate isomerases from chicken breast muscle and rabbit skeletal muscle, although the enzyme also has been purified from several other tissues (reviewed by Noltmann, 1972). Multiple forms of the rabbit muscle isomerase have been detected by electrophoresis by several investigators. Krietsch *et al.* (1971) separated three major and two minor components by chromatography on DEAE-cellulose. Because the isomerase from chicken muscle exhibits only one peak on electrophoresis, it was chosen for the crystallographic studies. Two simple purification schemes have been published for the isolation of pure triosephosphate isomerase from chicken muscle (Putnam *et al.*, 1972; McVittie *et al.*, 1972). Both allow purification with high yields using a few well-chosen purification steps (extraction, fractionation with ammonium sulphate, column chromatography and precipitation with ammonium sulphate).

Triosephosphate isomerase is a dimeric enzyme, with a molecular weight of about 48 000–54 000 for the dimer, or 24 000–27 000 for each of the two seemingly identical subunits (Noltmann, 1972; McVittie *et al.*, 1972). Corran and Waley (1973) have determined the sequence of the rabbit muscle enzyme. It contains 248 amino acids in each polypeptide chain, from which a molecular weight of 53 257 has been calculated for the dimer. The only variant forms they observed were due to partial deamidation of Asn and Gln residues. It is not yet clear if these forms are isoenzymes or artefacts of isolation. Each subunit contains five free cysteine residues and no dithiol bridges. In the chicken muscle enzyme, each subunit has one reactive but non-essential thiol group and three inaccessible thiols per subunit (Putnam *et al.*, 1972). The chicken muscle enzyme shows a high degree of homology to the rabbit muscle isomerase (87% of the residues are in identical positions) and possesses 247 residues per chain. There is no evidence from either modification experiments or kinetic examination, to suggest that there is co-operativity between the subunits of the chicken muscle enzyme.

Triosephosphate isomerase requires no cofactors whatsoever, and catalyzes a reaction involving only one substrate and one product. The elucidation of the probable mechanism of its action preceded the structural studies of the enzyme by quite a few years. The enzyme-catalyzed reaction has a non-enzymic counterpart in the base-catalyzed Lobry de Bruyn–Alberda van Ekenstein transformation (Speck, 1958), an alkaline isomerization which proceeds by an enolization:

$$
\begin{array}{ccccccc}
\text{HC}=\text{O} & & \text{HC}-\text{O}^- & & \text{HC}-\text{OH} & & \text{CH}_2\text{OH} \\
| & -\text{H}^+ & \| & & \| & +\text{H}^+ & | \\
\text{HCOH} & \rightleftharpoons & \text{C}-\text{OH} & \rightleftharpoons & \text{C}-\text{O}^- & \rightleftharpoons & \text{C}=\text{O} \\
| & & | & & | & & | \\
\text{R} & & \text{R} & & \text{R} & & \text{R}
\end{array}
$$

When dihydroxyacetone phosphate is incubated in tritiated water with triosephosphate isomerase, approximately 1 mol tritium is incorporated into the C^3 position of the substrate, indicating that

$$
\begin{array}{ccc}
\text{CH}_2\text{OH} & & \text{CHTOH} \\
| & \text{T}_2\text{O} & | \\
\text{C}=\text{O} & \xrightleftharpoons{} & \text{C}=\text{O} \\
| & \text{Isomerase} & | \\
\text{CH}_2\text{OPO}_3^{2-} & & \text{CH}_2\text{OPO}_3^{2-}
\end{array}
$$

hydrogen removal and addition is stereospecific (Bloom and Topper, 1956; Rieder and Rose, 1959). When the tritiated dihydroxyacetone phosphate is converted to 3-phosphoglycerate via glyceraldehyde-3-phosphate in the presence of the glyceraldehyde-3-phosphate dehydrogenase system, very little tritium is transferred to the C^2 of the product (Rieder and Rose, 1959; Knowles *et al.*, 1971). On the basis of this type of evidence, Rieder and Rose proposed a mechanism involving removal of one specific proton from the C^3 of dihydroxyacetone phosphate by a basic group on the enzyme with formation of an enzyme-bound enediol intermediate:

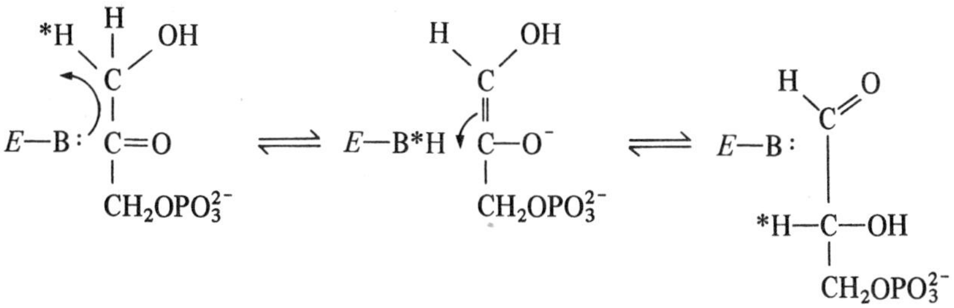

$$*H \overset{H}{\underset{C}{\diagdown \mid \diagup}} OH \qquad H \overset{}{\underset{C}{\diagdown \diagup}} OH \qquad H \overset{}{\underset{C}{\diagdown}} O$$

E—B: $\overset{\diagup}{C}$=O $\rightleftharpoons$ E—B*H $\overset{\diagup}{C}$—O⁻ $\rightleftharpoons$ E—B:

$\text{CH}_2\text{OPO}_3^{2-}$ $\text{CH}_2\text{OPO}_3^{2-}$ *H—C—OH

$\text{CH}_2\text{OPO}_3^{2-}$

Enediol

The intermediate can accept a proton at C^2 from the same basic group, which is now protonated, to form glyceraldehyde-3-phosphate. The low level of tritium transferred from C^3 to C^2 indicates that hydrogen transfer is not intramolecular (i.e. not a hydride transfer). The proton bound to the basic group of the enzyme presumably can exchange with water more rapidly than it can transfer the proton to the enediol intermediate.

The general base of the enzyme which accepts the proton from the substrate appears to be an extraordinarily reactive glutamate carboxyl. The first indication that glutamate is involved came from Hartman's use of chlorohydroxyacetone phosphate as an irreversible inhibitor of the isomerase and the identification of a substituted glutamate in the active site peptide (Hartman, 1971). Several other laboratories have since used active site-directed irreversible inhibitors to obtain similar results. The active-site residue is Glu-165 in the rabbit muscle triosephosphate isomerase (Corran and Waley, 1973), and occurs in a peptide containing the same Ala—Tyr—*Glu*—Pro—Val—Trp sequence as in the chicken muscle enzyme (de la Mare *et al.*, 1972).

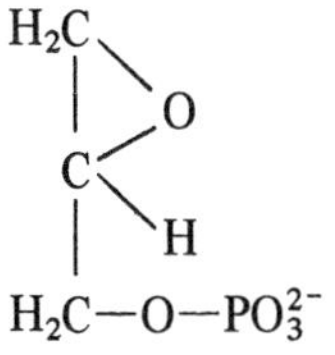

$$\text{H}_2\text{C} \diagdown$$
$$\mid \quad \diagdown \text{O}$$
$$\text{C} \diagdown$$
$$\mid \quad \diagdown \text{H}$$
$$\text{H}_2\text{C}—\text{O}—\text{PO}_3^{2-}$$

Glycidol phosphate
(2, 3–Epoxypropanol phosphate)

By measuring the rate of inactivation by the irreversible inhibitor glycidol phosphate, Waley (1972) has estimated that the pK_a of Glu-165 is about 6. This is in agreement with pH–initial velocity experiments done in both reaction directions with the chicken muscle isomerase (Plaut and Knowles, 1972). In contrast, Schray *et al.* (1973) have demonstrated that the rate of alkylation by glycidol phosphate under different reaction conditions is independent of pH over the range 5.5–10. This result suggests that the glutamate residue when alkylated does not have a pK_a value in this range. Using 3-chloroacetol sulphate, Hartman *et al.* (1975) have shown that the rabbit muscle isomerase is inactivated at a constant rate between pH values 5 and 8. The more stable yeast enzyme was tested below pH 5, and the pH dependence of its inactivation by chloroacetol sulphate indicates a pK_a of 3.9 for the group being inactivated.

As well as a basic γ-COO⁻ at the catalytic site, there appears to be an acidic (electrophilic) group, too (Rose, 1975; Webb and Knowles, 1974). This group, probably an amino residue, has not yet been identified. The enediol, single-base, single-electrophile mechanism proposed may be summarized as:

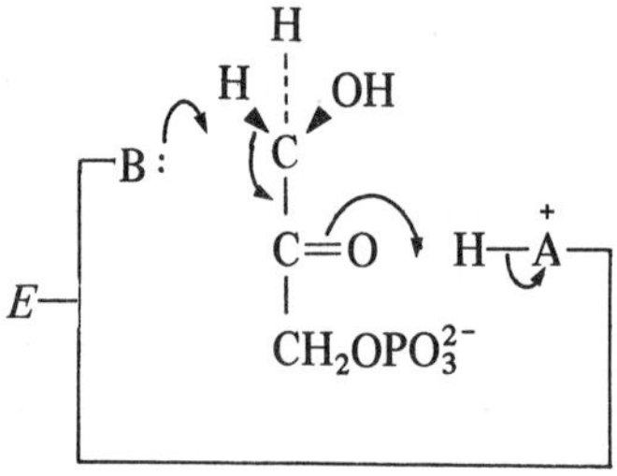

Knowles and his coworkers (1971) have used kinetic isotope studies to measure the rates of the various steps in the isomerase-catalyzed interconversion of dihydroxyacetone phosphate (DHAP) to glyceraldehyde-3-phosphate (GAP) (both forward and reverse reactions). These steps are shown below in a Cleland-type kinetic diagram:

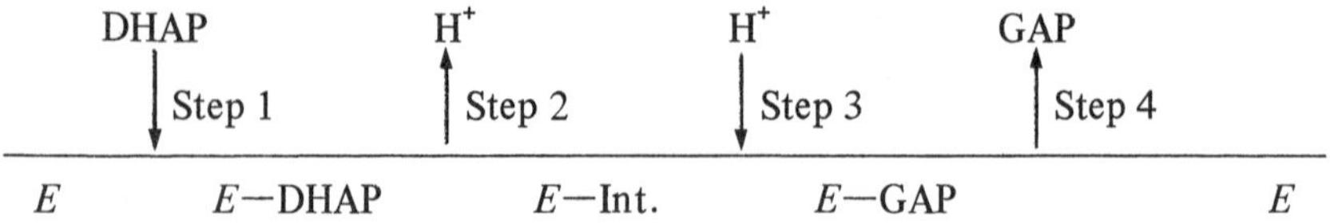

They used coupled enzymic assay systems to trap the product of the reaction, preventing any recycling in the presence of the isomerase. When the conversion of DHAP to GAP is performed in tritiated water under these irreversible conditions, there is appreciable radioactivity recovered in unreacted DHAP. This indicates that there is an obligatory intermediate, the enzyme-bound enediol (shown here as $E-$Int.), which can exchange with solvent. The protonation of the intermediate is not rate limiting, because k_H/k_T was only 1.3. This suggests that the release of GAP from the $E-$GAP complex is the rate-determining step in the forward direction.

In the reverse direction, using GAP and tritiated water, the specific activity of the product is only 11% of that of the solvent. This isotopic discrimination indicates that the rate of protonation of the enzyme-bound intermediate is slower than the rate of loss of DHAP from $E-$DHAP. The primary isotope effects using specifically deuterated 1-^{2}H-DHAP and 2-^{2}H-GAP were determined to be 3.0 : 1.0, respectively. In the forward direction, this is less than that expected (approximately 7) if deprotonation is the rate-limiting step. Therefore, proton abstraction from DHAP must be slightly faster than the release of GAP from the isomerase. The lack of a primary isotope effect in the reverse direction is as expected if the rate is determined by the rate of conversion of $E-$Int. to $E-$DHAP, the reverse of step 2. This is because the ^{2}H$^+$ from the labelled GAP has exchanged with ^{1}H$^+$ from the ^{1}H$_2$O solvent. The rate constant for binding of GAP to the enzyme has been calculated as $4 \times 10^8\,\text{M}^{-1}\text{s}^{-1}$, i.e. it is diffusion controlled. These results have been incorporated into a free energy–reaction co-ordinate profile for the reaction, and compared to that for the uncatalyzed enolization of DHAP and GAP (Hall and Knowles, 1975). The enzyme increases the rate of enolization by a factor of about 10^9. The highest energy barrier in the enzymic reaction has been found to be the diffusion of GAP to the enzyme, so the enzyme has reached maximal effectiveness as a catalyst!

Triosephosphate isomerase utilizes many of the catalytic effects described in Chapter 6. The enolization rate is catalyzed by proximity of the substrate to two appropriately oriented acid–base groups. The nucleophile and electrophile are in positions defining the stereospecificity of the reaction. Knowles's calculations demonstrate that the enzyme-bound transition state is stabilized more than that for the non-enzymic enolization. The reaction pathway is altered by the enzyme from a two-step process to a four-step process, with the energy barriers being quite similar for each of the four steps. The actual chemical catalysis steps have been so well refined during evolution that the limiting steps are the binding and release of a reactant.

REFERENCES

Alderton, G. and Fevold, H. L. (1946). *J. biol. Chem.* **164**, 1-5.

Banerjee, S. K., Holler, E., Hess, G. P. and Rupley, J. A. (1975). *J. biol. Chem.* **250**, 4355-4367.

Banner, D. W., Bloomer, A. C., Petsko, G. A. and Phillips, D. C. (1971). *Cold Spring Harb. Symp. quant. Biol.* **36**, 151-155.

Banner, D. W., Bloomer, A. C., Petsko, G. A., Phillips, D. C., Pogson, C. I., Wilson, I. A., Corran, P. H., Furth, A. J., Milman, J. D., Offord, R. E., Priddle, J. D. and Waley, S. G. (1975). *Nature, Lond.* **255**, 609-614.

Barnard, E. A. and Stein, W. D. (1959). *Biochem. J.* **71**, 19P-20P.

Blake, C. C. F., Koenig, D. F., Mair, G. A., North, A. C. T., Phillips, D. C. and Sarma, V. R. (1965). *Nature, Lond.* **206**, 757-761.

Blake, C. C. F., Mair, G. A., North, A. C. T., Phillips, D. C. and Sarma, V. R. (1967a). *Proc. R. Soc. B.* **167**, 365-377.

Blake, C. C. F., Johnson, L. N., Mair, G. A., North, A. C. T., Phillips, D. C. and Sarma, V. R. (1967b). *Proc. R. Soc. B.* **167**, 378-388.

Bloom, B. and Topper, Y. J. (1956). *Science, N.Y.* **124**, 982-983.

Canfield, R. E. (1963). *J. biol. Chem.* **238**, 2698-2707.

Carlisle, C. H., Palmer, R. A., Mazumdar, S. K., Gorinsky, B. A. and Yeates, D. G. R. (1974). *J. molec. Biol.* **85**, 1-18.

Corran, P. H. and Waley, S. G. (1973). *FEBS Letters* **30**, 97-99.

Crestfield, A. M., Stein, W. H. and Moore, S. (1963). *J. biol. Chem.* **238**, 2413-2428.

Dahlquist, F. W., Rand-Meir, T. and Raftery, M. A. (1969). *Biochemistry* **8**, 4214-4221.

de la Mare, S., Coulson, A. F. W., Knowles, J. R., Priddle, J. D. and Offord, R. E. (1972). *Biochem. J.* **129**, 321-331.

DeLorenzo, F., Goldberger, R. F., Steers, E., Jr., Givol, D. and Anfinsen, C. B. (1966). *J. biol. Chem.* **241**, 1562-1567.

Dunn, B. and Bruice, T. C. (1973). *Adv. Enzymol.* **37**, 1-60.

Eshdat, Y., Dunn, A. and Sharon, N. (1974). *Proc. natn. Acad. Sci. U.S.A.* **71**, 1658-1662.

Fevold, H. L. (1951). *Adv. Protein Chem.* **6**, 230.

Findlay, D., Herries, D. G., Mathias, A. P., Rabin, B. R. and Ross, C. A. (1962). *Biochem. J.* **85**, 152-153.

Flogel, M., Albert, A. and Biltonen, R. (1975). *Biochemistry* **14**, 2616-2621.

Ford, L. O., Johnson, L. N., Machin, P. A., Phillips, D. C. and Tjian, R. (1974). *J. molec. Biol.* **88**, 349-371.

Gross, E. and Witkop, B. (1966). *Biochem. biophys. Res. Commun.* **23**, 720-723.
Gundlach, H. G., Stein, W. H. and Moore, S. (1959). *J. biol. Chem.* **234**, 1754-1760.
Hall, A. and Knowles, J. R. (1975). *Biochemistry* **14**, 4348-4352.
Hartman, F. C. (1971). *Biochemistry* **10**, 146-154.
Hartman, F. C., LaMuraglia, G. M., Tomozawa, Y. and Wolfenden, R. (1975). *Biochemistry* **14**, 5274-5279.
Imoto, T. and Rupley, J. A. (1973). *J. molec. Biol.* **80**, 657-667.
Imoto, T., Johnson, L. N., North, A. C. T., Phillips, D. C. and Rupley, J. A. (1972). *In* "The Enzymes" (P. D. Boyer, ed.), Third edition, Vol. 7, pp. 665-868. Academic Press, New York.
Jollès, P. (1960). *In* "The Enzymes" (Boyer, P. D., Lardy, H. and Myrbäck, K., eds), Second edition, Vol. 4, pp. 431-445. Academic Press, New York.
Jollès, J., Juarequi-Adell, J., Bernier, I. and Jollès, P. (1963). *Biochim. biophys. Acta* **78**, 668-689.
Kartha, G., Bello, J. and Harker, D. (1967). *Nature, Lond.* **213**, 862-865.
Kenkare, U. W. and Richards, F. M. (1966). *J. biol. Chem.* **241**, 3197-3206.
Knowles, J. R., Leadlay, P. F. and Maister, S. G. (1971). *Cold Spring Harb. Symp. quant. Biol.* **36**, 157-164.
Krietsch, W. K. G., Pentchev, P. G. and Klingenburg, H. (1971). *Eur. J. Biochem.* **23**, 77-85.
Kunitz, M. and McDonald, M. R. (1953). *Biochem. Prep.* **3**, 9-19.
Kuramitsu, S., Ikeda, K., Hamaguchi, K., Fujio, H., Amano, T., Miwa, S. and Nishina, T. (1974). *J. Biochem.* **76**, 671-683.
Lin, T.-Y. and Koshland, D. E., Jr. (1969). *J. biol. Chem.* **244**, 505-508.
Markley, J. L. (1975). *Accts Chem. Res.* **8**, 70-80.
McVittie, J. D., Esnouf, M. P. and Peacocke, A. R. (1972). *Eur. J. Biochem.* **29**, 67-73.
Meadows, D. H., Jardetzky, O., Epand, R. M., Ruterjans, H. H. and Scheraga, H. A. (1968). *Proc. natn. Acad. Sci. U.S.A.* **60**, 766-772.
Noltmann, E. A. (1972). *In* "The Enzymes" (P. D. Boyer, ed.), Third edition, Vol. 6, pp. 271-354. Academic Press, New York.
Parsons, S. M. and Raftery, M. A. (1969). *Biochemistry* **8**, 4199-4205.
Parsons, S. M. and Raftery, M. A. (1972). *Biochemistry* **11**, 1623-1643.
Plaut, B. and Knowles, J. R. (1972). *Biochem. J.* **129**, 311-320.
Putnam, S. J., Coulson, A. F. W., Farley, I. R. T., Riddleston, B. and Knowles, J. R. (1972). *Biochem. J.* **129**, 301-310.
Rees, A. R. and Offord, R. E. (1972). *Biochem. J.* **130**, 965-968.
Richards, F. M. (1958). *Proc. natn. Acad. Sci. U.S.A.* **44**, 162-166.
Richards, F. M. and Vithayathil, P. J. (1959). *J. biol. Chem.* **234**, 1459-1465.
Richards, F. M. and Wyckoff, H. W. (1971). *In* "The Enzymes" (P. D. Boyer, ed.), Third edition, Vol. 4, pp. 647-806. Academic Press, New York.
Rieder, S. V. and Rose, I. A. (1959). *J. biol. Chem.* **234**, 1007-1010.
Roberts, G. C. K., Dennis, E. A., Meadows, D. H., Cohen, J. S. and Jardetzky, O. (1969). *Proc. natn. Acad. Sci. U.S.A.* **62**, 1151-1158.
Rose, I. A. (1975). *Adv. Enzymol.* **43**, 491-517.

Roxby, R. and Tanford, C. (1971). *Biochemistry* **10**, 3348-3352.

Schray, K. J., O'Connell, E. L. and Rose, I. A. (1973). *J. biol. Chem.* **248**, 2214-2218.

Secemski, I. I., Lehrer, S. S. and Lienhard, G. E. (1972). *J. biol. Chem.* **247**, 4740-4748.

Sharp, J. J., Robinson, A. B. and Kamen, M. D. (1973). *J. Am. Chem. Soc.* **95**, 6097-6108.

Smith, L. E. H., Mohr, L. H. and Raftery, M. A. (1973). *J. Am. Chem. Soc.* **95**, 7497-7500.

Smyth, D. G., Stein, W. H. and Moore, S. (1963). *J. biol. Chem.* **238**, 227-234.

Speck, J. C., Jr. (1958). *Adv. Carbohydrate Chem.* **13**, 63-103.

Sykes, B. D., Patt, S. L. and Dolphin, D. (1971). *Cold Spring Harb. Symp. quant. Biol.* **36**, 29-33.

Thomas, E. W., McKelvy, J. F. and Sharon, N. (1969). *Nature, Lond.* **222**, 485-486.

Timasheff, S. N. and Rupley, J. A. (1972). *Archs Biochem. Biophys.* **150**, 318-323.

van Eikeren, P. and Chipman, D. M. (1972). *J. Am. Chem. Soc.* **94**, 4788-4790.

Vernon, C. A. (1967). *Proc. R. Soc.* B. **167**, 389-401.

Waley, S. G. (1972). *Biochem. J.* **126**, 255-256.

Webb, M. R. and Knowles, J. R. (1974). *Biochem. J.* **141**, 589-592.

Weil, L. and Seibles, T. S. (1955). *Archs Biochem. Biophys.* **54**, 368-377.

Wetlaufer, D. B. and Ristow, S. (1973). *A. Rev. Biochem.* **42**, 135-158.

White, F. H., Jr. (1961). *J. biol. Chem.* **236**, 1353-1360.

Wyckoff, H. W., Hardman, K. D., Allewell, N. M., Inagami, T., Johnson, L. N. and Richards, F. M. (1967). *J. biol. Chem.* **242**, 3984-3988.

Wyckoff, H. W., Tsernoglou, D., Hanson, A. W., Knox, J. R., Lee, B. and Richards, F. M. (1970). *J. biol. Chem.* **245**, 305-328.

8 Coenzymes

In Chapter 1, I defined coenzymes as substances participating in enzymatic group transfer reactions by acting as transient carriers of mobile metabolic groups. The groups transferred from one metabolite to another range in size from electrons to large organic units. The traditional experimental approaches used in studying coenzymes are given in a review article by Huennekens (1956). He described coenzymes as being generally heat-stable, dialyzable, organic compounds required by some enzymes. Coenzymes alone can be catalysts, but their physiological catalysis is expressed only in conjunction with enzyme proteins. The classical proof of a coenzyme role for a compound is the resolution of the holoenzyme into its coenzyme and apoenzyme components, and the reconstitution of activity by addition to the apoenzyme of the suspected coenzyme to reform holoenzyme. Although I have divided coenzymes into cosubstrates and prosthetic groups, this distinction is arbitrary and depends mainly on the affinity of the apoenzyme for the coenzyme. Some coenzymes, such as biotin, or the flavin of succinic dehydrogenase, are covalently bound in their respective holoenzymes, while others such as pyridine nucleotides are not covalently bound and have K_m values sufficiently high to be considered as reaction substrates. As a rule of thumb, the K_m for a low molecular weight coenzyme varies as a function of its cellular concentration. Those coenzymes which are used in the fewest reactions (e.g. the vitamin B_{12} coenzymes in mammalian cells) occur in the lowest concentrations and have the smallest K_m values.

Even weakly bound coenzymes can be considered to be part of the active site of a coenzyme-requiring enzyme. With many dehydrogenases, NADH or NAD$^+$ must be bound to the apoenzyme before substrate can bind.* This is an indication that the enzyme and the

* In no instance has an ordered mechanism been found in which the substrate binds before NAD$^+$ or NADH.

coenzyme together are responsible for the observed catalysis. With this rationale that coenzymes are portions of the catalytic site, much effort has been applied to the examination of the chemical properties of coenzymes. The majority of the information obtained relates to the structures of the intermediate forms involved in group transfer by coenzymes. The detailed mechanisms of catalysis and the functional groups of the apoenzyme participating in the reactions are still to be discovered in many cases. However, the way in which coenzymes react covalently with the metabolic groups they transfer provides a great deal of the information we have about enzyme mechanisms.

In this chapter, I will talk about some of the low molecular weight coenzymes, and in the next about protein coenzymes. Table I is a list of a series of low molecular weight organic biochemicals that can be considered to be coenzymes. I have classified these coenzymes into group donors (strict cosubstrates), group-transferring coenzymes (prosthetic groups or coenzymes which bind several forms of the mobile groups) and oxidation–reduction coenzymes. Ten of these coenzymes are either B-vitamins (the water-soluble group of vitamins) or derivatives of B-vitamins. The B-vitamins are compounds, obtained from liver, plant or bacterial tissue, which cannot be synthesized by most mammals. They must be ingested in the diet, and usually are converted by the mammal into the coenzyme. This conversion, referred to as activation of the vitamin, often requires a reaction with ATP. Deficiency of a B-vitamin in an animal's diet may cause a characteristic disease, and the metabolic malfunctions observed in these vitamin deficiency diseases have indicated which metabolic roles are played by the individual B-vitamin coenzymes.

Here, I will describe four types of coenzymes—the pyridine nucleotides, flavins, pyridoxal phosphate and pterins—in some detail. Biotin and pantetheine are discussed in the next chapter, and the cobalamin (vitamin B_{12}) coenzymes in Chapter 11. A comprehensive description of coenzymes and their functions is given in Mahler and Cordes (1971) and should be consulted for information on those coenzymes not covered by this book.

PYRIDINE NUCLEOTIDES AND DEHYDROGENASES

The first coenzymes to be discovered were the pyridine (or nicotin-

Table I. Some low molecular weight coenzymes.

Coenzyme	Group donated or general function
A. *Group Donors*	
Adenosine-5'-phosphate (ATP)	Phosphate, pyrophosphate, tripolyphosphate, adenosyl, adenylyl; energy source
Phosphoguanidine compounds (e.g. phosphocreatine)	Phosphate
5-Phosphoribosyl-1-pyrophosphate (PRPP)	Ribose-5-phosphate
3'-Phosphoadenosine-5'-phosphosulphate (PAPS)	Sulphate
Uridine nucleotides (e.g. uridine diphosphate glucose)	Sugars
Cytidine nucleotides (e.g. cytidine diphosphate choline)	Alkyl phosphates
S-Adenosyl methionine	Methylation
Coenzyme M (2,2'-dithiodiethane sulphonate)	Methylation
B. *Group-Transferring Coenzymes*	
Pyridoxal phosphate[a]	Amino acid transformations
Coenzyme A[a]	Acyl groups
Tetrahydrofolate[a]	One-carbon groups
Biotin[a]	Carboxylation
Thiamine pyrophosphate (TPP)[a]	$-\overset{\mid}{\underset{\mid}{C}}-\overset{O}{\overset{\parallel}{C}}-$ cleavage
Polyprenol phosphate	Peptidoglycan components
Methyl cobalamin[a]	Methylation
Adenosyl cobalamin[a]	Rearrangement reactions
C. *Oxidation–Reduction Coenzymes*	
Pyridine nucleotides[a] (NAD$^+$, NADP$^+$)[a]	–
Flavin coenzymes[a] (FMN, FAD)[a]	–
Coenzyme Q (Ubiquinone)	–
Heme coenzymes	Electrons
Lipoate[a]	Acyl groups and hydrogen
Glutathione	–
Tetrahydrobiopterin	Reductant in hydroxylations

[a] These coenzymes are either B-vitamins or they have been derived from a B-vitamin.

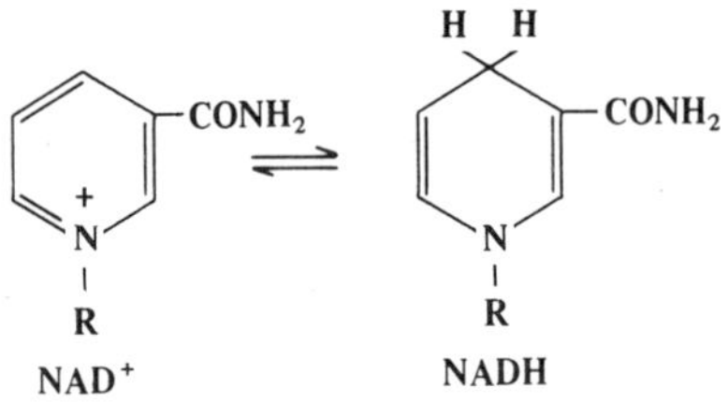

Fig. 1. The structure of nicotinamide adenine dinucleotide (NAD^+) is shown in (a). Nicotinamide adenine dinucleotide phosphate ($NADP^+$) has an additional phosphate group esterified to the $2'$-position of the adenosine moiety. For expediency, the structures of the oxidized and reduced pyridine nucleotide coenzymes are often abbreviated as shown in (b), with R representing adenosyl diphosphate ribose.

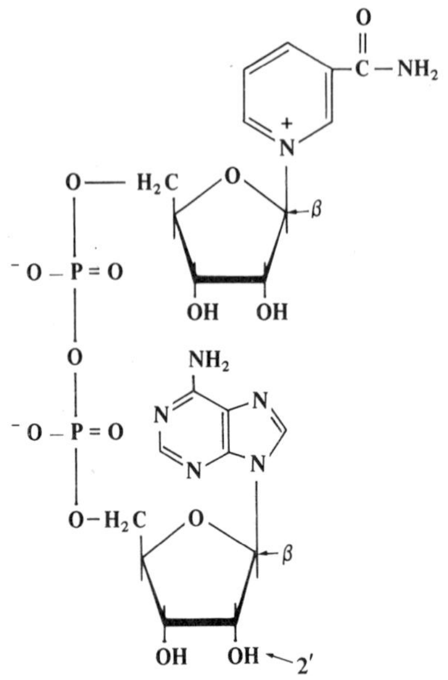

amide) nucleotides. These coenzymes are required to assist in oxidation–reduction reactions catalyzed by many dehydrogenases. The more abundant of the two coenzymes, nicotinamide adenine dinucleotide (NAD^+), was detected in 1904 as a heat-stable cofactor required for fermentation by yeast extracts. The oxidized forms of NAD^+ and $NADP^+$ (nicotinamide adenine dinucleotide phosphate) were purified and characterized in the 1930s in the laboratories of von Euler and Warburg. Both coenzymes contain one molecule of nicotinamide—the amide of the anti-pellagra vitamin, niacin or nicotinic acid—bound to ribose-5'-phosphate by a β-glycosidic bond to form nicotinamide mononucleotide (NMN) and a molecule of adenosine-5'-phosphate linked by a phosphodiester bond to the NMN (Fig. 1). NADP⁺ has an additional phosphomonoester group located at the 2'-position of the adenosine ribose. The α-nicotinamide analogues of NAD⁺ and NADP⁺ are inactive in enzymatic reactions. In dehydrogenation reactions, the pyridine nucleotides accept one hydrogen atom and an electron from the reduced substrate $(A H_2)$ and the second hydrogen removed from the substrate appears as a proton:

$$A H_2 + \text{NAD}^+ (\text{or NADP}^+) \rightleftharpoons A + \text{NADH} (\text{or NADPH}) + \text{H}^+.$$

The reduced forms of the coenzymes, NADH and NADPH, are stable and isolable compounds at neutral pH values. In general, NADH is formed in catabolism and is oxidized to yield energy, and NADPH is utilized for anabolic reductions.

The fundamental properties of the pyridine nucleotide coenzymes are described in reviews by Kaplan (1960), Bruice and Benkovic (1966) and Sund (1968). Five different systems of nomenclature have been used for NAD⁺ and NADP⁺, as listed below in chronological order:

Archaic	Cozymase	Phospho-cozymase
Archaic	Codehydrogenase I	Codehydrogenase II
Archaic	Coenzyme I	Coenzyme II
Current	Diphosphopyridine nucleotide (DPN⁺)	Triphosphopyridine nucleotide (TPN⁺)
Current	Nicotinamide adenine dinucleotide (NAD⁺)	Nicotinamide adenine dinucleotide phosphate (NADP⁺).

The names on the last line of the list are recommended and are used by most biochemists, but some major investigators still use the DPN⁺-TPN⁺ nomenclature.

Warburg's research during the mid 1930s showed that the nicotinamide ring was reversibly reduced in the action of NAD⁺ and NADP⁺, but it was 20 years later that the structure of the reduced form was elucidated. Until then, the 1,6-dihydro pyridine structure was favoured, from studies with model compounds. In the early 1950s, two sets of experiments provided the most definitive data we have on

H

the mechanism of oxidation and reduction of the pyridine nucleotides. One showed that the transfer of hydrogen to the nicotinamide ring of the coenzyme is direct and stereospecific, and the other demonstrated that the site of reduction was C^4.

Fisher *et al.* (1953), using alcohol dehydrogenase from yeast, showed that hydrogen was transferred directly from substrate to NAD^+. No isotope was incorporated into NADH when the oxidation of ethanol was performed in D_2O, but one atom of carbon-bound deuterium per molecule was transferred from C^1 labelled ethanol to NAD^+:

$$CH_3CD_2OH + NAD^+ \rightarrow CH_3CDO + NADD + H^+.$$

This reduction of NAD^+ had produced a new centre of asymmetry at the carbon accepting the deuterium. When the NADD was isolated and used to reduce acetaldehyde, the deuterium was quantitatively removed from the NADD:

$$CH_3CHO + NADD + H^+ \rightarrow CH_3CHDOH + NAD^+.$$

Since there are two hydrogens at the site of reduction in NADD (one protium and one deuterium), there must have been only one stereoisomer formed in the enzymic reduction of the NAD^+, because the same deuterium was removed in the enzymic reoxidation of the coenzyme. When chemically labelled reduced coenzyme was reoxidized in the same reaction system, only part of the deuterium was transferred to the alcohol. It has been found that dehydrogenases can be classified into two types (listed in Popják, 1970): those catalyzing the transfer of hydrogen to the A side of the nicotinamide ring, as does alcohol dehydrogenase, and those transferring hydrogen to the B side, to form the other stereoisomer.* The absolute configurations of the A and B forms of NADH have been deduced by Cornforth *et al.* (1966). Both the A- and B-deutero isomers of NADH were degraded by a procedure which converted their carbon atoms into succinic acid without altering the stereochemistry at C^4. When the two succinates were compared to a synthetic monodeuterosuccinate of known stereochemistry, it was concluded that A-specific dehydrogenases add hydrogen to the forward face of the nicotinamide ring as it is usually drawn, and B-specific dehydrogenases to

---

* X-ray crystallographic experiments with dehydrogenases indicate that the A- and B-side specificity is related to a $180°$ rotation about the nicotinamide—ribose glycosidic bond.

the face below the plane of the paper:

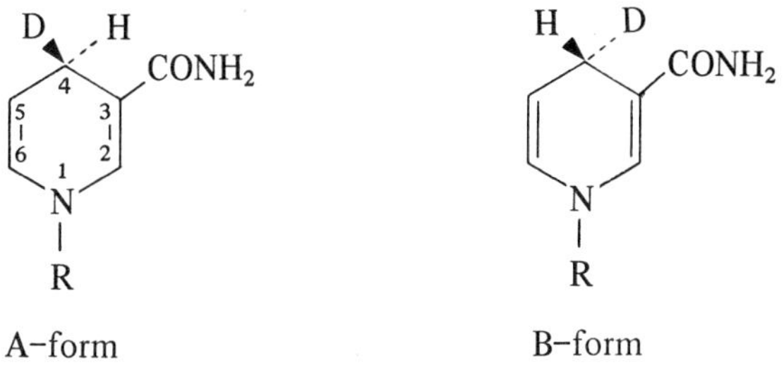

A-form   B-form

For a complete discussion of the stereochemistry of the reduction of NAD^+, I would refer the readers to the review by Popják (1970). The position of the nicotinamide ring which accepts the hydrogen during reduction was shown to be C^4 by Pullman *et al.* in 1954, in experiments based on the deuterium transfers done by Westheimer's group (Fisher *et al.*, 1953). NAD^+ was reduced in D_2O using sodium dithionite, and the labelled mixture of A- and B-forms was oxidized in the alcohol dehydrogenase reaction to form NAD^+ still containing some deuterium. The NAD^+ was enzymatically cleaved to nicotinamide, methylated with CH_3I, and the methyl nicotinamide oxidized with alkaline ferricyanide to the 2- and 6-pyridones (Fig. 2). In the formation of the two pyridones, there was no loss of deuterium. Therefore, since reduction at the *meta* position is unlikely and oxidation of the *ortho* positions had not caused loss of deuterium, the reduced form of NAD^+ must be the *para* or 1,4-dihydro pyridine

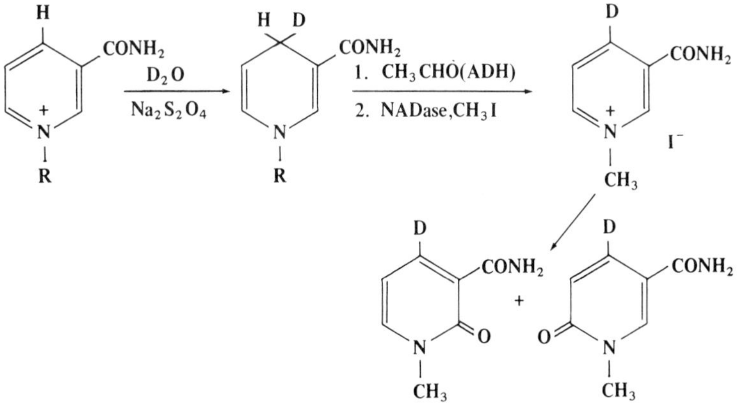

Fig. 2. Summary of the reactions used to show that NADH is a 1,4-dihydro nicotinamide derivative (Pullman *et al.*, 1954).

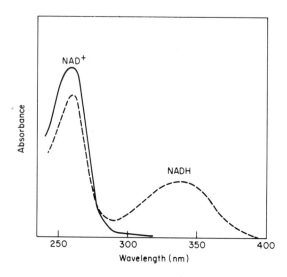

Fig. 3. Ultraviolet absorption spectra of NAD$^+$ and NADH over the pH range of 2–10.

compound. This structural assignment has since been verified by n.m.r. spectroscopy.

The absorption spectra of the oxidized and reduced forms of the pyridine nucleotide coenzymes (Fig. 3) are useful in experimental work. NAD$^+$ (and NADP$^+$) has a maximal absorption at 260 nm, due to absorption by adenine and by nicotinamide. Upon reduction, there is a decrease in absorbance at 260 nm and appearance of a band centred at 340 nm, from the formation of 1,4-dihydronicotinamide. This 340-nm band is used for many enzymatic assays, either direct assays for dehydrogenases or for enzymes which can be coupled to NADH-dependent enzymes (cf. p. 112). Although the extinction coefficients generally accepted are $\epsilon_{260} = 17 \times 10^3\,\mathrm{M}^{-1}\mathrm{cm}^{-1}$ for NAD$^+$ and $\epsilon_{340} = 6.22 \times 10^3\,\mathrm{M}^{-1}\mathrm{cm}^{-1}$ for NADH, measurements using highly purified materials show that these values should be 17.7×10^3 and 6.56×10^3, respectively (Silverstein, 1965; Gurr *et al.*, 1972). When NADH binds to some enzymes, its absorption maximum at 340 nm shifts to a lower wavelength so that the extent of formation of an enzyme–NADH complex can be measured. For example, with alcohol dehydrogenase, the maximum changes to 325 nm (cf. the use of this change on p. 138). NADH (but not NAD$^+$) fluoresces in the 440–460 nm region when excited at either 260 or

340 nm. This fluorescence also has been used for both kinetic experiments and to measure the binding of NADH to dehydrogenases.

A number of chemical reactions of pyridine compounds appear to be closely related to the reduction of NAD^+ and $NADP^+$. This is especially true for the addition of nucleophiles (X^-) to the coenzymes or to N-methyl nicotinamides:

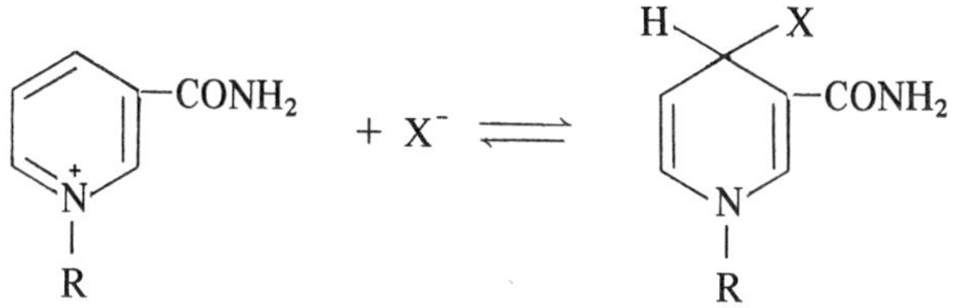

Most anions react with C^4 of the nicotinamide ring. When cyanide is mixed with NAD^+, an adduct having an absorption maximum at 327 nm is formed (Kaplan, 1960). Bisulphite reacts with NAD^+ in a similar manner. Borohydrides react with NAD^+ to give a mixture of 1,2-, 1,4- and 1,6-dihydronicotinamides.

The most convenient reducing agent for non-enzymic reduction of the pyridine nucleotide coenzymes is sodium dithionite, $Na_2S_2O_4$. NAD^+ is reduced to the active 1,4-NADH by dithionite via a transient yellow intermediate which can be trapped in alkaline solution. Yarmolinsky and Colowick (1956) suggested that this intermediate is a sulphinate adduct of NAD^+, a fact verified in 1966 by n.m.r. studies of the crystalline adduct (Biellmann and Callot, 1966; Caughey and Schellenberg, 1966). Dithionite consists of two SO_2^- fragments joined by a long and weak $-S-S-$ bond, which readily cleaves homolytically. Although it has been suggested that dithionite reduces NAD^+ by a free radical mechanism (Biellmann and Callot, 1968), it seems more likely that it reacts by nucleophilic addition. Kaplan's laboratory (Anderson et al., 1959) found that with NAD^+ and nine analogues there was a correlation between the ability of either cyanide or bisulphite to add to a nicotinamide compound and of dithionite to reduce the compound. As further support for a reduction involving nucleophilic addition, Kawai and Scrimgeour (1972) have shown that the formation of the yellow adduct is first-order with respect to $S_2O_4^{2-}$, not half-order as would be expected if an SO_2^- radical were reacting with NAD^+. Although dithionite can reduce many compounds by the one-electron pathway (cf. Lambeth and

Palmer, 1973), it reacts with nicotinamides by either direct addition of $HS_2O_4^-$ or a protonated sulphoxylate ion (HSO_2^-) formed from dithionite to the 4-position:

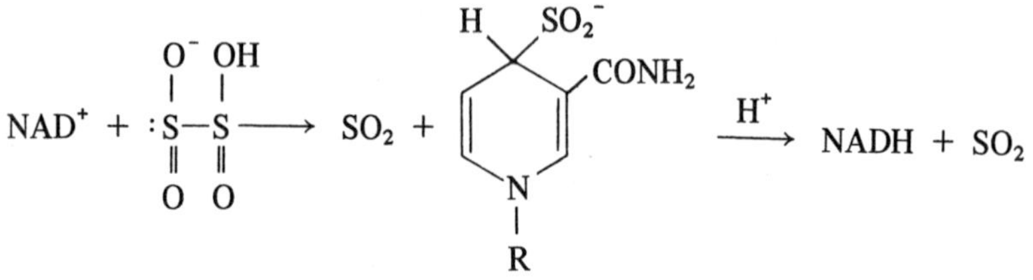

$$NAD^+ + \ :\!\!\overset{O^- \ OH}{\underset{O \ \ O}{S\!-\!S}}\!\! \longrightarrow \ SO_2 + \ \ \ \xrightarrow{H^+} \ NADH + SO_2$$

In a final step, the sulphinate group is replaced by a proton by hydrolysis.

Because of the number of non-enzymic reactions of NAD^+ and $NADP^+$ with nucleophiles and because there is little evidence to support free radical reactions of nicotinamides, most workers presume that oxidation and reduction involves the transfer of the equivalent of a hydride ion (i.e. a unit of two electrons and a proton). How the hydride equivalent—not necessarily a hydride ion itself—is transferred is still unknown. Hamilton (1971) has suggested several possible ionic mechanisms, stressing those which could be assisted by general acid–base catalysis. Of his mechanisms, the PHy (proton-hydride or $H^+ + H^-$) mechanism would have a chemical analogy in the nucleophilic addition reactions. The PP ($H^+ + H^+ + 2\ e$) mechanism, with both hydrogens transferred as protons and the two electrons transmitted by some other mechanism, is attractive as a general enzymic route because of its relation to many acid- and base-catalyzed reactions. Two possible ways of transferring two protons in a PP mechanism are by a covalent intermediate formed from the reductant and the substrate (PPC mechanism) and by a mechanism (PPM) which involves a metal ion. Reduction of NAD^+ by dithionite, involving a covalent intermediate, may be considered as a crude analogue of a PPC mechanism. Hamilton considered that either the PHy or PPC mechanism is likely for the oxidation of alcohols by NAD^+. The PHy pathway could be aided by the loss of the proton from the alcohol oxygen, with a resulting attraction between the alkoxide and the nicotinamide nitrogen:

$$PHy: \quad \begin{array}{c} \text{H} \\ \diagdown \diagup \\ \text{C} \\ \diagup \diagdown \\ \text{O} \\ | \\ \text{H} \end{array} \quad + \quad \begin{array}{c} \text{H} \quad \text{C}-\text{NH}_2 \\ \parallel \\ \text{O} \\ \diagdown \\ \overset{+}{\text{N}} \\ | \\ \text{R} \end{array} \quad \underset{\longleftarrow}{\overset{-\text{H}^+}{\rightleftharpoons}} \quad \begin{array}{c} \diagdown \\ \text{C}=\text{O} \\ \diagup \end{array} \quad + \quad \begin{array}{c} \text{H} \quad \text{H} \quad \text{C}-\text{NH}_2 \\ \parallel \\ \text{O} \\ \diagdown \\ \text{N} \\ | \\ \text{R} \end{array}$$

In lactate dehydrogenase, the PHy pathway applies and the imidazole of a histidine residue appears to be the general base (p. 216). Several model systems, such as reduction of the ketone trifluoromethyl-phenylcarbinol (Steffens and Chipman, 1971) and of a substituted methyl pyrimidine (Creighton *et al.*, 1973), proceed with direct hydrogen transfer from dihydronicotinamide to substrate, and kinetic experiments indicate that an intermediate of unknown nature is formed. However, studies on the non-enzymatic reduction by NADH of a series of 4-substituted 2,6-dinitrobenzene sulphonates points to a concerted transfer of the hydrogen and the electrons (Kurz and Frieden, 1975). There is now clear evidence from model and enzyme-catalyzed reactions indicating that pyridine nucleotides accept and donate hydrogens in simple dehydrogenase-catalyzed reactions by the PHy (i.e. $H^+ + H^-$) mechanism. In the oxidation of NADH in mitochondria, hydrogen probably is transferred directly from NADH to FMN. The mechanism for this important type of hydrogen transfer reaction is not yet known.

While bio-organic chemists are seeking plausible detailed mechanisms of action for NAD^+ and $NADP^+$, protein chemists are determining the structures of some pyridine nucleotide-dependent dehydrogenases. Sund (1968) has classified these enzymes into simple dehydrogenases, which require only the pyridine nucleotides as coenzymes, and complex dehydrogenases, which require a second coenzyme (e.g. flavin) for activity. Some of the simple dehydrogenases contain zinc as an essential component (cf. p. 140), and most are composed of more than one polypeptide subunit. The molecular weights of the dehydrogenases range from 15 000 for some dihydrofolate reductases to several million for glutamate dehydrogenase.

The three-dimensional structures of four simple dehydrogenases have been determined from high resolution X-ray crystallography experiments. These enzymes are the muscle lactate dehydrogenase from dogfish (Adams *et al.*, 1970; Rossmann *et al.*, 1971; Holbrook *et al.*, 1975), the cytoplasmic malate dehydrogenase from pig heart (Hill *et al.*, 1972; Webb *et al.*, 1973), horse liver alcohol dehydrogenase (Bränden *et al.*, 1973, 1975) and glyceraldehyde-3-phosphate dehydrogenase from lobster (Moras *et al.*, 1975). The first and last consist of four subunits and the other two are dimers. From the data now available, portions of tertiary structure of each of these dehydrogenases are similar, particularly at their coenzyme binding sites (the coenzyme-binding domain or nucleotide-binding unit). Lactate dehydrogenase is the most extensively studied dehydrogenase at present, and will be the only one described here. Its basic catalytic features are also present in the other simple dehydrogenases.

Lactate dehydrogenase (LDH) from dogfish muscle is a tetramer of molecular weight 140 000, having four identical subunits. It catalyzes the NAD^+-dependent interconversion of pyruvate and lactate. Almost all of the sequence of this LDH has been deduced (Taylor *et al.*, 1973; Holbrook *et al.*, 1975). The dehydrogenase has no bound metal and no dithiol bridges. One of the seven cysteine residues in each subunit reacts more readily with sulphydryl reagents (*p*-hydroxymercuribenzoate or iodoacetate) and has been thought essential for activity. Crystallographic studies of several forms of the dehydrogenase—including the apoenzyme, the LDH–NAD^+ binary complex, and the LDH–NAD^+–pyruvate ternary complex—have been extended to resolutions of 2.0 Å for the apoenzyme and of varying resolutions for individual complexes. Each subunit is a compact globulin, except for the 20 N-terminal amino acid residues which form an extended arm. Well over half of the residues participate in either helical or sheet configurations. The active site for each subunit is entirely within that subunit, in a large central cleft in which the adenosine group of NAD^+ is surrounded by hydrophobic residues and the two phosphate groups by charged or hydrophilic side chains. If substrates are added to crystals of the binary complex, or AMP or ADP to crystals of apoenzyme, the crystals shatter, suggesting that conformational changes occur upon binding of substrate and coenzyme. Although n.m.r. spectroscopy has shown that the NAD^+ molecule is in a folded configuration in free solution, the X-ray experiments indicate that it is bound to the cleft of LDH in an open or extended conformation. The cleft has access to solvent through an

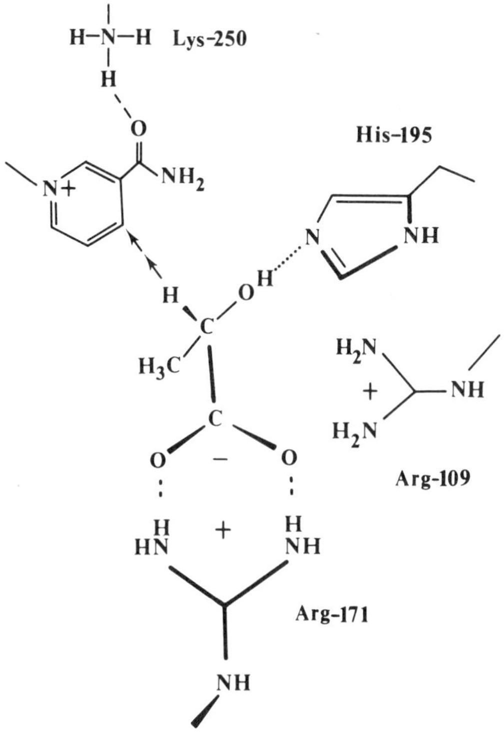

Fig. 4. Proposed structure of the lactate–NAD⁺-LDH active ternary complex catalytic site. Adapted from Adams *et al.* (1973).

opening made by a loop of residues 98–114. The binding of substrates to form the ternary complex causes a movement of this loop to a position over the catalytic centre pocket.

A summary of the sequence and X-ray data on LDH has been written recently (Adams *et al.*, 1973), rationalizing the mechanisms of coenzyme and substrate binding and catalysis. Kinetic studies (p. 132) indicate that the coenzyme binds before the substrate anion. The physical reason for this sequential binding has been discovered. If the loop were permanently in the position it occupies in the ternary complex, the phosphate groups of NAD⁺ or NADH could not approach their binding sites. The conformational change occurring on binding of the substrate to the enzyme–coenzyme complex may orient the nicotinamide so that hydride transfer is possible. It can be seen in the structural model that only side A of the nicotinamide ring can accept or donate hydrogen to the substrate. This is because the carboxamide group is hydrogen bonded to a Lys residue. The substrate is close to three ionizable side chains: the imidazole of His-195 and the guanidino groups of Arg-171 and Arg-109. Figure 4 shows the proposed alignment of L-lactate with these three active-site residues. The carboxylate of lactate or pyruvate is bound to Arg-171. His-195 accepts a proton from the hydroxyl of lactate, in the suggested PHy mechanism. The role of Arg-109 appears to be binding of the anionic substrates. The essential thiol, Cys-165, is not directly involved in binding of substrates, but its modification blocks His-195. Malate dehydrogenase also has a loop which covers its cleft upon binding of substrate, but this loop is not present in alcohol dehydrogenase. It will be intriguing to see the similarities and differences in binding and catalytic mechanisms of these and other dehydrogenases as they are elucidated.

NAD^+, acting as a prosthetic group and not a cosubstrate, is required by several epimerases which catalyze the inversion of configuration of asymmetric carbon atoms of sugar molecules (Glaser, 1972). Examination of one of these enzymes, UDP-galactose-4-epimerase, has shown that enzyme-bound NAD^+ is reduced and reoxidized during the reaction. This epimerase catalyzes the metabolic interconversion of galactose and glucose as their uridine nucleotide derivatives.* That NAD^+ is the transient hydrogen carrier for an intramolecular hydrogen transfer is suggested by:

1. lack of tritium incorporation into the sugar from tritiated water;

2. no loss of tritium from 4-labelled sugar during reaction; and

3. observation of a spectrum similar to that of NADH when the sugar substrate is added to the epimerase–NAD^+ complex.

Nelsestuen and Kirkwood (1971) have inactivated the enzyme by borohydride reduction of the enzyme-bound NAD^+ to NADH. Epimerase–NADH was reactivated by oxidation of the reduced pyridine nucleotide by several UDP-4-ketohexoses, providing strong support for an intramolecular redox mechanism (p. 92) which had

* See Chapter 13, p. 462 for the structure of uridine diphosphate galactose (UDPGal).

been proposed in 1957:

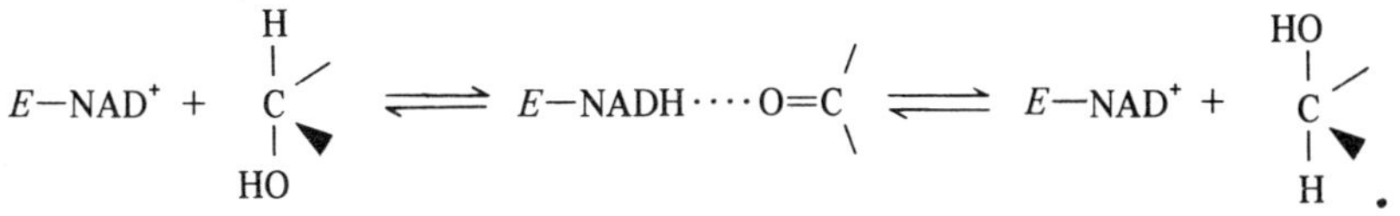

$$E-NAD^+ + \overset{H}{\underset{HO}{C}} \rightleftharpoons E-NADH\cdots O=C \rightleftharpoons E-NAD^+ + \overset{HO}{\underset{H}{C}}$$

UDPG	UDP–4–ketoglucose intermediate	UDPGal

The UDP-4-ketoglucose intermediate must exist in two different forms, with opposite faces of the ketosugar group towards the NADH. The mechanisms of the other NAD$^+$-dependent epimerases have not been clarified yet, but they have many characteristics similar to those of UDP-galactose-4-epimerase, so undoubtedly NAD$^+$ has a similar function with them.

NAD$^+$ is used as a specific substrate, not a coenzyme, in several reactions. Polynucleotide ligase catalyzes the joining of single-stranded breaks in double-helical DNA. The deoxypolynucleotide ends must have a free 3'-hydroxyl adjacent to a 5'-phosphate group. The ligase from E. coli requires NAD$^+$ as an energy source for the synthesis of the new phosphodiester bond (Richardson, 1969). Phage-induced and mammalian ligases use ATP rather than NAD$^+$ as the energy-providing substrate in this reaction. The first step in the joining reaction is the nucleophilic attack on the adenylyl phosphorus of NAD$^+$ by the ϵ-amino group of a lysine residue of the ligase to form an AMP derivative of the enzyme and release NMN (Gumport and Lehman, 1971). The adenylyl group is then transferred to the 5'-phosphate, and finally the 3'-hydroxyl replaces the AMP to complete the formation of the new phosphodiester bond:

1. $E + NAD^+ \longrightarrow E-AMP + NMN$

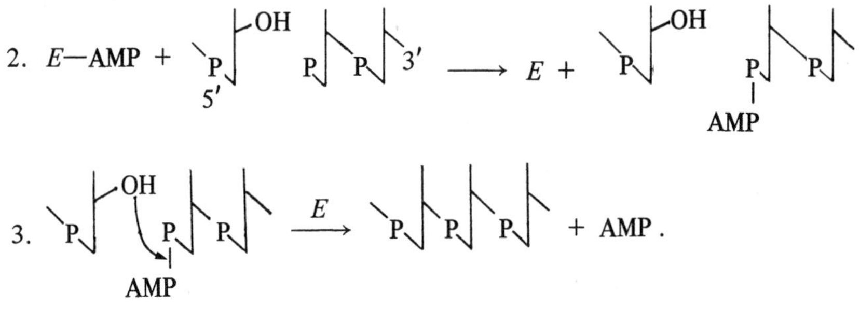

2. $E-AMP + \cdots \longrightarrow E + \cdots$

3. $\cdots \overset{E}{\longrightarrow} \cdots + AMP$.

An entirely different use of NAD^+ is as a substrate for the formation of polyadenosine diphosphate ribose (reviewed by Sugimura, 1973). The chromatin fraction of nuclei of animal cells contains a polymerase (poly(ADP—Rib) synthetase) which catalyzes the following reaction:

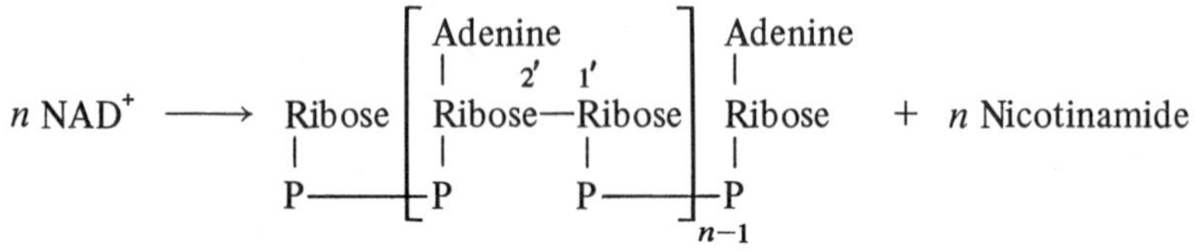

$$n\ NAD^+ \longrightarrow \left[\begin{array}{ccc} \text{Adenine} & & \\ | & 2' & 1' \\ \text{Ribose} & \text{Ribose—Ribose} \\ | & | & | \\ \text{P}\text{———}\text{P} & \text{P}\text{———} \end{array}\right]_{n-1} \begin{array}{c} \text{Adenine} \\ | \\ \text{Ribose} \\ | \\ \text{P} \end{array} + n\ \text{Nicotinamide}$$

The ribose units of this polymer are linked by $1',2'$-glycosidic bonds. Poly ADP-ribose appears to be covalently bound to nuclear proteins through serine phosphate bonds, and the substitution of these proteins may be used for their regulation. Although the number of known nuclear proteins capable of ADP-ribosylation in the presence of NAD^+ and poly(ADP—Rib) synthetase is increasing, the function of poly(ADP—Rib) in chromatin has not been established.

FLAVINS

The two flavin coenzymes, flavin mononucleotide (FMN) and flavin adenine dinucleotide (FAD), are derivatives of riboflavin, vitamin B_2.

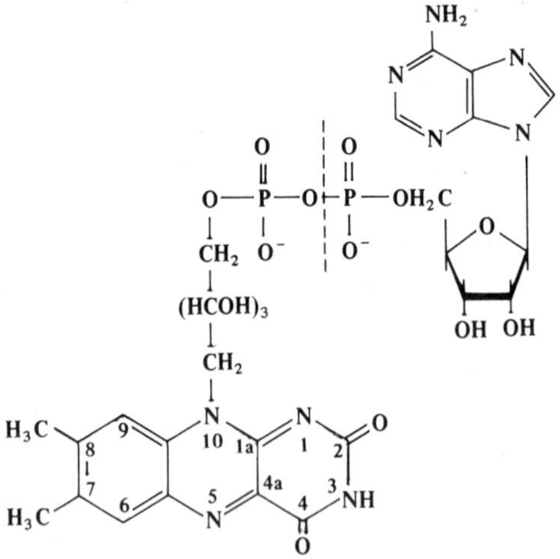

Fig. 5. Structure of flavin adenine dinucleotide (FAD). Flavin mononucleotide (FMN) consists of only the riboflavin-5'-monophosphate portion, shown by the dashed line.

They are prosthetic groups of flavoproteins, enzymes which catalyze oxidation–reduction reactions. Riboflavin contains a fused three-ring structure: a benzene ring, a pyrazine ring and a pyrimidine ring. This isoalloxazine nucleus contains two keto groups in the pyrimidine ring and is substituted with two methyl groups in the benzene ring and the alcohol D-ribitol at the N^{10} position. FMN is the 5-phosphate ester of riboflavin, and FAD contains an additional adenosine-5'-phosphate unit bound by a phosphodiester bond to FMN (Fig. 5).

An excellent appreciation of the chemistry of flavins can be obtained by studying the review written by Beinert in 1960, and the later articles by Ehrenberg and Hemmerich (1968) and Hemmerich *et al.* (1970) which supplement Beinert's review. The two heterocyclic rings of the riboflavin moiety are analogous to a pteridine-2,4-dione. This ring system can be reduced in two reversible steps, with formation first of a free radical semiquinone, and then of the fully reduced 1,5-dihydroflavin hydroquinone:

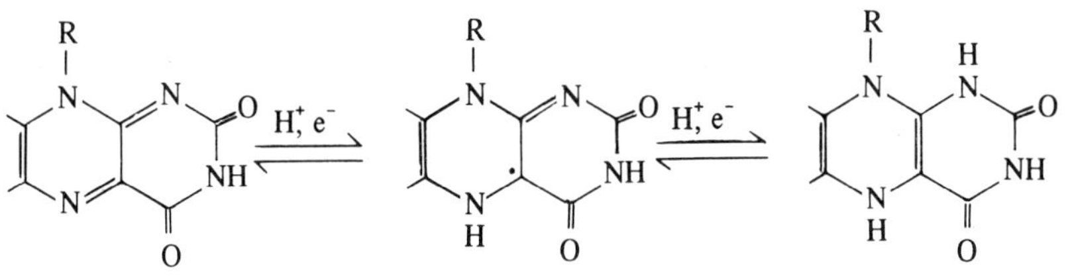

Flavoquinone	Flavosemiquinone	Flavohydroquinone
(FMN or FAD)	(FMNH· or FADH·)	(FMNH$_2$ or FADH$_2$)

The unpaired spin of the free radicals has been shown by e.s.r. experiments to be mainly distributed over the benzene and pyrazine rings, with the highest spin density being at the N^5 position (Müller *et al.*, 1971). The flavins are isolated in the oxidized or quinone form, and exist at that oxidation level under all but reducing conditions. Each oxidation state of the flavin can exist at different pH values in protonated, neutral and anionic forms, each with its characteristic absorption spectrum. Oxidized flavin can be reduced by many reducing agents, and the mechanism of reduction (i.e. whether or not the semiquinone is an intermediate) depends on the nature of the reductant. As we shall see, this versatility of reaction of flavins is of physiological importance.

The experimental approach used in studying the flavin coenzymes has been similar in many ways to that used for the pyridine nucleotide coenzymes, but the biochemistry of the flavins is much more complex. Simple inspection shows that the isoalloxazine ring system is more complicated than the nicotinamide ring. Absorption spectra can be used to advantage in flavoprotein studies, just as in the NAD^+-NADH interconversion, but the existence of three oxidation levels and the rapid autoxidation (spontaneous oxidation by O_2) of the reduced forms make experiments more difficult. The paramagnetism of the semiquinone free radical has enabled investigators to use e.s.r. spectrophotometry successfully, so that the radicals can be measured in the presence of high concentrations of the oxidized and reduced species. The greatest technical handicap for mechanistic studies is that direct transfer of hydrogen from substrate to FMN or FAD cannot be demonstrated because of the exchangeability of the nitrogen-bound protons with water. Because of all these experimental difficulties and the comparative complexity of the isoalloxazine ring, mechanistic interpretations are in a chaotic state at present.

The flavin coenzymes are prosthetic groups of their enzymes, not dissociable cosubstrates, acting as enzyme-bound redox intermediates between substrates (electron donors) and their oxidants (electron acceptors). Substrates for specific flavoproteins include reduced pyridine nucleotides, amino acids, fatty acyl CoA compounds and sugars. These reductants are not capable of reducing FAD or FMN by two one-electron steps, so reduction of the flavoprotein probably follows a PHy or PPC mechanism or a more complicated radical mechanism. It must be proved that any flavosemiquinones observed as possible intermediates in the oxidation of these substrates do not arise from reactions subsequent to the initial dehydrogenation. Compounds accepting electrons from reduced flavins vary with the flavoenzyme, and include O_2, cytochromes, electron-transferring flavoprotein (p. 285) and ferredoxin (p. 274). For some enzyme assays, artificial electron acceptors such as ferricyanide, phenazine methosulphate or dichlorophenol indophenol are used because they undergo an easily measured colour change upon reduction. Just as with reduction, the pathway of reoxidation of the reduced flavin depends on the nature of the reacting oxidizing agent, so it can be either one two-electron step or two one-electron steps.

Studies on the detailed mechanisms of reduction and oxidation of the flavin coenzymes are at much the same level as those with NAD^+, with the technical limitations mentioned above. The structures of the oxidized and reduced forms are now well documented, but the molecular processes by which the two hydrogens are added and removed have not yet been proven. Certainly within the next few years these mechanisms should be clarified, as current research is quite intense.

I will first describe the current state of knowledge on reduction of flavins, and then comment on the mechanisms of reoxidation. The two unsolved problems of flavin reduction are:

1. Do covalent flavin–substrate intermediates occur?
2. What is the site of substrate (or hydrogen) addition?

Homolytic mechanisms have been excluded because with most enzymes no free radical intermediates can be detected and the flavosemiquinone forms of the enzymes are inactive.

Oxidized flavins do not react as readily with most nucleophiles as do the nicotinamide compounds. Müller and Massey (1969) found that flavins can react reversibly with an excess of sulphite (SO_3^{2-}) to form adducts having spectra similar but not identical to 1,5-dihydroflavins. Because those spectra resemble that of a 5-acyl flavin analogue, they concluded that N^5 is the site of nucleophilic addition. However, Hevesi and Bruice (1973) have reported that with a substituted isoalloxazine, in which addition to N^5 is hindered, the C^{4a} sulphite adduct is formed. Therefore, depending on steric factors, either N^5 or C^{4a} adducts can be produced:

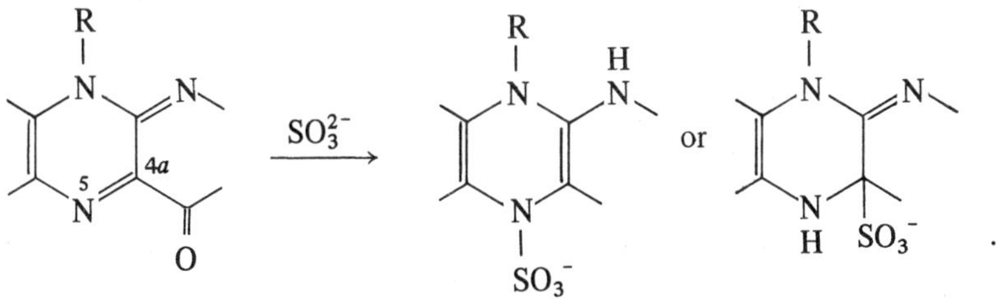

Demonstration of nonphotochemical nucleophilic addition to N^5 and to C^{4a} is relevant to either the PHy mechanism or to the suggestions by Hamilton (1971) and Hemmerich (Hemmerich and Jorns, 1972)

that covalent adducts are formed between substrates and flavo-coenzymes. Hamilton proposed that the alkoxide oxygen of the alcohol added to position C^{4a} of oxidized flavin, but this proposal

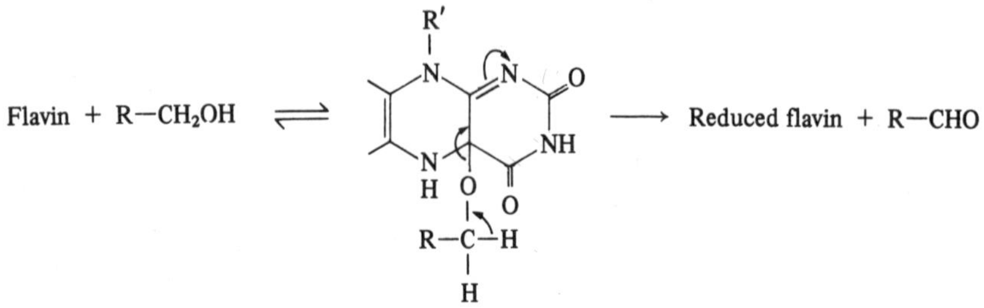

$$\text{Flavin} + \text{R-CH}_2\text{OH} \rightleftharpoons \qquad \qquad \longrightarrow \text{Reduced flavin} + \text{R-CHO}$$

has been criticized on several theoretical grounds. Based on many chemical reactions of flavins, Hemmerich has proposed a proton-carbanion mechanism in which the proton at the α-carbon of the substrate is first removed and the resultant carbanion adds to C^{4a}, then is shifted to N^5 to form a carbinolamine which can release the keto product (as $\text{R}-\overset{+}{\text{C}}\text{H}-\text{OH}$):

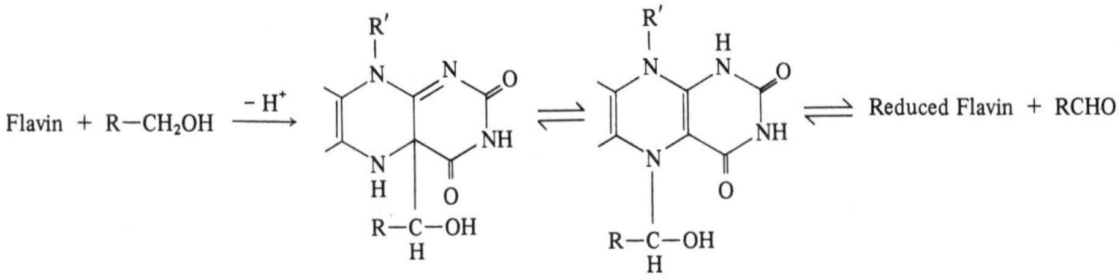

$$\text{Flavin} + \text{R-CH}_2\text{OH} \xrightarrow{-\text{H}^+} \qquad \rightleftharpoons \qquad \rightleftharpoons \text{Reduced Flavin} + \text{RCHO}$$

5-carbinolamine

Shinkai and Bruice (1973) have examined the kinetics of a model reaction, the anaerobic oxidation of a reduced flavin by ethyl pyruvate. They concluded that although a 5-carbinolamine is formed, it does not occur on the pathway to oxidized flavin. Other chemical studies have been performed since, but none of them has further clarified the interaction of alcohols with the $C^{4a}=N^5$ group of oxidized flavins.

The formation of carbanions from substrates during flavoprotein-catalyzed oxidations has been demonstrated in several experimental systems. For example, Walsh et al. (1972) showed that the acetylenic substrate-inactivator 2-hydroxy-3-butynoate forms a covalent adduct with the FMN of a bacterial lactate oxidase. Isotope studies showed that the inhibitor loses its α-hydrogen (i.e. forms a carbanion) in a step preceding adduct formation:

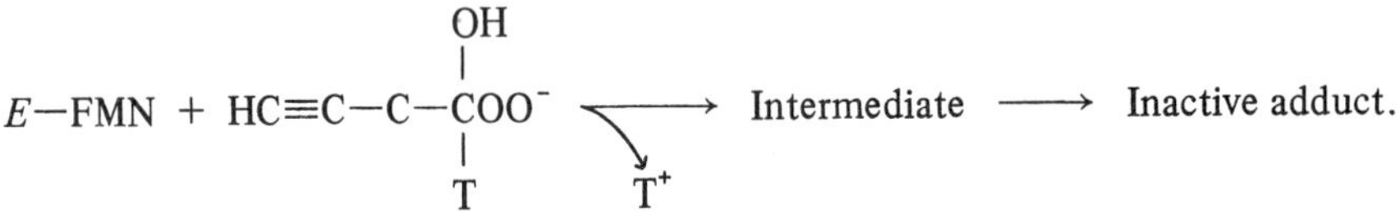

$$E-FMN + HC\equiv C-\underset{\underset{T}{|}}{\overset{\overset{OH}{|}}{C}}-COO^- \longrightarrow Intermediate \longrightarrow Inactive\ adduct.$$

Labelling with tritium on the acetylenic carbon (C^4) was used to show that the site of binding of the inhibitor was the FMN molecule, which was isolated from the denatured enzyme. Porter *et al.* (1973b) have shown that D-amino acid oxidase can oxidize the carbanion form of nitroethane ($H_3C-\bar{C}H-NO_2$). Cyanide can inhibit this oxidation, and in its presence a compound spectrally resembling 5-substituted dihydroflavin accumulates. Note that the formation of a carbanion intermediate has not yet been demonstrated with a natural substrate.

3-Dimethylamino-1-propyne ($HC\equiv C-CH_2-N(CH_3)_2$) is a potent inactivator of the FAD flavoprotein monoamine oxidase. After incubation of the oxidase with the irreversible inhibitor, followed by proteolytic digestion, Maycock *et al.* (1976) isolated a substituted flavin peptide from the inhibited enzyme. The FAD is in the form of an N^5 monosubstituted dihydroflavin, with the adduct having the structure:

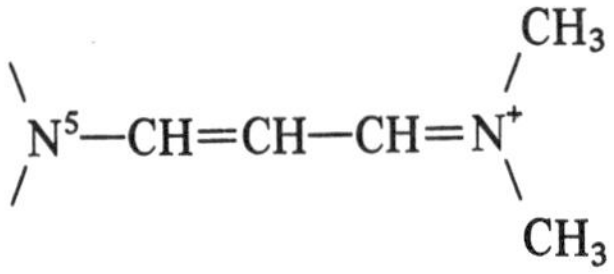

$$\underset{/}{\overset{\backslash}{N^5}}-CH=CH-CH=\overset{+}{N}\underset{\backslash}{\overset{/}{}}\overset{CH_3}{\underset{CH_3}{}}$$

The authors do not favour a carbanion mechanism for the formation of the covalent adduct. Nor do they consider the inhibitor–flavin adduct as support for the formation of N^5 adducts during oxidation, because the carbon of the inhibitor bound to the FAD is not that which normally undergoes oxidase-catalyzed reaction.

To overcome the impossibility of studying the transfer of hydrogen from substrates to flavin coenzymes (remember the N-bound hydrogens exchange with solvent rapidly), several investigators have synthesized and tested coenzymes in which N^5 of the isoalloxazine ring has been replaced by CH. Both deaza-FMN and deaza-FAD have

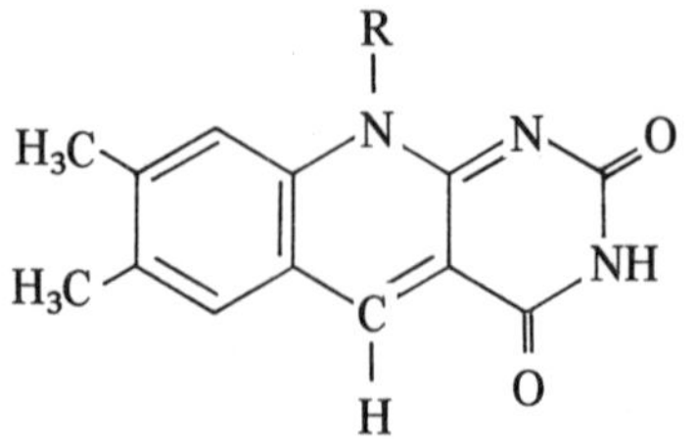

Deazaflavin coenzymes

been examined (Jorns and Hersh, 1974; Fisher and Walsh, 1974; Hersh and Jorns, 1975). Reduced deazaflavin coenzymes have been found to act as competent intermediates in flavoprotein-catalyzed oxidation reactions that have been studied. When substrates bearing tritium labelling at the appropriate positions are used, the hydrogen is transferred directly to the deazaflavin molecule. The reduced deazaflavins are more stable to autoxidation than are $FMNH_2$ and $FADH_2$, so they can be isolated. After isolation and aerobic reoxidation of ^3H-deaza-$FMNH_2$, non-exchangeable tritium remains in the oxidized deazaflavin. This suggests that there is a primary isotope effect in the non-enzymic oxidation, and that the tritium is bound at C^5 (Averill et al., 1975). No substrate–flavin covalent adducts have been detected when deazaflavins have replaced flavins as prosthetic groups, despite the expectation that they would accumulate because the C—C bond of a C^5 adduct would be stable. Results of all these experiments are consistent with stereospecific hydrogen transfer to position 5, suggesting a similar process with natural flavin coenzymes.

The mechanism of reduction of FMN and FAD by NADH is still an enigma. Few definitive experiments have been reported to distinguish between three possible mechanisms: a PHy mechanism with hydride addition at N^5 (or C^{4a}); a PPC covalent adduct mechanism; or charge–transfer complex formation between NADH and flavin. Brüstlein and Bruice (1972) synthesized a deazaflavin in which N^5 was replaced by a carbon atom. When this compound was reduced non-enzymatically by NADH in D_2O, no deuterium was bound to C^5, so that the hydrogen that was transferred there came directly from the 4-position of the NADH. This experiment has been criticized by Hemmerich (Hemmerich and Jorns, 1972) on the basis that N^5 is an essential component of flavins, and that substitution of a carbon for N^5 makes the deazaflavin a nicotinamide, not a flavin, model. However, the 5-deazaflavin coenzymes described above have been used as oxidants in several enzymic reactions including oxidation of ^3H-NADH, so Bruice's deazaflavin model system is applicable. Both N^5 and C^{4a} appear to be strongly electrophilic, and it is difficult to make the choice between these possible sites of hydrogen addition. Position 5 is the more probable site. Evidence is accumulating that complexes, possibly of the charge-transfer type, are formed between reduced nicotinamides and oxidized flavins (e.g., Porter *et al.*, 1973a; Blankenhorn, 1975), but the structures of the complexes are still to be determined. Complexes of this type have been detected as catalytic intermediates bound to several flavoproteins. No matter what the exact mechanism, a direct hydrogen transfer via N^5 is likely.

In summary, it appears that several mechanisms may exist for the reduction of FMN and FAD. These include carbanion formation for substrates with the structure

$$R-\overset{\displaystyle |}{\underset{\displaystyle |}{C}}-H,$$

with occurrence of intermediate covalent adducts at N^5 or C^{4a} and non-covalent hydrogen transfer such as that from NADH. It is not yet known how correct or extensive these mechanisms are, or whether different mechanisms (postulated but not yet tested) are utilized by enzymes.

There are several pathways possible for the reoxidation of reduced flavin coenzymes. Whereas both oxidation and reduction of pyridine nucleotides probably occur by a two-electron reaction, flavin redox reactions can occur by either one-electron or two-electron mechanisms. This versatility of the flavo-coenzymes is a result of the potential existence of the free radical form as a transient species, and allows FMN and FAD to join two-electron reductions to one-electron oxidations. For example, direct oxidation of NADH by the respiratory chain is not possible, but flavin intermediates—reduced by NADH—can be reoxidized by ferric atoms (either cytochromes or FeS proteins) via the semiquinone. Therefore, the flavin coenzymes are the pivotal chemicals that connect the obligatory two-electron and one-electron segments of the mitochondrial respiratory chain (Chapter 14, Fig. 6). Metalloflavoproteins also probably function as links between two-electron substrates and one-electron oxidants. In fact,

Rajagopalan and Handler (1968) have called these enzymes miniature electron-transport systems. Palmer and Massey (1968) have commented that there is no general pathway of catalysis by flavoproteins. Some enzymes seem to shuttle between the oxidized and fully reduced levels without detectable flavosemiquinone formation, some involve oxidation–reduction by transfer of only one electron (e.g. $FAD \rightleftharpoons FADH\cdot$ or $FADH\cdot \rightleftharpoons FADH_2$) and others utilize all three oxidation levels of the flavin. Examples of these three types of flavoproteins are glucose oxidase, flavodoxin and NADH dehydrogenase, respectively.

With several flavoproteins that use molecular oxygen as their electron acceptor, an enzyme-bound intermediate has been detected when O_2 is added to the reduced flavoprotein. An enzyme-bound intermediate postulated to be a C^{4a} peroxide of $FMNH_2$ has been isolated by Hastings et al. (1973) after oxidation of the reduced flavin in the presence of bacterial luciferase. These workers used an ethylene glycol–water mixture as solvent so that the enzyme-intermediate complex could be trapped and then separated from free flavin at $-20°$ C. The absorption spectrum of the luciferase–$FMNH_2$–O_2 complex (maximum absorption at 372 nm) is consistent with the hypothesis that $FMNH_2$ is oxygenated at position 4a. Entsch et al. (1974) have detected a similar intermediate (λ_{max} (approx.) = 385 nm) by stopped-flow spectrophotometry when O_2 is mixed with reduced p-hydroxybenzoate hydroxylase at low temperature. From molecular orbital calculations, Orf and Dolphin (1974) have suggested that the flavin peroxide may be formed from a dioxetane intermediate:

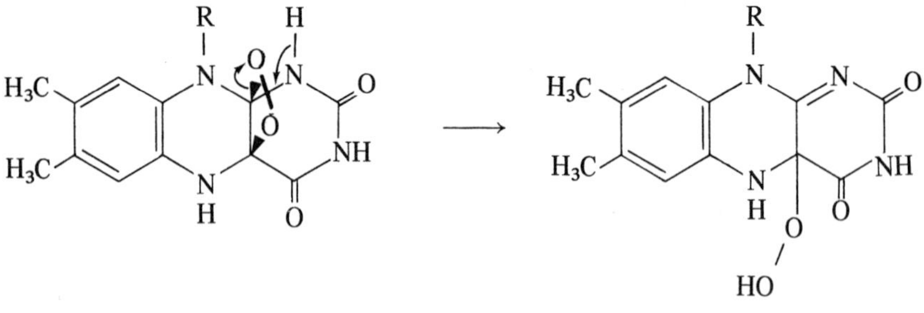

Dioxetane   4a–Peroxide

Massey et al. (1973) pointed out that the compounds formed by reaction of reduced flavins with O_2 could be converted to oxidized flavin by several routes. These could include direct formation of H_2O_2, as is observed in oxidase-catalyzed reactions, and formation of H_2O and a hydroxylated product in the presence of hydroxylases. Superoxide anion (O_2^{-}; see Chapter 11) and flavin semiquinone might be formed from oxygenated flavins bound to dehydrogenases. The possibility of addition of molecular oxygen across the $C^{1a}{=}C^{4a}$ double bond of reduced flavin coenzymes is high, and research to prove the structures of oxygenated reaction intermediates is now being performed.

It has been known for many years that flavin coenzymes are covalently attached to some flavoproteins. Singer and his coworkers have found that the site of attachment of peptides is the 8-methyl group (position 8α) of the flavin. For example, purified succinic dehydrogenase from beef heart mitochondria was digested with proteolytic enzymes to form flavin peptides. After acid hydrolysis of these peptides, a compound was isolated and shown to be *N*-3-histidyl-8α-flavin (Walker and Singer, 1970):

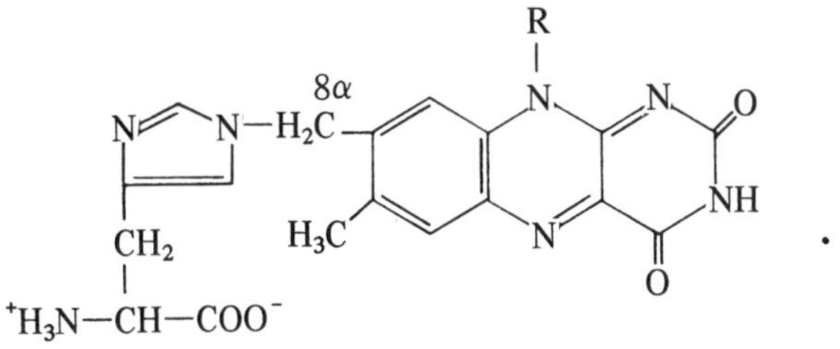

Although during the hydrolysis the flavin coenzyme was degraded to riboflavin, succinic dehydrogenase contains 8α-histidyl-FAD. By isolation and purification of a peptide with the composition

$$\begin{array}{c} \text{FAD} \\ | \\ \text{Ser—Gly—Gly—Cys—Tyr} \end{array},$$

it has been shown that monoamine oxidase has its FAD linked to cysteine by a thioether bond to the 8α-methylene group (Walker *et al.*, 1971). Several other examples of covalently bound FAD have been discovered and characterized (reviewed by Singer and Kenney, 1974; Singer and Edmondson, 1974).

PYRIDOXAL PHOSPHATE

One of the most interesting and mechanistically informative co-enzymes is pyridoxal-5'-phosphate (Braunstein, 1960; Jencks, 1969; Snell and Di Mari, 1970). This coenzyme is formed by metabolic oxidation of the 4-hydroxymethyl group of vitamin B_6 (pyridoxine) and ATP-mediated phosphorylation of the 5-hydroxymethyl group of the resulting pyridoxal:

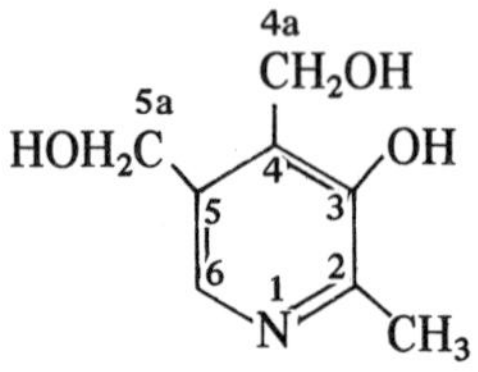

Pyridoxine 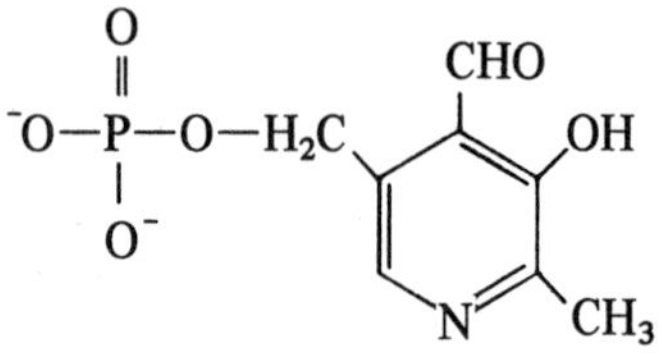   Pyridoxal phosphate (PLP)

Nutritional studies in the 1930s had suggested that vitamin B_6 was associated with the metabolic transformations of amino acids. Pyridoxal phosphate (PLP) has since been found to be a cofactor in more than a dozen different types of reactions of amino acids. These enzymic reactions include transamination, racemization, decarboxylation and side chain elimination or replacement (see Dunathan, 1971, for a complete list). Because transamination is the most common and best studied PLP-dependent reaction, I am restricting the present discussion to general properties of PLP and to this reaction. However, involvement of PLP in other reactions is described in other sections of the book.

The isolation of two metabolites of pyridoxine, an aldehyde form (pyridoxal) and an aminomethyl form (pyridoxamine)—which both

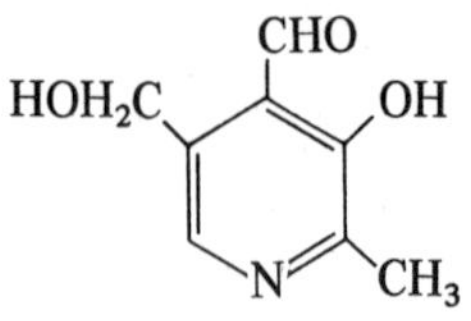

Pyridoxal 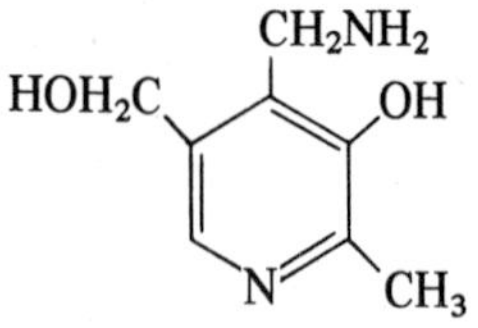   Pyridoxamine

have potent growth-promoting activity for lactic acid bacteria—along with the reversible non-enzymic interconversion of these two forms in the presence of glutamate and α-ketoglutarate (Snell, 1945), prompted the suggestion that these compounds have a catalytic role in biological transamination:

$$\text{Amino acid} + \underset{\underset{R}{|}}{CHO} \rightleftharpoons \text{Keto acid} + \underset{\underset{R}{|}}{CH_2NH_2}$$

Pyridoxal   Pyridoxamine

At about the same time, pyridoxal phosphate was synthesized and shown to act as the prosthetic group for several transaminases. Also, the 4-formyl group of PLP was implicated as a reactive centre by the inhibition of pyridoxal-dependent enzymatic reactions by addition of carbonyl reagents such as hydrazines. Schlenk and Fisher (1947) were the first to suggest that the pyridoxine-derived prosthetic groups of transaminases formed Schiff base *ES* complexes by dehydration reactions with keto acid and amino acid substrates. This combination of biological and chemical evidence in the early work is seen throughout the history of PLP biochemistry, and is most evident in the development of a generalized mechanism of action for PLP.

This general theory was proposed independently by Braunstein (Braunstein and Shemyakin, 1953) and Snell (Metzler *et al.*, 1954a). Braunstein's hypothesis was a rational explanation relating a number of seemingly unrelated enzyme-catalyzed reactions. Snell's conclusions were reached by examination of non-enzymic model reactions of amino acids, pyridoxal and di- or trivalent metal ions. Both approaches led to the same theory, with no major discrepancies. The theory suggests that the first reaction step for all substrates is the formation of a *Schiff base,* in this case an aldimine (i.e. the product of condensation of an aldehyde with an amine) which exists as at least two tautomers, I and II:

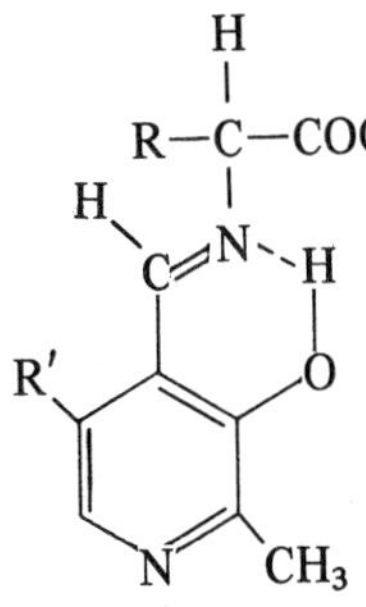

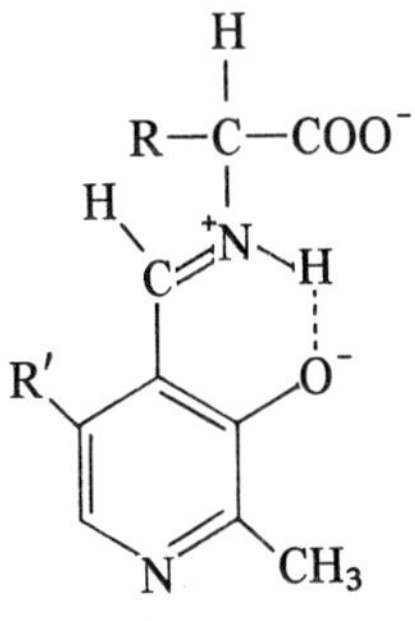

The Schiff base has absorption maxima at about 330 nm (form I) and 410 nm (form II), and has been observed in both enzymic and model systems. The cationic nitrogen of the imine and the conjugated system of double bonds which extends to the pyridine's nitrogen (probably protonated) withdraw electrons from the α-carbon of the amino acid, weakening the bonds to all three of its substituents. Release of the proton from the α-carbon can lead to racemization or transamination, release of the carboxylate can lead to decarboxylation, and release of the side-chain cation (R^+) to reactions such as the serine-glycine interconversion (p. 249). All of these reactions are made possible by the formation of an intermediate dihydropyridine or quinonoid compound of a ketimine. In transamination, the quinonoid intermediate is formed from the aldimine by removal of the α-hydrogen:

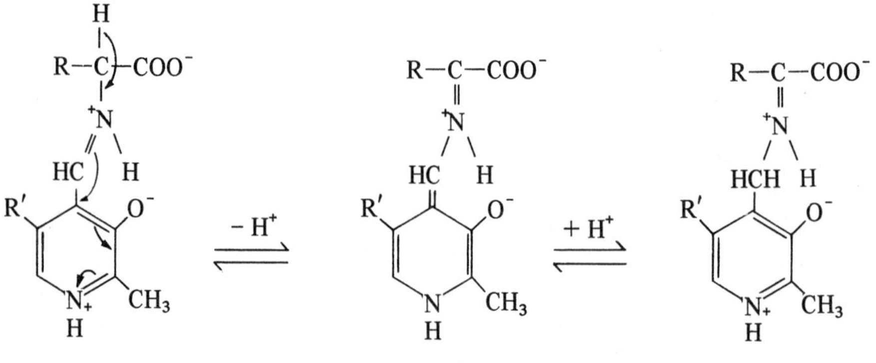

Aldimine   Quinonoid intermediate   Ketimine

Recently, Abbott and Bobrik (1973) have reported the first isolation of a 1,4-dihydropyridine tautomer. This compound was synthesized from pyridoxal and diethyl aminomalonate (EtO—CO—CHNH$_2$—CO—Et). It has a λ_{max} at 465 nm, which is comparable to the absorption spectra of intermediates which have been observed in several PLP-enzyme reactions (λ_{max} near 500 nm). The interconversion of the aldimine and the ketimine is the rate-determining step in model reactions, and also in some enzyme-catalyzed reactions. Auld and Bruice (1967) have shown that this interconversion can be catalyzed by general acid–base reagents, suggesting the presence of a basic group at the active site of PLP-requiring enzymes for assistance in release of the proton from the α-carbon.

The study of non-enzymatic model reactions has been of great assistance in the elucidation of the catalytic mechanisms of pyridoxine. Almost all pyridoxal-enzyme reactions have non-enzymic counterparts. The rate enhancement in the presence of the enzyme is at least 10^6, though. The non-enzymatic reactions whose stoichiometry was originally determined in Snell's laboratory required the presence of metal ion catalysts, and were performed at $100°$ C. Metal ions are not cofactors for most PLP-enzymes. Many refinements have been made to the early chemical experiments, to more closely approach the enzymic catalysis. For example, a model system for transamination not requiring metal ions has been devised, using 3-hydroxypyridine-4-aldehyde in place of pyridoxal, and its reactions proceed at much lower temperatures and allow measurement of reaction kinetics (Auld and Bruice, 1967). The apoenzyme in PLP-dependent reactions is not only responsible for the rate enhancement but it also performs the interrelated functions of stereospecificity and determination of cleavage of only one of the bonds of the substrate's α-carbon. During PLP-enzyme catalyzed reactions, the formation of the *ES* complexes (the stepwise intermediates) can be ascertained by observing the spectral changes occurring. By this technique, the enzymic and non-enzymic mechanisms have been found to be identical.

Often, PLP is bound tightly to its apoenzymes. Resolution into the coenzyme and apoenzyme may be effected by dialysis against an amino compound that can react with the aldehyde of PLP (e.g. cysteine or penicillamine). The reason for this effect is that the PLP is covalently bonded to the protein through its formyl group. Fischer *et al.* (1958) showed that PLP is bound to glycogen phosphorylase as a Schiff base involving the ε-amino group of a lysine residue. They reduced the holoenzyme with sodium borohydride to form a pyridoxylamine derivative:

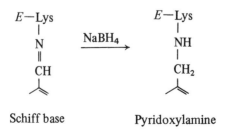

Schiff base   Pyridoxylamine

Hydrolysis of the reduced enzyme produced ϵ-N-pyridoxyllysine. Similar experiments with many other PLP-dependent enzymes have demonstrated that they too exist as internal imines with lysine in the absence of substrate, and that the first covalent reaction of a substrate with the enzyme is a *transimination* step:

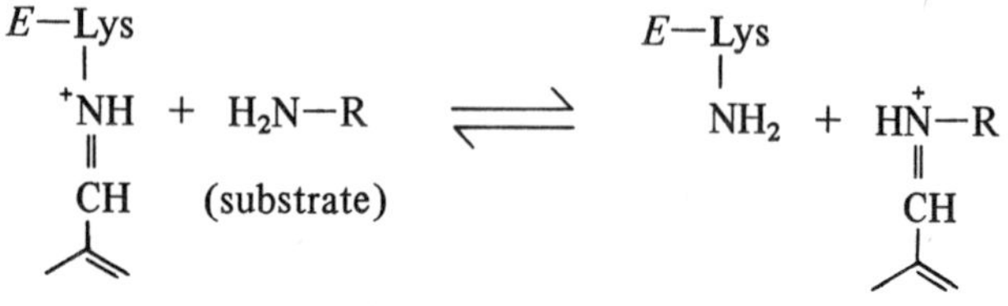

$$
\begin{array}{ccc}
E{-}\text{Lys} & & E{-}\text{Lys} \\
| & & | \\
{}^{+}\text{NH} \;+\; \text{H}_2\text{N}{-}\text{R} \;\;\rightleftharpoons\;\; & & \text{NH}_2 \;+\; \text{H}\overset{+}{\text{N}}{-}\text{R} \\
\| \quad\quad \text{(substrate)} & & \| \\
\text{CH} & & \text{CH}
\end{array}
$$

Internal imine   Schiff base
                                       with substrate

The Schiff base with the substrate also has been trapped by reduction with borohydride. Tobias and Kallen (1975) have concluded, based on results with a chemical model system, that transimination proceeds through a geminal diamine intermediate and not via formation of enzyme-bound PLP itself.

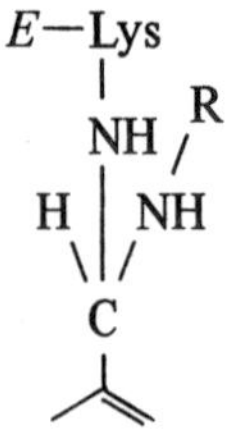

$$
\begin{array}{c}
E{-}\text{Lys} \\
| \\
\text{NH} \quad\! R \\
\quad\; \diagup \\
\text{H} \;\big|\; \text{NH} \\
\quad\diagdown\!\big|\!\diagup \\
\text{C} \\
\end{array}
$$

Geminal diamine

Ivanov and Karpeisky (1969) have summarized evidence which indicates that PLP is bound to apoenzymes through all of its functional groups. As well as the covalent bond at C^{4a}, ionic or hydrophobic bonds are formed by N^1, the 2-methyl group, the 3-phenolic group and the phosphate monoanion. These other interactions are necessary to bind PLP to the active site during the time in the catalytic reaction when the internal imine bond is broken. Anions such as Cl^-, SO_4^{2-} or PO_4^{3-} strongly inhibit the binding of PLP to apoenzymes, by competing with the phosphate group of PLP for an anion-binding pocket. The anion effect of sulphate may cause the commonly observed resolution of holoenzymes into apoenzymes and PLP in the presence of ammonium sulphate.

Aspartate Aminotransferase

The PLP-dependent enzyme which has received the most attention, and thus has the most completely delineated mechanism, is the aspartate aminotransferase from pig heart (also called glutamate aspartate transaminase). This enzyme (reviewed by Braunstein, 1973) catalyzes reversible transamination between glutamate and aspartate:

$$\begin{array}{c}
\text{COO}^-\\
|\\
\text{CH}_2\\
|\\
\text{CH}_2\\
|\\
\text{CH}-\overset{+}{\text{N}}\text{H}_3\\
|\\
\text{COO}^-
\end{array}
\quad + \quad
\begin{array}{c}
\text{COO}^-\\
|\\
\text{CH}_2\\
|\\
\text{C}=\text{O}\\
|\\
\text{COO}^-
\end{array}
\quad
\underset{\longleftarrow}{\overset{\text{Aminotransferase}}{\longrightarrow}}
\quad
\begin{array}{c}
\text{COO}^-\\
|\\
\text{CH}_2\\
|\\
\text{CH}_2\\
|\\
\text{C}=\text{O}\\
|\\
\text{COO}^-
\end{array}
\quad + \quad
\begin{array}{c}
\text{COO}^-\\
|\\
\text{CH}_2\\
|\\
\text{CH}-\overset{+}{\text{N}}\text{H}_3\\
|\\
\text{COO}^-
\end{array}$$

L–Glutamate   Oxaloacetate   α–Ketoglutarate   L–Aspartate
(Glu)   (OAA)   (α–KG)   (Asp)

This was the first pyridoxal phosphate enzyme to be purified extensively and obtained in quantities allowing use of substrate-level concentrations of enzyme (Jenkins $et\ al.$, 1959). The vitamin B_6 phosphate cofactors are bound firmly to the apoenzyme, but can be released by treatment with alkali. Jenkins and Sizer (1960) demonstrated that addition of glutamate to the aminotransferase caused a shift in the absorption spectrum, from that of the internal imine (E—PLP; $\lambda_{max} = 362\ \text{nm}$) to that of the pyridoxamine phosphate (E—PLP) form ($\lambda_{max} = 333\ \text{nm}$). The appearance of enzyme-bound pyridoxamine coincided with the formation of α-ketoglutarate. Proof of this E—PLP $\rightleftharpoons E$—PMP interconversion was obtained by Jenkins and D'Ari (1966), when they succeeded in preparing and purifying the PMP form of the aminotransferase. They also performed a spectrophotometric titration of the PMP form of the enzyme with α-ketoglutarate. The formation of aspartate from glutamate, then, is made up to two half-reactions:

$$\text{Glu} + E{-}\text{PLP} \rightleftharpoons \alpha{-}\text{KG} + E{-}\text{PMP}$$

$$\text{OAA} + E{-}\text{PMP} \rightleftharpoons \text{Asp} + E{-}\text{PLP}$$

$$\text{Sum:} \quad \text{Glu} + \text{OAA} \rightleftharpoons \alpha{-}\text{KG} + \text{Asp} \quad .$$

The donor amino acid (Glu) reacts with the aldehyde form of the enzyme to form an α-keto acid (α-KG) and methylamino form of the enzyme. Then the acceptor α-keto acid (OAA) reacts with the PMP–enzyme to produce Asp and the recycled PLP–enzyme. Kinetic studies of this reaction (Velick and Vavra, 1962; Henson and Cleland, 1964) are consistent only with a bireactant ping-pong mechanism:

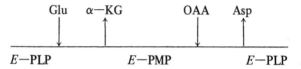

As an example of the results favouring a ping-pong kinetic mechanism, reciprocal plots for variable concentrations of Glu at fixed concentrations of OAA produce parallel lines (cf. p. 128). Parallel lines are also obtained with variable concentrations of Asp at fixed levels of α-KG.

The minimal mechanism of the first half-reaction involves:

1. non-covalent binding of the Zwitterion of glutamate to the enzyme to form the Michaelis complex;
2. nucleophilic addition of the free amino group of the glutamate anion to the PLP to form the aldimine (including transimination with the internal Schiff base);
3. the rate-limiting conversion of the aldimine to the ketimine via a negatively charged quinonoid intermediate (a stabilized carbanion); and
4. hydrolysis of the ketimine to give α-ketoglutarate and the pyridoxamine form of the enzyme (Fig. 6).

In the second half-reaction, the reverse pathway is followed with oxaloacetate in place of α-ketoglutarate.

The aspartate aminotransferase from the cytoplasm of pig heart has a molecular weight of 93 000. It consists of two identical subunits, each containing 412 amino acids and one molecule of PLP

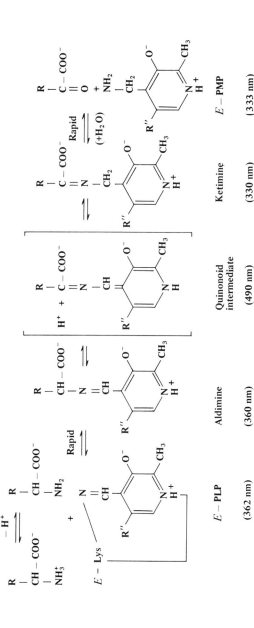

Fig. 6. A minimal mechanism for the aspartate aminotransferase-catalyzed reaction, showing the first half-reaction—the formation of keto acid No. 1 (R—CO—COO⁻). In the second half-reaction, this keto acid would be replaced by R′—CO—COO⁻. The absorption maxima assigned to each enzyme or *ES* species is indicated below its structure. All of the above compounds are bound to the enzyme, even though the enzyme is only shown as being present in *E*—PLP.

(Ovchinnikov *et al.*, 1973). The aminotransferase has no disulphide bridges, so that its globular shape is maintained entirely by non-covalent bonds. Morino and Watanabe (1969) found the sequence of the peptide containing the lysine residue forming the internal imine in the absence of substrate to be Ser—*Lys*—Asn—Phe. This sequence corresponds to residues 257–260 of the enzyme. The imidazole group of one of the eight histidine residues in each protomer is essential for activity. Several laboratories demonstrated that photo-oxidation of this histidine causes inactivation of aminotransferase activity. Peterson and Martinez-Carrion (1970) showed that the photo-oxidized enzyme is capable of forming the initial aldimine with aspartate and the ketimine with α-ketoglutarate, but could not cause the deprotonation needed to form the carbanionic quinonoid intermediate. They concluded that the histidine residue which is modified is the general-base catalyst. Yamasaki *et al.* (1975) have reinterpreted these results and presented evidence that the photo-oxidizable histidine is necessary for binding of the distal carboxylate group of the substrate. (The photo-oxidized enzyme can still catalyze the transamination of the poor substrate L-alanine.) They feel that the lysine residue originally in the internal imine is the general base that labilizes the substrate's α-hydrogen atom:

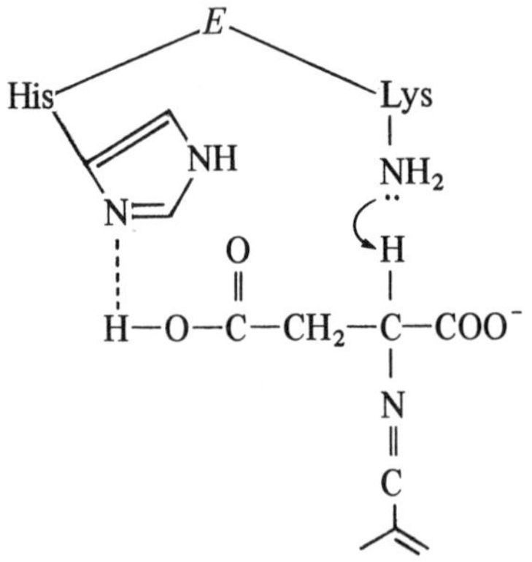

Many fine reviews of pyridoxal phosphate chemistry and enzymology are available (e.g. Bruice and Benkovic, 1966, their Chapter 8; Davis and Metzler, 1972; and the reviews cited above). There is not the space here to delve into the details of the enzymic mechanisms of all the PLP-mediated amino acid transformations. In all the reactions, a carbanion is formed—either at the α-carbon or the β-carbon of the amino acid—and stabilized by PLP. As discussed previously, the decision of which reaction is to occur (i.e. which bond is to be cleaved) is made by the enzyme. The one example discussed, transamination, exemplifies the relevance of organic-reaction mechanism studies to enzymic processes. Other enzymatic reactions, such as decarboxylation, racemization and eliminations can equally well be rationalized in logical chemical terms. One further example will be discussed at some length, that of serine transhydroxymethylase (p. 249). The principal importance of PLP from an academic viewpoint is the clearcut chemistry involved and the direct observation of transient *ES* complexes.

PTERIN COENZYMES

There are two coenzymes which contain a pterin functional moiety: *tetrahydrofolate,* a reduced derivative of the vitamin folic acid, and *tetrahydrobiopterin,* which can be synthesized by mammals. The former participates in transfer of one-carbon groups, and the latter is a redox cofactor. Workers in the pterin field are fortunate in having an excellent monograph (Blakley, 1969) for reference. Blakley's book fully documents the chemical, biological and enzyme-related properties of pteridine compounds. Pteridines are bicyclic ring compounds, containing a pyrimidine ring fused to a pyrazine ring. The naturally occurring pteridines are pterins, or 2-amino-4-oxo-pteridines:

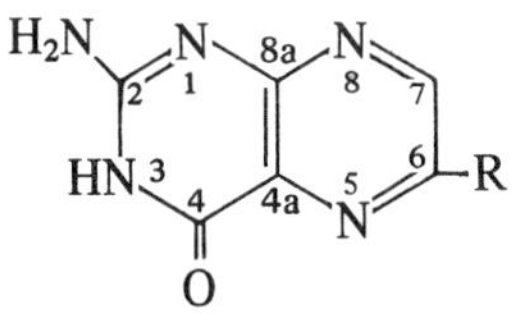

A pterin 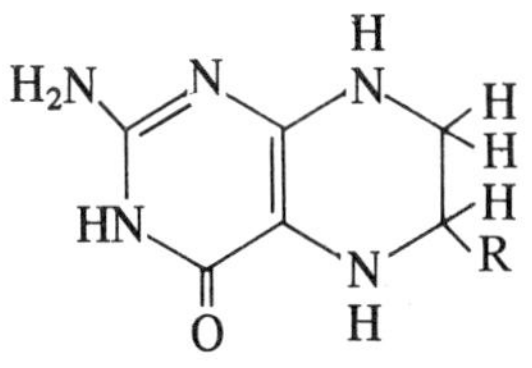   A tetrahydropterin

Folic acid and biopterin differ only in the R-group at C^6. Both are active as coenzymes only after the pyrazine ring is fully reduced; i.e. as their 5,6,7,8-tetrahydro forms. Although they are chemically similar, because their metabolic functions are so different they are discussed in separate sections. The pterin-requiring enzymes which I have described illustrate many basic principles of enzyme chemistry. In addition, an understanding of the enzymology of several of them is a prerequisite for the discussion of chemotherapeutic control of enzyme reactions (Chapter 15).

Tetrahydrofolate

Folic acid was isolated from extracts of yeast, spinach and liver in the 1940s, in a manner parallel to that of most B-vitamins. Purification methods were devised, utilizing assays of two responses to the vitamin: the growth of certain bacteria, and the prevention of anemia in chicks. The structure of folate was determined by degradation and synthesis as being a pterin substituted in the 6-position with a methylaminobenzoylglutamic acid:

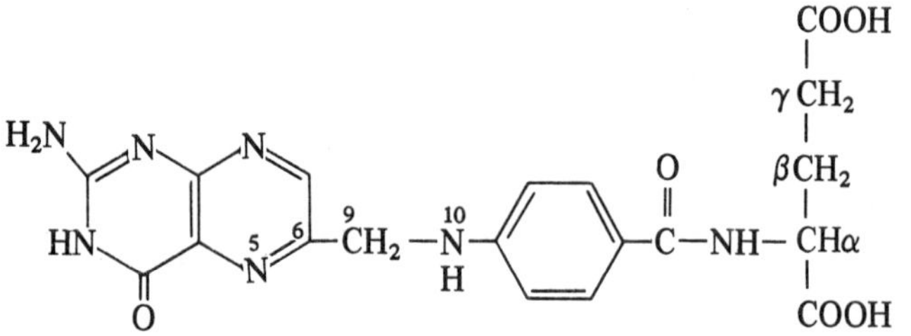

Folic acid exists as polyglutamates in which the glutamyl residue is linked to several other molecules of glutamate by γ-peptide bonds. The material commonly used in the laboratory is the monoglutamate, but Baugh and Krumdieck (1971) have synthesized a series of polyglutamyl folates by the solid phase synthetic method, providing well-defined samples of these compounds for future experiments.

The coenzyme form, tetrahydrofolate, is susceptible to rapid oxidative degradation. Enzymologists routinely add mild reducing agents such as ascorbate or 2-mercaptoethanol to assay mixtures to prevent this oxidation. Oxidation of tetrahydrofolate leads to the formation of a variety of products, with the proportions of these depending on the buffer composition and pH (Chippel and Scrimgeour, 1970). Tetrahydrofolate is prepared by catalytic hydrogenation of folic acid in acidic solution, with anaerobic removal of the platinum oxide catalyst (Hatefi et al., 1960). Reduction of folate by

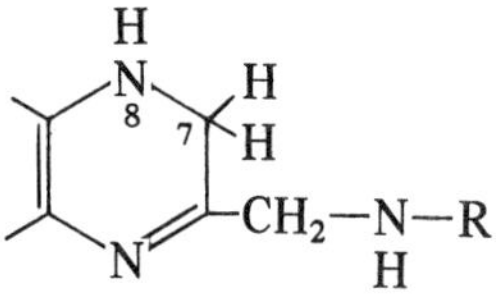

7, 8–dihydrofolate

dithionite at room temperature (Blakley, 1960) produces 7,8-dihydro-folate, which is much more stable than tetrahydrofolate but is also susceptible to oxidative degradation. The structure of 7,8-dihydro-folate was inferred from its chemical properties and from the knowledge that enzyme-catalyzed reduction produces a single diastereoisomer of tetrahydrofolate by introduction of a new asymmetric centre at C^6 (Mathews and Huennekens, 1960). This structural assignment was confirmed in 1963 by Pastore *et al.* by n.m.r. spectroscopy.

The final step in the biosynthesis of either tetrahydrofolate (or tetrahydrobiopterin) is the NADPH-dependent reduction of the pyrazine ring catalyzed by 7,8-dihydrofolate reductase. The dihydro-folate reductases of some cells can also catalyze the reduction of folate to tetrahydrofolate, although at much lower rates (e.g. 10% or less the rate of reduction of dihydrofolate). Therefore, the reactions catalyzed by dihydrofolate reductase can be written as:

$$7,8\text{–Dihydrofolate} + \text{NADPH} + \text{H}^+ \rightleftharpoons \text{Tetrahydrofolate} + \text{NADP}^+,$$

$$\text{Folate} + 2\,\text{NADPH} + 2\,\text{H}^+ \xrightarrow{\text{Slower}} \text{Tetrahydrofolate} + 2\,\text{NADP}^+.$$

Although 7,8-dihydrofolate is an obligatory intermediate in the reduction of folate in enzymic systems, it has not been proven to be so because of its much faster reduction. Dihydrofolate reductase is metabolically essential for three reasons:

1. it catalyzes the reduction of the end-product of vitamin biosynthesis by bacteria and plants, 7,8-dihydrofolate (Brown, 1971);
2. it catalyzes the reduction of exogenous folate; and
3. it functions as part of a cyclic enzyme system in the biosynthesis of thymidylate (TMP).

I

Because of its role in TMP synthesis, dihydrofolate reductase is one target site of the anti-folate drugs used in chemotherapy (Chapter 15).

Dihydrofolate reductase has been purified from many tissues and micro-organisms, and usually has a relatively low molecular weight (15 000–25 000 daltons). Most of the early work was performed with the reductase from chicken liver. Zakrzewski and Nichol (1960) demonstrated that one enzyme was responsible for the reduction of both folate and dihydrofolate. Principle evidence in reaching this conclusion was the constant ratio of activity with folate and dihydro-folate during purification, and heat inactivation causing loss of both the folate and dihydrofolate activity at the same rate. Both NADPH and NADH can act as reductants, but NADPH is preferable. The substrate specificity of the chicken liver reductase is dependent on the pH of the assay. Mathews and Huennekens (1963) reported that the pH–activity curve for the enzyme is biphasic, with optima at about pH 4.5 (major) and pH 7.5 (minor). At the higher pH optimum, only dihydrofolate is reduced and only NADPH can act as reducing agent. At pH 4.5, reduction of folate by NADPH and of dihydrofolate by NADH are appreciable, but reduction of dihydrofolate by NADPH is still the predominant reaction. This substrate specificity was verified by Kaufman and Gardiner (1966) when the enzyme was purified 8000-fold to homogeneity. Dihydrofolate reductases from many vertebrate sources show the remarkable property of activation in the presence of compounds which affect tertiary structure (salts, urea or guanidine) or of mercurials such as p-hydroxymercuribenzoate. These two types of reagents appear to cause increased catalytic activity by beneficial alterations in the conformation of the enzyme. Treatment with these reagents changes the pH–activity curve from biphasic to one having a single maximum. Many microbial dihydro-folate reductases show a single pH optimum, little or no activity with folate and minimal activations by denaturants or mercurials (cf. Table 5.1 in Blakley, 1969).

The purification of dihydrofolate reductases has been greatly aided by the development of affinity chromatography methods (p. 57). The first successful application of this technique involved preparation of a starch derivative to which the inhibitor ametho-pterin (4-amino-10-methylfolate, p. 518) was bound by a peptide bond to the amino group of a high molecular-weight soluble amino-ethylcellulose (Mell et al., 1968). Because of the strong binding of the enzyme inhibitor to the reductase ($K_i \approx 10^{-9}$ M), the reductase from a relatively crude preparation was specifically bound to the substituted starch and separated on a column of Sephadex G-75 from many contaminating proteins. After elution of the reductase from the high molecular-weight complex by addition of dihydrofolate, an enzyme having a 150-fold increase in specific activity was obtained. Soon after the solid-matrix method was introduced by Cuatrecasas, a half-dozen laboratories reported the preparation of amethopterin-amino-alkyl-agarose for rapid and efficient purification of dihydrofolate reductases. For example, the reductase from a mutant of *E. coli* has been purified by a batchwise procedure by addition of a cell extract to a slurry of amethopterin-aminoethyl-Sepharose. The affinity resin containing the enzyme was filtered and washed, and the reductase eluted with dihydrofolate at pH 8. After filtration through Sephadex G-75, solutions of homogeneous enzyme purified 1400-fold and in yields of over 40% were obtained (Poe *et al.*, 1972). Because of the possibility of obtaining much larger quantities of the reductase, amethopterin-resistant (i.e. dihydrofolate reductase enriched) bacteria often are being used as the source of the enzyme for physical experiments.

Reductase preparations from *Lactobacillus casei* can be separated by electrophoresis into two major fractions (Gundersen *et al.*, 1972). These are not isoenzymes or oligomers, but are the apoenzyme (form I) and an apoenzyme–NADPH non-covalent complex (form II). These two forms can be separated by chromatography on hydroxylapatite, as well as by polyacrylamide gel electrophoresis. The absorption spectra of forms I and II are compared in Fig. 7. Form I has a typical protein spectrum, but the additional absorption band at 340 nm for form II is due to the NADPH bound to it. Form I can be converted to form II by incubation with excess NADPH, and form II partially reverts to form I during electrophoresis. The NADPH appears to be bound to the catalytic site of the reductase. The binding of folates and pyridine nucleotides to the *L. casei* reductase has been examined by n.m.r. spectroscopy (Pastore *et al.*, 1974). Binary and ternary complexes have been tested (e.g. enzyme–dihydrofolate and enzyme–folate complexes can be prepared by elution from affinity columns using the appropriate folate compound). The inhibitor amethopterin has been used to form ternary enzyme–NADPH–amethopterin complexes. By studying the effect of order of addition of the ligands

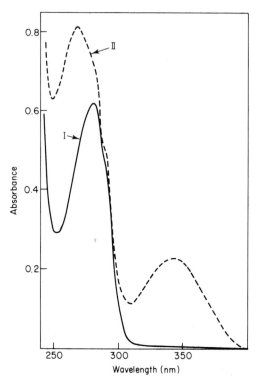

Fig. 7. Absorption spectra of forms I and II of dihydrofolate reductase isolated from amethopterin-resistant *L. casei.* The concentrations of the enzyme samples are both 2.9×10^{-5} M, and the pH is 7.0. From Gundersen *et al.* (1972), with permission.

to the reductase, Pastore *et al.* have concluded that the order of addition of ligands to the *L. casei* enzyme is random, not ordered. Neef and Huennekens (1975), after incubating a mammalian dihydrofolate reductase with substrates and inhibitors, succeeded in isolating by electrophoresis stable and well-defined binary and ternary complexes. The surprising stabilities of the reductase–ligand complexes suggest that observed multiple forms of other enzymes may be caused by binding of coenzymes or substrates.

Some chemical experiments which can be considered plausible models of the dihydrofolate reductase-catalyzed reaction have been performed. These all support the PHy mechanism proposed by Huennekens and Scrimgeour (1964). This mechanism proposes that the nitrogen of the $-C{=}N-$ bond undergoing reduction (N^8 of folate and N^5 of dihydrofolate) is protonated, displacing electrons toward the nitrogen and facilitating nucleophilic addition to the adjacent carbon. For example, reduction of 7,8-dihydrofolate would involve:

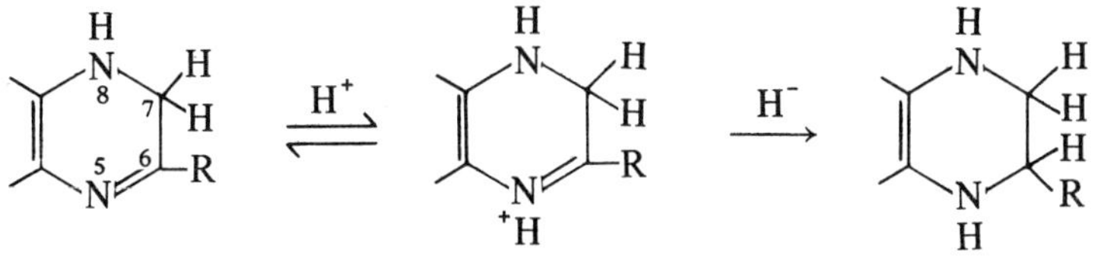

Dihydrofolate   Tetrahydrofolate

In agreement with this mechanism, one mole of tritium is incorporated at C^6 when dihydrofolate is reduced to tetrahydrofolate with NaB^3H_4 (Scrimgeour and Vitols, 1966). When folate is reduced with NaB^3H_4, first one mole of tritium is incorporated at C^7 to produce 7-^{3}H-7,8-dihydrofolate, and then a second mole of tritium is incorporated to give $6,7$-di^{3}H-5,6,7,8-tetrahydrofolate. Neither thermodynamic nor kinetic experiments are possible using borohydride, but the folate compounds reversibly form nucleophilic adducts similar to but much less stable than those of the nicotinamides. Addition of an excess of bisulphite to folate gives a compound similar in spectrum to dihydrofolate, and to dihydrofolate an adduct spectrally similar to tetrahydrofolate (Vonderschmitt et al., 1967). The dihydro form binds the nucleophile more tightly and more rapidly than does the aromatic folate form (Table II), giving both thermodynamic and kinetic reasons why 7,8-dihydrofolate is the preferred substrate in the enzyme-catalyzed reaction. Baker and Ho (1964) postulated that

Table II. Reaction of HSO_3^- with folate and dihydrofolate. Both the formation constants (Vonderschmitt et al., 1967) and rate constants (Kawai and Scrimgeour, 1972) for the reaction of HSO_3^- with the pterin were measured under anaerobic conditions at $31°$ C by spectrophotometric methods.

Compound	Formation constant	Rate constant
Folate	3 M^{-1}	4.4 M^{-1} s^{-1}
7,8-Dihydrofolate	83 M^{-1}	130.0 M^{-1} s^{-1}

a histidine residue is the general acid that protonates the nitrogen atoms, which would make dihydrofolate reductase similar to lactate dehydrogenase in that respect. The pK_a of the upper limb of the pH–activity curve of the mercurial-treated chicken liver reductase agrees with this suggestion.

Chemical modification studies with purified dihydrofolate reductases indicate that histidine, methionine and tryptophan residues are essential for ligand binding or for expression of catalytic activity. The sequences of several reductases have recently been published, and hopefully in the next few years more knowledge will be available pertaining to the specific amino acid residues participating in the catalytic process.

The chemical features of tetrahydrofolate which permit its metabolic function as a transient carrier of one-carbon groups are the optimally spaced secondary amino groups at N^5 and N^{10}. Whereas the mobile metabolic groups binding to PLP all bind to C^{4a}, the situation is much more complicated with tetrahydrofolate. The groups may bind either to N^5, to N^{10}, or to both nitrogen atoms. In considering these adducts of tetrahydrofolate, I shall abbreviate the structure of the "active site" of the coenzyme as shown below:

Tetrahydrofolate's
"active site"
(reactive portion of the molecule)

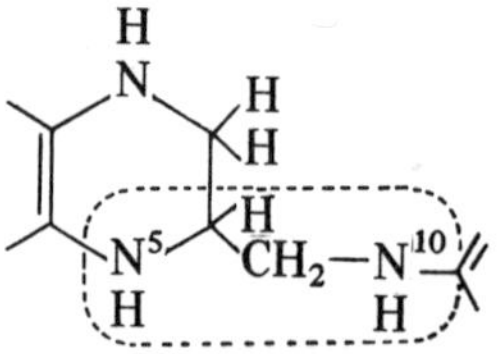

Abbreviation
(inside dotted line)

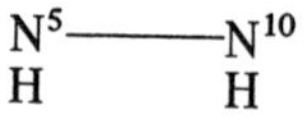

One-carbon groups at three different oxidation levels are bound to tetrahydrofolate. The abbreviated structures of the various covalent complexes of tetrahydrofolate are shown in Fig. 8, which outlines the metabolic pathways by which these complexes can be interconverted. Two of these C-1: tetrahydrofolate complexes are bridge-type five-membered rings involving both N^5 and N^{10}. These compounds, the formaldehyde adduct of tetrahydrofolate (methylenetetrahydrofolate) and one of three formate derivatives (methenyltetrahydrofolate), can be interconverted under the influence of methylene-

Oxidation level
of one – carbon
group

$$N^5 - N \longrightarrow \text{Methionine} \qquad CH_3OH$$
$$\underset{CH_3}{|} \quad H \qquad \text{Thymidylate}$$

NADH or NADPH

$$\underset{H}{N} - \underset{H}{N} \quad \xrightarrow{\text{HCHO}} \quad N \underset{CH_2}{\overset{}{\diagup}} N \quad \underset{2}{\rightleftharpoons} \quad \text{Serine} \qquad \text{HCHO}$$

Non – enzymic

1 $\Big\|$ HCOOH, ATP   NADPH $\Big\|$

$$\underset{H}{N} - \underset{\underset{CHO}{|}}{N^{10}} \quad \rightleftharpoons \quad N^5 \underset{\underset{CH}{\overset{+}{|}}}{\diagup} N^{10} \quad \rightleftharpoons \quad \underset{\underset{CHO}{|}}{N^5} - \underset{H}{N} \qquad \text{HCOOH}$$

$$N^5 - N$$
$$\underset{\|}{CH} \quad H$$
$$NH$$

Formyl
methionine   Purines

NADP⁺   Histidine $\longrightarrow$ FIGLU

$$CO_2 + \underset{H}{N} - \underset{H}{N} \qquad\qquad CO_2$$

Fig. 8. Interconversion of one-carbon : tetrahydrofolate covalent complexes, and some of the metabolic products and precursors of these complexes. The three numbered reactions are those catalyzed by: (1) formyltetrahydrofolate synthetase, (2) serine transhydroxymethylase, and (3) thymidylate synthetase.

tetrahydrofolate dehydrogenase:

5, 10–Methylenetetrahydrofolate   5, 10–Methenyltetrahydrofolate

Methylenetetrahydrofolate is a substrate for the synthesis of TMP, and also can be reduced by pyridine nucleotides in the presence of a flavoprotein to 5-methyltetrahydrofolate. The methyl derivative is used in the biosynthesis of methionine (Chapter 11) and thence for all the methylated compounds formed from S-adenosyl methionine. 5,10-Methenyltetrahydrofolate directly supplies C^8 of the purine structure, or C^2 after conversion to 10-formyltetrahydrofolate. 10-Formyltetrahydrofolate is also the donor of the C-1 unit in the synthesis of N-formylmethionyl-tRNA^{fMet}, which provides formyl-methionine, the amino acid initiating protein biosynthesis in most prokaryotes. 5-Formyltetrahydrofolate is thought to be a storage form of the coenzyme. 5-Formiminotetrahydrofolate is formed in the fermentation of purines by some bacteria and in the catabolism of histidine. A clinical test for folate deficiency is based on measure-ment of the level in urine of formiminoglutamate (FIGLU), the immediate precursor of formiminotetrahydrofolate in histidine breakdown. Normal subjects excrete negligible amounts (less than 20 mg) of FIGLU when fed a 14-g dose of histidine, while folate deficient patients can excrete close to 1 g of FIGLU. The enzymes catalyzing the transfer of one-carbon groups have recently been reviewed by Rader and Huennekens (1973), and the mechanisms of both enzymic and chemical model systems have been discussed by Benkovic and Bullard (1973).

In this chapter, I shall describe only three of the many enzymes involved in tetrahydrofolate-dependent metabolism: formyltetra-hydrofolate synthetase, serine transhydroxymethylase and thymidy-late synthetase (numbered 1–3 in Fig. 8). Formyltetrahydrofolate synthetase and serine transhydroxymethylase are the main points of entry of one-carbon compounds into the biosynthetic pathways requiring these C-1 units. Formate enters the C-1 pool as the free C-1 compound, whereas formaldehyde enters by the reversal of the transhydroxymethylase reaction. In fact, in most cells, formation of methylenetetrahydrofolate and glycine from serine is the major source of C-1 units. Free formaldehyde can react non-enzymically with tetrahydrofolate to reversibly form 5,10-methylenetetrahydrofolate, via a carbinolamine which undergoes acid-catalyzed dehydration to a rapidly cyclized cationic imine (Kallen and Jencks, 1966):

$$\text{N}\!-\!\!-\!\!-\text{N} + \text{HCHO} \rightleftharpoons \underset{\substack{|\\ \text{CH}_2\text{OH}}}{\text{N}^5\!-\!\!-\!\!-\underset{\text{H}}{\text{N}}} \underset{\substack{\\ }}{\overset{+\,\text{H}^+,\,-\,\text{H}_2\text{O}}{\rightleftharpoons}} \underset{\substack{\|\\ \text{CH}_2}}{\text{N}^+\!-\!\!-\!\!-\underset{\text{H}}{\text{N}}} \rightleftharpoons \underset{\substack{\\ \text{CH}_2}}{\text{N}\!-\!\!-\!\!-\text{N}}$$

| Tetrahydrofolate | Carbinolamine | Cationic imine | Methylene-tetrahydrofolate |

TMP synthetase catalyzes the transfer of the C-1 group of methylene-tetrahydrofolate to deoxyuridylate (dUMP) and, as well, reduction of the hydroxymethyl group to the methyl group of TMP. Because TMP is required for the synthesis of DNA, and thus for cell division, selective inhibition of TMP synthetase leads to death of rapidly growing cells (Chapter 15), a procedure which has been attempted for control of cancerous cells. These three enzymes not only exhibit some of the interesting facets of the biochemistry of tetrahydrofolate, but also demonstrate a diversity of properties representative of enzymes in general. Like most enzymes, they are larger than—and therefore not yet as well characterized as—lysozyme or ribonuclease. All three are oligomeric enzymes, and a brief description of their subunit interactions is a good introduction to the types of quaternary structures studied in Chapter 12.

Formyltetrahydrofolate Synthetase

Formyltetrahydrofolate synthetase (sometimes referred to as formate activating enzyme) catalyzes the entry of formate, produced by catabolic pathways, into the pool of one-carbon compounds covalently bound to tetrahydrofolate (Himes and Harmony, 1973). This enzyme has been detected in almost all cells tested. The reaction it catalyzes is:

$$\text{Formate} + \text{ATP} + \text{Tetrahydrofolate} \xrightleftharpoons{Mg^{2+}}$$

$$\text{10-Formyltetrahydrofolate} + \text{ADP} + P_i.$$

Despite the ubiquitous occurrence of the synthetase, it has been purified to homogeneity only from three species of *Clostridia*: *C. cylindrosporum, C. acidi-urici* and *C. thermoaceticum.* In these purine-fermenting bacteria, the synthetase does not play its normal role of formation of formyltetrahydrofolate, but instead catalyzes the reverse of the above reaction. Formyltetrahydrofolate is a product of purine degradation in these cells, and the synthetase-catalyzed reaction is believed to be a prime source of ATP for *Clostridia.* Despite an unfavourable equilibrium, the synthetase can efficiently catalyze this substrate-level phosphorylation *in vitro* (Curthoys and Rabinowitz, 1972). This alternate function of the synthetase in the clostridial cells requires the presence of large quantities of the enzyme (p. 45), probably formed by derepression. It is these elevated levels which have been the key to the preparation of crystalline formyltetrahydrofolate synthetase.

The equilibrium constant of the synthetase-catalyzed reaction is about 10^2, and so the enzyme can be used for preparation of 10-formyltetrahydrofolate. With excesses of the enzyme and two substrates, it can also be used as an analytical tool for determination of the third substrate. The assay for the enzyme is simple and rapid (formation, after acidification, of methenyltetrahydrofolate which absorbs at 355 nm; cf. pp. 109–110). The K_m values with the three clostridial enzymes for the three substrates are: formate, $2-7 \times 10^{-3}$ M; ATP (actually, MgATP), $1-6 \times 10^{-4}$ M; and tetrahydrofolate, $3-10 \times 10^{-4}$ M (Himes and Harmony, 1973). Only the C^6 diastereoisomer of tetrahydrofolate formed in the dihydrofolate reductase reaction is utilized as a substrate by the synthetase.

The synthetase has a molecular weight of 240 000 and consists of four monomeric units. These four subunits are identical when tested by SDS gel electrophoresis, peptide mapping and isoelectric focusing in 8 M urea (MacKenzie and Rabinowitz, 1971). The substrate-binding data of Curthoys and Rabinowitz (1972) suggest that one active site is located on each subunit, and that these active sites do not interact with each other.

Monovalent cations are necessary for the activity of the synthetase (Whiteley and Huennekens, 1962). Scott and Rabinowitz (1967) demonstrated that, in the absence of K^+, NH_4^+ or Rb^+, the synthetase dissociates into its 60 000 dalton monomers. Upon addition of KCl (50 mM) to the monomer, the enzyme reassociates and recovers its enzyme activity. The synthetase remains in the tetrameric form in the absence of NH_4^+ if substrates are present (Welch et al., 1968). Under these conditions, addition of effective monovalent cations increases the observed initial velocities several-fold, by lowering the K_m of formate from 50 mM to 5 mM. Therefore, the monovalent cations have two functions—prevention of dissociation and increase of apparent catalytic activity. Curthoys et al. (1972) have suggested that the monomer of the synthetase is catalytically inactive because it is unable to bind tetrahydrofolate or even the naturally occurring tetrahydropteroyl triglutamate, for which the enzyme has a 200-fold greater affinity ($K_s = 1.7 \times 10^{-6}$ M with the tetrameric form of synthetase). Harmony and Himes (1975) have examined the solvent isotope effect on the cation-induced reassociation of synthetase monomers. The overall stoichiometry of the reaction is:

$$4 \text{ Monomer} + 2 \text{ Me}^+ \rightleftharpoons \text{Tetramer--}(\text{Me}^+)_2 .$$

By substituting D_2O for H_2O, both the rate and extent of association are enhanced. Stabilization of quaternary structure by D_2O is generally attributed to strengthening of hydrophobic interactions (p. 488). However, the authors consider that their data are most consistent with the binding of a monovalent cation causing a change in bonding of a single proton in a monomer. Hydrophobic bonds *are* formed during the association, though.

Despite a great deal of research, the mechanism of the synthetase-catalyzed reaction is still unknown. No partial reactions have been unambiguously demonstrated, and the mechanism presently favoured is a concerted process first suggested by Himes and Rabinowitz in 1962:

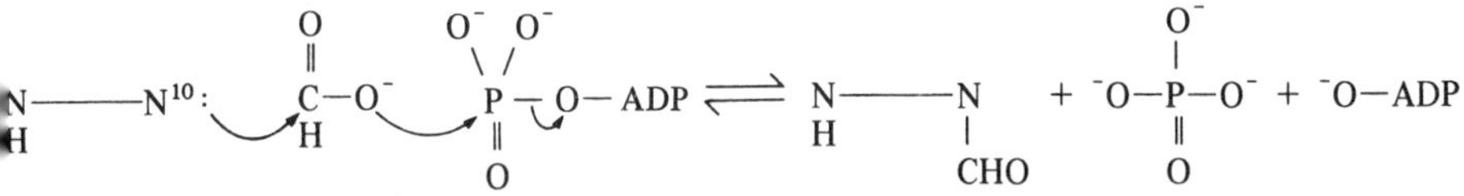

This mechanism is consistent with the kinetic data of Joyce and Himes (1966), which showed the kinetic mechanism to involve random binding of the substrates, with the binding of all three substrates before the release of any products. It is possible, though, that the reaction follows a stepwise mechanism involving a firmly-bound intermediate such as formylphosphate, with the binding of all substrates needed to provide the catalytically active conformation of the protein. Recent findings from e.s.r. and n.m.r. spectroscopy of the enzyme are compatible with either the concerted mechanism or active intermediates shielded by the active site (Buttlaire *et al.*, 1975). The pH-activity curve shows inflections at pH 6 and pH 9.2. The synthetase is inactivated by photo-oxidation of histidine residues, which suggests that the ionizable group at pH 6 is an imidazole general base which deprotonates N^{10} of tetrahydrofolate to produce the nucleophilic site for addition of formate.

Serine Transhydroxymethylase

The interconversion of L-serine and glycine is catalyzed by serine transhydroxymethylase (Chapter 8 in Blakley, 1969). Because this reaction required the presence of tetrahydrofolate, Blakley suggested in 1954 that 5,10-methylenetetrahydrofolate is a participant in the reaction:

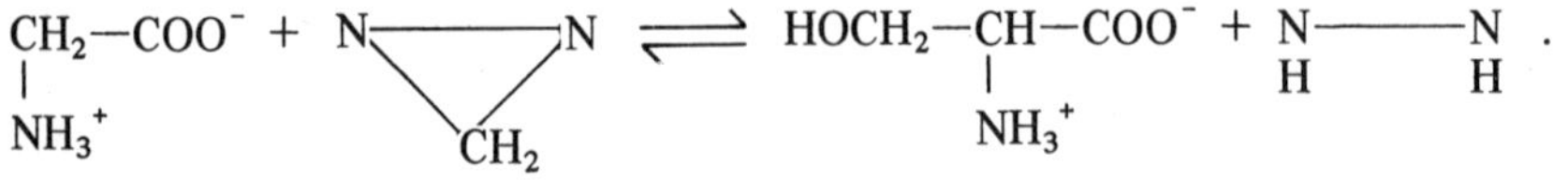

CH₂—COO⁻ + N N ⇌ HOCH₂—CH—COO⁻ + N—N
NH₃⁺
CH₂
NH₃⁺
H H

Glycine L-Serine

Addition of pyridoxal phosphate accelerates this reaction to a variable extent, depending on the amount of PLP bound to the particular enzyme preparation being examined. The purified serine transhydroxymethylase described below contains one mole of PLP per active site. Any description of the mechanism of the transhydroxymethylase reaction, therefore, must account for the roles of both coenzymes, PLP and tetrahydrofolate.

Metzler *et al.* (1954b) demonstrated that at 100° C in aqueous solution in the presence of pyridoxal and Al^{3+}, serine is cleaved to glycine and HCHO, as well as being deaminated to pyruvate. Under similar reaction conditions, a mixture of glycine and HCHO is converted to serine. They suggested that the reversible non-enzymic serine–glycine interconversion occurs by this mechanism (Metzler *et al.*, 1954a):

HOH₂C—CH—COO⁻
HO—C⁺ CH—COO⁻
H
H
H⁺
⁻CH—COO⁻
HC
N
Me²⁺
O
HOH₂C
CH₃
+ HCHO

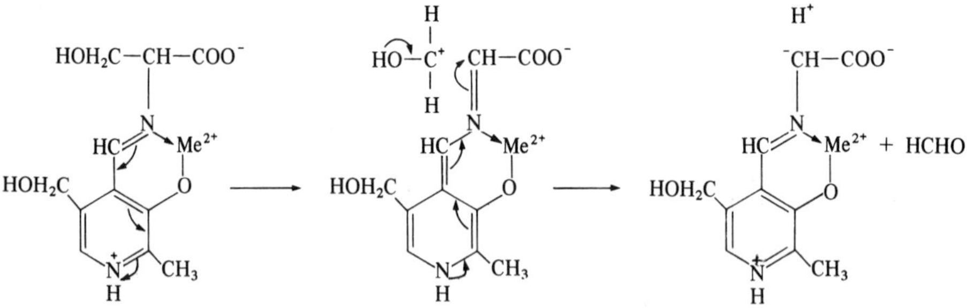

I II III

The cation of formaldehyde is released from the Schiff base of serine and pyridoxal (I) to produce the stabilized glycine carbanion (II). Subsequently, a proton is added (III) to the α-carbon of this intermediate, so that the glycine–pyridoxal Schiff base is produced. The mechanism of the reaction catalyzed by the transhydroxymethylase is almost identical. One major role of the enzyme is to allow only the aldol condensation reaction to occur (i.e. reversible cleavage of only the α-carbon—side-chain bond of the amino acid substrate).

Most of the mechanistic studies of the transhydroxymethylase have been performed using a homogeneous enzyme preparation from rabbit liver (Schirch and Mason, 1963). This enzyme has a molecular weight of 215 000 daltons and possesses four subunits (Martinez-Carrion et al., 1972). These subunits are identical, having glycine as their N-terminal residue and phenylalanine as their C-terminal (Schirch et al., 1973). The holoenzyme, a protonated internal imine, has an absorption maximum of 428 nm. ES complexes of serine transhydroxymethylase with several amino acids have been detected spectrally; for example, upon addition of glycine, major peaks are observed at 343 nm and 425 nm, with a minor peak at 495 nm (Schirch and Jenkins, 1964), indicating the existence of three enzyme–glycine complexes. Later, I shall discuss the use of these spectral bands for measurement of the formation and interconversion of ES complexes.

The stereochemistry of the serine transhydroxymethylase reaction and the existence of a carbanion intermediate have been discovered using isotopic-labelling experiments. Schirch and Jenkins (1964) demonstrated that the enzyme catalyzes an exchange between the α-hydrogen of glycine and the solvent. Jordan and Akhtar (1970) prepared glycine stereospecifically labelled with tritium at the α-carbon. The S hydrogen (H^S) at C^2 is lost when serine is formed, but the R hydrogen at C^2 is retained. Therefore, the overall reaction occurs with retention of configuration:

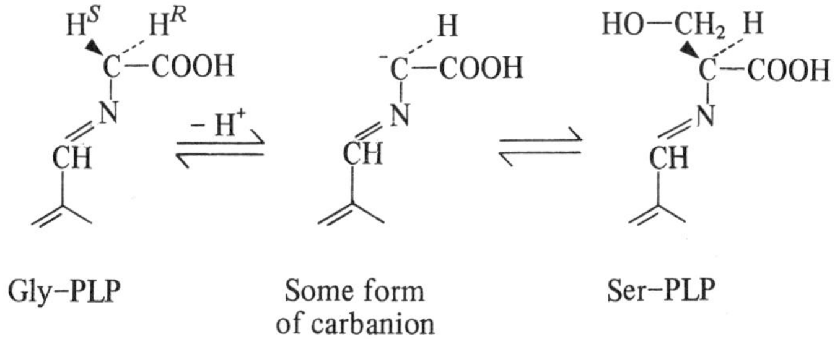

| Gly–PLP | Some form of carbanion | Ser–PLP |

In the absence of formaldehyde, the enzyme catalyzes the stereo-specific reprotonation of the carbanion to form Gly—PLP.

Serine transhydroxymethylase has a fairly wide specificity for the amino acid substrate (p. 30). The transfer reactions of most of these acids do not require tetrahydrofolate, and Chen and Schirch (1973a, b) have reported that even the reaction:

$$\text{Glycine} + \text{HCHO} \rightleftharpoons \text{Serine}$$

can occur in the absence of tetrahydrofolate, although at a greatly reduced rate. Their kinetic experiments demonstrated that glycine adds to the enzyme before formaldehyde, and suggest that there is a formaldehyde-binding site on the enzyme. The reaction of formaldehyde with the enzyme–glycine complex is shown clearly in Fig. 9, by

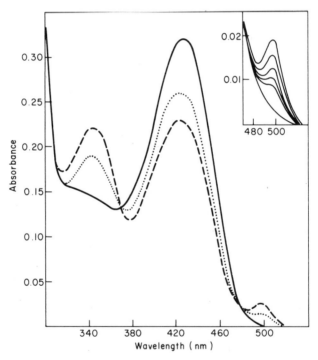

Fig. 9. Effect of glycine and formaldehyde on the absorption spectrum of serine transhydroxymethylase. The solid line depicts the spectrum of the enzyme (1.45 mg ml^{-1} at pH 8.1), the dashed line the spectrum after addition of glycine (to 91 mM) and the dotted line the spectrum after addition of HCHO (8.5 mM) to the enzyme–glycine complex. The inset figure shows the effect of several concentrations of HCHO on the absorption peak at 495 nm. The bottom line is free enzyme, the top line the enzyme–glycine complex, and the others are at 4.9, 9.7, 14.7 and 19.5 mM HCHO, reading downwards. From Chen and Schirch (1973b) with permission.

the quenching of the absorption maxima at 343 nm and 495 nm. Increasing levels of formaldehyde cause the peaks at 343 nm (the geminal diamine of PLP) and 495 nm (the enzyme–glycine anion) to shift to that absorbing at 425 nm (the Gly—PLP Schiff base). Chen and Schirch have postulated that the primary role of tetrahydrofolate is to catalyze the rate-limiting interaction of formaldehyde with the HCHO-binding site on the enzyme. Jordan and Akhtar (1970) had previously suggested a similar function, i.e. that methylenetetrahydrofolate acts solely as a carrier of the formaldehyde which reacts with the carbanion. When threonine is the substrate for the transhydroxymethylase, the aldol cleavage follows a mechanism in which acetaldehyde does not form a covalent linkage with the enzyme. This observation (Jordan *et al.*, 1976) supports direct reaction of the glycine carbanion with the reacting aldehyde, whether acetaldehyde or formaldehyde. Free formaldehyde might be generated from methylenetetrahydrofolate at the catalytic centre, under conditions allowing stereospecific reaction.

The assignments of *ES* structures to the absorption maxima (Fig. 9) of transhydroxymethylase–ligand complexes were made in the early 1960s (Schirch and Jenkins, 1964), with the exception of the 343 nm band. In 1971, O'Leary reported that reaction of 1,3-diaminopropane with PLP produced a compound absorbing at 330 nm and having the structure of a cyclic hexahydropyrimidine (a geminal diamine):

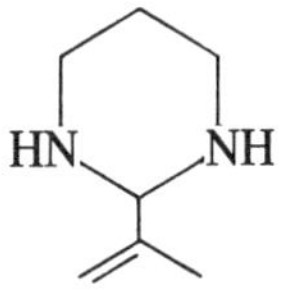

Diamine of PLP

This type of aldimine, formed by addition of a second amino group to C^{4a} of PLP, is now considered to be an obligatory intermediate in amino acid–PLP Schiff base formation from the E—PLP internal imine (cf. p. 232). The results of recent experiments in which rapid-kinetic techniques were used (Cheng and Haslam, 1972; Schirch, 1975) are consistent with the 343-nm chromophore being a geminal diamine. When glycine reacts with the enzyme, the first reaction is bimolecular.

$$E\text{-}428\,\text{nm} + \text{Gly} \rightleftharpoons E\text{-}343\,\text{nm} \rightleftharpoons E\text{-}425\,\text{nm} \rightleftharpoons E\text{-}495\,\text{nm}$$

This first reaction must be the addition of the anion of glycine to enzyme-bound PLP (428 nm) to form the geminal diamine (343 nm). Next, the diamine is converted to the Gly—PLP Schiff base (425 nm). Formation of the 495-nm complex from Gly—PLP is slow, but is stimulated by a factor of 1000 by tetrahydrofolate (Schirch, 1975). This result agrees with the observation that tetrahydrofolate stimulates the α-H : H_2O exchange (Schirch and Jenkins, 1964) but somewhat contradicts the conclusion that tetrahydrofolate is only a donor or acceptor of HCHO. The E-425 nm $\rightleftharpoons$ E-495 nm process is not simple, so tetrahydrofolate may cause an essential change in the conformation of the enzyme. The study of serine transhydroxymethylase has shown that the reaction is primarily a covalent catalysis involving PLP, with the reaction specificity controlled by the enzyme environment. Tetrahydrofolate interplays with PLP by presenting the HCHO group in the correct orientation, and possibly functions in an as yet unknown structural role. Above all, the reaction shows how well non-enzymic model systems (with both coenzymes) apply to the enzymic catalysis.

Thymidylate Synthetase

The final step in the biosynthesis of the pyrimidine nucleotide thymidylate (TMP) is the methylation of C^5 of the uracil derivative, deoxyuridylate (dUMP). This reaction, catalyzed by thymidylate synthetase, requires formaldehyde, dUMP and substrate amounts of tetrahydrofolate, but no extra reducing agent (Chapter 7 in Blakley, 1969; Friedkin, 1973). The stoichiometry of the *in vitro* reaction is:

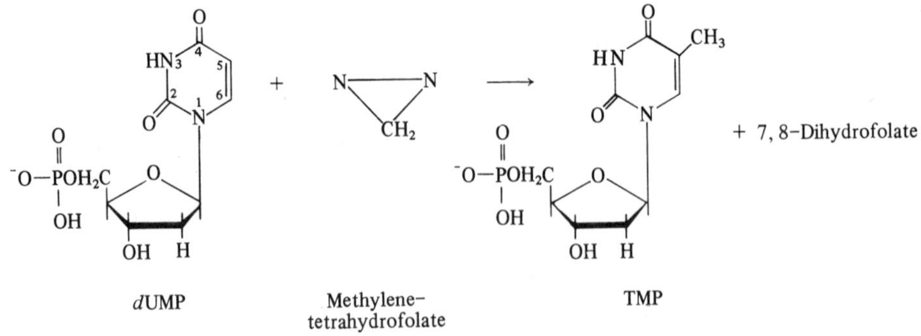

dUMP   Methylene–   TMP
            tetrahydrofolate

Table III. Transfer of tritium from tetrahydrofolate to TMP. Adapted from Pastore and Friedkin (1962).

Compound	Specific Activity (Ci mol^{-1})	
NADP^{3}H (A-side)	4.3	–
NADP^{3}H (B-side)	–	1.4
Tetrahydrofolate-^{3}H	4.2	0.03
TMP-^{3}H	3.4	0.03

Tetrahydrofolate is the one-carbon carrier in this process but it is also the hydrogen donor for reduction of the formaldehyde to the 5-methyl group of TMP. Pastore and Friedkin (1962) prepared NADPHs labelled at C^4 with tritium on the A-side and the B-side. They then reduced 7,8-dihydrofolate in the presence of dihydrofolate reductase, to form 6-^{3}H-tetrahydrofolate. Only the hydrogen from the A-side of NADPH was transferred in this reduction (Table III). When the labelled tetrahydrofolate was incubated with HCHO, dUMP, MgCl$_2$ and TMP synthetase, there was almost complete, and therefore direct, transfer of the tritium to TMP (first column, Table III). Chemical degradation established that the tritium was located in the methyl group of TMP, and other workers later showed that it was derived solely from C^6 of tetrahydrofolate. Friedkin proposed the formation of an intermediate (thymidylyltetrahydrofolate) having a methylene bridge between N^5 of tetrahydrofolate and C^5 of dUMP:

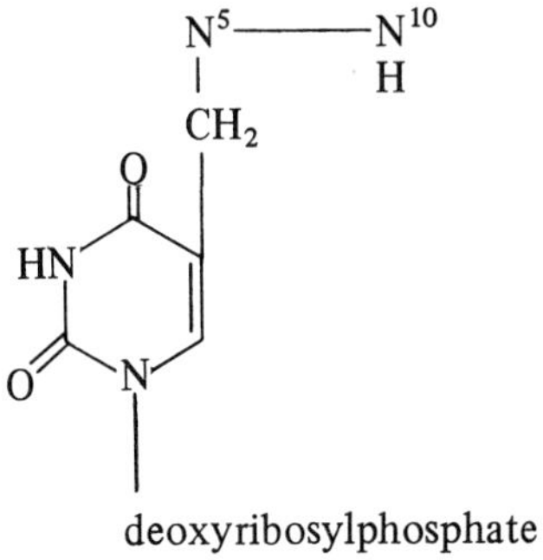

deoxyribosylphosphate

This intermediate has not been isolated or detected, but most mechanisms suggested for the synthetase include a bridged structure of this type.

The synthesis of TMP involved formation of methylenetetrahydrofolate (from HCHO *in vitro* and from the serine ⇌ glycine reaction *in vivo*), the synthetase-catalyzed reaction itself, and regeneration of tetrahydrofolate by the dihydrofolate reductase reaction. The sequence of these three reactions is considered to be a thymidylate synthesis cycle:

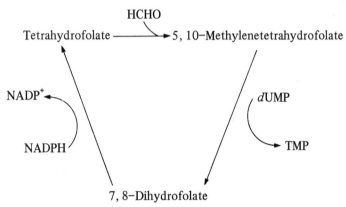

Experimentally, the TMP synthetase reaction is not reversible, and the equilibrium of the dihydrofolate reductase-catalyzed reaction lies in the direction of tetrahydrofolate formation. Rothman *et al.* (1973), by calorimetric experiments, have determined the thermodynamic parameters of each of the three reactions of the cycle

Table IV. Thermodynamic parameters for the three reactions in the biosynthesis of thymidylate. All values are for pH 7.4 and 25° C. The hydrated form of formaldehyde, $H_2C(OH)_2$, was considered the reactive species in reaction 1. The equilibrium constant and ΔG and ΔS values for reaction 2 are estimates. Data from Rothman *et al.* (1973).

Reaction	K (molar units)	ΔG (kcal mol^{-1})	ΔH (kcal mol^{-1})	ΔS (kcal mol^{-1} deg^{-1})
1. Formation of methylenetetrahydrofolate	3.2×10^4	-6.1	-25	-63
2. TMP synthetase reaction	2×10^{25}	-34	-9.7	$+81$
3. Reduction of dihydrofolate	5.6×10^4	-6.4	-12.6	-20.8

(Table IV). Their results show a modest decrease in enthalpy and a large increase in entropy (derived from opening of the 5-membered ring of methylenetetrahydrofolate) in the TMP synthetase reaction (reaction 2). The ΔH value for the sum of the three reactions, with the reduced folate compounds acting catalytically, is -47.1 kcal mol^{-1}, providing a large contribution to formation of TMP. Although the TMP synthetase reaction theoretically might be considered reversible, the thermodynamic measurements verify that experimentally it is irreversible.

There are three convenient assay procedures for analyzing for TMP synthetase activity. These assay techniques are instructive, because they show how to make use of some of the unique properties of an enzyme-catalyzed reaction. The simplest assay (Wahba and Friedkin, 1961) measures the increase in absorbance at 338 nm during the reaction, due to conversion of methylenetetrahydrofolate (λ_{max} at 295 nm) to 7,8-dihydrofolate (λ_{max} at 282 nm with an inflection at 306 nm). This assay is not sufficiently sensitive for use with preparations of low activity. In the conversion of dUMP to TMP, the hydrogen atom at C^5 of dUMP is replaced by a methyl group. Lomax and Greenberg (1967) have used the release of tritium from 5-^{3}H-dUMP into water (collected by distillation) as an assay. This assay is extremely sensitive, but is somewhat inconvenient. Santi et $al.$ (1974a) have recently devised a binding, rather than a kinetic, assay based on the specific covalent binding of radioactive 5-fluoro-2′-deoxyuridylate (FdUMP, Chapter 15) to TMP synthetase in the presence of methylenetetrahydrofolate. The $E-^3$H$-$FdUMP complex is trapped on nitrocellulose membranes under conditions allowing free ^{3}H$-$FdUMP to be washed away. This assay is 20 times more sensitive than the Wahba–Friedkin assay, and is especially useful for numerous routine assays, such as detection of activity during purification of the enzyme by column chromatography.

TMP synthetase has been purified extensively from mammalian tissues and bacterial cells. All the synthetases characterized to date have a molecular weight in the range 55 000–70 000 daltons, and are composed of two seemingly identical subunits. Elevated levels of synthetase activity (200- to 500-fold) have been found in sublines of $Lactobacillus$ $casei$ resistant to the anti-folates amethopterin or dichloroamethopterin, so that dihydrofolate reductase and TMP synthetase both can be isolated in large quantities from these cells.

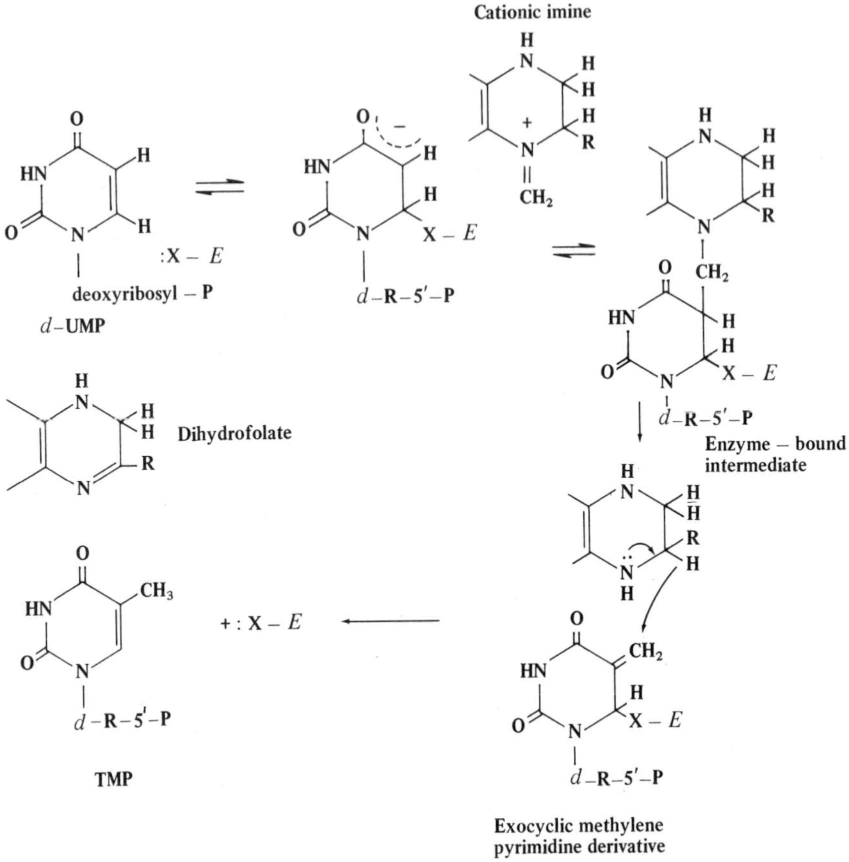

Fig. 10. Possible mechanism for the formation of dihydrofolate and TMP from methylenetetrahydrofolate and dUMP.

The TMP synthetase from *L. casei* has a molecular weight of 70 000, is composed of two subunits of 35 000 daltons (both with Met—Leu at their N-terminal; Loeble and Dunlap, 1972) and binds two molecules of FdUMP per molecule of enzyme dimer (Santi *et al.*, 1974b). TMP synthetase is one of only a few enzymes which can be inactivated by treatment with carboxypeptidase. When the C-terminal Val residue is selectively removed from only one of the two protomers, all activity is lost (Aull *et al.*, 1974). This suggests that the C-terminal is involved in catalysis, or, more probably, in maintenance of the native conformation.

The mechanism currently favoured for the TMP synthetase reaction is based on data from both enzymic and model reaction experiments, reviewed in Friedkin (1973) and Benkovic and Bullard (1973). The first step in the reaction is considered to be addition of a nucleophilic group (X) of the enzyme to C^6 of dUMP, followed by electrophilic attack by the cationic imine derived from methylenetetrahydrofolate (Fig. 10). The resulting intermediate differs from thymidylyltetrahydrofolate by its covalent linkage to the enzyme and by saturation of the 5,6-double bond of the pyrimidine. Several chemists have suggested that this intermediate undergoes the hydride shift necessary for reduction of the methylene bridge by cleavage to tetrahydrofolate and an enzyme-bound reactive exocyclic methylene compound which can abstract the hydrogen from C^6 of the tetrahydrofolate. The interaction of FdUMP with TMP synthetase seems to parallel the first few steps of the normal enzymic reaction. A nucleophilic group on the enzyme reacts covalently at C^6 of the pyrimidine, and methylenetetrahydrofolate is converted to a new chemical species, possibly the imine (Santi *et al.*, 1974b).

There is considerable controversy about the identity of the nucleophilic group of the enzyme which reacts at C^6 of the uracil of dUMP. Most workers regard the nucleophile to be a thiolate group of cysteine. This assignment is based on the requirement of exogenous RSH compounds for activity and stability of TMP synthetase, and on the loss of activity of the enzyme upon modification at one of the four —SH groups with *p*-hydroxymercuribenzoate (Dunlap *et al.*, 1971). Kalman (1971a, b) has elucidated the mechanism of a nonenzymic reaction in which the sulphydryl group of glutathione catalyzes the exchange of the hydrogen at C^6 of uridine. (This is a model system for the Lomax and Greenberg enzyme assay, p. 257.) His work is readily extended to give support to the essential role of a thiolate in the alkylation step catalyzed by TMP synthetase. A contrary conclusion has been reached by Santi, who has suggested that the hydroxymethyl group of a threonine residue of the enzyme is the active nucleophile. Data related to this conclusion and to the mechanism of inhibition of TMP synthetase by FdUMP are discussed in Chapter 15 (p. 524).

Tetrahydrobiopterin

The coenzyme tetrahydrobiopterin is structurally related to tetrahydrofolate, but serves an entirely different metabolic purpose—that of an oxidation–reduction cofactor. The only known reduction involving tetrahydrofolate is the synthesis of TMP, and in this reaction tetrahydrofolate is oxidized by a different mechanism than is tetrahydrobiopterin in its reactions. Tetrahydrobiopterin is a smaller molecule than tetrahydrofolate, with a dihydroxypropyl group at C^6. The naturally occurring isomer has an L-erythro side-chain. Many chemical studies and some enzymatic experiments have been performed with the analogous and more available synthetic 6-methyl- and 6,7-dimethyltetrahydropterins.

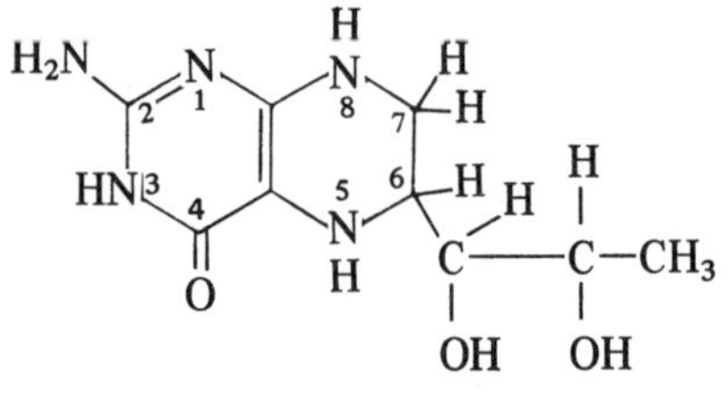

Tetrahydrobiopterin 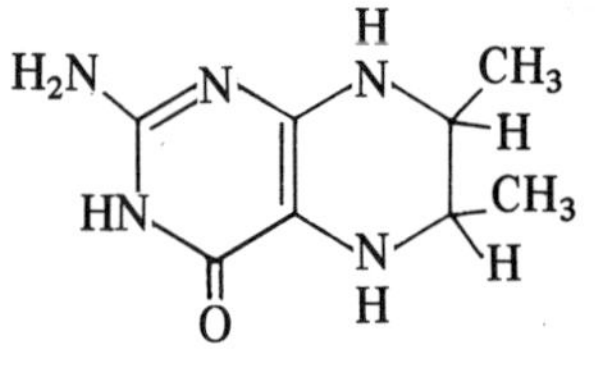   6,7–Dimethyltetrahydropterin

The metabolic role of tetrahydrobiopterin was discovered by Kaufman during his studies on the phenylalanine hydroxylating system (reviewed by Kaufman, 1971). The conversion of phenylalanine to tyrosine is a mixed function oxidation (i.e. a mono-oxygenase reaction in which one oxygen atom from O_2 is incorporated into the product) which Kaufman, in 1957, found to have the following stoichiometry:

$$\text{NADPH} + \text{H}^+ + \text{Phenylalanine} + O_2 \xrightarrow{\text{2 Enzymes}} \text{NADP}^+ + \text{Tyrosine} + H_2O.$$

The reaction is assayed by measuring either the production of tyrosine or the phenylalanine-dependent oxidation of NADPH. Kaufman observed that two enzymes are required to catalyze the hydroxylation. The more stable enzyme (E_2) occurred free of the other activity (E_1) in extracts of sheep liver. (In sheep liver, E_1 is not easily solubilized.) E_1 was then purified from rat liver, the most suitable source, using assay mixtures containing excess sheep liver E_2. During early kinetic studies, Kaufman noticed a lag in the oxidation of

NADPH which could be abolished by incubation of E_2 with NADPH. He correctly suspected that a cofactor was present in the E_2 fraction, and that it was reduced to its active form by NADPH in the presence of the crude E_2 fraction. Kaufman found that boiled extracts of liver contained cofactor activity, and that the chemical properties of the material were those of a 6-substituted pterin differing from folate. 6-Methyltetrahydropterin has high activity in replacing the cofactor. Tyrosine formation from phenylalanine required only O_2, tetrahydropterin and E_1 (which therefore is the phenylalanine hydroxylase), but is stimulated by the presence of E_2 and NADPH. Kaufman postulated that an oxidized pterin is formed in the hydroxylation and that E_2 catalyzes its cyclic reduction. While isolation of the naturally occurring cofactor proceeded, Kaufman used synthetic 6-methylpterins to ascertain the structure of the oxidized pterin, but because the oxidized pterin is very unstable, attempts at its isolation yielded only inactive 7,8-dihydropterin. Kaufman generated the intermediate by non-enzymic oxidation of tetrahydropterins. It has several distinctive properties:

1. titration of tetrahydropterin with oxidizing agents showed it is a dihydropterin;
2. it is active as a cofactor only after reduction;
3. it can be reduced by NADPH alone, or more rapidly by NADPH + E_2;
4. it can be reduced by 2-mercaptoethanol (7,8-dihydropterins cannot oxidize thiols); and
5. it is rapidly converted to a 7,8-dihydropterin under anaerobic conditions, so is probably a tautomer of 7,8-dihydropterin.

When the dihydro intermediate is reduced with chemically prepared NADP³H in the presence of E_2, no carbon-bound tritium is incorporated into the tetrahydropterin formed (Kaufman, 1964). This result rules out both the 5,8- and 5,6-dihydro structures, and favours a quinonoid structure for the intermediate, as proposed by Hemmerich. Almost simultaneously, Kaufman (1963) identified the cofactor from rat liver extracts as 7,8-dihydrobiopterin, so that part of the NADPH requirement is for reduction to tetrahydrobiopterin, catalyzed by 7,8-dihydrofolate reductase. The pathway for the phenylalanine hydroxylase system is shown in Fig. 11. The hydroxylation step is catalyzed by E_1, the phenylalanine hydroxylase, with the reducing

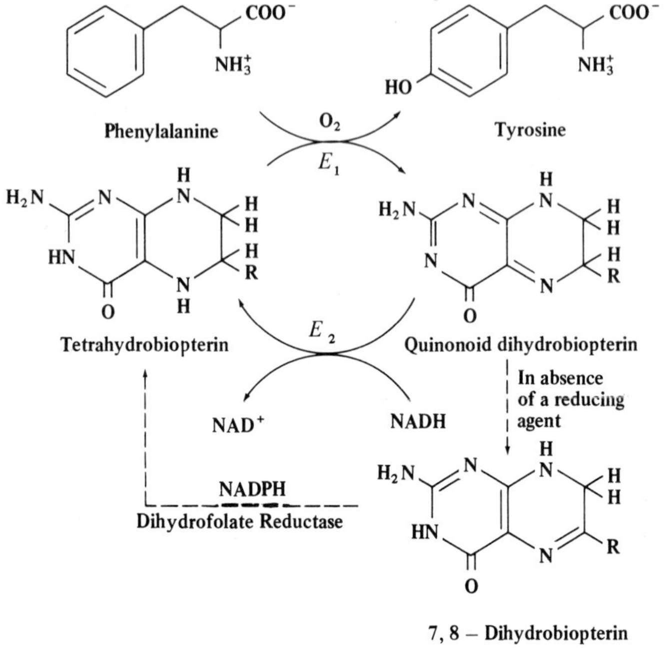

Fig. 11. Oxidation and reduction of tetrahydrobiopterin during the hydroxylation of phenylalanine. E_1 is phenylalanine hydroxylase and E_2 is quinonoid dihydropterin reductase. The pathway via 7,8-dihydrobiopterin, indicated by the dashed line, is a salvage pathway.

agent tetrahydrobiopterin being oxidized to the unstable quinonoid dihydrobiopterin. E_2, quinonoid dihydropterin reductase, catalyzes the reduction of the quinonoid dihydropterin by NADH or less effectively by NADPH (Nielsen *et al.*, 1969), recycling the unstable intermediate to active coenzyme. Dihydrofolate reductase and NADPH are used only in the initial reduction and to reactivate any 7,8-dihydro isomer formed by non-enzymatic degradation of the quinonoid compound.

Chemical characterization of the formation and isomerization of quinonoid dihydropterins has been achieved using anaerobic spectrophotometry and polarography, although instability of these compounds precludes isolation of the intermediate in solid form. Archer *et al.* (1972) have shown by stopped-flow spectrophotometry that oxidation of tetrahydropterins by Fe^{3+} occurs only after dissociation of N^5 of the cationic tetrahydropterin ($pK_a = 5.6$). The neutral tetra-

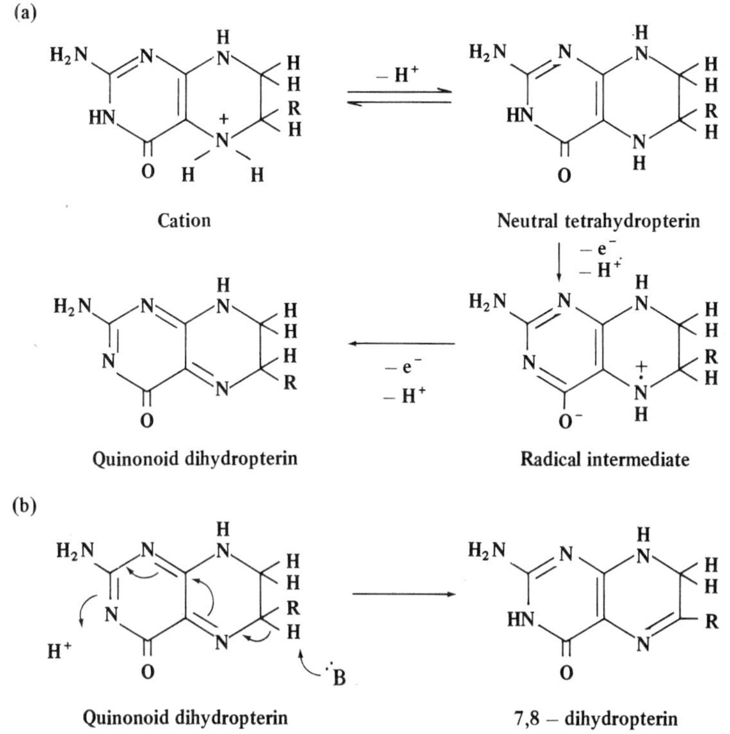

(a)

Cation   Neutral tetrahydropterin

Quinonoid dihydropterin   Radical intermediate

(b)

Quinonoid dihydropterin   7,8 – dihydropterin

Fig. 12. (a) Abbreviated mechanism for the rapid oxidation of tetrahydropterins by two molecules of Fe^{3+}. (b) Mechanism of the base-catalyzed tautomerization of quinonoid dihydropterin to 7,8-dihydropterin. Acid catalysis of this reaction occurs by a similar mechanism, in which a proton is donated to N^3 in a rapid step and the proton is removed from C^6 by base (B:) in the rate-determining step.

hydropterin is oxidized in two one-electron steps. This oxidation is extremely rapid, and is followed by the relatively slow isomerization (Fig. 12). Kinetic studies of the isomerization (Archer and Scrimgeour, 1970) indicate that this isomerization is catalyzed by both general acids and general bases. Primary kinetic isotope experiments using 6-^{2}H-quinonoid dihydropterin show that the cleavage of the C—H bond at position-6 is the rate-limiting step in the formation of 7,8-dihydropterins ($k_H/k_D = 10$–12). This latter finding proves Kaufman's conclusion that the quinonoid intermediate contains a hydrogen at C^6. Tetrahydropterins can be rapidly oxidized by molecular oxygen, but this reaction is favoured only at alkaline pH values, when the 3,4-cyclic amide group is ionized (Blair and Pearson,

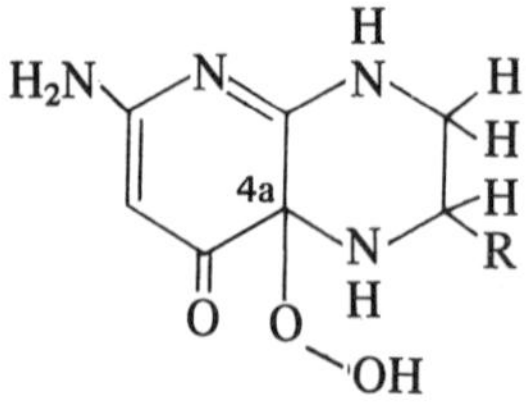

Possible hydroperoxide intermediate

1974). Autoxidation, first thought to proceed through a hydroperoxide intermediate, appears to be a free radical reaction like the oxidation by Fe^{3+} (Fig. 12). A hydroperoxide intermediate still might occur in the phenylalanine hydroxylase-catalyzed oxidation of tetrahydrobiopterin because non-enzymic and enzymic oxidations proceed by different mechanisms (Smith et al., 1975).

Phenylalanine hydroxylase has been purified to near homogeneity from rat liver. Interest in this enzyme is great because its absence or extremely low activity is the defect in phenylketonuria, the most common disorder of amino acid metabolism in humans. Untreated phenylketonuria leads to mental retardation, possibly by inhibition of pyruvate kinase (and energy production) in the brain by high levels of phenylalanine. The hydroxylase displays several features which must be explained when sufficient data are available for establishing a mechanism of action. It is an oligomeric enzyme, containing about one mole of ferric iron per subunit (Fisher et al., 1972). The e.p.r. signal of the iron disappears (probably by reduction to ferrous iron) when tetrahydropterin and phenylalanine are added, indicating that the iron participates in a catalytic step. When 7-methyltetrahydropterin is used in place of the 6-methyl compound, or when p-fluorophenylalanine replaces phenylalanine, the oxidation of tetrahydropterin exceeds the production of tyrosine. In the presence of lysolecithin (or chymotrypsin), the hydroxylase is completely uncoupled from hydroxylation, and catalyzes a product(tyrosine)-dependent oxidation of tetrahydropterin by oxygen, with formation of hydrogen peroxide (Fisher and Kaufman, 1973):

$$\text{Tetrahydropterin} + O_2 \xrightarrow{\text{Tyrosine}} \text{Quinonoid dihydropterin} + H_2O_2.$$

Peroxide formation suggests that a pterin hydroperoxide is formed. It is not yet known if the iron is reduced in this uncoupled reaction.

When tetrahydrobiopterin (but not either of the synthetic tetra-hydropterins) is the reducing substrate, the hydroxylase is stimulated by a phenylalanine hydroxylase-stimulating protein. This protein appears to be an enzyme which catalyzes the breakdown or release of a biopterin-containing intermediate in the hydroxylase reaction (Huang and Kaufman, 1973). During the hydroxylation reaction, tritium or deuterium at C^4 of phenylalanine is not replaced but migrates to the adjacent C^3 (Guroff *et al.*, 1967):

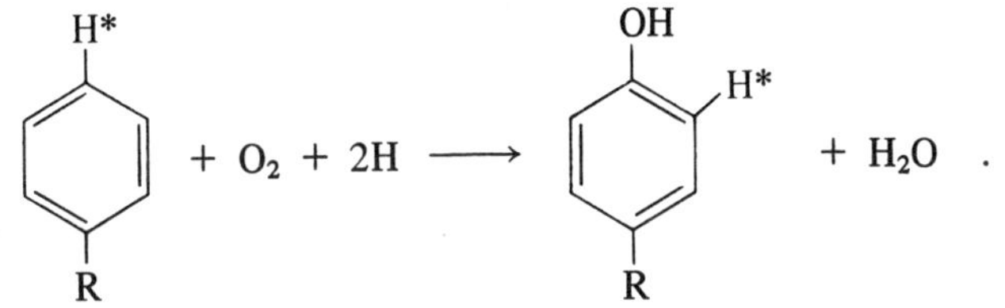

Chemical analogies suggest that the hydroxyl group is added as an electrophilic reactant (OH^+ or its equivalent) to form an arene oxide which could lead to the observed migration. The compound which

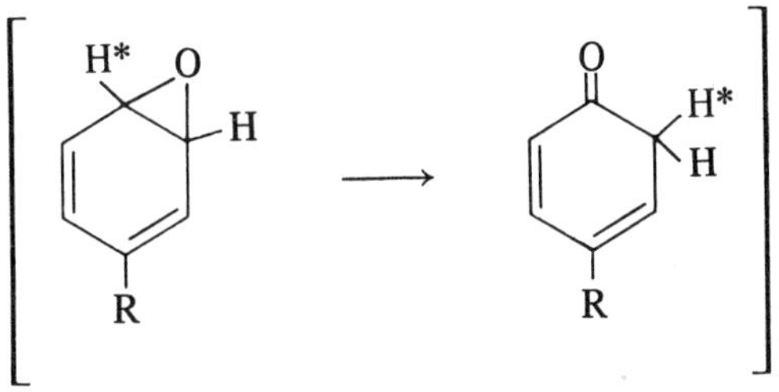

donates oxygen to phenylalanine is assumed to be the 4a-hydro-peroxide of tetrahydrobiopterin. In fact, Kaufman (1975) has succeeded in detecting a biopterin-containing intermediate in the early stages of tyrosine formation. He proposes that this intermediate is the 4a-hydroxytetrahydropterin formed from the hydroperoxide in this reaction:

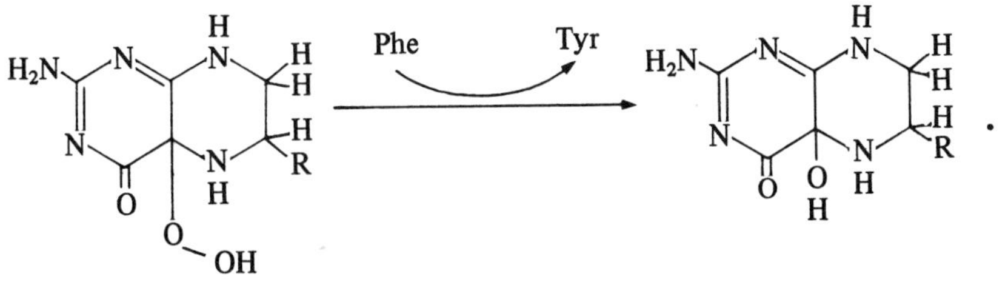

Although many non-enzymic hydroxylations have been studied and the chemistry of tetrahydropterin oxidation is considerably better understood than it was a few years ago, many additional experiments are needed.

The quinonoid dihydropterin reductase (E_2) is a widely distributed enzyme. It has been isolated (p. 73) in pure form from liver, brain and adrenal medulla (Cheema et al., 1973). The reductase from each of these tissues has a molecular weight of about 50 000, and is made up of two identical 25 000 dalton subunits. The high activity of the reductase in a wide variety of tissues is in keeping with its function in maintaining biopterin in the active tetrahydro form. It appears that the activity of the reductase is sufficient to catalyze reduction of the unstable quinonoid isomer and in this way minimize isomerization to 7,8-dihydrobiopterin. The reductase is a supplementary enzyme not only for phenylalanine hydroxylation, but also for other pterin-requiring hydroxylations such as the hydroxylation of tyrosine in adrenal medulla and brain, and of tryptophan in brain.

Despite the known function of quinonoid dihydropterin reductase as a supplementary enzyme for hydroxylation reactions, its full importance has only recently been discovered. An unusual type of phenylketonuria due to a deficiency of the reductase has been uncovered by Kaufman et al. (1975). The low phenylalanine diet that is used for the successful treatment of classical phenylketonuria did not prevent the onset of severe neurologic symptoms in the patient. Hydroxylation reactions in brain tissue are affected specifically because brain has no 7,8-dihydrofolate reductase activity. Therefore, neither the direct nor the salvage pathway (Fig. 11) is available for the reformation of tetrahydrobiopterin. In the absence of a good supply of tetrahydrobiopterin, a person afflicted with this rare inborn metabolic error cannot synthesize sufficient amounts of the neurotransmitters norepinephrine, dopamine and serotonin.

Examination of the chemistry of coenzymes has clearly established the premise that the functional group of a coenzyme is part of the catalytic site of its holoenzyme. For example, C^4 of NAD^+, the 4-formyl group of PLP, and the N^5 and N^{10} positions of tetrahydrofolate participate directly in group transfer reactions. These functional sites must interact with catalytic groups on the protein, such as His-195 of lactate dehydrogenase. Most coenzymes are covalent catalysts, with PLP and biotin carboxyl carrier protein (next chapter) being particularly good examples and logical progressions toward the discussion of covalent catalysis in Chapter 10. One last point worth noting is the versatility of some low molecular weight coenzymes, such as the variety of reactions in which ATP can be used, the dozen types of amino acid reactions in which PLP participates, the redox and non-redox reactions of NAD^+ and the interconversion of four oxidation levels of C-1 units in tetrahydrofolate-dependent reactions. These examples show how a limited number of coenzymes have been used for many purposes, just as a limited number of amino acids are used in the synthesis of proteins of diverse structure and function. In summary, coenzymes can be considered as supplying suitable catalytic sites, with their apoenzymes supplying binding sites and the other environmental characteristics essential for extremely fast reaction rates.

REFERENCES

Abbott, E. H. and Bobrik, M. A. (1973). *Biochemistry* **12**, 846-851.

Adams, M. J., Ford, G. C., Koekoek, R., Lentz, P. J., Jr., McPherson, A., Jr., Rossmann, M. G., Smiley, I. E., Schevitz, R. W. and Wonacott, A. J. (1970). *Nature, Lond.* **227**, 1098-1103.

Adams, M. J., Buehner, M., Chandrasekhar, K., Ford, G. C., Hackert, M. L., Liljas, A., Rossmann, M. G., Smiley, I. R., Allison, W. S., Everse, J., Kaplan, N. O. and Taylor, S. S. (1973). *Proc. natn. Acad. Sci. U.S.A.* **70**, 1968-1972.

Anderson, B. M., Ciotti, C. J. and Kaplan, N. O. (1959). *J. biol. Chem.* **234**, 1219-1225.

Archer, M. C. and Scrimgeour, K. G. (1970). *Can. J. Biochem.* **48**, 278-287.

Archer, M. C., Vonderschmitt, D. J. and Scrimgeour, K. G. (1972). *Can. J. Biochem.* **50**, 1174-1182.

Auld, D. S. and Bruice, T. C. (1967). *J. Am. Chem. Soc.* **89**, 2098-2106.

Aull, J. L., Loeble, R. B. and Dunlap, R. B. (1974). *J. biol. Chem.* **249**, 1167-1172.

Averill, B. A., Schonbrunn, A., Abeles, R. H., Weinstock, L. T., Cheng, C. C., Fisher, J., Spencer, R. and Walsh, C. (1975). *J. biol. Chem.* **250**, 1603-1605.

Baker, B. R. and Ho, B.-T. (1964). *J. Pharm. Sci.* **53**, 1457-1466.

Baugh, C. M. and Krumdieck, C. L. (1971). *Ann. N.Y. Acad. Sci.* **186**, 7-28.

Beinert, H. (1960). *In* "The Enzymes" (Boyer, P. D., Lardy, H. and Myrbäck, K., eds), Second edition, Vol. 2, pp. 339-416. Academic Press, New York.

Benkovic, S. J. and Bullard, W. P. (1973). *Prog. bioorg. Chem.* **2**, 133-175.

Biellmann, J. F. and Callot, H. (1966). *Tetrahedron Lett.* 3991-3996.

Biellmann, J. F. and Callot, H. (1968). *Bull. Soc. Chim. Fr.* 1154-1159.

Blair, J. A. and Pearson, A. J. (1974). *J. chem. Soc.* Perkin II. 80-88.

Blakenhorn, G. (1975). *Biochemistry* **14**, 3172-3176.

Blakley, R. L. (1960). *Nature, Lond.* **188**, 231-232.

Blakley, R. L. (1969). "The Biochemistry of Folic Acid and Related Pteridines." North-Holland, Amsterdam.

Brändén, C.-I., Eklund, H., Nordström, B., Boiwe, T., Söderlund, G., Zeppezauer, E., Ohlsson, I. and Åkeson, Å. (1973). *Proc. natn. Acad. Sci. U.S.A.* **70**, 2439-2442.

Brändén, C.-I., Jörnvall, H., Eklund, H. and Furugren, B. (1975). *In* "The Enzymes" (P. D. Boyer, ed.), Third edition, Vol. 11, pp. 103-190. Academic Press, New York.

Braunstein, A. E. (1960). *In* "The Enzymes" (Boyer, P. D., Lardy, H. and Myrbäck, K., eds), Second edition, Vol. 2, pp. 113-184. Academic Press, New York.

Braunstein, A. E. (1973). *In* "The Enzymes" (P. D. Boyer, ed.), Third edition, Vol. 9, pp. 379-481. Academic Press, New York.

Braunstein, A. E. and Shemyakin, M. M. (1953). *Biokhimiya* **18**, 393-411.

Brown, G. M. (1971). *Adv. Enzymol.* **35**, 35-77.

Bruice, T. C. and Benkovic, S. (1966). "Bioorganic Mechanisms", Vol. 2, pp. 301-349. Benjamin, New York.

Brüstlein, M. and Bruice, T. C. (1972). *J. Am. Chem. Soc.* **94**, 6548-6549.

Buttlaire, D. H., Reed, G. H. and Himes, R. H. (1975). *J. biol. Chem.* **250**, 261-270.

Caughey, W. S. and Schellenberg, K. (1966). *J. org. Chem.* **31**, 1978-1982.

Cheema, S., Soldin, S. J., Knapp, A., Hofmann, T. and Scrimgeour, K. G. (1973). *Can. J. Biochem.* **51**, 1229-1239.

Chen, M. S. and Schirch, L. (1973a). *J. biol. Chem.* **248**, 3631-3635.

Chen, M. S. and Schirch, L. (1973b). *J. biol. Chem.* **248**, 7979-7984.

Cheng, C.-F. and Haslam, J. L. (1972). *Biochemistry* **11**, 3512-3518.

Chippel, D. and Scrimgeour, K. G. (1970). *Can. J. Biochem.* **48**, 999-1009.

Cornforth, J. W., Cornforth, R. H., Donninger, C., Popják, G., Ryback, G. and Schroepfer, G. J., Jr. (1966). *Proc. R. Soc.* B **163**, 436-464.

Creighton, D. J., Hajdu, J., Mooser, G. and Sigman, D. S. (1973). *J. Am. Chem. Soc.* **95**, 6855-6857.

Curthoys, N. P. and Rabinowitz, J. C. (1972). *J. biol. Chem.* **247**, 1965-1971.

Curthoys, N. P., D'Ari Straus, L. and Rabinowitz, J. C. (1972). *Biochemistry* **11**, 345-349.

Davis, L. and Metzler, D. E. (1972). *In* "The Enzymes" (P. D. Boyer, ed.), Third edition, Vol. 7, pp. 33-74. Academic Press, New York.

Dunathan, H. C. (1971). *Adv. Enzymol.* **35**, 79-134.

Dunlap, R. B., Harding, N. G. L. and Huennekens, F. M. (1971). *Biochemistry* **10**, 88-97.

Ehrenberg, A. and Hemmerich, P. (1968). *In* "Biological Oxidations" (T. P. Singer, ed.), pp. 239-262. Interscience, New York.

Entsch, B., Massey, V. and Ballou, D. P. (1974). *Biochem. biophys. Res. Commun.* **57**, 1018-1025.

Fischer, E. H., Kent, A. B., Snyder, E. R. and Krebs, E. G. (1958). *J. Am. Chem. Soc.* **80**, 2906-2907.

Fisher, D. B. and Kaufman, S. (1973). *J. biol. Chem.* **248**, 4300–4304.

Fisher, H. F., Conn, E. E., Vennesland, B. and Westheimer, F. H. (1953). *J. biol. Chem.* **202**, 687–697.

Fisher, D. B., Kirkwood, R. and Kaufman, S. (1972). *J. biol. Chem.* **247**, 5161–5167.

Fisher, J. and Walsh, C. (1974). *J. Am. Chem. Soc.* **96**, 4345–4346.

Friedkin, M. (1973). *Adv. Enzymol.* **38**, 235–292.

Glaser, L. (1972). *In* "The Enzymes" (P. D. Boyer, ed.), Third edition, Vol. 6, pp. 355–380. Academic Press, New York.

Gumport, R. I. and Lehman, I. R. (1971). *Proc. natn. Acad. Sci. U.S.A.* **68**, 2559–2563.

Gundersen, L. E., Dunlap, R. B., Harding, N. G. L., Freisheim, J. H., Otting, F. and Huennekens, F. M. (1972). *Biochemistry* **11**, 1018–1023.

Guroff, G., Daly, J. W., Jerina, D. M., Renson, J., Witkop, B. and Udenfriend, S. (1967). *Science, N.Y.* **157**, 1524–1530.

Gurr, P. A., Bronskill, P. M., Hanes, C. S. and Wong, J. T. (1972). *Can. J. Biochem.* **50**, 1376–1384.

Hamilton, G. A. (1971). *Prog. bioorg. Chem.* **1**, 83–157.

Harmony, J. A. K. and Himes, R. H. (1975). *Biochemistry* **14**, 5379–5386.

Hastings, J. W., Balny, C., Le Peuch, C. and Douzou, P. (1973). *Proc. natn. Acad. Sci. U.S.A.* **70**, 3468–3472.

Hatefi, Y., Talbert, P. T., Osborn, M. J. and Huennekens, F. M. (1960). *Biochem. Prep.* **7**, 89–92.

Hemmerich, P. and Jorns, M. S. (1972). *In* "Enzymes: Structure and Function" (Drenth, J., Oosterbaan, R. A. and Veeger, C., eds), pp. 95–118. North-Holland, Amsterdam.

Hemmerich, P., Nagelschneider, G. and Veeger, C. (1970). *FEBS Letters* **8**, 69–83.

Henson, C. P. and Cleland, W. W. (1964). *Biochemistry* **3**, 338–345.

Hersh, L. B. and Jorns, M. S. (1975). *J. biol. Chem.* **250**, 8728–8734.

Hevesi, L. and Bruice, T. C. (1973). *Biochemistry* **12**, 290–297.

Hill, E., Tsernoglou, D., Webb, L. and Banaszak, L. J. (1972). *J. molec. Biol.* **72**, 577–591.

Himes, R. H. and Harmony, J. A. K. (1973). *Crit. Rev. Biochem.* **1**, 501–535.

Himes, R. H. and Rabinowitz, J. C. (1962). *J. biol. Chem.* **237**, 2915–2925.

Holbrook, J. J., Liljas, A., Steindel, S. J. and Rossmann, M. G. (1975). *In* "The Enzymes" (P. D. Boyer, ed.), Third edition, Vol. 11, pp. 191–292. Academic Press, New York.

Huang, C. Y. and Kaufman, S. (1973). *J. biol. Chem.* **248**, 4242–4251.

Huennekens, F. M. (1956). *In* "Currents in Biochemical Research" (D. E. Green, ed.), pp. 493–517. Interscience, New York.

Huennekens, F. M. and Scrimgeour, K. G. (1964). *In* "Pteridine Chemistry" (Pfleiderer, W. and Taylor, E. C., eds), pp. 355–373. Pergamon, New York.

Ivanov, V. I. and Karpeisky, M. Ya. (1969). *Adv. Enzymol.* **32**, 21–53.

Jencks, W. P. (1969). "Catalysis in Chemistry and Enzymology", pp. 133–146. McGraw-Hill, New York.

Jenkins, W. T. and D'Ari, L. (1966). *Biochem. biophys. Res. Commun.* 22, 376-382.

Jenkins, W. T. and Sizer, I. W. (1960). *J. biol. Chem.* 235, 620-624.

Jenkins, W. T., Yphantis, D. A. and Sizer, I. W. (1959). *J. biol. Chem.* 234, 51-57.

Jordan, P. M. and Akhtar, M. (1970). *Biochem. J.* 116, 277-286.

Jordan, P. M., El-Obeid, H. A., Corina, D. L. and Akhtar, M. (1976). *J. Chem. Soc. Commun.* 73-74.

Jorns, M. S. and Hersh, L. B. (1974). *J. Am. Chem. Soc.* 96, 4012-4014.

Joyce, B. K. and Himes, R. H. (1966). *J. biol. Chem.* 241, 5716-5731.

Kallen, R. G. and Jencks, W. P. (1966). *J. biol. Chem.* 241, 5851-5863.

Kalman, T. I. (1971a). *Ann. N.Y. Acad. Sci.* 186, 166-167.

Kalman, T. I. (1971b). *Biochemistry* 10, 2567-2573.

Kaplan, N. O. (1960). *In* "The Enzymes" (Boyer, P. D., Lardy, H. and Myrbäck, K., eds), Second edition, Vol. 3, pp. 105-169. Academic Press, New York.

Kaufman, B. T. and Gardiner, R. C. (1966). *J. biol. Chem.* 241, 1319-1328.

Kaufman, S. (1963). *Proc. natn. Acad. Sci. U.S.A.* 50, 1085-1093.

Kaufman, S. (1964). *J. biol. Chem.* 239, 332-338.

Kaufman, S. (1971). *Adv. Enzymol.* 35, 245-319.

Kaufman, S. (1975). *In* "Chemistry and Biology of Pteridines" (W. Pfleiderer, ed.), pp. 291-304. Walter de Gruyter, Berlin.

Kaufman, S., Holtzman, N. A., Milstein, S., Butler, I. J. and Krumholz, A. (1975). *New Engl. J. Med.* 293, 785-790.

Kawai, M. and Scrimgeour, K. G. (1972). *Can. J. Biochem.* 50, 1191-1198.

Kurz, L. C. and Frieden, C. (1975). *J. Am. Chem. Soc.* 97, 677-679.

Lambeth, D. O. and Palmer, G. (1973). *J. biol. Chem.* 248, 6095-6103.

Loeble, R. B. and Dunlap, R. B. (1972). *Biochem. biophys. Res. Commun.* 49, 1671-1677.

Lomax, M. I. S. and Greenberg, G. R. (1967). *J. biol. Chem.* 242, 109-113.

MacKenzie, R. E. and Rabinowitz, J. C. (1971). *J. biol. Chem.* 246, 3731-3736.

Mahler, H. R. and Cordes, E. H. (1971). "Biological Chemistry", Second edition, Chapter 8. Harper and Row, New York.

Martinez-Carrion, M., Critz, W. and Quashnock, J. (1972). *Biochemistry* 11, 1613-1615.

Massey, V., Palmer, G. and Ballou, D. (1973). *In* "Oxidases and Related Redox Systems" (King, T. E., Mason, H. S. and Morrison, M., eds), Vol. 1, pp. 25-43. University Park Press, Baltimore.

Mathews, C. K. and Huennekens, F. M. (1960). *J. biol. Chem.* 235, 3304-3308.

Mathews, C. K. and Huennekens, F. M. (1963). *J. biol. Chem.* 238, 3436-3442.

Maycock, A. L., Abeles, R. H., Salach, J. I. and Singer, T. P. (1976). *Biochemistry* 15, 114-125.

Mell, G. P., Whiteley, J. M. and Huennekens, F. M. (1968). *J. biol. Chem.* 243, 6074-6075.

Metzler, D. E., Ikawa, M. and Snell, E. E. (1954a). *J. Am. Chem. Soc.* 76, 648-652.

Metzler, D. E., Longenecker, J. B. and Snell, E. E. (1954b). *J. Am. Chem. Soc.* 76, 639-644.

Moras, D., Olsen, K. W., Sabesan, M. N., Buehner, M., Ford, G. C. and Rossmann, M. G. (1975). *J. biol. Chem.* **250**, 9137-9162.

Morino, Y. and Watanabe, T. (1969). *Biochemistry* **8**, 3412-3417.

Müller, F. and Massey, V. (1969). *J. biol. Chem.* **244**, 4007-4016.

Müller, F., Hemmerich, P. and Ehrenberg, A. (1971). *In* "Flavins and Flavoproteins" (H. Kamin, ed.), pp. 107-120. University Park Press, Baltimore.

Neef, V. G. and Huennekens, F. M. (1975). *Archs Biochem. Biophys.* **171**, 435-443.

Nelsestuen, G. L. and Kirkwood, S. (1971). *J. biol. Chem.* **246**, 7533-7543.

Nielsen, K. H., Simonsen, V. and Lind, K. E. (1969). *Eur. J. Biochem.* 497-502.

O'Leary, M. H. (1971). *Biochim. biophys. Acta.* **242**, 484-492.

Orf, H. W. and Dolphin, D. (1974). *Proc. natn. Acad. Sci. U.S.A.* **71**, 2646-2650.

Ovchinnikov, Yu. A., Egorov, C. A., Aldanova, N. A., Feigina, M. Yu., Lipkin, V. M., Abdulaev, N. G., Grishin, E. V., Kiselev, A. P., Modyanov, N. N., Braunstein, A. E., Polyanovsky, O-L. and Nosikov, V. V. (1973). *FEBS Letters* **29**, 31-34.

Palmer, G. and Massey, V. (1968). *In* "Biological Oxidations" (T. P. Singer, ed.), pp. 263-300. Interscience, New York.

Pastore, E. J. and Friedkin, M. (1962). *J. biol. Chem.* **237**, 3802-3810.

Pastore, E. J., Friedkin, M. and Jardetzky, O. (1963). *J. Am. Chem. Soc.* **85**, 3058-3059.

Pastore, E. J., Kisliuk, R. L., Plante, L. T., Wright, J. M. and Kaplan, N. O. (1974). *Proc. natn. Acad. Sci. U.S.A.* **71**, 3849-3853.

Peterson, D. L. and Martinez-Carrion, M. (1970). *J. biol. Chem.* **245**, 806-813.

Poe, M., Greenfield, N. J., Hirshfield, J. M., Williams, M. N. and Hoogsteen, K. (1972). *Biochemistry* **11**, 1023-1030.

Popják, G. (1970). *In* "The Enzymes" (P. D. Boyer, ed.), Third edition, Vol. 2, pp. 134-157. Academic Press, New York.

Porter, D. J. T., Blankenhorn, G. and Ingraham, L. L. (1973a). *Biochem. biophys. Res. Commun.* **52**, 447-452.

Porter, D. J. T., Voet, J. G. and Bright, H. J. (1973b). *J. biol. Chem.* **248**, 4400-4416.

Pullman, M. E., San Pietro, A. and Colowick, S. P. (1954). *J. biol. Chem.* **206**, 129-141.

Rader, J. I. and Huennekens, F. M. (1973). *In* "The Enzymes" (P. D. Boyer, ed.), Third edition, Vol. 9, pp. 197-223. Academic Press, New York.

Rajagopalan, K. V. and Handler, P. (1968). *In* "Biological Oxidations" (T. P. Singer, ed.), pp. 301-337. Interscience, New York.

Richardson, C. C. (1969). *A. Rev. Biochem.* **38**, 809-814.

Rossmann, M. G., Adams, M. J., Buehner, M., Ford, G. C., Hackert, M. L., Lentz, P. J., Jr., McPherson, A., Jr., Schevitz, R. W. and Smiley, I. E. (1971). *Cold Spring Harb. Symp. quant. Biol.* **36**, 179-191.

Rothman, S. W., Kisliuk, R. L. and Langerman, N. (1973). *J. biol. Chem.* **248**, 7845-7851.

Santi, D. V., McHenry, C. S. and Perriard, E. R. (1974a). *Biochemistry* **13**, 467-470.

K

Santi, D. V., McHenry, C. S. and Sommer, H. (1974b). *Biochemistry* **13**, 471–481.

Schirch, L. (1975). *J. biol. Chem.* **250**, 1939-1945.

Schirch, L. and Jenkins, W. T. (1964). *J. biol. Chem.* **239**, 3801-3807.

Schirch, L. and Mason, M. (1963). *J. biol. Chem.* **238**, 1032-1037.

Schirch, L., Edmiston, M., Chen, M. S., Barra, D., Bossa, F., Hinds, L. and Fasella, P. (1973). *J. biol. Chem.* **248**, 6456-6461.

Schlenk, F. and Fisher, A. (1947). *Archs Biochem.* **12**, 69-78.

Scott, J. M. and Rabinowitz, J. C. (1967). *Biochem. biophys. Res. Commun.* **29**, 418-423.

Scrimgeour, K. G. and Vitols, K. S. (1966). *Biochemistry* **5**, 1438-1443.

Shinkai, S. and Bruice, T. C. (1973). *J. Am. Chem. Soc.* **95**, 7526-7528.

Silverstein, E. (1965). *Analyt. Biochem.* **12**, 199-212.

Singer, T. P. and Edmondson, D. E. (1974). *FEBS Letters* **42**, 1-14.

Singer, T. P. and Kenney, W. C. (1974). *Vitamins and Hormones* **32**, 1-45.

Smith, J. R. L., Jerina, D. M., Kaufman, S. and Milstein, S. (1975). *J. Chem. Soc. Commun.* 881-882.

Snell, E. E. (1945). *J. Am. Chem. Soc.* **67**, 194-197.

Snell, E. E. and Di Mari, S. J. (1970). *In* "The Enzymes" (P. D. Boyer, ed.), Third edition, Vol. 2, pp. 335-370. Academic Press, New York.

Steffens, J. J. and Chipman, D. M. (1971). *J. Am. Chem. Soc.* **93**, 6694-6696.

Sugimura, T. (1973). *Prog. Nucleic Acid Res. molec. Biol.* **13**, 127-151.

Sund, H. (1968). *In* "Biological Oxidations" (T. P. Singer, ed.), pp. 603-639 and 641-705. Interscience, New York.

Taylor, S. S., Oxley, S. S., Allison, W. S. and Kaplan, N. O. (1973). *Proc. natn. Acad. Sci. U.S.A.* **70**, 1790-1794 and 2467.

Tobias, P. S. and Kallen, R. G. (1975). *J. Am. Chem. Soc.* **97**, 6530-6539.

Velick, S. F. and Vavra, J. (1962). *J. biol. Chem.* **237**, 2109-2122.

Vonderschmitt, D. J., Vitols, K. S., Huennekens, F. M. and Scrimgeour, K. G. (1967). *Archs Biochem. Biophys.* **122**, 488-493.

Wahba, A. J. and Friedkin, M. (1961). *J. biol. Chem.* **236**, PC11-PC12.

Walker, W. H. and Singer, T. P. (1970). *J. biol. Chem.* **245**, 4224-4225.

Walker, W. H., Kearney, E. B., Seng, R. L. and Singer, T. P. (1971). *Eur. J. Biochem.* **24**, 328-331.

Walsh, C. T., Schonbrunn, A., Lockridge, O., Massey, V. and Abeles, R. H. (1972). *J. biol. Chem.* **247**, 6004-6006.

Webb, L. E., Hill, E. J. and Banaszak, L. J. (1973). *Biochemistry* **12**, 5101-5109.

Welch, W. H., Irwin, C. L. and Himes, R. H. (1968). *Biochem. biophys. Res. Commun.* **30**, 255-261.

Whiteley, H. R. and Huennekens, F. M. (1962). *J. biol. Chem.* **237**, 1290-1297.

Yamasaki, M., Tanase, S. and Morino, Y. (1975). *Biochem. biophys. Res. Commun.* **65**, 652-657.

Yarmolinsky, M. B. and Colowick, S. (1956). *Biochim. biophys. Acta.* **20**, 177-189.

Zakrzewski, S. F. and Nichol, C. A. (1960). *J. biol. Chem.* **235**, 2984-2988.

9 Protein Coenzymes

We know that enzymes are proteins, and sometimes we tend to assume that all proteins are enzymes. Of course, we can immediately think of examples of non-enzymic proteins such as structural proteins, hemoglobin and antibodies. However, there is a group of proteins which do function in enzymatic reactions and, because they have no catalytic activity in themselves, are not enzymes. They could be called protein transfer factors, but I will call these molecules *protein coenzymes.*

Protein coenzymes can be categorized by their general properties. They have as their major component a protein molecule, to which a functional group is linked. This reactive function may be either an organic or inorganic prosthetic group, or can be part of the amino acid backbone of the protein. Protein coenzymes are not catalysts because they are not holoenzymes but are required for the action of other enzymes. Enzymes with firmly-bound prosthetic groups, such as pyridoxal phosphate enzymes, cannot be described as protein coenzymes because they are catalytically active. The protein coenzymes are low in molecular weight compared to most enzymes, but much higher in molecular weight than the coenzymes discussed in Chapter 8. Protein coenzymes, like the lower molecular weight coenzymes, function as carriers in either group or electron transfer. In addition, some of them—as with low molecular weight coenzymes—are involved in more than one metabolic reaction. They are more stable than most enzymes, and often can be easily purified. Cytochrome c (Dickerson and Timkovich, 1975) is the protein coenzyme with which we are most familiar. It has no enzyme activity but transfers electrons by connecting the cytochrome c reductase and cytochrome oxidase enzyme systems. The heme which undergoes the $Fe^{2+} \rightleftharpoons Fe^{3+}$ oxidation and reduction is the functional group of cytochrome c. The molecular weight of mammalian cytochrome c is about 12 000, well below that of most enzymes. It is stable and relatively easily isolated from mitochondrial enzyme fractions.

In this chapter, I have chosen to discuss some of the more recently discovered protein coenzymes. These include the oxidation–reduction coenzymes which contain non-heme iron as their functional centre, group transfer proteins that have derivatives of B-vitamins at their active centres and transport proteins that are essential in enzyme-catalyzed transport reactions. Other transport proteins, such as intrinsic factor, are not discussed, but might also be classed as a type of coenzyme. Modifier proteins such as α-lactalbumin or repressors similarly approach coenzyme status, but since they act by means other than group transfer, they have been arbitrarily excluded.

OXIDATION-REDUCTION PROTEINS

There are four types of protein coenzymes that participate in oxidation–reduction reactions. Besides the hemoproteins such as cytochromes, there are the iron–sulphur proteins, the flavoprotein coenzymes and the thiol coenzymes. Most of the latter three types of coenzymes are well-characterized considering their recent discovery. Elucidation of the mechanisms of their action will greatly aid the understanding of complicated enzymes which have the same functional groups. The first protein coenzymes we shall examine are the iron–sulphur (FeS) proteins—the ferredoxins and the rubredoxins (reviewed by Orme-Johnson, 1973 and Lovenberg, 1973). These two coenzymes are important in their own right but study of their chemistry should also provide information applicable to the FeS proteins of mitochondria.

Ferredoxins

The name *ferredoxin* was coined for a brown iron-containing protein which was isolated from extracts of the anaerobic bacterium *Clostridium pasteurianum* (Mortenson et al., 1962). This protein is essential for several bacterial metabolic reactions, including nitrogen fixation and conversion of pyruvate to acetyl phosphate. A rather similar protein had been isolated from spinach during the 1950s. Called "methaemoglobin reducing factor" by Davenport et al. (1952) and

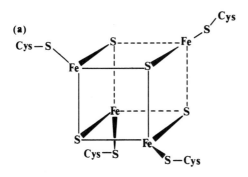

Fig. 1. Presumed structures of the iron–sulphur clusters of bacterial (a) and plant (b) ferredoxins.

"photosynthetic pyridine nucleotide reductase" by San Pietro and Lang (1958), it was found to function in electron transport in photosynthesis. Since 1962, both the bacterial and plant iron-containing proteins have been named ferredoxins. Although these ferredoxins have many similarities, they are sufficiently different to constitute two distinct classes of ferredoxins.

Both the bacterial and plant ferredoxins contain equimolar quantities of iron and acid-labile sulphur. The sulphur is released as H_2S when HCl is added to the protein. These sulphur atoms and the thiol groups of cysteine are the ligands binding the iron in ferredoxins (hence the former name, non-heme iron proteins). Both the plant and bacterial ferredoxins have reduction potentials of about −0.4 volts at pH 7, making them powerful reducing agents. They have characteristic absorption bands between 350 and 600 nm which are diminished upon reduction of the ferredoxin. Most of the brown bacterial ferredoxins such as the clostridial type contain eight moles each of iron, sulphur and cysteine and have a molecular weight of about 6000. The red chloroplast ferredoxins are twice as large but contain only two moles each of iron and acid-labile sulphur. After a decade of research, it is now thought that the iron in bacterial ferre-

Table I. Purification of ferredoxin from *C. pasteurianum*. The ferredoxin activity was measured by its ability to reactivate a preparation of pyruvate synthase, freed from ferredoxin by chromatography on DEAE-cellulose. From the data of Mortenson (1964).

Step	Specific Activity (units/mg protein)
Extract	2.6
Acetone-treated solution	54
DEAE-cellulose eluate	307
Recrystallized material	468

doxins is in two clusters of four iron atoms each (Fig. 1a) and that in the plant ferredoxins it is in a single cluster of two irons (Fig. 1b).

Early work on the bacterial ferredoxins was summarized by Valentine in 1964. The purification of ferredoxin from *C. pasteurianum* is a simple procedure. Mortenson (1964) described a three-step purification based on the small size, the acidic nature and the stability of ferredoxin (Table I). The crude extract of the bacteria is mixed with an equal volume of cold acetone, precipitating most proteins but leaving the ferredoxin in solution. The supernatant solution containing the ferredoxin is poured onto a column of DEAE-cellulose, where the ferredoxin remains at the top of the column during elution with 0.2 M Tris, pH 8. The ferredoxin is eluted with 0.5 M Tris, pH 8. (During this elution, a slower-moving band of red material can be observed with some *Clostridia*. This even more acidic protein is rubredoxin, discussed below.) The eluted ferredoxin is crystallized by precipitation with ammonium sulphate. The purest ferredoxin from *C. pasteurianum* is recovered in the 75–90% saturation fraction. Well over half the total ferredoxin activity in the crude extracts is obtained as crystalline material. DEAE-cellulose columns have been used to separate ferredoxin from many crude enzyme preparations, since at an ionic strength of about 0.1 the ferredoxin is retained as a dark brown band and other proteins are eluted in a large peak. Mayhew (1971) has devised a scheme for purifying ferredoxin without using acetone to denature other proteins. He used several column-chromatography steps and obtained a highly purified ferredoxin by a rapid and mild procedure. The ferredoxin which is isolated by these two procedures is in the more stable oxidized form (designated ferredoxin$_{ox}$ below).

Although crystalline preparations of ferredoxin from a number of bacteria were available in 1964, the analysis of the protein was far from simple. Lovenberg *et al.* (1963) showed that ferredoxins from five species of *Clostridium* had similar molecular weights by sedimentation velocity experiments, and contained about seven moles each of iron and acid-labile sulphur. Amino acid analysis of the *C. pasteurianum* ferredoxin indicated a molecular weight of 6000 and the presence of about seven half-cystines and a relatively high number of glutamic and aspartic residues. The cysteine residues were not present as either free sulphydryls or as disulphides, so presumably they were bonded to the iron. Hong and Rabinowitz (1970) re-evaluated the iron and acid-labile sulphide contents of clostridial ferredoxin, and

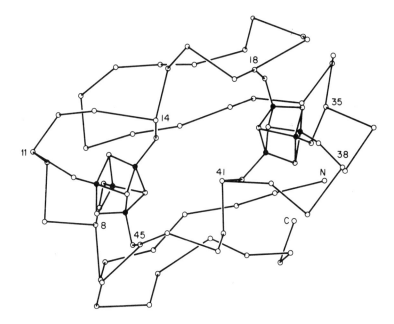

Fig. 2. Plot of the positions of the α-carbon atoms, the cysteine sulphur atoms and the iron atoms (solid circles) of the ferredoxin isolated from *P. aerogenes*. N signifies the amino-terminal, C is the carboxy-terminal, and the numerals show the positions of the α-carbons of the eight cysteine residues. Prepared by P. N. Lewis using the co-ordinates published by Adman *et al.* (1974) and the OR TEP computer program.

by careful analysis under suitable conditions, they obtained a value of eight moles each for iron, sulphur and cysteine, which agrees with data now available from X-ray crystallography.

X-ray crystallographic experiments with ferredoxin were delayed because of difficulty in obtaining crystals of the correct type and stability. Finally, in 1973, Adman *et al.* succeeded in determining the structure of the ferredoxin from *Peptococcus aerogenes*, based upon X-ray diffraction experiments at 2.8 Å resolution (Fig. 2). The four iron and four sulphur atoms in each cluster are at alternate corners of a distorted cube-shaped structure. A cysteine sulphur atom which extends out diagonally from the cubes binds to each of the iron atoms. The cysteine residues which co-ordinate the iron atoms are Cys-8, -11, -14 and -45 for one iron, and Cys-18, -35, -38 and -41 for the other. The two iron-sulphur clusters appear to be identical in shape. They are within the molecule, covered by hydrophobic side chains, and separated by about 12 Å. Jensen suggested that the clusters might contact electron donors or acceptors via two tyrosine residues which have edges exposed to the solvent. However, the tyrosine residues do not participate in electron transfer because a bacterial ferredoxin devoid of tyrosine retains full biological activity (Lode *et al.*, 1974). Other paths for transfer of electrons must be considered.

Difficulty in the analysis of ferredoxins extended to the nature and reactivity of the iron–sulphur clusters. The sulphur is thought to be in the form of sulphide, but the formal valence of both the iron and the sulphur (if any can be ascribed) is unknown. Even as simple a question as the number of electrons accepted per molecule of bacterial ferredoxin was in question for many years, but we now know that bacterial ferredoxins can accept two electrons *in vitro*. The oxidized form of ferredoxin can release half of its iron as Fe^{3+} and half as Fe^{2+} upon addition of mercurials. When reduced, two more of the irons released are ferrous (Sobel and Lovenberg, 1966). Sobel and Lovenberg also found by four other techniques that two electrons were transferred per molecule of ferredoxin. Titration by dithionite monitored by measuring the e.p.r. signals of the iron confirmed the stoichiometry of two electrons for the bacterial ferredoxins, and demonstrated the presence of two different iron signals arising sequentially during reduction (Orme-Johnson and Beinert, 1969a). Eisenstein and Wang (1969) have presented evidence suggesting that the two iron–sulphur clusters have reduction potentials differing from each other by 0.03 volts. Packer *et al.* (1975) have shown by n.m.r. studies that the midpoint redox potentials of the ferredoxins from *C. acidi-urici* and *C. pasteurianum* differ by 0.05 volts, with the difference being attributed to differences in protein structure. They verified that one Fe_4S_4 cluster in the *C. pasteurianum* ferredoxin is harder to reduce, but found that both clusters have the same E'_0 in the *C. acidi-urici* ferredoxin.

Carter *et al.* (1972) have found that the structures of the iron–sulphur clusters of oxidized ferredoxin and of both oxidized and reduced high-potential iron protein are geometrically similar. High-potential iron protein (HIPIP), isolated from *Chromatium vinosum*, has a single 4 iron, 4 sulphur and 4 cysteine cluster, like the dual clusters in bacterial ferredoxins. However, its reduction potential is +0.35 volts, almost 0.8 volts different from the ferredoxins. These workers suggested that three oxidation states might exist for the Fe_4S_4 iron–sulphur clusters: one state (C^-) as in reduced ferredoxin, one state (C) as in oxidized ferredoxin or reduced HIPIP, and one state (C^+) as in oxidized HIPIP. Cammack (1973) has been able to reversibly reduce HIPIP with dithionite in 80% dimethyl sulphoxide to the C^- state (i.e. "super-reduced" to a compound with properties similar to reduced ferredoxin). The high concentration of dimethyl sulphoxide is necessary to effect a change in protein conformation, which in turn seems to permit the iron–sulphur group to accept a full complement of electrons. When clostridial ferredoxin is treated with ferricyanide, an e.p.r. signal consistent with formation of super-oxidized (C^+) ferredoxin can be observed. It is thought that, in formal valence terms, a C^+-state cluster contains three Fe^{3+} and one Fe^{2+} atoms, a C-state cluster contains two Fe^{3+} and two Fe^{2+} atoms, and a C^- state cluster one Fe^{3+} and three Fe^{2+} atoms.

The physiological uses of ferredoxin in bacteria are based upon its ability to accept and donate these electrons. Ferredoxin is the electron carrier in the reduction of pyridine nucleotides, of nitrogen to ammonia (p. 370) and of protons to H_2 (the hydrogenase reaction). It also is necessary for a number of reversible carboxylation reactions (reviewed by Buchanan, 1972). These ferredoxin-linked carboxylations of acyl coenzyme A esters occur only in anaerobic and photosynthetic bacteria. They lead to the biosynthesis of amino acids in most cases. Each of the reactions can easily be reversed to produce reduced ferredoxin. A widespread reaction is catalyzed by pyruvate synthase (also called pyruvate-ferredoxin oxidoreductase):

Pyruvate + Ferredoxin$_{ox}$ + CoASH $\rightleftharpoons$ Acetyl CoA + CO$_2$ + Ferredoxin$_{red}$.

This enzyme has been purified from *C. acidi-urici* by Uyeda and Rabinowitz (1971), who found that thiamine pyrophosphate (TPP) and an iron–sulphur chromophore are prosthetic groups of the enzyme. The sequence of reactions seems to be transfer of an acetaldehyde group from pyruvate to form hydroxyethyl-TPP and CO$_2$. The active acetaldehyde reacts with coenzyme A to form acetyl CoA and at the same time reduces the iron–sulphur group of pyruvate synthase, which can then reduce ferredoxin. In *Clostridium,* this reaction is used for breakdown of pyruvate formed by fermentation, but in other anaerobic bacteria it is used for the synthesis of pyruvate and hence the amino acid alanine. Ferredoxin-linked carboxylation of acetyl CoA and succinyl CoA can be utilized by photosynthetic bacteria for a cyclic CO$_2$ assimilation pathway. The mechanism of light-induced reduction of ferredoxin is not as well understood in these cases as is that in plant chloroplasts described below.

Reduced ferredoxin is such a strong reducing agent that it may also carry out non-enzymatic reductions in bacterial cells. For example, Wright and Anderson (1958) described an enzyme system in extracts of *C. sticklandii* which reduced folate to 7,8-dihydrofolate using substrate amounts of pyruvate and coenzyme A as reductants. Scrimgeour *et al.* (1967) found that the reaction was a manifestation of the pyruvate synthase reaction, in which reduced ferredoxin formed from pyruvate oxidation was acting as the reducing agent for non-enzymatic reduction of folate.

Plant-type ferredoxins have been isolated from the chloroplasts of a wide variety of plants, from algae to higher plants. Like the bacterial ferredoxins, they are stable and easily purified in most cases. Because they can be replaced as cofactors in the photoreduction of NADP$^+$ in chloroplasts by bacterial ferredoxins, they inherited the name ferredoxin, too. As I mentioned above, these ferredoxins are larger proteins than the bacterial ferredoxins and contain a single 2-iron reactive centre. The sequences of the 97 amino acids of spinach and four other chloroplast ferredoxins have been determined (Rao and Matsubara, 1970) but no X-ray crystallographic data are yet available. Physical chemical measurements of the properties of the iron

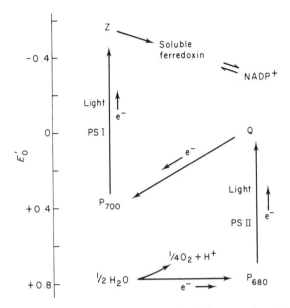

Fig. 3. Abbreviated diagram showing the path of electrons in photosynthesis in plants. Water is the ultimate donor and $NADP^+$ the final acceptor of electrons. Two electron-transporting photosystems, PS I and PS II, are arranged in series to transfer electrons (e^-) from water to $NADP^+$. The chlorophyll pigment of photosystem I, P_{700}, reduces the primary electron acceptor Z (either membrane-bound ferredoxin or an as yet unidentified redox factor) and oxidizes the unidentified reducing agent Q. Compound Z passes its reducing equivalents to $NADP^+$ via soluble ferredoxin. Reduced ferredoxin can also be oxidized by an electron-transport chain which leads back to P_{700}, with a simultaneous conversion of ADP and P_i to ATP. Oxidative phosphorylation also occurs in association with the reduction of P_{700} by reduced compound Q via a central electron-transport chain.

of plant ferredoxins can be explained by the iron–sulphur–cysteine cluster shown in Fig. 1b (Brintzinger *et al.*, 1966). In this binuclear cluster, both iron atoms are thought to be high-spin ferric in oxidized ferredoxin, and one high-spin ferric and one high-spin ferrous in the reduced state. Thus, the plant ferredoxins can accept only one electron per mole (Orme-Johnson and Beinert, 1969b).

The role of ferredoxin in plants is that of an electron carrier in photosynthesis (cf. review by Bishop, 1971). Photosynthesis can be described as the formation of a stable reducing agent, NADPH, and of an energy source, ATP, both of which are needed for the fixation of CO_2 into sugar. The energetics of the photosynthetic process are somewhat analogous to a reverse of the respiratory chain in mito-chondria, but because of their greater complexity the reactions in the chloroplast are even less clearly understood. Figure 3 traces the path of electrons from water to $NADP^+$. Light energy converts two specialized chlorophyll molecules (P_{680} and P_{700}) to excited states of much lower reduction potentials. These chlorophyll molecules, with reduction potentials in the ground state of about $+0.8$ and $+0.4$ volts, boost electrons in two steps (through photosystems I and II) to a reduction potential sufficiently negative to lead to reduction of $NADP^+$ (Lehninger, 1975). An electron is passed from the light-energized chlorophyll molecules to an electronegative electron acceptor, leaving positively charged chlorophyll. The positive chloro-phyll oxidizes H_2O in photosystem II and one of the reducing agents (indirectly, the reduced form of compound Q in Fig. 3) in photo-system I. Ferredoxin accepts electrons from photosystem I, the membrane-bound chlorophyll–polypeptide complex of lower reduc-tion potential. Reduced ferredoxin then reduces $NADP^+$ in a reaction catalyzed by a flavoprotein called ferredoxin-$NADP^+$ reductase (Shin et al., 1963):

$$2 \text{ Ferredoxin}_{red} + H^+ + NADP^+ \rightleftharpoons 2 \text{ Ferredoxin}_{ox} + NADPH .$$

Arnon and his associates (e.g. Arnon, 1965) have often commented that, because ferredoxin is the most electronegative compound that has been isolated from cells, it must be the primary physiological acceptor of electrons in photosynthesis. Indirect evidence suggested that there is an additional cofactor having a reduction potential lower than -0.4 volts. This electron acceptor still has not been identified. Malkin and Bearden (1971) reported that broken spinach chloroplasts from which all the soluble ferredoxin had been removed showed an e.p.r. signal characteristic of plant ferredoxins when reduced under illumination. They postulated that this membrane-bound ferredoxin is the primary acceptor of electrons from P_{700}. Based upon potentiometric titration observations, Ke et al. (1973) were able to detect the presence of three iron–sulphur proteins or centres in plant chloroplast membranes. The redox potentials of these electron carriers fall into two regions, both being highly nega-tive (-0.53 and -0.58 volts). After a harsh extraction, using meth-anol and acetone, one of the bound ferredoxins has been isolated (Malkin et al., 1974). This purified ferredoxin is quite different from the soluble plant ferredoxin. It contains four atoms of iron and four inorganic sulphides, it has spectral properties similar to bacterial ferredoxin, and it cannot substitute for soluble ferredoxin in the reaction catalyzed by ferredoxin-NADP$^+$ reductase. In fact, spectral experiments suggest that only the Fe_4S_4 type of iron–sulphur proteins are bound to spinach chloroplast membranes. When all the FeS centres of the membrane are chemically reduced, P_{700} can still undergo reduction upon illumination at low temperatures, suggesting the presence of yet another electron acceptor. Evans *et al.* (1975) have proposed that this compound, having an e.p.r. spectrum different from ferredoxins, is the primary acceptor of electrons from P_{700}. Until this compound is identified and its role verified, and the function of the membrane-bound ferredoxins fully explained, the pathway of electron transport from P_{700} to the soluble ferredoxin will be in doubt.

Rubredoxins

A group of bright red proteins called rubredoxins are also iron-sulphur coenzymes. Rubredoxins from anaerobic bacteria are the simplest of the iron–sulphur proteins and have been the most completely characterized. Unfortunately, the specific function of these rubredoxins is not known, although they can replace ferredoxins as electron carriers in some reactions.

Lovenberg and Sobel (1965) crystallized rubredoxin from *Clostridium pasteurianum* after separating it from ferredoxin by ammonium sulphate precipitation of the ferredoxin. The clostridial rubredoxin is red in the oxidized form in which it is isolated, and colourless when reduced. It has an E_0' value of −0.06 volts and a molecular weight of about 6000. It contains only one mole of iron and transfers only one electron per molecule. Since there is no acid-labile sulphur in rubredoxin, Bachmeyer *et al.* (1967) suggested that the four cysteine residues form the iron-binding site. The iron seems to stabilize the conformation of the protein as well as accept and donate the electron transferred.

The crystal structure of rubredoxin from *C. pasteurianum* has been determined in Jensen's laboratory (Watenpaugh *et al.*, 1971). The process of structure refinement by calculation from the 2 Å map has been applied to rubredoxin (the first time this method has

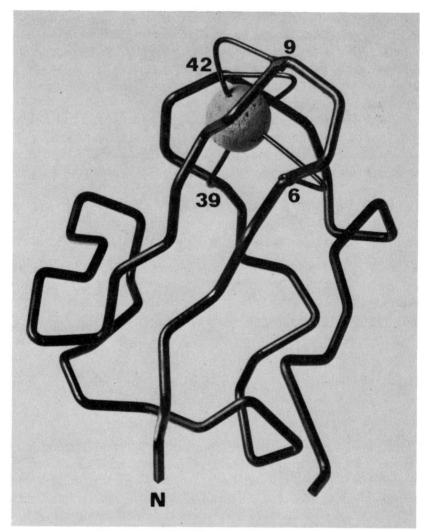

Fig. 4. Wire model of the backbone of rubredoxin. The large sphere represents the atom of iron, and the positions of the α-carbons of the four cysteine residues are indicated by their respective numbers. N indicates the amino terminus of the polypeptide chain.

been used with a protein), resulting in a structure with 1.5 Å resolution. Figure 4 shows the structure of a rubredoxin molecule. The iron atom is bonded to the cysteine residues at positions 6, 9, 39 and 42. The Fe—S bond lengths and angles indicate that there is distortion from a tetrahedral configuration, with one bond about 0.3 Å shorter than the other three. This distortion is probably caused by the constraints placed on the iron–sulphur bonds by the primary structure of the protein.

A larger and more complicated red protein coenzyme, also called rubredoxin, has been isolated from the aerobic bacterium *Pseudomonas oleovorans*. Although structural data on this protein is incomplete, a specific function has been assigned. The pseudomonad rubredoxin is one of three protein components of an enzyme complex which catalyzes the hydroxylation of octane and the ω-position of some fatty acids:

$$n\text{–Octane} + \text{NADH} + \text{H}^+ + \text{O}_2 \rightarrow n\text{–Octanol} + \text{NAD}^+ + \text{H}_2\text{O} .$$

The other two proteins are enzymes: the first enzyme is an NADH-rubredoxin reductase and the second a mixed function oxidase, ω-hydroxylase (Peterson *et al.*, 1967). When the pseudomonad rubredoxin is isolated, it usually contains one mole of iron per mole of rubredoxin (molecular weight 19 000; Lode and Coon, 1971). The protein contains 10 half-cystine residues and can readily bind a second atom of iron. It is assumed that each iron is bound by four cysteine residues, as in the iron–sulphur cluster of the clostridial rubredoxin. An anomaly with the pseudomonad rubredoxin is that both forms of the molecule (1-Fe and 2-Fe) can accept one electron per atom of iron, but both have equal enzymatic activity in the hydroxylation reaction at the same protein concentration.

Flavoprotein Coenzymes

Several flavoproteins can be classified as protein coenzymes. A high-molecular weight electron carrier, the electron-transferring flavoprotein (ETF), occurs in mammalian heart and liver mitochondria, and also can be isolated from the anaerobic bacterium *Peptostreptococcus elsdenii*. A lower-molecular weight flavoprotein, flavodoxin, has been isolated from quite a few micro-organisms and has been extensively studied. Blue-green algae contain a somewhat similar FMN-containing protein, phytoflavin (Smillie and Entsch, 1971). All of these flavoprotein coenzymes serve as electron-carriers in redox reactions. The latter two seem to be interchangeable with ferredoxin in function.

The electron-transferring flavoprotein contains FAD as a prosthetic group (Beinert, 1963). As a component of the matrix or "soluble" phase of mitochondria, ETF accepts electrons from fatty acyl CoA dehydrogenases (e.g. butyryl CoA dehydrogenase) and transfers them to the respiratory chain, presumably at the coenzyme Q site. Also, ETF is required to couple the oxidation of sarcosine (N-methyl glycine) to the respiratory chain. Hoskins (1966) reported that an ETF from monkey liver mitochondria mediates electron transfer in the oxidation of both sarcosine and butyryl CoA, but this protein may possess different binding sites for each of the reduced dehydrogenases. The occurrence of ETF in many mammals and also in *Mycobacterium* suggests that it has a general role in mitochondrial electron transport. Although various ETF proteins had been obtained in fairly pure forms, until recently little was known about their physical properties. The electron-transferring protein of *P. elsdenii* has a molecular weight of 75 000 and is composed of two non-identical subunits (Whitfield and Mayhew, 1974). Hall and Kamin (1975) have purified ETF from pig liver mitochondria by a new purification procedure. They reported that ETF has a molecular weight of 55 000–60 000 daltons, and is composed of two subunits, each of which contains one mole of FAD. In addition, a membrane-bound iron–sulphur flavoprotein has been partially purified from beef heart mitochondria, and has been shown to be rapidly reduced by the fatty acyl dehydrogenase and ETF system (Ruzicka and Beinert, 1975). Thus, three flavoproteins seem to be required for electron transfer from substrates to the main electron transfer system:

$$\text{Butyryl CoA} \rightarrow \underset{\underset{\text{FAD}}{|}}{\text{Acyl CoA dehydrogenase}} \rightarrow \underset{\underset{\text{FAD}}{|}}{\text{ETF}} \rightarrow \underset{\underset{\text{FAD}}{|}}{\text{FeS}} \rightarrow \text{Coenzyme Q.}$$

Study of the flavin–flavin interactions between ETF and those enzymes which can reduce it (three acyl dehydrogenases and sarcosine dehydrogenase) will define the role of the protein coenzyme in metabolism.

The flavoprotein coenzyme flavodoxin was first isolated from *Clostridium pasteurianum* grown in an iron-deficient medium designed to suppress the biosynthesis of ferredoxin (Knight and Hardy, 1966). Flavodoxin can replace ferredoxin in the hydrogenase, pyruvate synthase and nitrogen fixation reactions, but with slightly lower effectiveness. Under similar growth conditions, flavodoxins have now been isolated from other micro-organisms including *P. elsdenii*. The *P. elsdenii* flavodoxin contains one mole of bound FMN, which can be reduced to both the semiquinone and fully reduced states (Mayhew *et al.*, 1969). The E'_0 value for the fully reduced/semiquinone couple is -0.37 volts and probably is the reaction by which flavodoxin functions when it replaces ferredoxin *in vivo*. Phytoflavin also has a very low E'_0 (-0.45 volts) for this couple (Entsch and Smillie, 1972). With a molecular weight of about 15 000, flavodoxin is the smallest flavoprotein that has been isolated, its crystals having been suitable subjects for X-ray structure studies. The structures of the oxidized form of *Desulfovibrio vulgaris* flavodoxin to 2.0 Å resolution (Watenpaugh *et al.*, 1973) and of clostridial flavodoxin to 1.9 Å resolution (Burnett *et al.*, 1974) have been reported. Both flavodoxins are oblate spheroids with a planar FMN molecule partially buried near the surface at one end of the flavodoxin. In both types of flavodoxin, the α-carbon atoms of the polypeptide backbone are in approximately equivalent positions, but there are differences in the flavin binding sites, especially in the regions surrounding the isoalloxazine ring. The avid binding of FMN to the flavodoxins is not due to covalent bonds but to multiple non-covalent bonds. With the flavin ring being located at the surface, interaction with electron donor and acceptor proteins could readily be envisaged, except that it is the pyrimidine ring of the flavin that is buried most deeply and the dimethylbenzene moiety that is in contact with solvent. The semiquinone form of the clostridial flavodoxin has a similar conformation to the oxidized form, so large changes in crystal structure do not occur upon reduction. Two possible mechanisms for electron transfer to and from flavodoxin are conformational changes during binding to the proteins with which flavodoxin interacts, allowing exposure of N^5 of the flavin, or electron transfer via the dimethylbenzene unit.

Thiol Proteins

The final group of oxidation–reduction protein coenzymes to be discussed have thiol groups at their active centres. Thioredoxins are low molecular weight proteins which cycle between their dithiol and disulphide forms. The two sulphur atoms involved in the transfer of reducing power are present as neighbouring cysteine residues in thioredoxin. The low-molecular weight thiol protein which is one of the coenzymes in the glycine decarboxylation reaction also shuttles between the dithiol and disulphide forms, but has lipoic acid as a prosthetic group.

Thioredoxin was discovered in extracts of *Escherichia coli* by Reichard and his colleagues during their isolation of the enzyme system which converts ribonucleotides to deoxyribonucleotides (Reichard, 1968). This enzyme and its allosteric control are described in Chapter 13. Four proteins are responsible for the conversion of ribonucleoside diphosphates to deoxyribonucleoside diphosphates:

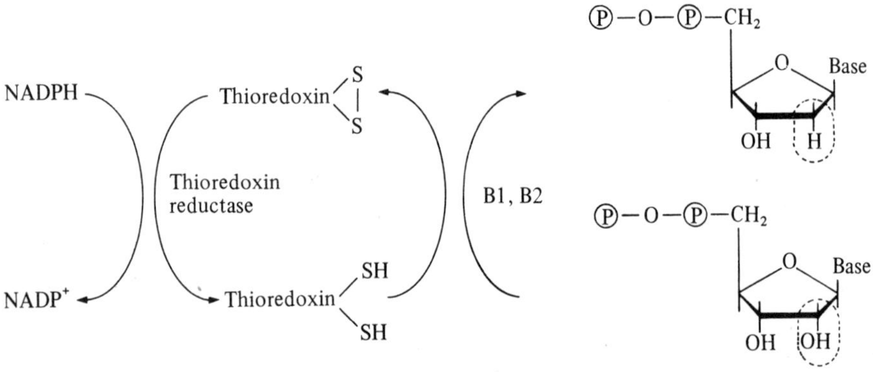

Thioredoxin reductase (protein 1) catalyzes the reduction of the disulphide bridge of thioredoxin (protein 2) by NADPH. The reduced protein coenzyme then reacts with the ribonucleotide under the influence of ribonucleotide reductase, which is composed of two subunits, B1 and B2 (proteins 3 and 4).

Thioredoxins are heat-stable acidic proteins, with molecular weights of 10 000–13 000. They have been isolated from several bacteria, yeast, rat liver tissue and tumour tissue. The thioredoxin from *E. coli* (Holmgren, 1968) is a polypeptide with 108 amino acid residues. The N-terminal half of the molecule is acidic, and the C-terminal portion contains many hydrophobic residues. Its reduction potential is −0.26 volts, making it a slightly better reducing agent than a monothiol. The two cysteines which undergo the oxidation and reduction are at positions 32 and 35, and are in a sequence

$$-\text{Trp}-\overline{\text{Cys}-\text{Gly}-\text{Pro}-\text{Cys}}-\text{Lys}- .$$

The same sequence occurs at the active centre in the two thioredoxins isolated from *Saccharomyces cerevisiae* (Hall *et al.*, 1971). However, the thioredoxin synthesized by *E. coli* infected with T4 bacteriophage has

$$-\overline{\text{Cys}-\text{Val}-\text{Tyr}-\text{Cys}}-$$

as its active disulphide ring (Sjöberg and Holmgren, 1972). The only homology seen among the structures is the formation of the 14-membered ring in the oxidized or disulphide state. Completion of X-ray diffraction studies to 2.8 Å resolution show that the disulphide group active centre is a prominent protrusion, not in a cleft (Holmgren *et al.*, 1975). Since both thioredoxin reductase and ribonucleotide reductase presumably have active site clefts, the finding that thioredoxin is a "male" protein is quite reasonable, and in keeping with my description of it as a protein coenzyme.

The enzyme thioredoxin reductase has been purified to homogeneity by Thelander (1967). It has a molecular weight of 66 000 and contains two molecules of FAD. It has an absolute specificity for oxidized thioredoxin as electron acceptor, and the K_m for NADPH is about 400 times lower than that for NADH. Zanetti and Williams (1967) and Thelander (1968) found that it took four moles of NADPH to completely reduce one mole of thioredoxin reductase. This inferred that there were two groups being reduced, the FAD molecules and another redox function. This second group was shown to be a disulphide. The oxidized enzyme has no —SH groups which can react with alkylating agents, but the enzyme reduced by substrate reacts with four moles of alkylating compounds. Thelander (1968) observed that the molecular weight of thioredoxin reductase decreased to about 32 000 in 6 M urea or 6 M guanidine. This indicated that the enzyme has two subunits. It appears that the two subunits, each containing a disulphide and one FAD, are identical. There is only one N-terminal (glycine) in the reductase. The number of tryptic peptides formed (22) is close to that expected for identical subunits, because there are 52 potential sites of cleavage by trypsin per 66 000 daltons. The sequence of the cystine peptide is

$$-\text{Ala}-\overline{\text{Cys}-\text{Ala}-\text{Thr}-\text{Cys}}-\text{Asp}-\text{Gly}-\text{Phe}-$$

(Thelander, 1970). The finding of only one cystine peptide also supports the identical nature of the two subunits. This active disulphide reacts directly with the sulphur of thioredoxin. It is noteworthy that, as in thioredoxin, there are only two amino acids separating the two half-cystine residues.

Is thioredoxin a coenzyme for more than one reaction in any or all cells? This is an intriguing and unanswered question. Indirect evidence suggests that a general role for thioredoxin, not just the specific function in conversion of ribose to deoxyribose, is possible. Elford *et al.* (1970) assayed for the levels of ribonucleotide reductase and of thioredoxin–thioredoxin reductase in rat liver, working through a series of rat hepatomas of varying growth rate, and regenerating and foetal rat liver. The specific activity of ribonucleotide reductase showed a close relationship to the rate of tissue growth, over a several thousand-fold range. The thioredoxin–thioredoxin reductase complex was present at comparable levels in all the tissues, though, suggesting it may have other functions. Thioredoxin and its reductase can act as electron carriers in the reduction of methionine sulphoxide to methionine and of sulphate (as PAPS, 3'-phosphoadenosine-5'-phosphosulphate) to sulphite in the presence of NADPH and the specific reductases (Porqué *et al.*, 1970). The occurrence of the two thioredoxins of differing amino acid sequence in yeast (Hall *et al.*, 1971) might also indicate several functions for thioredoxin. Future experiments elucidating the role of thioredoxins will be of great interest.

The oxidation of glycine to carbon dioxide, ammonia and formaldehyde by *Peptococcus glycinophilus* also requires four proteins,

$$\text{Glycine} + \text{Tetrahydrofolate} + \text{NAD}^+ \xrightarrow{\text{4 proteins}}$$

$$\text{Methylene tetrahydrofolate} + CO_2 + NH_4^+ + \text{NADH,}$$

one of which is a low molecular weight acidic thiol protein called P_2 (Klein and Sagers, 1966). The disulphide form of P_2 is reduced in the oxidation of glycine and in turn the dithiol form reduces NAD^+ to NADH. Motokawa and Kikuchi (1969) have purified a dithiol protein with similar activity, which they have called "hydrogen carrier protein", from rat liver mitochondria. The other three proteins in the bacterial system are P_1, a pyridoxal phosphate-containing protein which catalyzes the decarboxylation of glycine to form a reduced

P_2; P_3, a flavoprotein which catalyzes the reduction of NAD^+ by reduced P_2; and P_4, needed for transfer of the remaining methylene group of glycine to tetrahydrofolate and the release of ammonia. The first step in glycine oxidation is the formation of the glycine–P_1 PLP Schiff base:

$$
\begin{array}{c}
\text{CHO} \\
| \\
P_1
\end{array}
+ \; H_2N-CH_2-COO^- \;\rightleftharpoons\;
\begin{array}{c}
N-CH_2-COO^- \\
\| \\
CH \\
| \\
P_1
\end{array}
$$

In the presence of the dithiol protein coenzyme, P_2, decarboxylation of the Schiff base can occur. It was shown in 1967 (Baginsky and Huennekens) that P_3 has many of the properties of lipoyl dehydrogenase and can be replaced by purified preparations of lipoyl dehydrogenase. P_2 contains three $-SH$ groups per molecular weight of about 12 000. Only one of these is cysteine. The reducible disulphide of the protein has been identified as lipoic acid, 6,8-dithio-octanoate (Robinson *et al.*, 1973). The manner in which lipoic acid is bound to the electron carrier P_2 and whether lipoate is also present in hydrogen carrier protein from rat liver are still unknown. Motokawa and Kikuchi (1974) have detected an intermediate in the reversible cleavage of glycine, which they suggest is a $-CH_2NH_2$ group (derived from C^2 of glycine) bound to sulphur of the hydrogen carrier protein. They observed a similar intermediate with a bacterial enzyme system. From their results, the role of P_4 seems to be catalysis of the interconversion of the proposed intermediate and the dithiol form of P_2:

$$
\begin{array}{c}
\text{SH} \\
/ \\
_2 \\
\backslash \\
S-CH_2-NH_2
\end{array}
+ \; \text{Tetrahydrofolate} \;\overset{P_4}{\rightleftharpoons}\;
\begin{array}{c}
\text{SH} \\
/ \\
P_2 \\
\backslash \\
\text{SH}
\end{array}
+ \; \text{Methylenetetrahydrofolate} \; + \; NH_3 \,.
$$

Presence in mitochondria of a functional glycine cleavage enzyme system is essential for glycine catabolism in humans. When this metabolic pathway does not operate, glycine accumulates in abnormal amounts in body fluids. This inborn metabolic error, called nonketotic hyperglycinemia, causes severe mental retardation and early death. It is not yet known which protein component(s) of the cleavage system is defective or absent in this genetic disease.

Protein *A*, a component of glycine reductase, is a low molecular weight, acidic and heat stable protein component participating in the reductive deamination of glycine by extracts of *Clostridium sticklandii* (Stadtman, 1966). It apparently possesses one or more thiol groups essential for activity as demonstrated, in part, by its inactivation by iodoacetamide. The glycine reductase system is a multi-protein complex, bound to the cell membrane. It can use several electron donors, including a dimercaptan substrate such as dithiothreitol, to cleave glycine:

$$R(SH)_2 + {}^+H_3N-CH_2-COO^- \xrightarrow{\text{3 proteins}} CH_3COO^- + NH_4^+ + R(S)_2 .$$

Of the three proteins needed to catalyze the above reaction, two have been purified close to homogeneity and one partially (Turner and Stadtman, 1973). Protein A has received the most attention. It has been found by sucrose density-gradient sedimentation to have a molecular weight of about 12 000. Its activity drastically decreases in extracts of bacteria from cultures allowed to grow for prolonged periods, as though a trace nutrient were required for its synthesis. This nutrient has turned out to be selenite, and experiments with radioactive selenium showed that one gram atom of this element is covalently bound to each molecule of protein A. Although the chemical form of the selenium is not yet known, it may be an —SeH compound. The selenium possibly is a redox-active group, analogous to the sulphur atoms of the dithiol protein coenzymes. Turner and Stadtman have suggested that protein A may also be a component of the clostridial formate dehydrogenase, because this enzyme complex also contains selenium (Andreesen and Ljungdahl, 1973). Further experiments may prove correct the suspicion that protein A is a selenoprotein having the role of an electron carrier (Stadtman, 1974).

GROUP TRANSFER PROTEINS

There are two enzyme systems which catalyze the biosynthesis of fatty acids, and each of these enzymes has a protein coenzyme associated with it. These protein coenzymes are similar in one major respect—they both have the active portion of a low molecular-weight covalently-bound coenzyme as their active site. The two enzymes required for fatty acid synthesis are acetyl CoA carboxylase and fatty acid synthetase. Acetyl CoA carboxylase catalyzes the formation of malonyl CoA from acetyl CoA and carbon dioxide:

$$CH_3CO-SCoA + HCO_3^- + ATP \xrightarrow{\text{3 proteins}}$$

$$^-OOCCH_2CO-SCoA + ADP + P_i.$$

Biotin carboxyl carrier protein (BCCP, or sometimes just CCP) is the protein coenzyme for this reaction. Fatty acid synthetase is a high molecular weight complex which utilizes for most of the steps protein-bound acyl groups rather than coenzyme A esters to catalyze the multi-step synthesis of palmitic acid:

$$\text{Acetyl CoA} + 7\text{ Malonyl CoA} + 14\text{ NADPH} + 14\text{ H}^+ \xrightarrow{\text{Fatty acid synthetase}}$$

$$CH_3(CH_2)_{14}COOH + 7\text{ CO}_2 + 14\text{ NADP}^+ + 8\text{ CoASH} + 6\text{ H}_2O.$$

The protein which binds the acyl groups during the elongation of the acyl chain is called acyl carrier protein. For discussions with more detail about these enzymes than given here, the reader should consult the reviews by Moss and Lane (1971) and Volpe and Vagelos (1973), and turn to p. 472 in Chapter 14.

Biotin Carboxyl Carrier Protein

Several laboratories observed in the 1950s that bicarbonate participates in the biosynthesis of long-chain fatty acids from acetyl CoA. In 1958, Wakil showed that bicarbonate combines with acetyl CoA to form malonyl CoA, the first isolated intermediate in fatty acid biosynthesis. Wakil and Gibson (1960) demonstrated that an enzyme fraction which catalyzes the carboxylation of acetyl CoA contains biotin. The activity of this enzyme fraction is inhibited by avidin, a protein from egg white which specifically binds to biotin. These data suggested that an enzyme–biotin–CO_2^- complex is an intermediate in the formation of malonyl CoA. Subsequently, acetyl CoA carboxylase was purified from several tissues, and the carboxylase from *E. coli* has been carefully examined in the laboratories of Lane and Vagelos. This enzyme can dissociate into three proteins which are functionally distinct (Fig. 5). Biotin carboxylase catalyzes the carboxylation of the carboxyl carrier protein (BCCP)

$$ATP + HCO_3^- + BCCP \underset{\xrightarrow{\text{Biotin carboxylase}}}{\rightleftharpoons} BCCP-CO_2^- + ADP + P_i$$

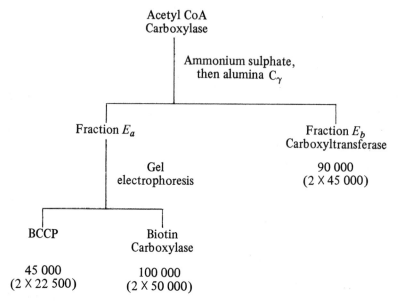

Fig. 5. Separation of acetyl CoA carboxylase from *E. coli* into its three protein components (Alberts and Vagelos, 1972). The molecular weights of each of the proteins and their subunits are shown on the diagram.

and carboxyltransferase catalyzes the transfer of CO_2 from the carboxylated protein coenzyme to acetyl CoA

$$BCCP{-}CO_2^- + \text{Acetyl CoA} \xrightleftharpoons{\text{Carboxyltransferase}} \text{Malonyl CoA} + BCCP.$$

BCCP was purified by growing large batches (approximately 100 lb) of *E. coli* in the presence of ^{3}H-biotin and isolating the radioactive fractions (Alberts and Vagelos, 1972). Fall and Vagelos (1972) devised a purification procedure which prevents limited proteolysis of BCCP during isolation and thus gives only one fraction with BCCP activity. This protein has a molecular weight of 22 500 but readily dimerizes. It contains one mole of biotin per 22 500 daltons. In all biotin proteins examined, the biotin is covalently-bound by its valeric acid side-chain to the ϵ-amino group of a lysine residue of the polypeptide, as shown in the structure at the top of p. 295.

The carboxyl group is attached to the 1′-position of the biotin when carboxybiotin is isolated (stabilized in the form of its methyl ester), and it has been concluded that carboxylation occurs at this position

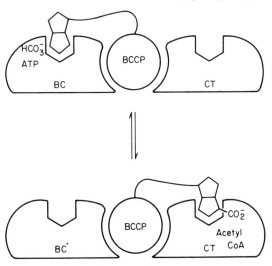

(Moss and Lane, 1971). Model studies (e.g. Bruice and Hegarty, 1970) had raised the possibility that carboxylation occurs at the apparently more reactive 2'-oxygen, with a spontaneous rearrangement of the O-acyl compound to an N-acyl compound occurring during the methylation. Guchhait et al. (1974) have ruled out O-carboxylation of the biotin by demonstrating that authentic 1'-N-carboxybiotin is a substrate for carboxyltransferase and biotin carboxylase, and that the carboxybiotin compound generated in the enzymatic reaction is indistinguishable from 1'-N-carboxybiotin.

Since the two half-reactions of malonyl CoA formation occur at active sites on different subunits, BCCP must be involved in an inter-subunit transfer of a biotin-bound carboxyl group. Figure 6 illustrates

Fig. 6. Possible mechanism for translocation of carboxy-BCCP between the subunits of acetyl CoA carboxylase. In the upper drawing, the biotin molecule is bound to the active site of biotin carboxylase (BC), and in the lower diagram it has moved—carrying the carboxyl group—to the active site of carboxyltransferase (CT). Adapted from Guchhait et al. (1971).

this type of oscillatory function for the carrier protein. Kinetic evidence to support a similar two-site mechanism has been obtained with other biotin-requiring enzyme systems.

Another biotin-requiring enzyme which has its biotin present in a small protein subunit is transcarboxylase (Wood, 1972). Transcarboxylase, found only in propionic acid bacteria, catalyzes the reversible transfer of a carboxyl group from methylmalonyl CoA (MMCoA) to pyruvate:

$$
\underset{\text{MMCoA}}{\overset{\overset{\displaystyle COO^-}{\underset{|}{|}}}{CH_3-CH-CO-SCoA}} + \underset{\text{Pyr}}{CH_3-\overset{\overset{\displaystyle O}{\|}}{C}-COO^-} \rightleftharpoons \underset{\text{PrCoA}}{CH_3-CH_2-\overset{\overset{\displaystyle O}{\|}}{C}-SCoA} + \underset{\text{OAA}}{^-OOC-CH_2-\overset{\overset{\displaystyle O}{\|}}{C}-COO^-}
$$

The intact enzyme has a molecular weight of 790 000 and is composed of six molecules each of three polypeptides, arranged as three different types of subunits: 12 S, 5 S and 1.3 S (Green et al., 1972). The 12 S subunit is a central hexamer, and the 6 S subunits are three dimers on the periphery of the complex. The 1.3 S carboxyl carrier protein component has a molecular weight of about 12 000 and contains one mole of covalently-bound biotin. The transfer of the carboxyl group from MMCoA to pyruvate does not involve free CO_2. It proceeds by these partial reactions:

1. MMCoA + Biotinyl subunit $\underset{\Longleftarrow}{\overset{12\,S}{\Longrightarrow}}$

   CO_2^-—Biotinyl subunit + propionyl CoA (PrCoA)

2. CO_2^-—Biotinyl subunit + Pyr $\underset{\Longleftarrow}{\overset{5\,S}{\Longrightarrow}}$ Biotinyl subunit + OAA .

The two half-reactions are catalyzed by dissimilar subunits of transcarboxylase (Chuang et al., 1975). The keto acids (Pyr and OAA) bind to 5 S subunits, and the coenzyme A esters to the 12 S subunits. The biotin-containing subunit not only carries the active —CO_2 from one catalytic centre to the other but also is essential for association of the 5 S and 12 S subunits.

Kinetic studies had foreshadowed the results of the later structural studies. Northrop (1969) measured both the initial-velocity and product-inhibition kinetics for transcarboxylase. Initial-velocity patterns indicated, as expected, a ping-pong kinetic pattern with an intermediate enzyme–biotin–CO_2^- complex:

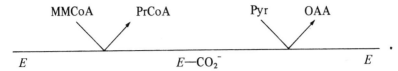

In a ping-pong mechanism, pyruvate would be a competitive inhibitor of propionyl CoA and a mixed noncompetitive inhibitor of oxalo-acetate. However, Northrop observed the opposite inhibition pattern: the CoA esters were competitive with each other, and the keto acids were also competitive with each other. This indicated that the acyl CoA esters combine with one site of the enzyme and the keto acids with another, and the biotin protein acts as a carboxyl carrier between these two sites. These and other kinetic data led to the postulate, now confirmed, that the two catalytic sites are on different subunits. The anomalous kinetic mechanism observed has been termed a modi-fied ping-pong mechanism, or—more correctly—a two-site substituted enzyme mechanism.

Initially, the same sort of two-site substituted enzyme kinetic mechanism was indicated for pyruvate carboxylase (e.g. Barden *et al.*,

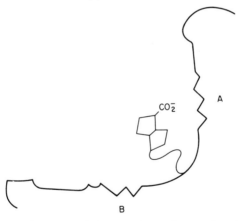

Fig. 7. Diagrammatic structure of the two-site active centre of a pyruvate carb-oxylase subunit, based on the early observation of two-site substituted enzyme kinetics. A is the site of partial reaction (1), the ATP–ADP site. B is the site of partial reaction (2), the keto acid site. These two sites would be far enough apart to allow binding of both types of substrates simultaneously, but close enough to allow biotin (probably bound to the side chain of a lysine) to rapidly reach both sites. Later non-ping-pong kinetic results do not eliminate this type of structure.

1972). This enzyme is a tetramer, with one mole of biotin present in each subunit. The reaction catalyzed by pyruvate carboxylase also has two partial reactions, the second of which is identical to the second step of the transcarboxylase reaction:

1. E–biotin + $MgATP^{2-}$ + HCO_3^- $\rightleftharpoons$ E–biotin–CO_2^- + $MgADP^-$ + P_i

2. E–biotin–CO_2^- + Pyruvate $\rightleftharpoons$ E–biotin + Oxaloacetate .

As with transcarboxylase, the kinetic evidence was explained by assuming that there is a separate catalytic site for each partial reaction and that the biotin moiety moves between these two sites (Fig. 7). It was assumed that, in pyruvate carboxylase, both sites are present on the same subunit. The kinetic mechanism for pyruvate carboxylase is considerably more complicated than originally envisaged. Using carboxylase from pig liver, Warren and Tipton (1974) have found that the binding of substrates and release of products follows this Theorell–Chance type pathway:

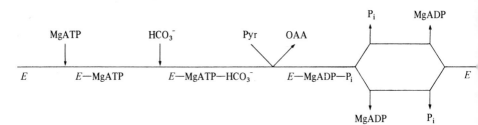

where E indicates an enzyme molecule having the effectors K^+, acetyl CoA and Mg^{2+} already bound. They did not observe any feature rendering the mechanism ping-pong in nature. MgADP and P_i are released after the binding of pyruvate. This suggests that the carboxylation of the biotin prosthetic group of pyruvate carboxylase traps MgADP and P_i at their binding sites. Warren and Tipton showed that each pyruvate carboxylase protomer contains three polypeptide chains, but only one molecule of biotin. Perhaps pyruvate carboxylase resembles acetyl CoA carboxylase and transcarboxylase in possessing a BCCP-like polypeptide in each active subunit.

Acyl Carrier Protein

Fatty acid synthetase catalyzes the repetitive condensation of 2-carbon units derived from malonyl CoA with acetate or the longer developing fatty acid, and the reduction of the β-keto groups which are formed. Lynen (1961) found that thiol groups were essential for the catalytic activity of the yeast fatty acid synthetase and suggested that the acyl intermediates were bound to the $-SH$ groups of the enzyme. A heat-stable thiol-containing protein required for the synthesis of fatty acids was isolated from *E. coli*. This protein, the acyl carrier protein (ACP), binds as thioesters both the substrates and all the intermediates of the biosynthesis. For example, the condensation of the acetyl CoA primer and the first molecule of malonyl CoA is:

$$
\underset{\text{CH}_3-\text{C}-\text{S}-\text{ACP}}{\overset{\text{O}}{\|}} + \underset{\text{CH}_2-\text{C}-\text{S}-\text{ACP}}{\overset{\text{HOOC}\quad\text{O}}{|\quad\|}} \rightleftharpoons \underset{\text{CH}_3-\text{C}-\text{CH}_2-\text{C}-\text{S}-\text{ACP}}{\overset{\text{O}\quad\quad\text{O}}{\|\quad\quad\|}} + \text{ACP}-\text{SH} + \text{CO}_2
$$

The acyl$-$S$-$ACP substrates are formed by transesterification from their CoA esters. The only $-SH$ group which the *E. coli* ACP contained was 4-phosphopantetheine (a portion of CoA), linked by a phosphodiester bond to a serine hydroxyl (Majerus *et al.*, 1965; Pugh and Wakil, 1965):

$$
\underset{\underset{-\text{Asp}-\text{Ser}-\text{Leu}-}{\underset{|}{\underset{\text{OH}}{|}}}{\overset{\text{O}}{\|}}\text{O}-\text{P}-\text{O}-\text{CH}_2-\underset{\overset{|}{\underset{\text{CH}_3}{}}}{\overset{\overset{\text{CH}_3}{|}}{\text{C}}}-\text{CH}-\overset{\overset{\text{OH}}{|}}{}\underset{}{\overset{\overset{\text{O}}{\|}}{\text{C}}}-\text{NH}-\text{CH}_2-\text{CH}_2-\overset{\overset{\text{O}}{\|}}{\text{C}}-\text{NH}-\text{CH}_2-\text{CH}_2-\text{SH}.
$$

Acyl carrier proteins, with molecular weights ranging from 8600 to 16 000, have now been isolated from bacteria, plants, yeast and mammalian liver. ACP is a component of all fatty acid synthetases tested.

The complete amino acid sequence of the acyl carrier protein from *E. coli* was established by Vanaman *et al.* (1968). The molecule has 77 amino acid residues and the pantetheine is attached to serine 36 (Fig. 8). A polypeptide having the 74 N-terminal residues of the *E. coli* ACP has been synthesized (Hancock *et al.*, 1972) by the solid phase method. Histidine causes complications in the synthetic procedure, and the first His is at position 75. After 4-phosphopantetheine was introduced enzymatically, the 74-residue peptide (ACP-

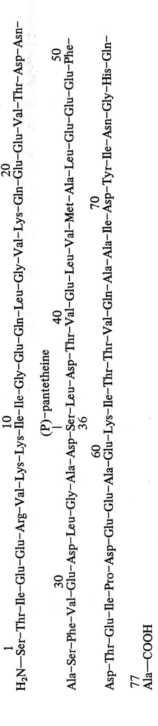

1
H₂N—Ser–Thr–Ile–Glu–Glu–Arg–Val–Lys–Lys–Ile–Gly–Glu–Gln–Leu–Gly–Val–Lys–Gln–Glu–Val–Thr–Asp–Asn–

(P)–pantetheine
|
36
Ala–Ser–Phe–Val–Glu–Asp–Leu–Gly–Ala–Asp–Ser–Leu–Asp–Thr–Val–Glu–Leu–Val–Met–Ala–Leu–Glu–Glu–Phe–

Asp–Thr–Glu–Ile–Pro–Asp–Glu–Glu–Ala–Glu–Lys–Ile–Thr–Thr–Val–Gln–Ala–Ala–Ile–Asp–Tyr–Ile–Asn–Gly–His–Gln–

77
Ala—COOH

Fig. 8. Amino acid sequence of acyl carrier protein from *E. coli* (Vanaman *et al.*, 1968).

(1–74)) had high biological activity in an enzyme system measuring malonyl pantetheine–CO_2 exchange. A number of analogues of the ACP have been synthesized, enabling some structure–function relationships to be established (Hancock *et al.*, 1973). The C-terminus does not participate directly in the activity of ACP, but when the amino acids in the N-terminal region (positions 1–6, making ACP-(2–74) to ACP-(7–74)) were serially omitted from the synthetic peptides, a large decrease in activity was observed. This indicates that the N-terminal portion of ACP helps in maintaining an active structure for ACP.

An acyl carrier protein has been isolated from citrate lyase, a multi-enzyme complex which catalyzes the cleavage of citrate to acetate and oxaloacetate in several bacteria. The cleavage occurs in two steps, an acyl exchange reaction and an acyl lyase reaction:

1. E–ACP–acetyl + Citrate ⇌ E–ACP–citryl + Acetate

2. E–ACP–citryl ⇌ E–ACP–acetyl + Oxaloacetate .

Acetylated citrate lyase is the active form of the enzyme. An enzyme is present in the bacteria to catalyze the ATP-dependent addition of acetate, to activate the des-acetyl citrate lyase. Continuous substrate turnover regenerates new acetyl enzyme. After treatment of ^{14}C-acetyl-labelled citrate lyase from *Klebsiella aerogenes* with SDS, Dimroth *et al.* (1973) were able to isolate all of the radioactivity in one protein fraction. This fraction was purified and found to be an ACP of molecular weight of about 10 000. Experiments the previous year had shown that citrate lyase contains phosphopantetheine. Srere and his colleagues (Carpenter *et al.*, 1975) have performed a careful analysis of the lyase enzyme complex. They found that the complex (520 000 dalton molecular weight) is composed of three types of subunits, with molecular weights of 54 500, 32 000 and 11 000 (the ACP). The largest subunit catalyzes reaction (1), the formation of citryl–ACP, and the 32 000 dalton subunit catalyzes reaction (2), the cleavage of citryl–ACP to acetyl–ACP (Dimroth and Eggerer, 1975). The number of each type of subunit in the enzyme complex has not yet been accurately determined, but there appear to be about six copies of each. Extracts of *K. aerogenes* contain two ACPs, one specific for fatty acid synthesis and the other specific for citrate lyase activity. The ACP associated with the citrate lyase is *chemically* different from the fatty acid synthetase also, notably in the composition of its prosthetic group, which contains adenine, phosphate and sugars as well as pantetheine.

Phosphopantetheine is also bound to a high molecular weight enzyme required for the biosynthesis of the antibiotic Gramicidin S (Gilhuus-Moe *et al.*, 1970) and the enzyme which catalyzes the biosynthesis of 6-methyl salicylic acid (Lynen, 1972). Lipmann (1973) has reported the isolation of an acyl carrier protein from the enzyme system catalyzing nonribosomal biosynthesis of antibiotics, and suggests that the thiol-linked formation of polypeptides has evolved from fatty acid synthesis.

TRANSPORT PROTEINS

Examination of the mechanisms of transport of solutes across cellular membranes suggests that some solutes interact with specific proteins present in the membrane during their entry into (or release from) the cell (reviewed by Kaback, 1970). The strongest evidence implicating these proteins in transport comes from bacterial mutants which are unable to synthesize the solute-specific binding proteins and are thus also unable to transport or concentrate the respective solutes. Of the transport proteins characterized to date, the histidine-containing phosphocarrier protein (HPr) required for transport of carbohydrates across the membranes of some bacteria (Roseman, 1972) shows the closest similarity to the other protein coenzymes which we have discussed. Whether some or all other transport proteins can be classified as coenzymes will depend on the elucidation of the mechanisms of their action (i.e. do they interact directly with enzymes?), which for many, especially those in higher organisms, is a few years hence.

In 1964, a heat- and acid-stable protein was isolated from *E. coli* in Roseman's laboratory (Kundig *et al.*, 1964) which, along with two enzyme fractions, was needed for the phosphorylation of several sugars. This protein was phosphorylated by phosphoenol pyruvate (PEP), in the presence of enzyme fraction I. When the protein was phosphorylated with ^{32}P-phosphate, the ^{32}P was released immediately in acid but was retained after treatment with 0.1 N NaOH for 1 h at 25° C. The hydrolysis of phosphohistidine follows this pattern (phosphoserine is stable in acid) and indeed, upon alkaline hydrolysis of the labelled protein, ^{32}P-phosphohistidine was detected. This protein, HPr, was the substrate which phosphorylated the sugar, under the influence of enzyme fraction II. Although the protein components of this phosphotransferase system were purified extensively, no biological significance for this reaction was clear in 1964.

The phosphocarrier protein HPr has been purified to homogeneity from *E. coli* (Anderson *et al.*, 1971), *Staphylococcus aureus* (Simoni *et al.*, 1973) and *Salmonella typhimurium* (Roseman, 1972), and the PEP-dependent phosphotransferase system has been found in three other micro-organisms. The HPr from *E. coli* has a molecular weight of about 9500, and contains two moles of histidine per mole of protein. It is not as heat-stable as was originally believed. Three forms of the *E. coli* protein, differing by the amount of ammonia estimated after hydrolysis (9, 8 and 6-7 moles of ammonia per 2 moles His) could be isolated. The most active form (containing 9 moles of NH$_3$) was prepared by a relatively gentle series of purification steps (e.g. at 0-4° C, pH 5.5-7.6). The less active forms, artefacts caused by deamidation during their isolation, were prepared by a procedure involving among other steps heating to 95° C, acidification to pH 2.4, several 24-h dialysis steps and concentration in a rotary evaporator at 35° C. Harsh steps such as these were not used in the purification of the *S. aureus* HPr, and it was shown to be a single protein band in gel electrophoresis. It has a molecular weight of about 9000 (9200 from sedimentation equilibrium, 9000 from SDS gels and 8630 from its amino acid analysis). There is only one histidine residue in the HPr obtained from *S. aureus*. Both the *E. coli* and *S. aureus* HPr's bind one mole of phosphate, apparently to the N^1 position of histidine. Phosphohistidine isolated from HPr is hydrolyzed at approximately the same rate as synthetic 1-phosphohistidine and much faster than synthetic 3-phosphohistidine. The binding of phosphate to N^1 of the single histidine residue of the *S. aureus* HPr has been confirmed by n.m.r. spectroscopy (Schrecker *et al.*, 1975).

The isolation of HPr was based solely on its properties in the enzymatic transfer of phosphate from PEP to a variety of sugars. It is now known that the phosphotransferase system is coupled to translocation of the sugar phosphate acceptor into the bacterial cell. Enzyme I catalyzes the phosphorylation of HPr by PEP, a reaction common to the phosphorylation of several sugars, and fraction II

L

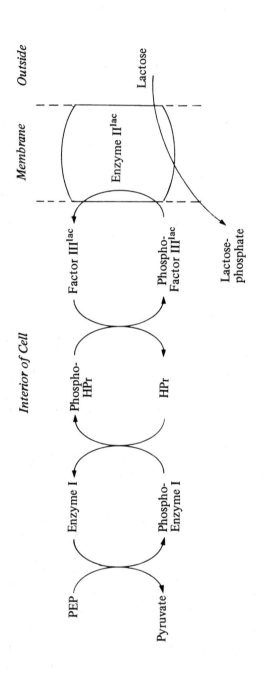

Fig. 9. The coupling of the transfer of phosphate from phosphoenol pyruvate (PEP) to lactose with the transport of the sugar into *S. aureus* (Simoni *et al.*, 1973). Enzyme I, HPr, and Factor III transfer the phosphoryl group from PEP to lactose. Factor III^{lac} and Enzyme II^{lac} are specific for the transport and phosphorylation of lactose.

contains the sugar-specific proteins. Experiments (Roseman, 1972) with the Enzyme II fractions from *E. coli* and *S. aureus* have shown that two proteins in this fraction are required for transfer of phosphate from phospho-HPr to any sugar, and that at least one of these proteins is bound to the cellular membrane. Specific proteins for the transport of glucose, mannose and fructose into *E. coli* and lactose into *S. aureus* have been examined. For example, enzymatic and genetic results support the scheme shown in Fig. 9 for phosphorylation of lactose as it crosses the cell membrane (Simoni *et al.*, 1973 and accompanying papers). Factor III accepts a phosphate group from phospho-HPr and transfers this phosphate to lactose during the existence of a phospho-Factor III-Enzyme II-lactose complex. Thus, HPr is a protein coenzyme carrying a phosphoryl group from the phosphorylated form of Enzyme I to any of several sugar-specific membrane-bound kinases.

The protein coenzymes which I have discussed in this chapter do not include all that are known, but are a representative sampling of them. The diversity of the group transfer reactions in which they participate indicates their importance in metabolism. An appreciation of the function of both the low molecular weight coenzymes and the protein coenzymes will aid in understanding the mechanisms of the enzymes described in the following chapter, all of which utilize covalent catalysis as one mode of their function.

REFERENCES

Adman, E. T., Sieker, L. C. and Jensen, L. H. (1973). *J. biol. Chem.* **248**, 3987–3996.

Alberts, A. W. and Vagelos, P. R. (1972). *In* "The Enzymes" (P. D. Boyer, ed.), Third edition, Vol. 6, pp. 37–82. Academic Press, New York.

Anderson, B., Weigel, N., Kundig, W. and Roseman, S. (1971). *J. biol. Chem.* **246**, 7023–7033.

Andreesen, J. R. and Ljungdahl, L. G. (1973). *J. Bacteriol.* **116**, 867–873.

Arnon, D. I. (1965). *Science, N.Y.* **149**, 1460–1470.

Bachmeyer, H., Piette, L. H., Yasunobu, K. T. and Whiteley, H. R. (1967). *Proc. natn. Acad. Sci. U.S.A.* **57**, 122–127.

Baginsky, M. L. and Huennekens, F. M. (1967). *Archs Biochem. Biophys.* **120**, 703–711.

Barden, R. E., Fung, C.-H., Utter, M. F. and Scrutton, M. C. (1972). *J. biol. Chem.* **247**, 1323–1333.

Beinert, H. (1963). *In* "The Enzymes" (Boyer, P. D., Lardy, H. and Myrbäck, K., eds), Second edition, Vol. 7, pp. 467–476. Academic Press, New York.

Bishop, N. I. (1971). *A. Rev. Biochem.* **40**, 197–226.

Brintzinger, H., Palmer, G. and Sands, R. H. (1966). *Proc. natn. Acad. Sci. U.S.A.* **55**, 397–404.

Bruice, T. C. and Hegarty, A. F. (1970). *Proc. natn. Acad. Sci. U.S.A.* **65**, 805–809.

Buchanan, B. B. (1972). *In* "The Enzymes" (P. D. Boyer, ed.), Third edition, Vol. 6, pp. 193–216. Academic Press, New York.

Burnett, R. M., Darling, G. D., Kendall, D. S., LeQuesne, M. E., Mayhew, S. G., Smith, W. W. and Ludwig, M. L. (1974). *J. biol. Chem.* **249**, 4383–4392.

Cammack, R. (1973). *Biochem. biophys. Res. Commun.* **54**, 548–554.

Carpenter, D. E., Singh, M., Richards, E. G. and Srere, P. A. (1975). *J. biol. Chem.* **250**, 3254–3260.

Carter, C. W., Jr., Kraut, J., Freer, S. T., Alden, R. A., Sieker, L. C., Adman, E. and Jensen, L. H. (1972). *Proc. natn. Acad. Sci. U.S.A.* **69**, 3526–3529.

Chuang, M., Ahmad, F., Jacobson, B. and Wood, H. G. (1975). *Biochemistry* **14**, 1611–1619.

Davenport, H. E., Hill, R. and Whatley, F. R. (1952). *Proc. R. Soc. B,* **139**, 346–358.

Dickerson, R. E. and Timkovich, R. (1975). *In* "The Enzymes" (P. D. Boyer, ed.), Third edition, Vol. 11, pp. 397–547. Academic Press, New York.

Dimroth, P. and Eggerer, H. (1975). *Proc. natn. Acad. Sci. U.S.A.* **72**, 3458–3462.

Dimroth, P., Dittmar, W., Walther, G. and Eggerer, H. (1973). *Eur. J. Biochem.* **37**, 305–315.

Eisenstein, K. K. and Wang, J. H. (1969). *J. biol. Chem.* **244**, 1720–1728.

Elford, H. L., Freese, M., Passamani, E. and Morris, H. P. (1970). *J. biol. Chem.* **245**, 5228–5233.

Entsch, B. and Smillie, R. M. (1972). *Archs Biochem. Biophys.* **151**, 378–386.

Evans, M. C. W., Sihra, C. K., Bolton, J. R. and Cammack, R. (1975). *Nature, Lond.* **256**, 668–670.

Fall, R. R. and Vagelos, P. R. (1972). *J. biol. Chem.* **247**, 8005–8015.

Gilhuus-Moe, C. C., Kristensen, T., Bredesen, J. E., Zimmer, T. L. and Laland, S. G. (1970). *FEBS Letters* **7**, 287–290.

Green, N. M., Valentine, R. C., Wrigley, N. G., Ahmad, F., Jacobson, B. and Wood, H. G. (1972). *J. biol. Chem.* **247**, 6284–6298.

Guchhait, R. B., Moss, J., Sokolski, W. and Lane, M. D. (1971). *Proc. natn. Acad. Sci. U.S.A.* **68**, 653–657.

Guchhait, R. B., Polakis, S. E., Hollis, D., Fenselau, C. and Lane, M. D. (1974). *J. biol. Chem.* **249**, 6646–6656.

Hall, C. L. and Kamin, H. (1975). *J. biol. Chem.* **250**, 3476–3486.

Hall, D. E., Baldesten, A., Holmgren, A. and Reichard, P. (1971). *Eur. J. Biochem.* **23**, 328–335.

Hancock, W. S., Prescott, D. J., Marshall, G. R. and Vagelos, P. R. (1972). *J. biol. Chem.* **247**, 6224–6233.

Hancock, W. S., Marshall, G. R. and Vagelos, P. R. (1973). *J. biol. Chem.* **248**, 2424-2434.

Holmgren, A. (1968). *Eur. J. Biochem.* **6**, 475-484.

Holmgren, A., Söderberg, B.-O., Eklund, H. and Bränden, C.-I. (1975). *Proc. natn. Acad. Sci. U.S.A.* **72**, 2305-2309.

Hong, J.-S. and Rabinowitz, J. C. (1970). *J. biol. Chem.* **245**, 4982-4987.

Hoskins, D. D. (1966). *J. biol. Chem.* **241**, 4472-4479.

Kaback, H. R. (1970). *A. Rev. Biochem.* **39**, 561-598.

Ke, B., Hansen, R. E. and Beinert, H. (1973). *Proc. natn. Acad. Sci. U.S.A.* **70**, 2941-2945.

Klein, S. M. and Sagers, R. D. (1966). *J. biol. Chem.* **241**, 197-205.

Knight, E., Jr. and Hardy, R. W. F. (1966). *J. biol. Chem.* **241**, 2752-2756.

Kundig, W., Ghosh, S. and Roseman, S. (1964). *Proc. natn. Acad. Sci. U.S.A.* **52**, 1067-1074.

Lehninger, A. L. (1975). "Biochemistry", Second edition, Chapter 22. Worth, New York.

Lipmann, F. (1973). *Accs Chem. Res.* **6**, 361-367.

Lode, E. T. and Coon, M. J. (1971). *J. biol. Chem.* **246**, 791-802.

Lode, E. T., Murray, C. L. and Rabinowitz, J. C. (1974). *Biochem. biophys. Res. Commun.* **61**, 163-169.

Lovenberg, W. (1973). (ed.) "Iron-Sulfur Proteins", Vols 1 and 2. Academic Press, New York.

Lovenberg, W. and Sobel, B. E. (1965). *Proc. natn. Acad. Sci. U.S.A.* **54**, 193-199.

Lovenberg, W., Buchanan, B. B. and Rabinowitz, J. C. (1963). *J. biol. Chem.* **238**, 3899-3913.

Lynen, F. (1961). *Fedn Proc. Fedn Am. Socs exp. Biol.* **20**, 941-951.

Lynen, F. (1972). *In* "Enzymes: Structure and Function" (Drenth, J., Oosterbaan, R. A. and Veeger, C., eds), pp. 177-200. North-Holland, Amsterdam.

Majerus, P. W., Alberts, A. W. and Vagelos, P. R. (1965). *J. biol. Chem.* **240**, 4723-4726.

Malkin, R. and Bearden, A. J. (1971). *Proc. natn. Acad. Sci. U.S.A.* **68**, 16-19.

Malkin, R., Aparicio, P. J. and Arnon, D. I. (1974). *Proc. natn. Acad. Sci. U.S.A.* **71**, 2362-2366.

Mayhew, S. G. (1971). *Analyt. Biochem.* **42**, 191-194.

Mayhew, S. G., Foust, G. P. and Massey, V. (1969). *J. biol. Chem.* **244**, 803-810.

Mortenson, L. E. (1964). *Biochim. biophys. Acta.* **81**, 71-77.

Mortenson, L. E., Valentine, R. C. and Carnahan, J. E. (1962). *Biochem. biophys. Res. Commun.* **7**, 448-452.

Moss, J. and Lane, M. D. (1971). *Adv. Enzymol.* **35**, 321-442.

Motokawa, Y. and Kikuchi, G. (1969). *Archs Biochem. Biophys.* **135**, 402-409.

Motokawa, Y. and Kikuchi, G. (1974). *Archs Biochem. Biophys.* **164**, 634-640.

Northrop, D. B. (1969). *J. Biol. Chem.* **244**, 5808-5819.

Orme-Johnson, W. H. (1973). *A. Rev. Biochem.* **42**, 159-204.

Orme-Johnson, W. H. and Beinert, H. (1969a). *Biochem. biophys. Res. Commun.* **36**, 337-344.

Orme-Johnson, W. H. and Beinert, H. (1969b). *J. biol. Chem.* **244**, 6143-6148.
Packer, E. L., Sternlicht, H., Lode, E. T. and Rabinowitz, J. C. (1975). *J. biol. Chem.* **250**, 2062-2072.
Peterson, J. A., Kusunose, M., Kusunose, E. and Coon, M. J. (1967). *J. biol. Chem.* **242**, 4334-4340.
Porqué, P. B., Baldesten, A. and Reichard, P. (1970). *J. biol. Chem.* **245**, 2371-2374.
Pugh, E. L. and Wakil, S. J. (1965). *J. biol. Chem.* **240**, 4727-4733.
Rao, K. K. and Matsubara, H. (1970). *Biochem. biophys. Res. Commun.* **38**, 500-506.
Reichard, P. (1968). "The Biosynthesis of Deoxyribose." John Wiley, New York.
Robinson, J. R., Klein, S. M. and Sagers, R. D. (1973). *J. biol. Chem.* **248**, 5319-5323.
Roseman, S. (1972). *In* "Metabolic Pathways" (L. E. Hokin, ed.), Third edition, Vol. 6, pp. 41-89. Academic Press, New York.
Ruzicka, F. J. and Beinert, H. (1975). *Biochem. biophys. Res. Commun.* **66**, 622-631.
San Pietro, A. and Lang, H. M. (1958). *J. biol. Chem.* **231**, 211-229.
Schrecker, O., Stein, R., Hengstenberg, W., Gassner, M. and Stehlik, D. (1975). *FEBS Letters* **51**, 309-312.
Scrimgeour, K. G., Vitols, K. S., Norris, M. L. and Pushkar, H. J. (1967). *Archs Biochem. Biophys.* **119**, 159-166.
Shin, M., Tagawa, K. and Arnon, D. I. (1963). *Biochem. Z.* **338**, 84-96.
Simoni, R. D., Nakazawa, T., Hays, J. B. and Roseman, S. (1973). *J. biol. Chem.* **248**, 932-940.
Sjöberg, B.-M. and Holmgren, A. (1972). *J. biol. Chem.* **247**, 8063-8068.
Smillie, R. M. and Entsch, B. (1971). *In* "Methods in Enzymology" (A. San Pietro, ed.), Vol. 23, pp. 504-514. Academic Press, New York.
Sobel, B. E. and Lovenberg, W. (1966). *Biochemistry* **5**, 6-13.
Stadtman, T. C. (1966). *Archs Biochem. Biophys.* **113**, 9-19.
Stadtman, T. C. (1974). *Science, N.Y.* **183**, 915-922.
Thelander, L. (1967). *J. biol. Chem.* **242**, 852-859.
Thelander, L. (1968). *Eur. J. Biochem.* **4**, 407-422.
Thelander, L. (1970). *J. biol. Chem.* **245**, 6026-6029.
Turner, D. C. and Stadtman, T. C. (1973). *Archs Biochem. Biophys.* **154**, 366-381.
Uyeda, K. and Rabinowitz, J. C. (1971). *J. biol. Chem.* **246**, 3111-3125.
Valentine, R. C. (1964). *Bact. Rev.* **28**, 497-517.
Vanaman, T. C., Wakil, S. J. and Hill, R. L. (1968). *J. biol. Chem.* **243**, 6420-6431.
Volpe, J. J. and Vagelos, P. R. (1973). *A. Rev. Biochem.* **42**, 21-60.
Wakil, S. J. (1958). *J. Am. Chem. Soc.* **80**, 6465.
Wakil, S. J. and Gibson, D. M. (1960). *Biochim. biophys. Acta.* **41**, 122-129.
Warren, G. B. and Tipton, K. F. (1974). *Biochem. J.* **139**, 297-329.
Watenpaugh, K. D., Sieker, L. C., Herriott, J. R. and Jensen, L. H. (1971). *Cold Spring Harb. Symp. quant. Biol.* **36**, 359-367.

Watenpaugh, K. D., Sieker, L. C. and Jensen, L. H. (1973). *Proc. natn. Acad. Sci. U.S.A.* **70**, 3857-3860.

Whitfield, C. D. and Mayhew, S. G. (1974). *J. biol. Chem.* **249**, 2801-2810.

Wood, H. G. (1972). *In* "The Enzymes" (P. D. Boyer, ed.), Third edition, Vol. 6, pp. 83-115. Academic Press, New York.

Wright, B. E. and Anderson, M. L. (1958). *Biochim. biophys. Acta.* **28**, 370-375.

Zanetti, G. and Williams, C. H., Jr. (1967). *J. biol. Chem.* **242**, 5232-5236.

10 Covalent Catalysis

The principles of covalent catalysis of non-enzymatic reactions were introduced in Chapter 4, and two examples of this process were discussed there—one involving a Schiff base intermediate (p. 93) and the other an acyl imidazole intermediate (p. 101). The reactions of many of the coenzymes can be considered as parallel to the covalent functions of the enzymes which we shall now examine. For example, the reactions of enzyme-bound pyridoxal phosphate resemble the reactions of enzymes which have amino acid residues capable of forming Schiff bases with substrates. The characterization of covalent intermediates in enzyme-mediated reactions is striking proof of the existence of *ES* complexes. These intermediates are not just simple Michaelis complexes, though. They are formed by covalent reaction of a portion of the substrate molecule with a reactive group present in the enzyme.

The most satisfactory evidence implicating a covalent *ES* intermediate in a reaction is the isolation and characterization of the intermediate after reaction of substrate with enzyme. The isolated intermediate should be *kinetically competent* (i.e. react in the presence of acceptor at least as fast as the overall reaction occurs). Because most *ES* intermediates are extremely reactive, they can seldom be isolated without some sort of *stabilization* or chemical trapping technique. Often, stabilization is achieved by using artificial substrates which react more slowly than natural substrates. It is then necessary to show that these substrates react by the same mechanism as do the natural substrates. Trapping of enzyme-bound groups can be either by a chemical change such as a reduction, or by rapid denaturation of the enzyme.

Kinetic experiments of several types—ping-pong behaviour, a "burst," or demonstration of a common intermediate—can indicate involvement of a covalent step in catalysis. The observation of *ping-*

pong kinetics can be taken as an indication of the formation of a covalent intermediate. However, not all enzymes that function by formation of covalent *ES* intermediates show simple ping-pong kinetics. The presence of an initial *"burst"* of product formation under suitable conditions (cf. p. 158) suggests that a portion of the substrate is still bound to the enzyme after release of the leaving group A.

$$A—B + E \xrightarrow{\text{fast}} A + B—E \xrightarrow{\text{slow}} B + E$$

After its rapid formation, the intermediate composed of the enzyme and the remaining part of the substrate (B—*E*) undergoes a slower cycle of decomposition and reformation in the steady state. Sometimes several substrates can donate the same group to the enzyme, or several acceptors can react with the *ES* intermediate:

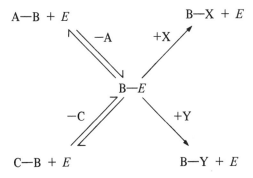

The existence of the *common B—E intermediate* in two or more reactions can be demonstrated by several types of kinetic experiments. If the decomposition of the intermediate is the rate-determining step, then both A—B and C—B would produce the same maximal velocity for the formation of B—X. If the formation of B—*E* is the slower step, then the products B—X and B—Y should be formed from A—B at similar rates. The common intermediate should be partitioned at a constant ratio between two acceptors, no matter which step is rate limiting. For example, at fixed concentrations of a mixed solution of X and Y, any substrate transferring B to the enzyme should give the same ratio of B—X and B—Y.

Exchange reactions, most often using an isotopically labelled substrate or product, have commonly been used as a test for the occurrence of a covalent *ES* intermediate. For example, in a direct displacement reaction such as (1), incorporation of labelled A (*A) into A—B in the absence of acceptor X would not be expected. If B—E were a covalent intermediate, as in the two-step reaction (2), then *A could react with B—E when X is absent to form *A—B, by partial reaction (2a).

1. A—B + X $\xrightarrow{E}$ A + B—X

2a. A—B + E ⇌ A + B—E

2b. B—E + X ⇌ B—X + E

Addition of *B would give no exchange by either mechanism. Observation of an exchange reaction is valid evidence for a covalent intermediate if *A cannot bind to the site normally occupied by X, and *X cannot bind to the site occupied by A (Bell and Koshland, 1970).

After surveying the mechanistic data available for transferases, Spector (1973) has speculated that transferases use only the two-step or double displacement mechanism. He extended this premise by considering all enzymes to catalyze transferase reactions of some type, so that all enzymes ought to catalyze their reactions in discrete steps with transient formation of covalent ES intermediates. I would agree that many transferase-catalyzed reactions proceed by formation of covalent intermediates (most, considering the holoenzymes of the coenzymes) and discrete steps occur in most reactions. However, there is no compelling reason why all enzymes should use utilize covalent catalysis. As noted in Chapter 4 (p. 93), the catalytic role of a covalent intermediate is severely restricted. Non-covalent forces stabilize intermediates in many reactions, and acid–base catalyzed, i.e. ionic, reactions can hardly be termed "covalent" (I do not consider a protonated enzyme to be a covalent ES complex). Covalent ES intermediates may occur for reasons other than catalysis (e.g. retention of stereochemistry, or preservation of bond energy), but covalent ES intermediates are not essential to enzymatic mechanisms. The number of proven cases of covalent catalysis is not yet large, but it is growing.†

† Occasionally, the number shrinks. For example, the "phosphorylenzyme" prepared from phosphoglycerate kinase and ATP has been shown to be an enzyme–1,3-diphosphoglycerate complex (Johnson et al., 1975). The kinetic mechanism for this enzyme is a sequential pathway with phosphoryl transfer between substrates within a ternary complex.

Table 1. Examples of covalent *ES* intermediates.

Enzyme	Reacting Residue	Covalent Intermediate
Chymotrypsin	Serine ($-$OH)	Acyl serine
Glyceraldehyde-3-phosphate dehydrogenase	Cysteine ($-$SH)	Acyl cysteine
Succinyl CoA synthetase	Histidine (imidazole)	Phosphohistidine
Alkaline phosphatase	Serine ($-$OH)	Phosphoserine
Acetoacetate decarboxylase	Lysine ($-$NH$_2$)	Schiff base
Histidine decarboxylase	Pyruvic acid ($-\overset{\overset{\text{O}}{\|}}{\text{C}}-$)	Schiff base
Sucrose phosphorylase	Carboxyl	Glucosyl enzyme

Bell and Koshland (1971) have compiled a useful list of 60 well-established covalent *ES* intermediates, and summarized the evidence upon which their existence is founded. These intermediates include pyrophosphoryl, phosphoryl, acyl (both proteolytic and non-proteolytic), Schiff base and carboxy ester compounds. In Table I, I have listed seven enzymes operating by covalent catalytic mechanisms. These enzymes exemplify seven different uses of covalent catalysis, and their properties will be outlined in this chapter. Chymotrypsin is the best-studied of the proteases and esterases having a nucleophilic serine residue as the covalent acceptor of an acyl group from the substrate. A closely allied group of enzymes (e.g. papain and ficin) utilize the thiol group of cysteine as the nucleophilic reagent for the hydrolysis. With glyceraldehyde-3-phosphate dehydrogenase, a cysteine residue is used to preserve energy in a multi-step oxidative reaction leading to formation of ATP from ADP and P$_i$. Both succinyl CoA synthetase and alkaline phosphatase form phosphoenzyme intermediates, but the former is a phosphohistidine enzyme and the latter a phosphoserine enzyme. Schiff-base intermediates are formed by substrates with both acetoacetate decarboxylase and histidine decarboxylase. However, these enzymes (or their substrates, if preferred) differ in the group supplied by each: a nucleophilic amino group by the enzyme and a carbonyl group by the substrate for acetoacetate decarboxylase; and an electrophilic carbonyl group by the enzyme and an amino group by the substrate for histidine decarboxylase. Finally, a carboxyl group of sucrose phosphorylase is the transient carrier of the glucosyl moiety of sucrose. In this case, the formation of a covalent *ES* intermediate assures that the glucose-1-phosphate formed from sucrose is the α-anomer. Examination of research using these enzymes will show how the experimental approaches described above have been applied.

CHYMOTRYPSIN

My first example of an enzyme using covalent catalysis in its action is chymotrypsin. Chymotrypsin is one of several proteases, called *serine proteases,* which contain an essential serine group which is acylated during catalysis. Other serine proteases are trypsin, thrombin, elastase and subtilisin. There are also some serine esterases, such as acetylcholinesterase, which have similar properties. Chymotrypsin is synthesized as its inactive precursor, the zymogen chymotrypsinogen, in the pancreas. It is stored there in protein granules and released when needed into the duodenal region of the small intestine via the pancreatic ducts. In the intestine, the zymogen is converted to chymotrypsin, which aids in the digestion of proteins.

The mechanism of action of chymotrypsin has been more extensively studied than that of any other enzyme. Luckily, excellent reviews of the literature are available (Bruice and Benkovic, 1966; Hess, 1971; Zeffren and Hall, 1973). The data permitting the elucidation of the three-dimensional structures of both chymotrypsinogen (Kraut, 1971) and chymotrypsin (Blow, 1971) also have been reviewed. As we shall soon see, these data have given much insight into the mechanism of chymotrypsin.

Two forms of chymotrypsinogen occur in bovine pancreas, chymotrypsinogen A and chymotrypsinogen B, differing substantially in their amino acid sequences. The A form has been studied more fully, and is the only isoenzyme to be considered here. The complete amino acid sequence of chymotrypsinogen A was published by Hartley in 1964, with two minor corrections being reported two years later (Hartley and Kauffman, 1966; Meloun *et al.*, 1966). A final correction, replacement of Asn-102 by an Asp, was made after the X-ray crystallography data became available.

The specificity of chymotrypsin is primarily for peptide bonds in which the carbonyl group belongs to an aromatic amino acid (tryptophan, tyrosine or phenylalanine). However, peptide bonds having leucine, methionine, asparagine or glutamine on the N-terminal side also can be hydrolyzed at a slower rate. Synthetic amide or ester substrates are used for most kinetic studies. Those synthetic substrates which are amides or esters of aromatic amino acids (such as *N*-acetyl-L-tyrosine ethyl ester) are called *specific substrates*, while other substrates (such as *p*-nitrophenyl acetate) are called *nonspecific substrates*.

The activation of chymotrypsinogen to chymotrypsin occurs by cleavage of only one peptide bond, that between Arg-15 and Ile-16. After this one change, the molecule has full enzymatic activity. This

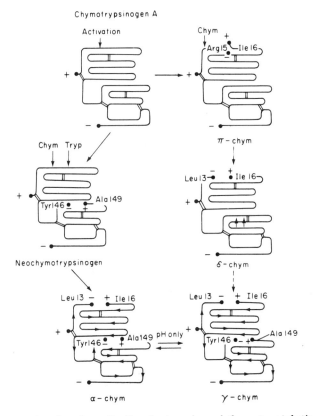

Fig. 1. Diagram showing the activation by trypsin and the autocatalytic reactions in the conversion of chymotrypsinogen A to α-chymotrypsin (from Blow, 1971). The pathway on the left represents "slow activation," which occurs in the presence of low concentrations of trypsin. The pathway on the right is the "fast activation" which occurs at higher concentrations of trypsin. The fine double lines represent cystine bridges in the proteins.

cleavage is catalyzed by trypsin (which has specificity for peptide bonds of arginine and lysine), and under physiological conditions is the first modification of chymotrypsinogen. Several bonds of chymotrypsinogen (involving the carboxyl groups of Leu-13, Tyr-146 and Asn-148) are susceptible to autocatalytic digestion. Because these bonds are on the surface of the zymogen, they are easily accessible to other molecules of chymotrypsin (Kraut, 1971). Most of the peptide bonds that would be susceptible in denatured chymotrypsin are not hydrolyzed in the native protein because they are buried. α-Chymotrypsin is the form which is most easily obtained in homogeneous crystals, and as shown in Fig. 1, it contains three peptide chains (residues 1-13, 16-146 and 149-245, the C-terminal of chymotrypsinogen) held together by disulphide and non-covalent bonds.

Due to the great amount of research done with chymotrypsin, many findings of general application have emerged. The structure and mechanism of action of chymotrypsin are similar to those of other serine proteases, but information gained from the studies on chymotrypsin also applies to many other types of enzymes. The inhibition by active site-directed inhibitors, the "burst" phenomenon and the detailed X-ray studies on chymotrypsin are examples of well-characterized properties of chymotrypsin which have already augmented our knowledge of basic enzymology.

Di-isopropyl phosphofluoridate (or di-isopropyl fluorophosphonate, DFP) is one of a group of toxic organophosphorus compounds,

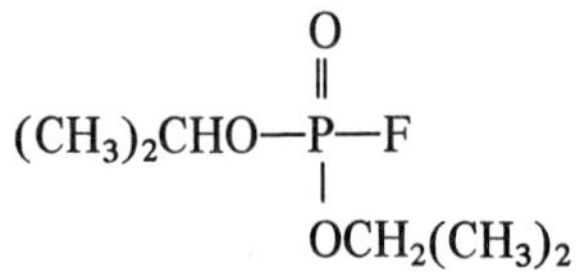

$$
\begin{array}{c}
O \\
\parallel \\
(CH_3)_2CHO\!-\!P\!-\!F \\
\mid \\
OCH_2(CH_3)_2
\end{array}
$$

DFP

called "nerve gases" because of their ability to inhibit cholinesterases. One mole of DFP reacts with one mole of chymotrypsin to irreversibly inactivate the enzyme (Balls and Jansen, 1952). Chymotrypsinogen reacts with DFP, but only at an extremely slow rate. Using ^{32}P-DFP, Schaffer et al. (1953) were able to isolate ^{32}P-phosphoserine from a hydrolysate of inactivated chymotrypsin. By establishing the sequence of two ^{32}P-labelled peptides obtained by enzymatic hydrolysis of ^{32}P-DFP-treated enzyme, Oosterbaan et al. (1958) found that the ^{32}P-dialkyl phosphoryl group was bound only to the serine present in the sequence Gly-Asp-*Ser*-Gly-Gly-Pro-Leu. When the complete sequence of chymotrypsin was made available, this serine was identified as Ser-195. Because DFP completely inhibits chymotrypsin by reacting with only one of the 27 serine residues in the molecule, it is considered to be an active site-directed covalent inhibitor, with Ser-195 being part of the active site. Inactivation by DFP is used as a test for active-site serine hydroxyl groups, and about a dozen proteases or esterases react in a manner similar to chymotrypsin. In view of the extreme toxicity of DFP, rigid precautions should be taken to prevent its inhalation, ingestion or contact with the skin (Cohen *et al.*, 1967).

Two pieces of kinetic evidence have aided in the elaboration of a minimal kinetic expression for the hydrolyses catalyzed by chymotrypsin. Hartley and Kilby (1954) observed a biphasic release of the *p*-nitrophenol from *p*-nitrophenyl acetate or *p*-nitrophenyl ethyl carbonate (p-NO_2—C_6H_4—O—CO—OC_2H_5). There was a rapid

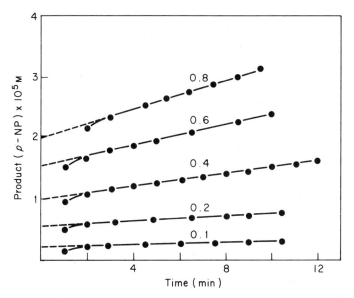

Fig. 2. Biphasic curves showing the formation of *p*-nitrophenol (*p*-NP) from 5×10^{-4} M *p*-nitrophenyl ethyl carbonate in the presence of chymotrypsin. The reactions were conducted at $20°$ C in 0.066 M phosphate buffer, pH 7.6 containing 5% isopropanol (to suppress spontaneous hydrolysis) at five concentrations of chymotrypsin. The concentrations of chymotrypsin (in mg ml^{-1}) are indicated on the curves. Adapted from Hartley and Kilby, 1954.

release of about one mole (1.1–1.2 moles) of p-nitrophenol per mole of enzyme, followed by a slower linear release of the phenol (Fig. 2). Hartley and Kilby proposed that this kinetic "burst" was caused by rapid acylation of the enzyme, followed by rate-limiting hydrolysis of the acylated active centre. They proposed a kinetic expression of this type:

$$E + A \rightleftharpoons EA \xrightarrow{\text{Rapid}} EQ \xrightarrow[\text{Slower}]{H_2O} E + Q.$$
$$\searrow P$$

To measure the kinetics of the "initial burst," stopped-flow spectrophotometry was required (Spencer and Sturtevant, 1959). Because the stability of the acyl-enzyme intermediate varies with the structure of the acyl group and also is greater at low pH, McDonald and Balls (1957) succeeded in crystallizing an acyl-enzyme, trimethylacetyl chymotrypsin, by precipitation with ammonium sulphate at pH 4. The ^{14}C-acetyl group from radioactive p-nitrophenyl acetate reacts with the same serine residue as does DFP (Oosterbaan et al., 1962). The second piece of kinetic evidence came from demonstration of a common intermediate in the hydrolysis of specific ester substrates. Steady-state kinetic experiments showed that the methyl, ethyl and p-nitrophenyl esters of N-acetyl-L-phenylalanine are hydrolyzed in the presence of chymotrypsin at the same maximal velocity (Zerner et al., 1964) as are the esters of N-acetyl-L-tryptophan (Table II). These data also support the mechanistic scheme involving binding, then acylation, and a rate-limiting deacylation of

Table II. Comparison of the rates of chymotrypsin-catalyzed hydrolysis of some specific substrates. Maximal velocities are expressed in terms of the catalytic constant k_{cat} (cf. p. 116). From the data of Zerner et al. (1964).

Substrate	k_{cat} (s^{-1})	
	N-Acetyl-L-Phenylalanine Derivative	N-Acetyl-L-Tryptophan Derivative
Ethyl ester	63	27
Methyl ester	58	28
p-Nitrophenyl ester	77	30

the EQ intermediate, as written above. Using rapid-reaction techniques, the rates of formation and decomposition of intermediates and the release of products have been measured for a number of non-specific substrates. Results indicate that two ES complexes occur before formation of the acyl enzyme (Hess, 1971). Based on the pH dependence of chymotrypsin-catalyzed hydrolysis of anilide substrates, Caplow (1969) has suggested that these are the Michaelis (EA) complex and a tetrahedral intermediate (EA') formed by attack of Ser-195 on the substrate:

$$E + A \rightleftharpoons EA \rightleftharpoons EA' \overset{}{\underset{P}{\searrow}} \longrightarrow EQ \rightarrow E + Q.$$

Subsequent experiments with both enzymic and non-enzymic reaction systems have provided evidence in favour of nucleophilic addition of the ionized hydroxyl of Ser-195 to the substrate with formation of the transient tetrahedral intermediate or transition state, followed by release of the leaving group to form acylated chymotrypsin:

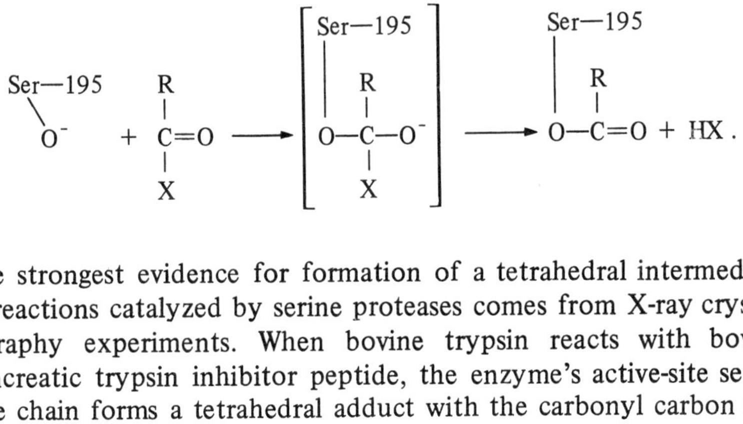

The strongest evidence for formation of a tetrahedral intermediate in reactions catalyzed by serine proteases comes from X-ray crystallography experiments. When bovine trypsin reacts with bovine pancreatic trypsin inhibitor peptide, the enzyme's active-site serine side chain forms a tetrahedral adduct with the carbonyl carbon of a specific lysine residue of the inhibitor (Rühlmann et al., 1973). Crystallographic experiments with subtilisin–substrate analogue complexes, discussed below, virtually prove the existence of the tetrahedral intermediate.

As well as Ser-195, four other amino acid residues have been implicated in the mechanism of action of chymotrypsin. The left limb of the pH v. activity curve for chymotrypsin (p. 120) has a midpoint at pH 7, indicating that a group with a pK_a of 7 (possibly an unprotonated imidazole of histidine) is essential in the deacylation step. Stopped-flow experiments showed a similar pH effect for the acylation step. In addition photo-oxidation (to which histidine residues are sensitive) inactivates chymotrypsin. Shaw and his co-workers (Ong *et al.*, 1965) synthesized TPCK, an analogue of a specific substrate, *N*-tosylphenylalanine ethyl ester.

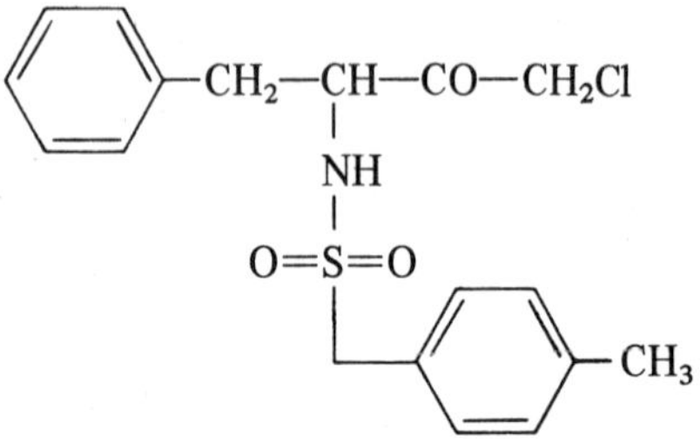

TPCK, L–1–tosylamino–2–phenylethylchloromethyl ketone

Like DFP, TPCK reacts only with the active enzyme. TPCK alkylates only one of the two histidine residues in chymotrypsin, His-57. Hess and his colleagues (Oppenheimer *et al.*, 1966) showed that the α-amino group of Ile-16 (formed during the activation process) is responsible for the right-hand part of the pH activity curve, the mid-point of about pH 8.5. They acetylated all the free amino groups of the zymogen, and then showed that upon activation to acetylated δ-chymotrypsin the only free amino group formed was that of Ile-16. The essential roles of two aspartate residues have been deduced from the structure obtained by X-ray crystallography experiments. The carboxyl group of Asp-194 forms an ion pair with the protonated Ile-16 α-amino group and disruption of this ion pair (e.g. by loss of the proton on the amino group) causes the enzyme to change into a catalytically inactive conformation. The carboxyl of Asp-102 inter-acts with the imidazole ring of His-57. The functions of these five essential residues: Ser-195, His-57, Ile-16, Asp-194 and Asp-102, should become clearer from a consideration of the three-dimensional structure of α-chymotrypsin. Obviously, not just these five amino acids are essential for the activity of chymotrypsin, but the roles of other functional side chains have not yet been clarified.

High-resolution electron density maps have been obtained for both α-chymotrypsin (Blow, 1971) and chymotrypsinogen A (Kraut, 1971). The enzyme and the zymogen have very similar structures.

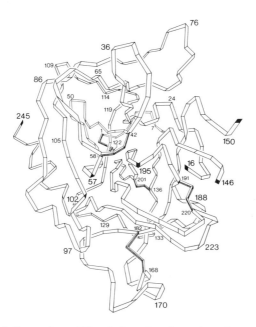

Fig. 3. Shape of the polypeptide chain of α-chymotrypsin, as shown by the position of the α-carbon atoms. From Blow (1971).

There are important but not extensive rearrangements when the four peptide bonds are broken by activation. Figure 3 shows the ellipsoid shape of chymotrypsin. As observed with other proteins, the hydrophobic residues tend to be located in the interior of the molecule. The only charged groups in the interior are the α-amino group of Ile-16 and the carboxyls of Asp-102 and Asp-194. In chymotrypsinogen, Asp-194 is hydrogen bonded to the side chain of His-40. Upon activation, when Ile-16 is protonated and pairs up with Asp-194, the substrate-binding cavity is formed. Comparison of the structures of α-chymotrypsin and its complexes with formyl-L-tryptophan and formyl-L-phenylalanine show that the aromatic groups of these virtual substrates are imbedded in a hole, bounded by residues 189–192 on one side and residues 213–217 on the other, located in a position appropriately close to the catalytic centre.

In the three-dimensional structure, Blow *et al.* (1969) noticed that Ser-195, His-57 (both at the surface) and Asp-102 (buried) were in a configuration favouring hydrogen bonding between them as shown below:

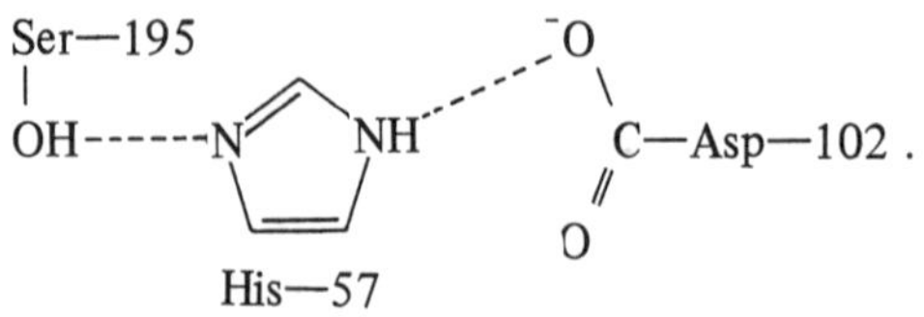

Ser—195
|
OH------N⟍⟋NH --------- ⁻O
⟍
His—57   C—Asp—102 .
∥
O

They suggested that this "charge relay system" conducts electrons from the buried carboxyl group to the surface, and in this way makes the hydroxyl group of Ser-195 a powerful nucleophile at the pH optimum. Fersht and Sperling (1973) proposed that the carboxyl group of Asp-102 is ionized at pH values even below those at which chymotrypsin is active (pK_a of 3 or less) and they considered the pK near 7 to be caused primarily by the imidazole of His-57. These latter conclusions on assignment of pK_a values must be modified, though, on the basis of the experiments on α-lvtic protease described below. Polgár and Bender (1969) have suggested that the charge relay system has an alternative role in catalysis, the stabilization of the imidazolium ion of His-57 by a negative charge on Asp-102 to prevent premature release of its proton. This suggestion also would appear to be ruled out now. The "charge relay system" occurs in chymotrypsinogen, but the zymogen has little or no catalytic activity because of a lack of a suitable substrate-binding site (Wright, 1973). The defective binding site in chymotrypsinogen is reflected by a low binding affinity of the zymogen for inhibitors and by a rate of ester hydrolysis which is 10^6–10^7 times slower than that in the presence of chymotrypsin (Gertler et al., 1974).

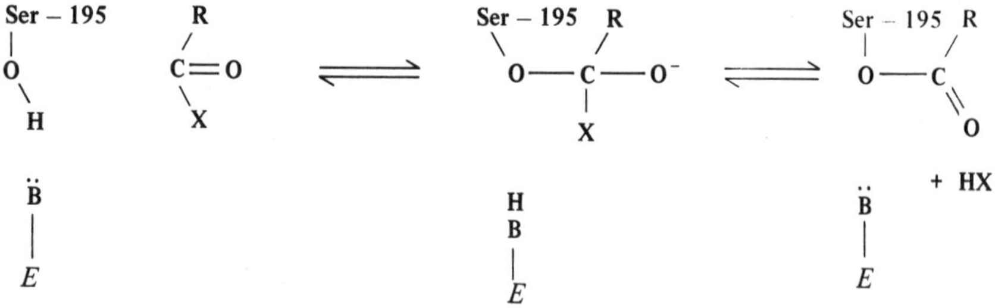

Ser – 195   R
| ⟋
O   C=O
⟍   ⟍
H   X

B̈
|
E

⇌

Ser – 195   R
⟍   ⟋
O—C—O⁻
|
X

H
B
|
E

⇌

Ser – 195 R
| ⟋
O—C
⟍
O

B̈   + HX
|
E

Fig. 4. Postulated acid–base catalytic function in catalysis of hydrolysis by chymotrypsin. The basic group (made up of residues His-57 and Asp-102) removes a proton from the active-site serine, and donates a proton to the group (—X) leaving the tetrahedral intermediate formed by addition of the serine hydroxyl to the substrate.

The steps in the hydrolysis of a peptide (or ester) bond, as shown in Fig. 4, appear to be:

1. removal of a proton from the hydroxyl of Ser-195 by the other two components of the charge relay system (His-57 and Asp-102);
2. formation of a tetrahedral intermediate by nucleophilic attack of the alkoxide of the serine on the carbonyl carbon of the peptide bond; and
3. formation of the acyl-serine intermediate by loss of an amine (or alcohol), $H-X$.

In the deacylation of the covalent intermediate, water replaces the leaving group $H-X$ in a reversal of these steps. I have already discussed (p. 166) Bender's assignments of various catalytic effects—change of pathway to covalent alcoholysis, proximity and acid–base catalysis—to the chymotrypsin-catalyzed reaction. To these effects must be added stabilization of the transition state by specific hydrogen bonding (see below).

Other serine proteases have structural and mechanistic analogies to chymotrypsin. Sequence and X-ray data have shown that chymotrypsin, trypsin (Keil, 1971; Stroud et al., 1971) and elastase (Hartley and Shotton, 1971) have clusters of homologous sequences and very similar tertiary structures, and contain the Ser-His-Asp hydrogen-bond complex at their active centres. Like chymotrypsinogen, both trypsinogen and proelastase are activated by cleavage of a single bond in the N-terminal region. In all of these enzymes, the newly formed N-terminal amino group is important for activity, presumably for rearrangement to the catalytically active conformation (Kurosky and Hofmann, 1972). In trypsin, the side-chain cavity contains an aspartate residue at its base, explaining the specificity for lysine and arginine. The specificity pocket of elastase is obstructed by two side chains, so that bulky aromatic amino acid peptide bonds cannot be hydrolyzed.

Subtilisin, a serine protease from *Bacillus,* and α-chymotrypsin have different overall three-dimensional structures, but the composition of the catalytic sites of both are identical and part of the specificity cavity of subtilisin is the same as that in chymotrypsin. Examination by X-ray crystallography of the structure of polypeptide–subtilisin complexes has suggested plausible models for the Michaelis complex and the tetrahedral and acyl intermediates (Robertus et al., 1972).

Subtilisin has a groove or channel on the surface capable of holding a polypeptide chain at least six amino acid residues long. The stereochemistry of the virtual substrate–enzyme complexes would allow formation of several new hydrogen bonds in the tetrahedral intermediate. In particular, an asparagine residue and an NH group seem capable of binding the oxyanion group (Fig. 5). Stabilization of the tetrahedral intermediate would allow it to proceed more readily toward the transition state leading to the formation of the acyl enzyme. Kraut's laboratory have extended their studies to include solution of the structures of complexes in which subtilisin has reacted covalently with boronic acid transition-state analogues (Matthews *et al.*, 1975). The adducts with both benzeneboronic acid (C_6H_5—$B(OH)_2$) and 2-phenylethaneboronic acid (C_6H_5—CH_2—B—$(OH)_2$) have the structure:

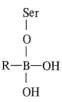

and one oxygen atom is hydrogen bonded to the oxyanion binding site. These observations strongly support the mechanism previously postulated, including the stabilization of the transition state. This stabilization effect should apply equally to catalysis by chymotrypsin.

α-Lytic protease isolated from *Myxobacter 495* is another serine protease closely related to the mammalian proteases. It is a very stable protein containing the Ser-His-Asp charge relay grouping. α-Lytic protease contains only one histidine residue, and by growing the bacteria in the presence of a 2-^{13}C-histidine, Hunkapiller *et al.* (1973) have achieved the first specific ^{13}C enrichment of a single atom in an enzyme. They then examined, by [13]C n.m.r. titrations, the properties of the peak assigned to C^2 of the active site histidine. This residue becomes protonated only below pH 4, but its chemical shift is affected by a neighbouring group having a pK_a of 6.7, probably Asp-102. The suggestion that Asp-102 is a weaker acid than His-57 agrees with the postulated effects of relatively nonpolar environments on the perturbed pK_a values of His-57 (lowered from 6.4) and Asp-102 (raised from 4.5) in chymotrypsin. The carboxyl group of Asp-102 is the ultimate base accepting the proton from the serine

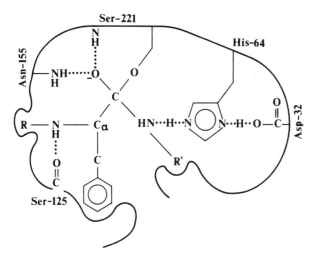

Fig. 5. Schematic structure of the postulated tetrahedral intermediate formed during hydrolysis of a polypeptide by subtilisin. Ser-221, His-64 and Asp-32 form a "charge relay system" identical to that in chymotrypsin. The tetrahedral intermediate is stabilized by formation of hydrogen bonds with Asn-155 and the backbone —NH— of Ser-221 and with Ser-125. None of these hydrogen bonds can be formed in the Michaelis complex. Therefore, the intermediate (and probably also the transition state) is bound more tightly than either product or substrate. Donation of the proton from His-64 produces the leaving group, $R'NH_2$, and the acyl enzyme. Adapted from Robertus *et al.* (1972).

hydroxyl, and a neutral carboxylic acid-imidazole system is formed, not an ion pair:

The imidazole group insulates the carboxyl from water, assuring its higher pK_a, and relays the proton from serine. The Ser-His-Asp catalytic unit would be more efficient than a simpler Ser-His couple because there is no charge separation with the triad as there would be with a Ser-His diad. Hunkapiller *et al.* conclude their article by saying that the high pK_a of Asp-102, the stabilization of the tetrahedral intermediate by hydrogen bonding and the lack of charge separation during formation of the intermediate probably account for much of the catalytic efficiency of serine proteases.

Present studies can explain the specificities of several of the serine proteases. Covalent catalysis and acid–base catalysis have been established in the action of these enzymes. No induced fit has been observed with chymotrypsin. If the transition states are indeed bound more tightly than the substrate, as suggested by the work on subtilisin, then slight distortions of the substrates as they pass along the reaction co-ordinate may occur. No large distortions have yet been observed for the binding of substrates. The lack of proof for a mechanism for even this well-studied enzyme explains why it will be many years before mechanistic enzymology will be a dead science.

GLYCERALDEHYDE-3-PHOSPHATE DEHYDROGENASE

Glyceraldehyde-3-phosphate dehydrogenase catalyzes one of the more complicated reactions in glycolysis. In collaboration with a second enzyme, phosphoglycerate kinase, it catalyzes the oxidation of glyceraldehyde-3-phosphate, reduction of NAD^+ and formation of ATP from ADP and P_i:

(1)

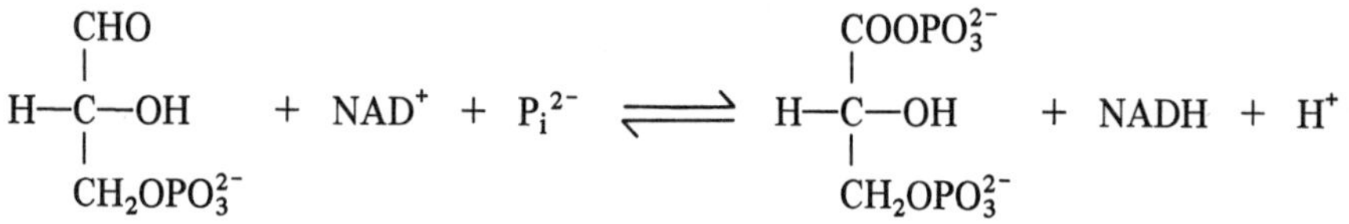

$$\begin{array}{c} CHO \\ | \\ H-C-OH \\ | \\ CH_2OPO_3^{2-} \end{array} + NAD^+ + P_i^{2-} \rightleftharpoons \begin{array}{c} COOPO_3^{2-} \\ | \\ H-C-OH \\ | \\ CH_2OPO_3^{2-} \end{array} + NADH + H^+$$

Glyceraldehyde–3–phosphate   1,3–Diphosphoglycerate

(2)

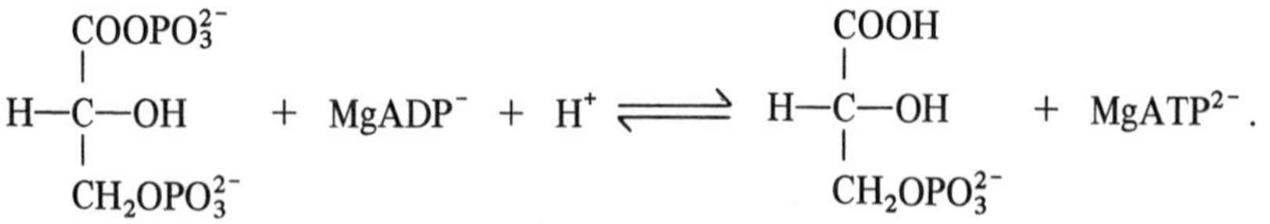

$$\begin{array}{c} COOPO_3^{2-} \\ | \\ H-C-OH \\ | \\ CH_2OPO_3^{2-} \end{array} + MgADP^- + H^+ \rightleftharpoons \begin{array}{c} COOH \\ | \\ H-C-OH \\ | \\ CH_2OPO_3^{2-} \end{array} + MgATP^{2-}.$$

3–Phosphoglycerate

This type of reaction, in which a compound of high negative free energy of hydrolysis is formed in one step and its energy transferred to ADP to form ATP in a second, is called *substrate-level phosphorylation*. Reaction (1) requires the participation of both NAD^+ and inorganic phosphate, producing the energy-rich mixed anhydride, 1,3-diphosphoglycerate. Covalent catalysis is important in this reaction, because an acyl intermediate composed of 3-phosphoglycerate bound to a cysteine residue of the dehydrogenase conserves the energy obtained in the oxidation.

Most of the mechanistic and structural research of glyceraldehyde-3-phosphate dehydrogenase has been performed using the crystalline enzyme obtained from rabbit muscle (p. 45) although crystalline dehydrogenases from other animals, birds and plants have been found to be quite similar. Determinations of the molecular weight of the rabbit muscle dehydrogenase produced a range of values from about 100 000–150 000 daltons, and for many years a value of about 120 000 was assumed. Fox and Dandliker (1956) reported that the molecular weight was 136 000–140 000. They applied several physical techniques, and obtained the same value with all the methods. Later research has verified their data. The glyceraldehyde-3-phosphate dehydrogenase from both rabbit and pig muscle is composed of four subunits of identical primary structure, each with a molecular weight of 36 000 ± 1000 (Harris and Perham, 1965; Harrington and Karr, 1965). This enzyme differs from most dehydrogenases by having tightly bound NAD^+ when isolated. The number of moles of pyridine nucleotide bound depends on the isolation procedure. There is a decrease in the binding constant for the addition of each subsequent molecule of NAD^+ (negative co-operativity). This and other interesting subunit effects of this enzyme are discussed briefly in Chapter 12 (p. 425). Here, I will concentrate on the mechanism of the reaction catalyzed by glyceraldehyde-3-phosphate dehydrogenase.

In the 1930s, evidence was obtained to indicate that glyceraldehyde-3-phosphate dehydrogenase is inhibited by iodoacetic acid (ICH_2COOH), and elaboration of these experiments verified that the —SH group of cysteine is essential for activity. For example, Holzer and Holzer (1952) and Segal and Boyer (1953) found that the dehydrogenase is specifically protected from inactivation by iodoacetic acid in the presence of its substrate, D-glyceraldehyde-3-phosphate. It was postulated that the essential —SH group reacts with the aldehyde group of the substrate to form an acyl enzyme and NADH. Krimsky and Racker (1955) prepared both the acetyl and 3-phosphoglyceryl derivatives from NAD⁺-free glyceraldehyde-3-phosphate dehydrogenase by reacting the enzyme with acetyl phosphate or 1,3-diphosphoglycerate. These acyl enzymes either could form hydroxamic acids when treated with NH_2OH, or could oxidize NADH, presumably with the formation of the thiol-ester derivatives of the enzyme. The enzyme was reacted with two different markers, ^{14}C-iodoacetate and p-nitrophenyl ^{14}C-acetate, and the radioactive substituted peptides were isolated from trypsin digests (Harris et al., 1963). The sequences of the respective peptides are:

$$^{14}CH_2COOH$$
$$|$$
Ile–Val–Ser–Asn–Ala Ser–Cys–Thr–Thr–Asn–Cys–Leu–Ala–Pro–Leu–Ala–Cys

$$^{14}CO—CH_3$$
$$|$$
and   Lys–Ile–Val–Ser–Asn–Ala–Ser–Cys

It is clear that both reagents labelled the same cysteine residue, which must be the active-site —SH group (now known to be Cys-149).

Additional evidence for a multi-step covalent mechanism came from studies of partial reactions catalyzed by the dehydrogenase. Velick and Hayes (1953), using substrate-level concentrations of enzyme, observed rapid reduction of NAD⁺ when glyceraldehyde-3-phosphate was added. Segal and Boyer (1953), in similar experiments, found that this "burst" of NADH formation was equivalent to the participation of two acyl sites per molecule of dehydrogenase, even with excess substrate present. Addition of phosphate after the original reduction of NAD⁺ provided another increment of NADH formation, as the phosphate accepted the acyl group from the substituted enzyme. Koeppe et al. (1956) demonstrated the ability of the acylated enzyme to be a valid intermediate by finding that the initial velocity for formation of NADH was the same with or without phosphate, and that the cleavage of the acyl-enzyme by phosphate is as rapid as the overall reaction. With a few minor modifications, the mechanism postulated by Segal and Boyer in 1953 is still accepted. In this mechanism, the aldehyde substrate reacts with the active —SH to form a hemithioacetal. The enzyme-bound NAD⁺ is then reduced and the hemithioacetal is oxidized to the energy-rich thiol ester. Segal and Boyer suggested that in the third step, NAD⁺ in solution would be reduced by the bound NADH. Actually, NADH is released from the dehydrogenase in this step and replaced by NAD^+. Finally, the acyl group is transferred from the thiol ester-enzyme to phosphate:

1. $RCHO + HS-E-NAD^+ \rightleftharpoons RCHOH-S-E-NAD^+$

2. $RCHOH-S-E-NAD^+ \rightleftharpoons RCO-S-E-NADH (+ H^+)^*$

3. $RCO-S-E-NADH + NAD^+ \rightleftharpoons RCO-S-E-NAD^+ + NADH$

4. $RCO-S-E-NAD^+ + P_i \rightleftharpoons HS-E-NAD^+ + RCO-P.$

Jencks (1969) pointed out that the addition of the $-S^-$ nucleophile to the aldehyde aids in removal of electrons from the electrophilic carbonyl group, and in the reverse direction facilitates reduction of the acyl group. He has aptly termed the formation of the acylated nucleophile by oxidation of the nucleophilic adduct "oxidative activation."

As a result of kinetic and structural studies reported in the past few years, the mechanism of glyceraldehyde-3-phosphate dehydrogenase is more fully documented. Steady-state (Duggleby and Dennis, 1974) and rapid (Harrigan and Trentham, 1973 and 1974) kinetic measurements are in accord with a ping-pong mechanism. Both a stable $E-NAD^+$ holoenzyme and a stable $S-$acyl enzyme are on the reaction pathway. The free enzyme is not catalytically active; NAD^+ must be bound to the enzyme for reaction of either the substrate or the product with Cys-149. The basic kinetic mechanism is:

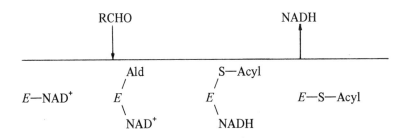

* The proton is retained by a basic group on the enzyme for release during deacylation (step 4).

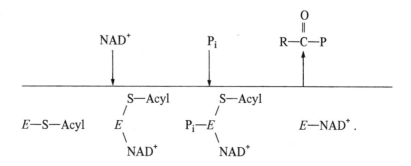

The $E-NAD^+$ holoenzyme reacts with 3-phosphoglyceraldehyde (RCHO) to form an $E-Ald-NAD^+$ ternary complex which re-arranges into the thioester and reduced pyridine nucleotide. The NADH is released, replaced by NAD^+, and then phosphorolysis of the acyl enzyme leads to formation of $E-NAD^+$ and the acyl phosphate, 1,3-diphosphoglycerate (RCO$-$P).

X-ray crystallography experiments have provided a three-dimensional structure for the lobster glyceraldehyde-3-phosphate dehydrogenase holoenzyme at 2.9 Å resolution (Moras *et al.*, 1975). At this resolution, chemical groups but not isolated atoms can be identified. Although the four polypeptide chains of the tetrameric enzyme are known to be identical, small deviations in conformation between the two classes of chains have been observed. In half the subunits, the nicotinamide ring of the NAD^+ is closer to the sulphur of Cys-149 than in the other two subunits. It is assumed that a charge transfer complex is formed between the nicotinamide and the sulphur. His-176 probably removes a proton from Cys-149, and the negatively charged sulphur adds to the aldehyde group of the substrate. The thio-hemiacetal tetrahedral intermediate (Fig. 6) is positioned correctly by the enzyme, allowing the hydride to be transferred to C^4 of the NAD^+. The proton sequestered by His-176 would not be released until the phosphorolysis step, when the proton is released and the imidazole group forms an ion-pair with the sulphur atom of Cys-149 (Polgár, 1975), which also participates in the charge transfer complex. Further crystallographic experiments will help explain the mechanism in more detail and will provide the information needed to fully explain the negative co-operativity of NAD^+-binding by the enzyme.

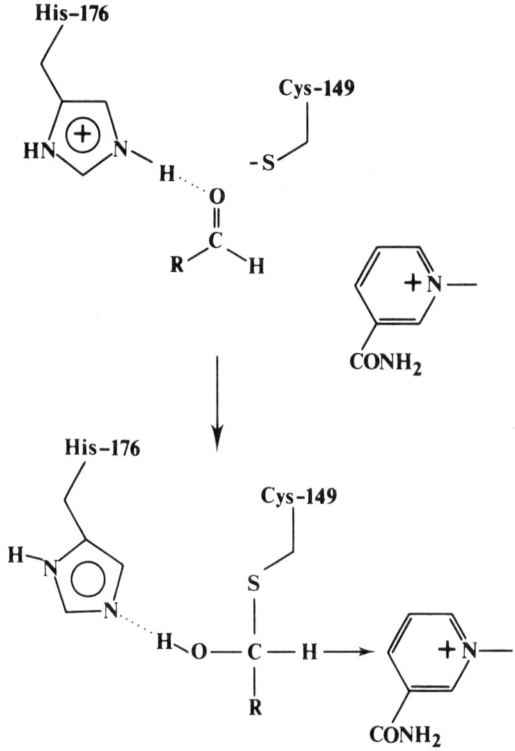

Fig. 6. Steps leading to the formation of the thioester intermediate during the redox step in the glyceraldehyde-3-phosphate dehydrogenase-catalyzed reaction. Based on the crystallographic experiments of Moras *et al.* (1975).

SUCCINYL COENZYME A SYNTHETASE

Covalent phosphoryl-enzyme intermediates have been detected in a number of phosphoryl cleavage or transfer reactions. Three functional groups of proteins—the imidazole of histidine, the hydroxyl group of serine and the carboxyl group of aspartate or glutamate— all have been implicated as specific sites of phosphoryl binding in different enzymes. As mentioned in the discussion of phosphocarrier protein (HPr, p. 302), the effect of pH on the stability of a phospho-enzyme can help to identify the group reacting with the phosphate. Phosphohistidine proteins are stable in alkali and phosphoserine proteins in acid. Phosphoacyl proteins would be labile under both acidic and basic conditions. Examples of both phosphohistidine and phosphoserine intermediates are discussed below. The phospho-enzyme form of $(Na^+ + K^+)$-ATPase is thought to be an acyl phosphate (p. 497).

Of the phosphohistidine enzymes (Bridger, 1973), the first to be isolated was succinyl CoA synthetase (reviewed by Bridger, 1974). This enzyme catalyzes the following reversible reaction:

$$\text{Succinyl CoA} + \text{NDP} + P_i \xrightleftharpoons{\text{Mg}^{2+}} \text{NTP} + \text{Succinate} + \text{CoASH}.$$

This reaction is part of the citric acid cycle, and supplies both energy and the next metabolite of the cycle, succinate. In the forward direction, the potential energy of the thiolester is transferred to the new anhydride grouping of the nucleoside triphosphate, so it is another example of substrate-level phosphorylation. In *E. coli*, the synthetase-catalyzed reaction operates primarily in the reverse direction to maintain the supply of the anabolite succinyl CoA. The discovery of a phosphorylated form of succinyl CoA synthetase was the result of experiments designed to detect a covalent intermediate in oxidative-phosphorylation in mitochondria. Beef liver mitochondria carrying out oxidative phosphorylation were incubated with $^{32}P_i$ for a half minute and then mixed with a cold solution of urea containing NH_4OH. Column chromatography of the resulting solution gave a protein fraction containing bound ^{32}P. The protein-bound radio-activity was alkali stable and acid labile. Hydrolysis of the labelled peptide, using both alkali and enzyme treatment, produced a compound identical to the more stable 3-phospho isomer of phospho-histidine (DeLuca *et al.*, 1963):

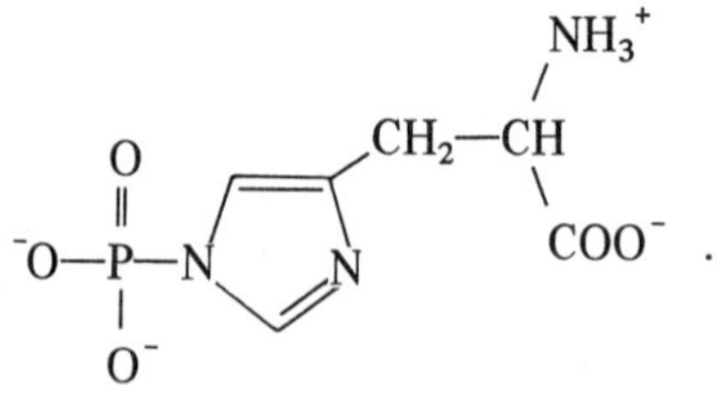

Isolation of this ^{32}P-labelled protein from mitochondria led to the logical suggestion (Boyer, 1963) that it was an intermediate in oxidative phosphorylation. Only a year later, though, Boyer's own laboratory identified the protein as succinyl CoA synthetase (Mitchell

et al., 1964). Thus, the phosphorylated imidazole group is an intermediate in a substrate-level phosphorylation, not in the electron
transport chain coupled oxidative-phosphorylation.

Most further experiments have been performed using succinyl CoA
synthetases from *E. coli* and pig heart. I will describe work done
with the former. Rapid-mixing experiments (Bridger *et al.*, 1968)
showed that phosphoenzyme is formed initially at a rate sufficiently
rapid to account for its being an intermediate in the overall reaction.
In addition, in the forward direction, the steady-state level of phosphoenzyme is attained before the steady-state level of ATP formation is
established. Therefore, the phosphoenzyme qualifies as an obligatory
covalent intermediate. One anomaly that Boyer and his colleagues
observed and explained was the slow rate of a partial reaction,
$ADP \rightleftharpoons ATP$ exchange catalyzed by the enzyme in the absence of
other substrates. They found that the exchange rate is stimulated
when other substrates are present (e.g. succinyl CoA gave maximal
stimulation). They termed the phenomenon in which the active site
is fully active only when all its binding sites are occupied *substrate
synergism.* Wang *et al.* (1972) have succeeded in isolating and sequencing a 12-residue active-site phosphopeptide from tryptic hydrolysates
of ^{32}P-labelled synthetase, prepared and purified under phosphoryl-
stabilizing conditions. Succinyl phosphate is formed from succinate
and the phosphoryl-enzyme intermediate (Nishimura, 1967), and it
appears that enzyme-bound succinyl phosphate is the other intermediate in a three-step mechanism (Hildebrand and Spector, 1969),
written here for the reverse direction:

1. $E + ATP \rightleftharpoons P{\sim}E + ADP$

2. $P{\sim}E + Succinate \rightleftharpoons Succinyl\ phosphate{-}E$

3. $Succinyl\ phosphate{-}E + CoASH \rightleftharpoons E + Succinyl\ CoA + P_i$.

Kinetic measurements did not indicate the ping-pong mechanism
expected for an enzyme reaction utilizing covalent catalysis. Instead,
a partially random sequential pattern was observed, as shown on
p. 334 (Moffet and Bridger, 1973). The release of ADP only after
formation of succinyl CoA is consistent with the substrate synergism
of succinyl CoA synthetase.

The succinyl CoA synthetase of *E. coli* has a molecular weight of
about 140 000 and is composed of two polypeptide chain types,

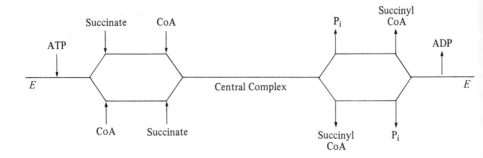

called α and β. In concentrated solution, the predominant species is the $\alpha_2\beta_2$ tetramer, but at low concentrations the enzyme can dissociate to an $\alpha\beta$ form (Krebs and Bridger, 1974). Teherani and Nishimura (1975) also demonstrated the close association of α and β subunit chains, using gel electrophoresis and crosslinking reagents such as dimethylsuberimidate (p. 71). The only species observed in their experiments were monomers, $\alpha\beta$ (the principal crosslinked product), $\alpha_2\beta$ and $\alpha_2\beta_2$. Note the absence of dimers of like subunits. The most likely arrangement of seven possible quaternary structures is

The smaller α subunit (29 500 daltons) contains the histidine residue which is phosphorylated during catalysis. Neither subunit alone can catalyze the overall reaction, but the α subunit can slowly catalyze its own phosphorylation by ATP (Pearson and Bridger, 1975). Therefore, it probably contains the ATP binding site. Release of ^{32}P from the phosphoenzyme in the presence of succinate and CoA requires both subunit types. Preliminary evidence suggests that the β chain is the site of binding of the succinyl phosphate intermediate. Thus, each subunit of the $\alpha\beta$ protomer may catalyze a separate partial reaction by a two-site mechanism such as that of transcarboxylase (p. 296). The phosphoryl group would be transferred from the phosphorylated α subunit to the nearby β subunit–succinate complex. Bridger has concluded that the active site of succinyl CoA synthetase is close to the point of contact of the α and β subunits in the $\alpha_2\beta_2$ oligomer.

ALKALINE PHOSPHATASE

Alkaline phosphatase is a hydrolase having a pH optimum in the range of pH 8-10 (cf. reviews by Reid and Wilson, 1971; Fernley, 1971). It has a broad substrate specificity, catalyzing the hydrolysis of most phosphomonoesters:

$$R—O—\overset{\overset{\displaystyle O}{\|}}{\underset{\underset{\displaystyle O^-}{|}}{P}}—O^- + H_2O \rightleftharpoons ROH + P_i.$$

Its activity has been detected in many tissues and cells, but neither the biological function nor the significance of the alkaline pH optimum is known for any alkaline phosphatase.

The formation of a phosphoserine covalent enzyme intermediate has been established for the alkaline phosphatase-catalyzed reaction. Cohn (1949) showed that in the hydrolysis of glucose-1-phosphate catalyzed by intestinal alkaline phosphatase the O—P bond rather than the C—O bond is broken. The basis for this conclusion is the incorporation of ^{18}O from labelled water into the inorganic phosphate produced. In the absence of a phosphate acceptor, alkaline phosphatase catalyzes the exchange of oxygen between water and inorganic phosphate (Stein and Koshland, 1952), supporting a mechanism in which a nucleophilic group on the enzyme displaces the —OR group of the substrate. Engström (1961) demonstrated that ^{32}P-phosphate can be covalently incorporated into purified alkaline phosphatase from calf intestine at pH 5. When acid-inactivated phosphoprotein is hydrolyzed in HCl, the ^{32}P-label is located in phosphoserine. Therefore, the serine hydroxyl appears to be the nucleophilic group at the active site of alkaline phosphatase.

With the availability of large amounts of the phosphatase from *E. coli* (p. 44), much of the research on the mechanism shifted to use of the inducible enzyme from the bacterium. In *E. coli* cells, alkaline phosphatase activity occurs only in the periplasmic space, between the membrane and the cell wall. Purification of the phosphatase was facilitated by this cellular distribution, which allows removal of the cell wall to release phosphatase (p. 46). Most of the bacterial enzymes are retained by the spheroplasts (the plasma membrane and the contents it surrounds). Schwartz and Lipmann (1961) isolated

M

phosphoserine and phosphopeptides from ^{32}P-phosphate treated *E. coli* phosphatase. They showed by electrophoresis of a tryptic hydrolysate that a single serine residue is labelled. Kinetic experiments (the observation that the enzyme catalyzes the hydrolysis of a wide variety of esters at similar rates, and the "burst" of alcohol formation during hydrolysis) are compatible with a phosphoryl-enzyme mechanism. Supplementary evidence for this mechanism came from transferase experiments, in which alcoholic phosphate acceptors were added to compete with water for the phosphoryl group. The ratio of the amount of phosphorylation of the acceptors to the amount of P_i formed is independent of the nature of the substrate (Neumann, 1969; Barrett *et al.*, 1969). If the reactive intermediate in the transferase reaction were a reversible $E-RO-P (=O)-(OH)_2$ complex, then the ratio of products would vary with the structure of the leaving group, $-OR$. The results prove that the reaction occurs in two steps:

1. $E + RO-P(=O)(OH)_2 \rightarrow E-P + ROH$

2. $E-P + H_2O \rightarrow E + P_i$

    (or $E-P + R'OH \rightarrow R'O-P(=O)(OH)_2 + E$ in the transferase reaction).

Pre-steady state experiments show that above pH 8 the rate-limiting step in the overall reaction is a rearrangement of the original Michaelis complex before formation of the phosphoenzyme (Trentham and Gutfreund, 1968). The elucidation of the physical nature of this step will be needed to explain the enzyme's mechanism.

The alkaline phosphatase of *E. coli* is an oligomer composed of two identical subunits. It appears that the dimer contains four atoms of Zn^{2+} and two atoms of Mg^{2+}. Removal of the zinc leads to loss of catalytic activity. Replacement of the zinc by other divalent metals (e.g. Co^{2+} or Cd^{2+}) produces an enzyme with lower maximal activity. The initial-velocity kinetics of alkaline phosphatase follow a classical hyperbolic pattern. However, reaction of only one active site under certain conditions has been observed in many experiments. This is an example of half-site reactivity (p. 425). Lazdunski (Lazdunski *et al.*, 1971) has suggested that during the catalysis the phosphorylation of one subunit by non-covalently bound organic phosphate accompanies dephosphorylation of the other phosphoserine. His "flip-flop" mechanism postulates that there is interaction between the two subunits which restricts covalent phosphorylation to one site. As with other half-site mechanisms, there is insufficient evidence available to fully explain the lower reactivity of one active site. X-ray crystallographic experiments in progress should aid greatly in the study of alkaline phosphatase, by providing data on the subunit interaction and also the roles of the metal ions in substrate positioning and activation.

ACETOACETATE DECARBOXYLASE

Acetoacetate decarboxylase is one of several enzymes in which a single lysine residue is the site of substitution during covalent catalysis. The mechanism of action of this enzyme, crystallized from *Clostridium acetobutylicum,* has been studied by Westheimer and his co-workers (reviewed by Fridovich, 1972). The decarboxylase has a molecular weight of 340 000, and is composed of 12 seemingly identical subunits. The enzyme-catalyzed formation of acetone and CO_2 proceeds by formation of a Schiff base formed by reaction of the β-carbonyl group of the substrate with the ϵ-amino group of the lysine residue, followed by decarboxylation. The mechanism proposed for this decarboxylation (Tagaki and Westheimer, 1968) is:

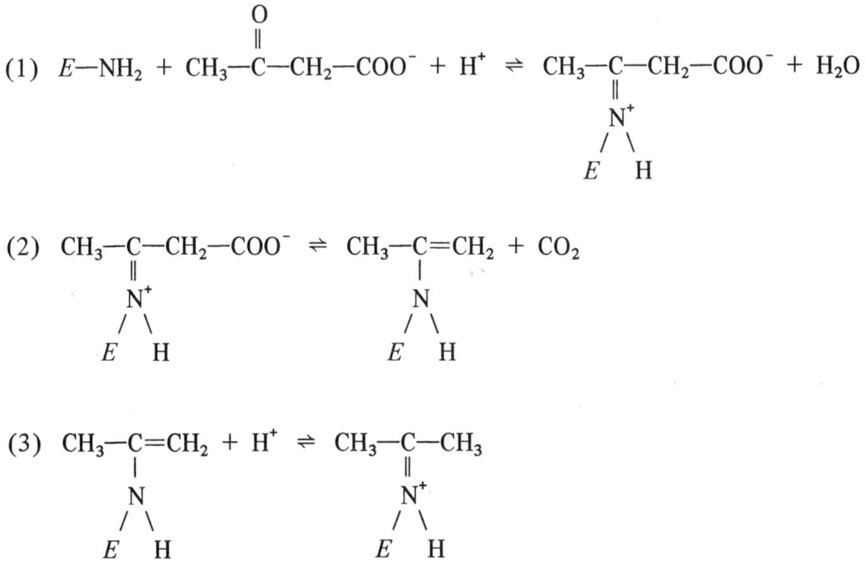

$$\text{(1)} \quad E{-}NH_2 + CH_3{-}\overset{\overset{\textstyle O}{\|}}{C}{-}CH_2{-}COO^- + H^+ \;\rightleftharpoons\; CH_3{-}\underset{\underset{\textstyle E \quad H}{\underset{\textstyle / \;\backslash}{N^+}}}{\overset{\textstyle \|}{C}}{-}CH_2{-}COO^- + H_2O$$

$$\text{(2)} \quad CH_3{-}\underset{\underset{\textstyle E \quad H}{\underset{\textstyle / \;\backslash}{N^+}}}{\overset{\textstyle \|}{C}}{-}CH_2{-}COO^- \;\rightleftharpoons\; CH_3{-}\underset{\underset{\textstyle E \quad H}{\underset{\textstyle / \;\backslash}{N}}}{\overset{\textstyle |}{C}}{=}CH_2 + CO_2$$

$$\text{(3)} \quad CH_3{-}\underset{\underset{\textstyle E \quad H}{\underset{\textstyle / \;\backslash}{N}}}{\overset{\textstyle |}{C}}{=}CH_2 + H^+ \;\rightleftharpoons\; CH_3{-}\underset{\underset{\textstyle E \quad H}{\underset{\textstyle / \;\backslash}{N^+}}}{\overset{\textstyle \|}{C}}{-}CH_3$$

$$(4) \quad CH_3-\underset{\underset{\underset{E \quad H}{/ \ \backslash}}{\overset{+}{N}}}{\overset{\|}{C}}-CH_3 + H_2O \; \rightleftharpoons \; CH_3-\overset{\overset{O}{\|}}{C}-CH_3 + H^+ + E-NH_2 \, .$$

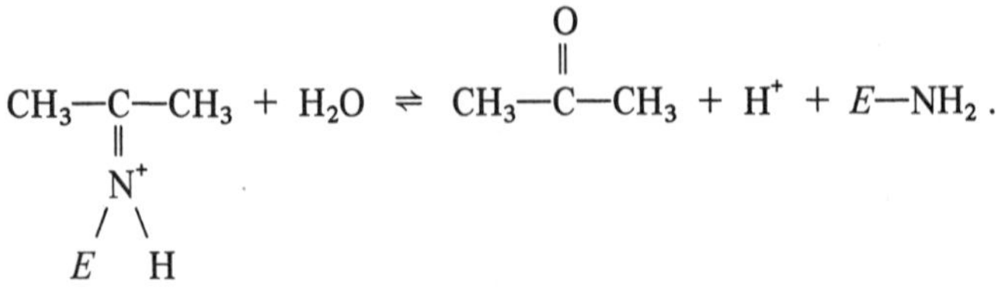

The Schiff base is formed in step 1, decarboxylated in step 2 to give the enamine of acetone, which is protonated in step 3. In step 4, the iminium ion–enzyme complex is hydrolyzed to produce acetone and regenerated enzyme. This mechanism is based on evidence from both exchange and trapping experiments. It was originally postulated on the basis of enzyme-dependent exchange of the ^{18}O-labelled carbonyl oxygen of acetoacetate (reaction 1) and of acetone (reaction 4) with water (Hamilton and Westheimer, 1959). Sodium borohydride has been used to trap the Schiff base (Warren et al., 1966) as was done with pyridoxal phosphate-substrate Schiff bases. Borohydride does not inactivate the decarboxylase unless acetoacetate (or under some conditions, acetone) is also present. When ^{14}C-acetoacetate was used in these experiments, acid hydrolysis of the reduced ES complex gave a single labelled spot on paper chromatograms. This compound was identified as ε-N-isopropyllysine

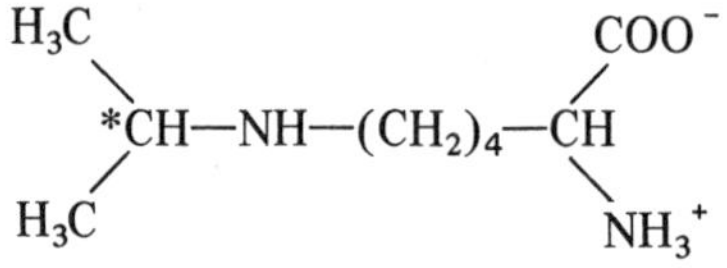

$$\underset{H_3C}{\overset{H_3C}{\diagdown}} {}^*CH-NH-(CH_2)_4-\underset{NH_3^+}{\overset{COO^-}{\diagup}} CH$$

indicating that it was formed from the protonated Schiff base of acetone with the enzyme.

The four-step mechanism above requires a nucleophilic (unprotonated) lysine group for reaction to occur. Schmidt and Westheimer (1971) found that the pK_a of the essential lysine is 5.9, a decrease of about four pH units from that of an ordinary lysine. This pK_a value allows a large fraction of the enzyme molecules to have the essential lysine group in an unprotonated form at the pH_{opt} (approximately pH 6) for the enzymic reaction. Comparison of the rates of model reactions with the enzymatic reaction indicates that the enzyme accelerates both the formation of the Schiff base and its decarboxylation. It has already been shown that the effectiveness of nucleophilic catalysis is due to increased reactivity and instability of the Schiff base (p. 93). Here, the decarboxylation is enhanced by attraction of electrons from the carboxyl toward the nitrogen of the Schiff base. Isotope effects, comparing the rates of decarboxylation of ^{12}C- and ^{13}C-carboxyl labelled acetoacetate, show that the decarboxylation step is rate limiting (O'Leary and Baughn, 1972). Recent n.m.r. experiments (Hammons *et al.*, 1975) have shown that reaction 4 consists of two separate steps—removal of a proton from the iminium ion by a catalytic group and hydrolysis of the acetone imine. The hydrolysis is rapid in comparison to the proton exchange, but proton exchange is faster than decarboxylation.

HISTIDINE DECARBOXYLASE

There is also a group of enzymes which contain an electrophilic prosthetic group capable of forming Schiff base intermediates with nucleophilic groups of substrates. The reactions involved are somewhat similar to pyridoxal phosphate-requiring reactions, but protein-bound α-keto acids—pyruvate or α-ketobutyrate—are the reactive functions. The first enzyme demonstrated to have pyruvate at its active site was D-proline reductase (Hodgins and Abeles, 1967, 1969), in which the pyruvate seems to be bound by an amide linkage. Two other enzymes, histidine decarboxylase (Riley and Snell, 1968) and adenosyl methionine decarboxylase (Wickner *et al.*, 1970), have been found to contain pyruvate. Urocanase from *Pseudomonas putida* (George and Phillips, 1970) and serine-threonine dehydratase (Kapke and Davis, 1975) contain α-ketobutyrate. As an example of this class of enzymes, I shall examine histidine decarboxylase.

When histidine decarboxylase was isolated in crystalline form from *Lactobacillus,* Rosenthaler *et al.* (1965) noticed that the spectrum of the enzyme was that of a simple protein (i.e. there was no absorption above 300 nm). No pyridoxal phosphate was detectable by microbiological assay and addition of pyridoxal phosphate to assay mixtures had no effect on enzyme activity. These data led them to conclude that, unlike other amino acid decarboxylases, this enzyme does not use pyridoxal phosphate as a cofactor. Histidine decarboxylase is inhibited by reduction with $NaBH_4$ or by reaction with phenylhydrazine, reagents which also inhibit pyridoxal phosphate-dependent enzymes. Riley and Snell (1968) demonstrated the presence of pyruvic acid in the enzyme by isolating tritiated lactic acid from decarboxylase that had been reduced with NaB^3H_4 and then hydrolyzed. When ^{14}C-phenylhydrazine was added to histidine decarboxylase, and the substituted enzyme submitted to proteolytic digestion, the phenylhydrazone of pyruvoylphenylalanine was formed. Therefore, the essential carbonyl group is a pyruvic acid molecule bound by an amide bond to an N-terminal phenylalanine residue of the enzyme:

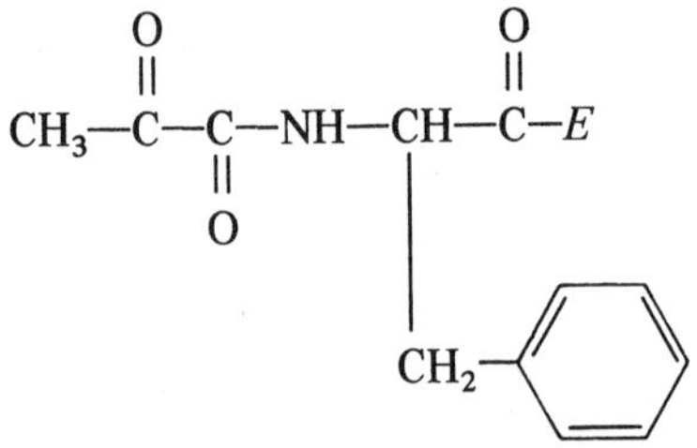

$$CH_3-\overset{\overset{O}{\|}}{C}-\overset{\overset{\|}{O}}{\underset{\underset{O}{\|}}{C}}-NH-\underset{\underset{CH_2-\hexagon}{|}}{CH}-\overset{\overset{O}{\|}}{C}-E$$

The enzyme appears to have ten subunits, five with a molecular weight of 29 700, containing all the pyruvoyl residues, and five sub-units with a molecular weight of 9000 (Riley and Snell, 1970). Tracer experiments showed that serine is the precursor of the pyruvate. The pyruvate is possibly formed by deamination of an N-terminal serine residue of the larger subunit. With urocanase, threonine is converted to α-ketobutyrate by a similar reaction (Lynch and Phillips, 1972).

The pyruvoyl residue of histidine decarboxylase has a role similar to that of the pyridoxal phosphate of other decarboxylases. Recsei and Snell (1970) showed that the enzyme forms a Schiff base with the substrate, histidine, by reducing a mixture of decarboxylase and ^{14}C-histidine with borohydride, and isolating N^2-(1-carboxyethyl)-^{14}C-histidine after hydrolysis. That the product, histamine, forms a Schiff base was shown by a similar experiment. The Schiff bases are presumably intermediates in a catalytic process, such as that shown in Fig. 7.

Another type of electrophilic centre prosthetic group occurs in two enzymes which catalyze the elimination of ammonia from amino acids. Both histidine ammonia lyase (Givot et al., 1969) and phenyl-alanine ammonia lyase (Hanson and Havir, 1972) contain a bound form of dehydroalanine (2-aminoacrylic acid, $^+H_3N-C(=CH_2)COO^-$).

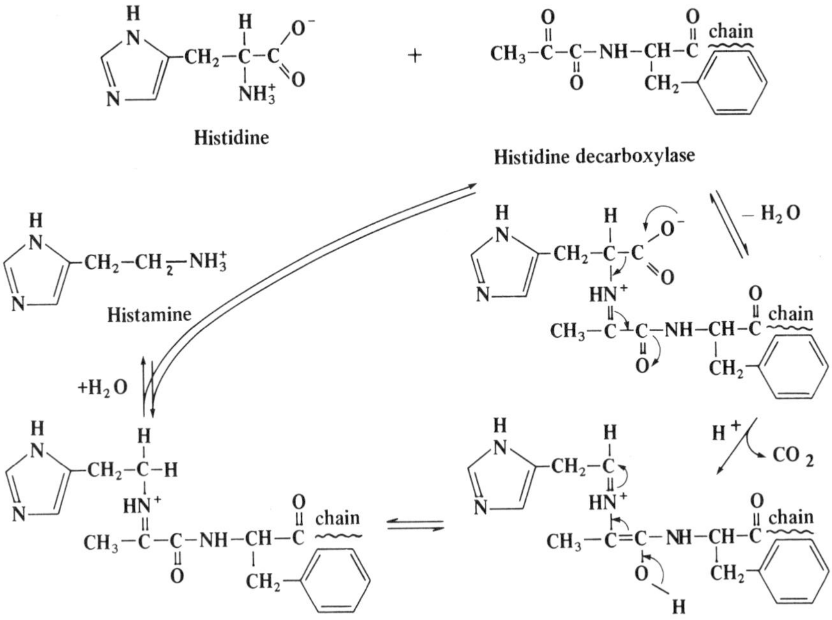

Fig. 7. Postulated mechanism of action of histidine decarboxylase showing the function of enzyme-bound pyruvate in this process. From Boeker and Snell (1972).

The exact structures of the active sites and mechanisms of action of these enzymes should be elucidated soon.

SUCROSE PHOSPHORYLASE

Although it was proposed in 1947 that the sucrose phosphorylase reaction proceeds via a covalent *ES* intermediate, it has only been in the past few years that a protein-bound reactant has been isolated. Sucrose phosphorylase from *Pseudomonas saccharophila* catalyzes the reversible phosphorolysis of sucrose (written here as glucosyl-fructoside):

Glucosyl–fructoside + P_i ⇌ Glucose–1–phosphate + Fructose.

Doudoroff *et al.* (1947) discovered that the phosphorylase catalyzes the incorporation of ^{32}P-phosphate into glucose-1-phosphate in the absence of fructose. This finding suggested that an enzyme–glucose covalent complex is formed, and that this complex acts as a glucosyl

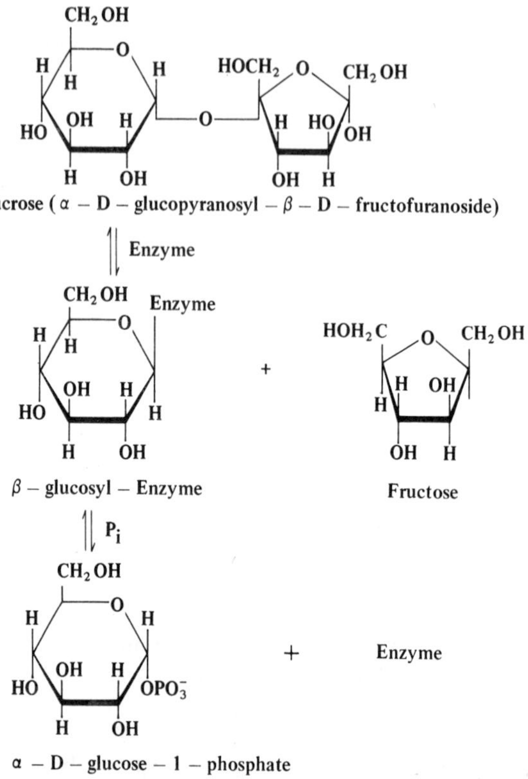

Fig. 8. Stereochemistry of the glucose unit during the phosphorolysis of sucrose catalyzed by sucrose phosphorylase.

donor to either a sugar acceptor or to P_i. Further evidence in support of this mechanism came from the observation that, in both sucrose and glucose-1-phosphate, the glucose was in the α-anomeric configuration. The attack of the enzyme on the glycosidic bond of sucrose is probably a backside displacement to form the β-glucosyl-enzyme intermediate. When the product is formed in the second step, inversion to the α-configuration occurs (Koshland, 1954). In this double displacement reaction, inversion of configuration occurs twice, giving retention of the α-configuration in the overall reaction (Fig. 8).

These experiments were amongst the most definitive of the early research on covalent catalysis. The validity of the double displacement mechanism based on them has been substantiated by recent kinetic and isolation experiments performed by Abeles and his collaborators (reviewed by Mieyal and Abeles, 1972). Sucrose phosphorylase was purified close to homogeneity, and bisubstrate kinetic experiments were performed (Silverstein *et al.*, 1967). Initial-velocity measurements with varying concentrations of substrates (e.g. sucrose at five levels of phosphate) gave parallel lines (except that inhibition is observed at high concentrations of glucosyl acceptor), indicating a modified ping-pong mechanism consistent with the proposed mechanism. Incubation of sucrose phosphorylase with sucrose in the absence of P_i should lead to accumulation of a glucose-enzyme complex, but because water can act as an acceptor, the glucose–enzyme complex can be isolated only after denaturation with hot methanol. One mole of glucose is bound per mole of enzyme. Voet and Abeles (1970) devised two additional procedures for isolating inactivated glucose–enzyme complexes. Treatment of a mixture of ^{14}C-sucrose and enzyme with $NaIO_4$ gives a complex containing covalently bound glucose. This complex can react slowly with sugar acceptors, indicating that an active form of glucose is present. Proteolytic digestion of an acid-denatured complex produces glucosyl-peptides, from which β-glucose can be isolated. The base lability of the glucosyl-enzyme and peptides suggested that the intermediate is an ester formed from glucose and a free carboxyl of the enzyme. Release of glucose from the acid-denatured complex makes a carboxyl group available for a carboxyl-specific substitution reaction, providing supporting evidence for an ester linkage (Mieyal and Abeles, 1972).

Mieyal and Abeles have presented a tentative mechanism for sucrose phosphorylase (Fig. 9). They have postulated that an acid group on the enzyme aids in the dissociation of the glucosidic bond of sucrose, leaving a carbonium ion at C^1. This carbonium ion would be stabilized either by formation of a strained ester bond or an ion pair by interaction with the carboxyl group on the enzyme. This mechanism, including the involvement of a carboxyl group, is quite similar to that of lysozyme (p. 179).

The mechanistic studies on the seven enzymes just discussed indicate that detectable covalent *ES* intermediates do occur. When grouped with the reactions in which covalent coenzyme–substrate intermediates exist, covalent enzymic catalysis accounts for a signifi-

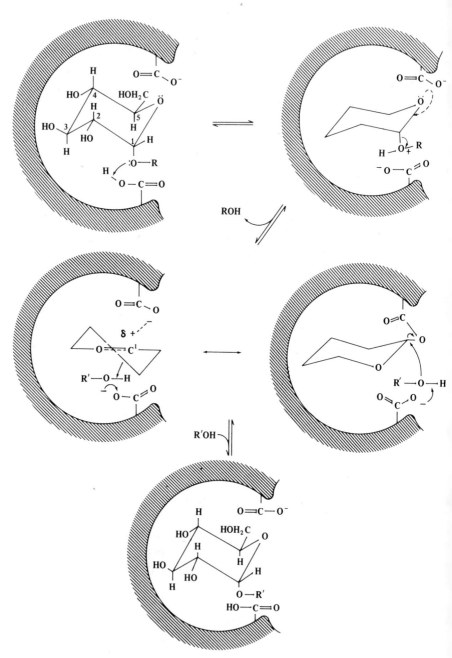

Fig. 9. Mechanism for the action of sucrose phosphorylase as postulated by Mieyal and Abeles (1972).

cant fraction of all enzymic reactions. Probably the strength of these covalent bonds formed by reaction with natural substrates varies, from stable and isolable intermediates to short-lived distorted bonds and strong ion pairs in the intermediate. At the other extreme would be the many enzymes which act by accepting and donating protons or whose *ES* intermediates have no covalent character whatsoever.

REFERENCES

Balls, A. K. and Jansen, E. F. (1952). *Adv. Enzymol.* **13**, 321-343.

Barrett, H., Butler, R. and Wilson, I. B. (1969). *Biochemistry* **8**, 1042-1047.

Bell, R. M. and Koshland, D. E., Jr. (1970). *Biochem. biophys. Res. Commun.* **38**, 539-545.

Bell, R. M. and Koshland, D. E., Jr. (1971). *Science, N.Y.* **172**, 1253-1256.

Blow, D. M. (1971). *In* "The Enzymes" (P. D. Boyer, ed.), Third edition, Vol. 3, pp. 185-212. Academic Press, New York.

Blow, D. M., Birktoft, J. J. and Hartley, B. S. (1969). *Nature, Lond.* **221**, 337-340.

Boeker, E. A. and Snell, E. E. (1972). *In* "The Enzymes" (P. D. Boyer, ed.), Third edition, Vol. 6, p. 244. Academic Press, New York.

Boyer, P. D. (1963). *Science, N.Y.* **141**, 1147-1153.

Bridger, W. A. (1973). *PAABS Revista* **2**, 83-125.

Bridger, W. A. (1974). *In* "The Enzymes" (P. D. Boyer, ed.), Third edition, Vol. 10, pp. 581-606. Academic Press, New York.

Bridger, W. A., Millen, W. A. and Boyer, P. D. (1968). *Biochemistry* **7**, 3608-3616.

Bruice, T. C. and Benkovic, S. J. (1966). "Bioorganic Mechanisms", Vol. 1, Chapter 2. W. A. Benjamin, New York.

Caplow, M. (1969). *J. Am. Chem. Soc.* **91**, 3639-3645.

Cohen, J. A., Oosterbaan, R. A. and Berends, F. (1967). *In* "Methods in Enzymology" (C. H. W. Hirs, ed.), Vol. 11, pp. 686-702. Academic Press, New York.

Cohn, M. (1949). *J. biol. Chem.* **180**, 771-781.

DeLuca, M., Ebner, K. E., Hultquist, D. E., Kreil, G., Peter, J. B., Moyer, R. W. and Boyer, P. D. (1963). *Biochem. Z.* **338**, 512-525.

Doudoroff, M., Barker, H. A. and Hassid, W. Z. (1947). *J. biol. Chem.* **168**, 727-732.

Duggleby, R. G. and Dennis, D. T. (1974). *J. biol. Chem.* **249**, 167-174.

Engström, L. (1961). *Biochim. biophys. Acta.* **52**, 49-59.

Fernley, H. N. (1971). *In* "The Enzymes" (P. D. Boyer, ed.), Third edition, Vol. 4, pp. 417-447. Academic Press, New York.

Fersht, A. R. and Sperling, J. (1973). *J. molec. Biol.* **74**, 137-149.

Fox, J. B., Jr. and Dandliker, W. B. (1956). *J. biol. Chem.* **218**, 53-57.

Fridovich, I. (1972). *In* "The Enzymes" (P. D. Boyer, ed.), Third edition, Vol. 6, pp. 255-270. Academic Press, New York.

George, D. J. and Phillips, A. T. (1970). *J. biol. Chem.* **245**, 528-537.

Gertler, A., Walsh, K. A. and Neurath, H. (1974). *Biochemistry* **13**, 1302-1310.
Givot, I. L., Smith, T. A. and Abeles, R. H. (1969). *J. biol. Chem.* **244**, 6341-6353.
Hamilton, G. A. and Westheimer, F. H. (1959). *J. Am. Chem. Soc.* **81**, 6332-6333.
Hammons, G., Westheimer, F. H., Nakaoka, K. and Kluger, R. (1975). *J. Am. Chem. Soc.* **97**, 1568-1572.
Hanson, K. R. and Havir, E. A. (1972). *In* "The Enzymes" (P. D. Boyer, ed.), Third edition, Vol. 7, pp. 154-162. Academic Press, New York.
Harrigan, P. J. and Trentham, D. R. (1973). *Biochem. J.* **135**, 695-703.
Harrigan, P. J. and Trentham, D. R. (1974). *Biochem. J.* **143**, 353-363.
Harrington, W. F. and Karr, G. M. (1965). *J. molec. Biol.* **13**, 885-893.
Harris, J. I. and Perham, R. N. (1965). *J. molec. Biol.* **13**, 876-884.
Harris, I., Meriwether, B. P. and Park, J. H. (1963). *Nature, Lond.* **198**, 154-157.
Hartley, B. S. (1964). *Nature, Lond.* **201**, 1284-1287.
Hartley, B. S. and Kauffman, D. L. (1966). *Biochem. J.* **101**, 229-231.
Hartley, B. S. and Kilby, B. A. (1954). *Biochem. J.* **56**, 288-297.
Hartley, B. S. and Shotton, D. M. (1971). *In* "The Enzymes" (P. D. Boyer, ed.), Third edition, Vol. 3, pp. 323-373. Academic Press, New York.
Hess, G. P. (1971). *In* "The Enzymes" (P. D. Boyer, ed.), Third edition, Vol. 3, pp. 213-248. Academic Press, New York.
Hildebrand, J. G. and Spector, L. B. (1969). *J. biol. Chem.* **244**, 2606-2613.
Hodgins, D. and Abeles, R. H. (1967). *J. biol. Chem.* **242**, 5158-5159.
Hodgins, D. S. and Abeles, R. H. (1969). *Archs Biochem. Biophys.* **130**, 274-285.
Holzer, H. and Holzer, E. (1952). *Z. Physiol. Chem.* **291**, 67-86.
Hunkapiller, M. W., Smallcombe, S. H., Whitaker, D. R. and Richards, J. H. (1973). *Biochemistry* **12**, 4732-4743.
Jencks, W. P. (1969). "Catalysis in Chemistry and Enzymology", Chapter 2. McGraw-Hill, New York.
Johnson, P. E., Abbott, S. J., Orr, G. A., Sémériva, M. and Knowles, J. R. (1975). *Biochem. biophys. Res. Commun.* **62**, 382-389.
Kapke, G. and Davis, L. (1975). *Biochemistry* **14**, 4273-4276.
Keil, B. (1971). *In* "The Enzymes" (P. D. Boyer, ed.), Third edition, Vol. 3, pp. 249-275. Academic Press, New York.
Koeppe, O. J., Boyer, P. D. and Stulberg, M. P. (1956). *J. biol. Chem.* **219**, 569-583.
Koshland, D. E., Jr. (1954). *In* "The Mechanism of Enzyme Action" (McElroy, W. D. and Glass, B., eds), pp. 608-641. Johns Hopkins Press, Baltimore.
Kraut, J. (1971). *In* "The Enzymes" (P. D. Boyer, ed.), Third edition, Vol. 3, pp. 165-183. Academic Press, New York.
Krebs, A. and Bridger, W. A. (1974). *Can. J. Biochem.* **52**, 594-598.
Krimsky, I. and Racker, E. (1955). *Science, N.Y.* **122**, 319-321.
Kurosky, A. and Hofmann, T. (1972). *Can. J. Biochem.* **50**, 1282-1296.
Lazdunski, M., Petitclerc, C., Chappelet, D. and Lazdunski, C. (1971). *Eur. J. Biochem.* **20**, 124-139.

Lynch, M. C. and Phillips, A. T. (1972). *J. biol. Chem.* **247**, 7799-7805.

Matthews, D. A., Alden, R. A., Birktoft, J. J., Freer, S. T. and Kraut, J. (1975). *J. biol. Chem.* **250**, 7120-7126.

McDonald, C. E. and Balls, A. K. (1957). *J. biol. Chem.* **227**, 727-736.

Meloun, B., Kluh, I., Kostka, V., Morávek, L., Prusik, Z., Vaněček, J., Keil, B. and Šorm, F. (1966). *Biochim. biophys. Acta.* **130**, 543-546.

Mieyal, J. J. and Abeles, R. H. (1972). *In* "The Enzymes" (P. D. Boyer, ed.), Third edition, Vol. 7, pp. 515-532. Academic Press, New York.

Mitchell, R. A., Butler, L. G. and Boyer, P. D. (1964). *Biochem. biophys. Res. Commun.* **16**, 545-550.

Moffet, F. J. and Bridger, W. A. (1973). *Can. J. Biochem.* **51**, 44-55.

Moras, D., Olsen, K. W., Sabesan, M. N., Buehner, M., Ford, G. C. and Rossmann, M. G. (1975). *J. biol. Chem.* **250**, 9137-9162.

Neumann, H. (1969). *Eur. J. Biochem.* **8**, 164-173.

Nishimura, J. S. (1967). *Biochemistry* **6**, 1094-1099.

O'Leary, M. H. and Baughn, R. L. (1972). *J. Am. Chem. Soc.* **94**, 626-630.

Ong, E. B., Shaw, E. and Schoellmann, G. (1965). *J. biol. Chem.* **240**, 694-698.

Oosterbaan, R. A., Kunst, P., van Rotterdam, J. and Cohen, J. A. (1958). *Biochim. biophys. Acta.* **27**, 549-563.

Oosterbaan, R. A., van Adrichem, M. and Cohen, J. A. (1962). *Biochim. biophys. Acta.* **63**, 204-206.

Oppenheimer, H. L., Labouesse, B. and Hess, G. P. (1966). *J. biol. Chem.* **241**, 2720-2730.

Pearson, P. H. and Bridger, W. A. (1975). *J. biol. Chem.* **250**, 8524-8529.

Polgár, L. (1975). *Eur. J. Biochem.* **51**, 63-71.

Polgár, L. and Bender, M. L. (1969). *Proc. natn. Acad. Sci. U.S.A.* **64**, 1335-1342.

Recsei, P. A. and Snell, E. E. (1970). *Biochemistry* **9**, 1492-1497.

Reid, T. W. and Wilson, I. B. (1971). *In* "The Enzymes" (P. D. Boyer, ed.), Third edition, Vol. 4, pp. 373-415. Academic Press, New York.

Riley, W. D. and Snell, E. E. (1968). *Biochemistry* **7**, 3520-3528.

Riley, W. D. and Snell, E. E. (1970). *Biochemistry* **9**, 1485-1491.

Robertus, J. D., Kraut, J., Alden, R. A. and Birktoft, J. J. (1972). *Biochemistry* **11**, 4293-4303.

Rosenthaler, J., Guirard, B. M., Chang, G. W. and Snell, E. E. (1965). *Proc. natn. Acad. Sci. U.S.A.* **54**, 152-158.

Rühlmann, A., Kukla, D., Schwager, P., Bartels, K. and Huber, R. (1973). *J. molec. Biol.* **77**, 417-436.

Schaffer, N. K., May, S. C., Jr. and Summerson, W. H. (1953). *J. biol. Chem.* **202**, 67-76.

Schmidt, D. E., Jr. and Westheimer, F. H. (1971). *Biochemistry* **10**, 1249-1253.

Schwartz, J. H. and Lipmann, F. (1961). *Proc. natn. Acad. Sci. U.S.A.* **47**, 1996-2005.

Segal, H. L. and Boyer, P. D. (1953). *J. biol. Chem.* **204**, 265-281.

Silverstein, R., Voet, J., Reed, D. and Abeles, R. H. (1967). *J. biol. Chem.* **242**, 1338-1346.

Spector, L. B. (1973). *Bioorg. Chem.* **2**, 311-321.

Spencer, T. and Sturtevant, J. M. (1959). *J. Am. Chem. Soc.* **81**, 1874-1882.

Stein, S. S. and Koshland, D. E., Jr. (1952). *Archs Biochem. Biophys.* **39**, 229-230.

Stroud, R. M., Kay, L. M. and Dickerson, R. E. (1971). *Cold Spring Harb. Symp. quant. Biol.* **36**, 125-140.

Tagaki, W. and Westheimer, F. H. (1968). *Biochemistry* **7**, 901-905.

Teherani, J. A. and Nishimura, J. S. (1975). *J. biol. Chem.* **250**, 3883-3890.

Trentham, D. R. and Gutfreund, H. (1968). *Biochem. J.* **106**, 455-460.

Velick, S. F. and Hayes, J. E., Jr. (1953). *J. biol. Chem.* **203**, 545-562.

Voet, J. G. and Abeles, R. H. (1970). *J. biol. Chem.* **245**, 1020-1031.

Wang, T., Jurášek, L. and Bridger, W. A. (1972). *Biochemistry* **11**, 2067-2070.

Warren, S., Zerner, B. and Westheimer, F. H. (1966). *Biochemistry* **5**, 817-823.

Wickner, R. B., Tabor, C. W. and Tabor, H. (1970). *J. biol. Chem.* **245**, 2132-2139.

Wright, H. T. (1973). *J. molec. Biol.* **79**, 1-23.

Zeffren, E. and Hall, P. L. (1973). "The Study of Enzyme Mechanisms", Chapters 5 and 9. Wiley-Interscience, New York.

Zerner, B., Bond, R. P. M. and Bender, M. L. (1964). *J. Am. Chem. Soc.* **86**, 3674-3679.

11 Metals and Enzymes

Roughly one-quarter of the enzymes known require metallic cations to achieve their full catalytic activity. Metal cofactors (abbreviated as M) are analogous to organic coenzymes in that some are mobile cofactors and others are anchored prosthetic groups. The Michaelis complexes of these enzymes (EMS) contain one mole of bound metal per catalytic site. Enzymes containing firmly bound metal atoms, retained during isolation and located at the active site, are often called *metalloenzymes*. Other enzymes, termed *metal-activated enzymes,* show absolute requirements for or are stimulated by the addition of metal ions. Mildvan (1970) has suggested a stability constant of $10^8 \, M^{-1}$ for the enzyme–metal complex as an arbitrary division between these two types of metal-requiring enzymes. The strength of complexation or the mobility of the metal ions in enzyme reactions also can be related to the action of the ions in the reactions, as I will show with the examples in this chapter.

Although the roles of metals in enzyme reactions have been studied for many years, the artificial title "bioinorganic chemistry" has become popular only within the past few years, during a time in which more and more inorganic chemists have turned their talents to the challenges of enzymology. The sophistication of their experiments and the success of their work have produced chemical explanations for many metal-requiring enzymic reactions, approaching the work of mechanistic chemists on the physical organic chemistry of enzyme-related reactions. Because of the overlap of inorganic, organic and enzyme chemistry in this field of bioinorganic research, I prefer the description *co-ordination chemistry of enzyme-catalyzed reactions.* Readers not conversant with co-ordination chemistry should read the text by Basolo and Johnson (1964), which is an excellent introduction to the chemistry of co-ordination compounds.

Co-ordination compounds or *metal complexes* are compounds with a central metal ion surrounded by a cluster of ions or molecules.

Modern co-ordination chemistry started with the proposal of a theory of co-ordination in 1893 by Werner, based on the properties of a series of complexes of $CoCl_3$ with ammonia. In more modern terminology, Werner postulated that:

1. most elements show two types of valence—oxidation state and co-ordination number;
2. an element tends to satisfy both types of valence; and
3. co-ordination bonds are directed toward specific geometric configurations.

The groups co-ordinated to the central metal atom are called *ligands*. Ligands, said to be in the co-ordination sphere of the metal, donate free electron pairs to the metal ion. Therefore, ligands are Lewis bases and metal ions in complexes are Lewis acids. A ligand molecule capable of binding to two or more positions of the central atom is called a *chelate* ligand. For example, α,α'-bipyridine can form chelate

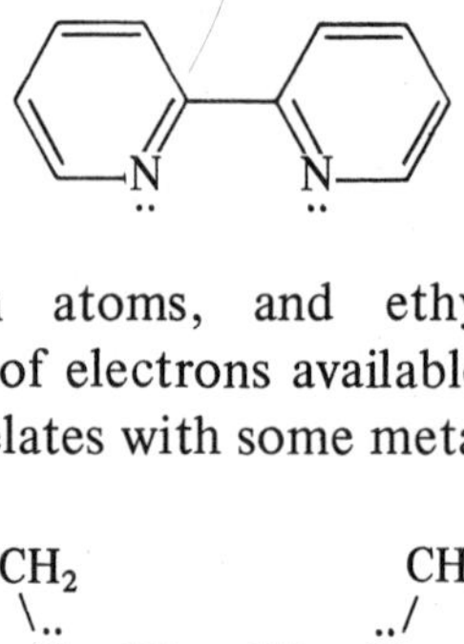

compounds with iron atoms, and ethylenediaminetetraacetate (EDTA), with six pairs of electrons available (a hexadentate ligand), can form very stable chelates with some metal ions.*

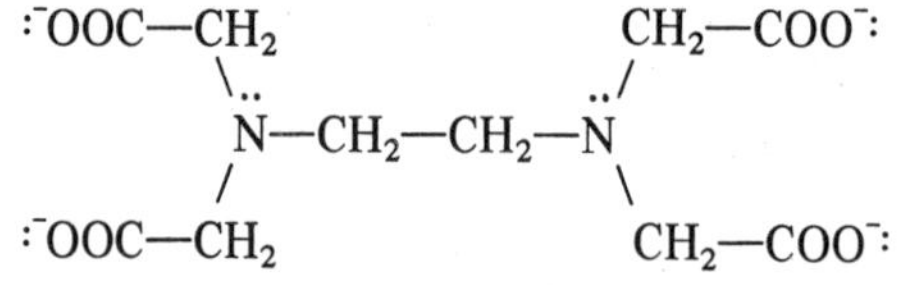

The co-ordination properties of a metal ion can only be understood in terms of its electronic structure. Electrons in an atom occupy successive energy levels, with 2 electrons in the first or lowest level, 8 in the second, 18 in the third and 32 in the fourth. Each of these levels may have sublevels, s, p, d and f. Of these, the s sublevel is lowest in energy and first to be filled with electrons. An s sublevel

---

* The greater stability of chelates over other complexes is due almost entirely to a large increase in entropy during their formation (e.g. $Mg(H_2O)_6^{2+}$ + EDTA $\rightleftharpoons$ MgEDTA + 6 H_2O, where one EDTA has replaced 6 H_2O).

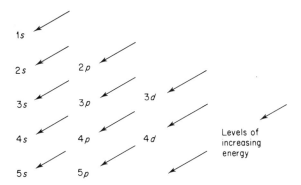

Fig. 1. Usual order of occupancy of atomic orbitals by electrons. The orbital of lowest energy level is 1s, followed by 2s, 2p and so on, as indicated by the arrows.

may contain 2 electrons in one *orbital,* and this s orbital has a spherical shape. A p sublevel can accept 6 electrons in three orbitals, with each p orbital being dumb-bell-shaped and at 90° to each other p orbital. Any d sublevels can be occupied by 10 electrons in five orbitals, four of which are cloverleaf-shaped and one is dumb-bell. Fourteen electrons can be accommodated in an f sublevel. The energy levels of successive electronic sublevels are indicated diagrammatically in Fig. 1. The difference in energy between the 1s and 2s sublevels is large, but as higher sublevels are reached the energy difference becomes smaller. The electronic structure of an atom can be written with circles for each orbital, and an arrow for each electron, as shown for nitrogen (atomic number 7):

This structure can be abbreviated as $1s^2 \, 2s^2 \, 2p^3$. Electrons in the 2p sublevel of N occupy separate orbitals because of mutual repulsion. The direction of the arrows in the circles indicates the direction of spin of the electrons, usually the same in any unfilled sublevel (Hund's rule).

   In discussing complexation of metals with enzymes, I will be referring to stereochemical and to magnetic properties of the complexes. The common co-ordination numbers are 4 and 6. A co-ordination number of 4 usually corresponds to either a tetrahedral or a square planar complex, while a co-ordination number of 6 produces an octahedral complex. The bonds between ligands and metal which form these complexes may involve combinations of orbitals, *hybrid orbitals*, which blend to give the stable molecular structures. For example, unco-ordinated Co^{3+} (atomic number 27, $27 - 3 = 24$ electrons, and co-ordination number 6) would have the electronic distribution $1s^2\ 2s^2\ 2p^6\ 3s^2\ 3p^6\ 3d^4\ 4s^2$, with the sublevel of highest energy, $3d$ (Fig. 1), containing only four of the ten electrons it could hold. When Co^{3+} reacts with $6\ NH_3$ to form the $[Co(NH_3)_6]^{3+}$ complex, the cobalt accepts six pairs of electrons, one from each molecule of ammonia. Pauling's valence bond theory describes the structure of the complex as covalent bonds between the six ligand molecules and six hybrid orbitals (d^2sp^3) of the cobalt—two at the $3d$ sublevel, one at $4s$ and three at $4p$ (the next probable sublevel, from Fig. 1). These six bonds are directed toward the corners of an octahedron, with an electronic distribution:

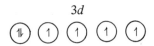

The magnetic properties of transition–metal complexes are a manifestation of the pairing or lack of pairing of electrons in their complexes. In $[Co(NH_3)_6]^{3+}$, all the electrons are forced into pairing, so the complex is *diamagnetic* (not attracted to a magnet) and termed *low-spin*. Some ligands form $4s\ 4p^3\ 4d^2$ complexes with Co^{3+}, leaving only six electrons at the $3d$ level:

$$3d$$

Outer-orbital complexes of this type are *paramagnetic* (weakly attracted to a magnet) and called *high-spin* complexes.

Williams (1970) has related the strengths of formation of complexes by metallic cations to their activity in biological systems. He classified the monovalent cations Na^+ and K^+ as charge-carriers of high mobility (i.e. forming weak complexes), the divalent cations Mg^{2+} and Ca^{2+} as semi-mobile structure formers, and the transition-metal cations as immobile metals (prosthetic groups), used either as acid or redox catalysts. I have used his classification of decreasing mobility to develop this chapter on the co-ordination chemistry of enzymes. Groups or individual examples of metal-requiring enzymes are discussed, starting with K^+-requiring enzymes and continuing with enzymes that employ Mg^{2+} and then the trace elements Cu, Zn, Mo, Fe and Co as cofactors. In similarity with the preceding four chapters, only a few enzymes can be examined, but these are representative in their diversity of properties. Metal-containing enzymes and coenzymes discussed in previous chapters include alcohol dehydrogenase, phenylalanine hydroxylase and ferredoxin, and other metalloenzymes occur in the later chapters on complex enzymes.

METAL-ACTIVATED ENZYMES

The presence of weakly-bound activator ions is essential for the activity of many enzymes. These cations belong to either the alkali metal series (K^+ and Na^+) or the alkaline earths (Ca^{2+} and Mg^{2+}). K^+, Na^+, Ca^{2+} and Mg^{2+} constitute over 99% of the metal content of the human body. Living cells maintain the necessary concentrations of these four ions by selectively accumulating K^+ and Mg^{2+} in intracellular fluids and transporting Na^+ and Ca^{2+} out of the cells into the extracellular fluids. In keeping with the higher intracellular concentrations of K^+ and Mg^{2+}, most cation-activated enzymes require one of these two ions rather than either Na^+ or Ca^{2+}. For this reason, I will discuss only the mechanisms of action of K^+ and Mg^{2+} with metal-activated enzymes.

Monovalent Cations (K^+)

Due to the fact that K^+ is used as a counter-ion in the preparation of buffers, its requirement (either absolute or stimulatory) often goes unnoticed. However, over 60 enzymes are known which are dependent on K^+ for full activity (Suelter, 1970). Despite the occurrence of so many enzymes requiring K^+, its mode of action is not understood fully. The first K^+-activated reaction to be examined in detail was that catalyzed by rabbit muscle pyruvate kinase. Kachmar and Boyer (1953) found that in the absence of K^+ the enzyme had 1.5% or less of its maximal activity. Rb^+ or NH_4^+ could also stimulate activity, but Li^+ and Na^+ inhibited the activating effect of K^+. Kachmar and Boyer concluded that the ions combined with a negatively charged site on the enzyme, possibly to cause a favourable displacement of its structure. Their work has led to the generally held concept that K^+ is required to stabilize an active conformation of the enzyme molecule, but is catalytically passive (e.g. Evans and Sorger, 1966). An alternative proposal, formulated by Suelter (1970), is that K^+ is necessary for the formation of a functional ternary complex (*EMS*). He noted that most K^+-activated reactions involve either phosphoryl transfer or elimination reactions proceeding through a keto-enol tautomer, and therefore proposed that the K^+ ion is a bridge between the enzyme and the enolate ion of the substrate. Either theory is compatible with the high concentrations of K^+ needed for maximal activity and its property of forming weak co-ordination compounds.

Kachmar and Boyer (1953) suggested that the stimulating activity of monovalent cations is related to their ionic radii. In 1968 and 1969, several laboratories reported that thallous ion (Tl^+) could replace K^+. Williams (1970) commented that Tl^+ and/or NH_4^+ often are equally or more effective than K^+. Tl^+ has approximately the same ionic radius as K^+, but forms complexes about ten-fold greater in affinity. With many enzymes, both the K_m and the maximal velocity are dependent on the ionic radius of the activating cation, as can be seen from the effect of monovalent ions on the activity of IMP dehydrogenase (Table I). Here, Tl^+ shows both the highest

Table I. Effect of monovalent cations on the activity of IMP dehydrogenase from *Bacillus subtilis*. From the data of Wu and Scrimgeour (1973).

Cation	Unhydrated ionic radius (Å)	K_m (mM)	V (mμmol min^{-1})
Tl^+	1.40	5.4	42
K^+	1.33	6.3	24
NH_4^+	1.48	7.4	16
Rb^+	1.48	9.0	11
Cs^+	1.69	11.2	9
Na^+	0.95	100	5.0
Li^+	0.60	450	4.5

affinity (lowest K_m) and the greatest V, followed closely by K^+ and NH_4^+. The very low stimulations by Na^+ and Li^+ show that they do not fit properly into the site of co-ordination of the K^+.

Pyruvate kinase catalyzes the formation of pyruvate from phospho-enolpyruvate (PEP), in one of the last steps of glycoloysis:

$$
\begin{array}{c}
COO^- \\
| \\
C-OPO_3^{2-} + H^+ + ADP \\
\| \\
CH_2
\end{array}
\longrightleftharpoons
\begin{array}{c}
COO^- \\
| \\
C=O + ATP. \\
| \\
CH_3
\end{array}
$$

The kinase from muscle (reviewed by Kayne, 1973) is a tetramer having a molecular weight of 230 000 daltons. Both a monovalent cation and a divalent cation (Mg^{2+} under physiological conditions, or Mn^{2+}) are needed for activity. $^{205}Tl^+$, the most abundant isotope of thallium, exhibits an n.m.r. spectrum. Reuben and Kayne (1971) have studied, under a variety of conditions, the n.m.r. spectra of $^{205}Tl^+$ bound to pyruvate kinase. From their data, they calculated that the Tl^+-Mn^{2+} separation distance is 8.2 Å in the E–Mn^{2+} complex, and 4.9 Å in the E–Mn^{2+}–PEP complex. The Tl^+, therefore, is located near the active site, supporting Suelter's postulate of catalytic involvement. Nowak and Mildvan (1972) have measured the effect of K^+ on the binding of PEP and some of its analogues to pyruvate kinase. They reported that K^+ raises the affinity of E–Mn^{2+} only for substrate or for analogues possessing a carboxylate group, suggesting that enzyme-bound K^+ forms a bridge complex with the carboxyl group of PEP and alters the enzyme's conformation to a catalytically active form. Similar results have been found when $CH_3NH_3^+$ is used as a cation probe for magnetic resonance measurements (Nowak, 1973). Therefore, the monovalent cation probably activates pyruvate kinase by playing an essential role in formation of an active *EMS* ternary complex. Kinetic and physical studies of the effect of a series of alkylamines on pyruvate kinase have clarified the mechanism of activation by monovalent cations (Nowak, 1976). The activation process involves two conformation changes. The first, caused by binding of the ion to the enzyme, is necessary but not sufficient to form the active ternary complex conformation. The cation-activated complex then undergoes a conformation change when the substrate is bound to form the catalytically active enzyme conformation. In the absence of a monovalent cation of suitable size, the crucial active conformation is not achieved.

In Chapter 8 (p. 248), I described the K^+-dependent reassociation of the subunits of formyltetrahydrofolate synthetase. Because dissociation and complete reactivation can be achieved, the synthetase can be used to study the effect of monovalent cations on quaternary structure. Reassociation to the active tetramer occurs in the presence of any of several cations, with the order of effectiveness being related to ionic radius:

$$NH_4^+ > Tl^+ > Rb^+ \sim K^+ > Cs^+ > Na^+ > Li^+$$

(Harmony et al., 1974). The four most effective cations have ionic radii between 1.33 Å (K^+) and 1.48 Å (Rb^+). This selectivity for cation diameter is within the 0.2 Å limit suggested by Williams (1970) for biological systems generally. The cations might either neutralize charges on the surface, preventing electrostatic repulsion of subunits, or cause a conformation change needed for the association of subunits.

Present data all suggest that monovalent cations form weakly-bonded complexes with anionic groups of apoenzymes to confer conformational changes, at either the ternary or quaternary structure level, and also possibly aid in substrate binding. Extension of current research should soon clarify the roles of K^+.

Alkaline Earth Cations (Mg^{2+})

Ca^{2+} and Mg^{2+} form complex bonds of moderate strength, which fall between the weak bonds of the more abundant K^+ and Na^+ and the strong bonds of the trace metals. They are analogous to K^+ and Na^+ in preferring oxygen (such as in carboxylate or phosphate) as a ligand atom. Ca^{2+} and Mg^{2+} both have a co-ordination number of 6, so form octahedral complexes. In general, enzymes requiring calcium for activity are extracellular enzymes, as expected from the higher concentration of Ca^{2+} in extracellular fluids. For example, Ca^{2+} is required in the activation of the zymogens trypsinogen from pancreas and prothrombin in blood. Calcium is present in salivary and pancreatic α-amylases, making these calcium metalloenzymes. The role of the atom of calcium in amylase appears to be structural, providing a catalytically active conformation and a protein structure resistant to proteolytic degradation (Vallee *et al.*, 1959).

A variety of enzymes, including all the kinases, require Mg^{2+} for activity. Kinetic experiments indicate that a Mg–ATP chelate is the actual substrate for many kinase-catalyzed reactions. Physico-chemical experiments have provided more meaningful data on the co-ordination sphere of the divalent cation, and thus its mechanism of action. Metal-activated enzymes can form three types of ternary *ES* complexes (Mildvan, 1970):

E—S—M	E—M—S	M—E—S
Substrate Bridge	Metal Bridge	Enzyme Bridge

The substrate bridge is not possible with metalloenzymes. Examples of all three types of co-ordination scheme are known. Binding studies can indicate which of the three types of co-ordination occurs. One of the most useful techniques is the measurement of the enhancement of the proton relaxation rate (PRR) of water, applied to enzyme studies by Cohn (1963). In these experiments, the diamagnetic Mg^{2+} is replaced by the paramagnetic probe Mn^{2+}, which has five unpaired electrons and thus a large magnetic moment. Because of its similar chemistry, Mn^{2+} can replace Mg^{2+} in most magnesium-requiring enzyme reactions. In solutions of $Mn(H_2O)_6^{2+}$, the rotation and tumbling of protons and water in the hydration sphere of Mn^{2+} is much faster than when the manganese–water complex is bound to a macromolecule (the enzyme). The proton relaxation time of the n.m.r. of this water is measured with a pulsed apparatus. The enhancement factor, ϵ, is defined as

$$\epsilon = \frac{\text{Relaxation rate in presence of complexing agent}}{\text{Relaxation rate in absence of complexing agent}}.$$

The key data obtained are the effects of enzyme and of substrate on the water bound to Mn^{2+} (i.e. information on the co-ordination sphere of the Mn^{2+}). When the PRR enhancements of the binary E-Mn^{2+} mixture (ϵ_b) and the ternary E-Mn^{2+}–substrate complex (ϵ_T) are compared, the following behaviours are observed:

E—S—M	E—M—S	M—E—S
$\epsilon_b < \epsilon_T$	$\epsilon_b > \epsilon_T$	$\epsilon_b = \epsilon_T$.

Substrate bridge complexes give little or no enhancement over $Mn(H_2O)_6^{2+}$ in the binary mixture, but a large enhancement when the enzyme is added. In metal bridge complexes, the binary complex shows enhancement over aqueous Mn^{2+}, but because addition of substrate replaces some of the co-ordinated water, the enhancement of the ternary complex is less than that of the binary complex. The enzyme bridge complexes show PRR enhancement in the $E-Mn^{2+}$ mixture, but addition of substrate provides no change in enhancement. Assignments based on PRR measurements can be verified by e.p.r. spectroscopy of the Mn^{2+} complexes, by the effect of enzyme-bound Mn^{2+} on the relaxation rate of substrates, or by X-ray crystallography of ternary complexes (Mildvan, 1970). It is sometimes possible to estimate the distance between the Mn^{2+} probe and various functional groups by measuring the effect of the metal ion on the n.m.r. signal of the functional groups.

Most kinases form $E-S-M$ complexes, and metalloenzymes and some enzymes catalyzing enolization and elimination reactions form $E-M-S$ complexes. There are fewer examples of enzyme bridge $(M-E-S)$ complexes. The metal in these latter enzymes may stabilize a catalytically active conformation, as calcium does in amylase. I will now briefly describe one example each of the $E-S-M$ and $E-M-S$ metal-activated enzymes.

Creatine kinase is typical of the kinases using a substrate bridge complex (Cohn, 1970). Creatine kinase catalyzes the reversible phosphorylation of creatine, utilizing the magnesium–ATP chelate as the phosphate donor.

$$MgATP^{2-} + Creatine\ (Cr) \;\rightleftharpoons\; MgADP^- + Phosphocreatine\ (PC) + H^+$$

The creatine kinase reaction is used for storage of energy in muscle tissue in the form of a phosphoguanidine, phosphocreatine (ΔG for hydrolysis $= -10\,kcal\,mol^{-1}$). Properties of the enzyme have been reviewed by Watts (1973). The kinetic mechanism for the reaction is rapid equilibrium random, with formation of E–MgADP–Cr and E–MgATP–PC dead-end complexes (Morrison and James, 1965; Morrison and Cleland, 1966):

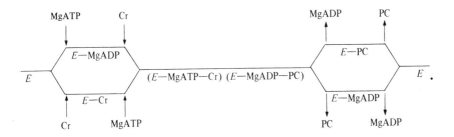

When nucleotide substrates are added to the E–Mn^{2+} mixture, large PRR enhancements of water are observed (Table II). These data provide evidence for the formation of a ternary complex of E–Nucleotide–Mn^{2+} structure. Addition of creatine or phosphocreatine to the enzyme–Mn^{2+} mixture gives no significant PRR enhancement, so no ternary E–Creatine–Mn^{2+} complex is formed. Data from e.p.r. spectra indicate that the ligands of MnADP are unchanged when the enzyme is added, ruling out simultaneous binding of the metal to the enzyme and ADP. Two substrates bind to creatine kinase, not just the adenine nucleotide. The structure of the abortive creatine + ADP complex appears to be:

$$
\begin{array}{c}
\text{ADP—Mn} \\
/ \\
\text{Creatine kinase} \\
\backslash \\
\text{Creatine}
\end{array}
$$

Although creatine has no effect on the PRR of the enzyme–Mn^{2+} mixture, it does affect that of the E–ADP–Mn complex, with a slower water exchange in the quaternary complex. Probably there is

Table II. PRR enhancements for complexes of nucleotides and creatine kinase with Mn^{2+}. From the computer-calculated data presented by Cohn (1970).

Mixture	ϵ_b	ϵ_T
E-Mn	1.5	–
ADP-Mn	1.6	–
E-ADP-Mn	–	19.6
ATP-Mn	1.6	–
E-ATP-Mn	–	15.2

a conformational change upon binding of creatine, but no inter-action between the Mn^{2+} and the guanidino substrate. Cohn *et al.* (1971) have used magnetic resonance and paramagnetic data to estimate interatomic distances between the manganese atom and the substrates. Mn^{2+} is attached to the α- and β-phosphates of both enzyme-bound ADP and ATP, and not to the phosphoryl group that is transferred from ATP to creatine. The creatine and ATP are oriented in a manner favourable for reversible transfer of the phos-phoryl group. Although the function of Mg^{2+} or Mn^{2+} in the reaction has not been proven, the Mg–ATP chelate is the true substrate, with the cation at least assisting in the orientation of the phosphate bonds of ADP and ATP. Mg^{2+} may also make the β-phosphate a better leaving group, by withdrawing electrons.

Binding studies, performed using the PRR–Mn^{2+} technique, have shown that in enolase the divalent cation co-ordinates the substrate to the active site of the enzyme (E–M–S complex formation). Enolase catalyzes the reversible hydration of phosphoenol pyruvate to 2-phosphoglycerate (PGA):

$$
\begin{array}{ccc}
\text{COO}^- & & ^1\text{COO}^- \\
| & & | \\
\text{C}-\text{O}-\text{PO}_3^{2-} + \text{H}_2\text{O} \rightleftharpoons & & \text{H}^2\text{C}-\text{O}-\text{PO}_3^{2-} \\
\| & & | \\
\text{CH}_2 & & ^3\text{CH}_2\text{OH}
\end{array}
$$

$$\text{PEP} \qquad\qquad\qquad \text{PGA}$$

It is a dimeric enzyme, with Mg^{2+} having a stabilizing influence on association of two identical subunits (Wold, 1971). Reassociation to the active dimer requires two moles of Mg^{2+} per mole of enolase dimer, and an additional two moles of Mg^{2+} may be bound upon addition of substrate. Therefore, the dimer can be represented as

$$S-Mg-E-(Mg)_2-E-Mg-S$$

with both subunits having catalytic activity. The geometry of an active enolase–Mn^{2+}–(H_2O)–PEP analogue complex has been deter-mined by nuclear relaxation experiments (Nowak *et al.*, 1973). The distances from the Mn^{2+} to the phosphorus atom and to the carboxyl group of the methylene analogue of PEP

$$
\begin{array}{c}
\text{COO}^- \\
| \\
\text{C}-\text{CH}_2-\text{PO}_3^{2-} \\
\| \\
\text{CH}_2
\end{array}
$$

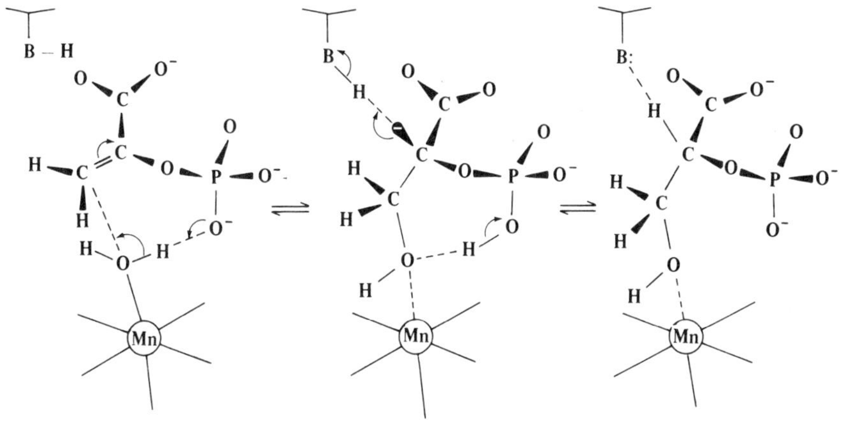

Fig. 2. Mechanism proposed for the function of enzyme-bound Mn^{2+} (and therefore Mg^{2+}) at the active site of enolase. From Nowak *et al.* (1973), with permission.

were too great for direct co-ordination of Mn^{2+}, but are consistent with co-ordination through a molecule of water. Isotope exchange and isotope rate experiments by Dinovo and Boyer (1971) have shown that an enzyme-bound carbanion (X^-) is formed by the rate-limiting hydroxyl addition to PEP:

$$E-PEP \xrightarrow{\text{OH}^-} E-PEP-OH^- \rightleftharpoons E-X^- \xrightarrow{\text{H}^+} E-PGA .$$

From the geometry of the ternary complex and from Boyer's data, Mildvan has proposed that the divalent cation activates the water molecule, so that a co-ordinated OH^- adds to C^3 of PEP (Fig. 2). In the reverse direction, the 3-hydroxyl group of PGA would bind to the Mn^{2+}. The phosphate group possibly acts as a general base in the hydration of PEP.

The above discussions of Mg^{2+} and K^+ point out the difficulties in determining the co-ordination environment and function of these weakly-bound cations. Experimental evidence lies increasingly in the direction of catalytic as well as bridging or structural roles for these ions.

METALLOENZYMES

Due to the weak affinity of the cations to metal-activated enzymes, physical and kinetic measurements often must be carried out with high concentrations of the cations (the excesses are needed for saturation of the metal-binding sites). Metalloenzymes are somewhat easier to characterize because of the properties of their firmly attached metal ions. Coleman (1971) has described both the applicable chemical properties of metals occurring in metalloenzymes and the mechanistic features of a great many metalloenzymes. Here, I shall discuss enzymes utilizing copper, zinc and molybdenum, and the cobalt-coenzymes of vitamin B_{12}. Finally, I will mention the role of iron in the oxygen-carrier protein, hemoglobin, as an introduction to co-operative effects between protein subunits.

In discussing low molecular weight coenzymes (p. 203), I noted that the K_m for a coenzyme is dependent on the amount of coenzyme available in the cell. By similar reasoning, one would expect essential trace metal ions in living organisms to be bound tightly to their apoenzymes. This indeed is true. There is a delicate balance between nutritional deficiencies and toxicities from slight excesses of metals. In Australia, some livestock were found to be diseased because of copper deficiency, while others in another part of the country had jaundice from copper poisoning. In humans, Cu^{2+} can inhibit $-SH$ enzymes or excess Zn^{2+} can cause gastrointestinal irritation. Selective chelating agents such as EDTA (for iron), 2,3-dimercaptopropanol (for Hg^{2+} or Pb^{2+}), or penicillamine (for Cu^{2+}) may be administered as drugs to remove inhibitory metals. In enzyme experiments, inhibition by chelating agents can indicate the presence of a metal ion in an enzyme. The sequestering agent may either remove the ion or may bind at the active site, blocking catalysis. Direct analysis of the purified active enzyme for metal content provides much more satisfactory evidence for the involvement of a metal, and mechanistic studies as described below are even more instructive.

Superoxide Dismutase (Cu)

Copper occurs as a cofactor in a number of oxidative metalloenzymes. As isolated, most of these enzymes contain cupric copper. For example, in both cytochrome oxidase and dopamine β-hydroxylase, the enzymes contain Cu^{2+} which cycles between the Cu^{2+} and Cu^+ oxidation states during catalytic action. One recently discovered exception to this one-electron redox pattern is the mechanism of galactose oxidase (Dyrkacz *et al.*, 1976). With this enzyme, the one atom of copper per protein molecule oscillates between the trivalent and monovalent levels:

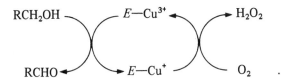

A copper metalloenzyme now receiving a great deal of attention is superoxide dismutase (Fridovich, 1972, 1975). This enzyme catalyzes the dismutation of a reactive form of oxygen, the superoxide anion, O_2^-:

$$2\,O_2^- + 2\,H^+ \;\longrightarrow\; H_2O_2 + O_2 \;.$$

The discovery and characterization of the dismutase has occurred since 1969, based on observations that during the oxidation of xanthine by molecular oxygen in the presence of xanthine oxidase, an extremely unstable species capable of initiating free-radical chain reactions was produced. Knowles *et al.* (1969) provided the first unequivocal evidence that this compound was O_2^-, formed by a one-electron reduction of O_2 by reduced xanthine oxidase. They detected the enzymically-generated free radical by a rapid-freezing e.p.r. technique. McCord and Fridovich (1969) reported that a previously unsuspected enzymatic activity could catalyze the removal of O_2^- generated by the xanthine oxidase system. This new enzyme, superoxide dismutase, was purified and shown to be identical to a copper protein, erythrocuprein, previously isolated without knowledge of its activity.

Beef erythrocyte superoxide dismutase is a blue-green protein containing two atoms of Cu^{2+} and two atoms of Zn^{2+} per molecule of enzyme. It has a molecular weight of 31 300, and is composed of two subunits of equal size. An inactive apoenzyme containing the Zn^{2+} but little Cu^{2+} has been prepared. This apoenzyme regains its spectral, electrophoretic and catalytic properties when copper is added to it (Rotilio *et al.*, 1972b). The zinc, then, is not involved in the catalytic activity, but appears to stabilize the enzyme's structure. A metal-free apoenzyme has been prepared, using stabilization at pH 3.8 (Beem *et al.*, 1974). Reconstitution with copper alone gives variable but appreciable activity (20–80% of the Cu–Zn holoenzyme), confirming the catalytic role of Cu^{2+} and the lack of a direct catalytic role for Zn^{2+}. Full activity is restored when zinc is added to the copper-enzyme derivative. Superoxide dismutase has been shown to operate by alternate reduction and reoxidation of the bound copper (Rotilio *et al.*, 1972a; Klug-Roth *et al.*, 1973). The copper is reduced by one superoxide ion, then reoxidized by the next. The reduction of about 40% of the Cu^{2+} by O_2^- formed by pulse radiolysis of O_2 was observed by measuring either the decrease in the Cu^{2+} e.p.r. signal or the Cu^{2+} absorption band at 650 nm. Equations fitting the observed oxidation and reduction are:

$$(1) \quad E\text{—}Cu^{2+} + O_2^- \longrightarrow E\text{—}Cu^+ + O_2$$

$$(2) \quad E\text{—}Cu^+ + O_2^- \xrightarrow{2\,H^+} E\text{—}Cu^{2+} + H_2O_2 \,.$$

Superoxide dismutase demonstrates half-site reactivity, with the reduction of one copper atom rendering the other copper less reactive. The fully reduced enzyme can be obtained by chemical reduction, though.

The structure of bovine superoxide dismutase has been determined recently. Steinman *et al.* (1974) have published the complete amino acid sequence. The two subunits have identical structures, with acetyl-alanine at the N-terminal and lysine at the C-terminal (position 151). The two polypeptide chains are associated through strong non-covalent interactions. X-ray studies show that these are principally hydrophobic side-chain interactions. Richardson *et al.* (1975) have reported the crystal structure of the dismutase at 3 Å resolution. The dominant feature of the polypeptide backbone is a cylindrical barrel made up of eight antiparallel β chains. Both subunits have the same conformation. The two Cu atoms are at opposite ends of the dimer, with a Zn atom close to each. Each Cu atom is complexed to four histidine imidazoles, in a slightly distorted square plane. The copper atom is positioned so that it is accessible to solvent. The Zn ligands are three histidines and one aspartate, and the shape at the zinc complex is approximately tetrahedral.

Superoxide dismutase has now been purified from many tissues and cells. The blue-green copper–zinc dismutase has been isolated from eukaryotes, including plants, fungi and yeasts, and the cytoplasm of birds and mammals. A second type of dismutase, a pinkish manganese metalloprotein, has been isolated from bacteria and from liver mitochondria (Weisiger and Fridovich, 1973). *Escherichia coli* also contains a second superoxide dismutase, which has iron rather than manganese as its metal ion. The structural and functional properties of the mitochondrial and bacterial manganese enzymes are similar, but quite distinct from the family of Cu^{2+}–Zn^{2+} enzymes (Steinman and Hill, 1973). It is believed that superoxide dismutase is a defensive enzyme, preventing oxygen toxicity by removal of O_2^- generated *in vivo*. In agreement with this suggestion, it is primarily organisms which metabolize oxygen that contain detectable amounts of dismutase (Fridovich, 1972). (A few obligate anaerobic bacteria possess superoxide dismutase activity, but all respiring cells contain it.) Although O_2^- is unstable, Fridovich (1975) has estimated that the dismutase increases the rate of decay of O_2^- in liver at pH 7.4 by at least 10^9 times. Besides this interesting physiological role, superoxide dismutase is often used as a tool to test for the involvement of O_2^- in enzymic reactions. Inhibition of some hydroxylation reactions by addition of dismutase suggests that O_2^- may be an intermediate generated by reduction of O_2 by iron in the hydroxylases. Similar evidence has indicated that O_2^- may be formed in the autoxidation of reduced flavins (p. 227).

Carboxypeptidase A (Zn)

Zinc is an essential cofactor for the enzymes alcohol dehydrogenase, carbonic anhydrase and carboxypeptidase A, as well as maintaining the active structure in several other enzymes (e.g. superoxide dismutase). Carboxypeptidase, one of the first enzymes to be crystallized, is the most extensively characterized metalloenzyme and the only one for which a clearcut mechanism of action is available for the zinc. It is an exopeptidase which catalyzes the hydrolysis of the peptide bond adjacent to a C-terminal free carboxylate ion. Carboxypeptidase A shows a preference for C-terminal aromatic or branched aliphatic amino acids. Because there are several comprehensive reviews describing this enzyme (Hartsuck and Lipscomb, 1971; Quiocho and Lipscomb, 1971), I will summarize just the chemistry of the zinc and other active site components (Lipscomb, 1974). Carboxypeptidase A from beef pancreas is a single polypeptide chain with 307 amino acid residues and one atom of zinc. Inhibition of carboxy-peptidase by metal-binding reagents such as cysteine, sulphide and cyanide (Smith and Hanson, 1949) indicated that the enzyme is a metalloprotein. Quantitative emission spectrographic analysis of recrystallized carboxypeptidase showed 0.96 moles of zinc per mole of enzyme, with only minute quantities of any other metals (Vallee and Neurath, 1954). Many experiments have been performed on the specificity and the physical chemistry of carboxypeptidase. X-ray crystallographic results from Lipscomb's laboratory were available before the complete amino acid sequence (Bradshaw *et al.*, 1969) was determined. Now, these two results have been combined, providing a great deal of information about the catalytic site. The molecule is an ellipsoid of about $50 \times 42 \times 38$ Å. The active site, comprising about one-quarter of the molecule, contains the co-ordinated Zn^{2+}, a surface groove for binding of polypeptides and a pocket for binding the side chain of the C-terminal residue of the polypeptide substrate. Completion of a three-dimensional model has identified the four ligands bound to the Zn^{2+} as His-69, Glu-72, His-196 and H_2O. These ligands form distorted tetrahedral bonds with the zinc. Crystallo-graphic results show that in the Gly–L-Tyr complex with carboxy-peptidase, the carbonyl oxygen of the peptide bond being cleaved replaces the water molecule co-ordinated to Zn^{2+}.

Vallee (1964) proposed a mechanism for carboxypeptidase A which included the fundamental principles now favoured, although the functional groups of the enzyme and the ligands of the zinc were not known at that time. He proposed

1. a binding area on the protein for the C-terminal side-chain;
2. a positive charge on the protein to bind the carboxylate of the substrate;
3. co-ordination of the carbonyl group of the substrate peptide bond to Zn^{2+}; and
4. base (nucleophilic catalyst) and acid (proton donor) groups on the enzyme.

The 2 Å structure of the Gly–Tyr complex of carboxypeptidase sub-stantiates these general principles (Hartsuck and Lipscomb, 1971), but is ambiguous about the role of the basic group in the active-site cleft. Binding of the poorly hydrolyzed substrate causes several conformational changes in the protein (p. 31). (The scissile bond of

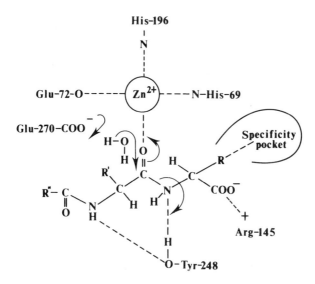

Fig. 3. Diagrammatic representation of the structure of the *ES* complex formed between carboxypeptidase A and a tripeptide, as suggested by the data of Lipscomb and his coworkers (Hartsuck and Lipscomb, 1971) and by other chemical studies on the enzyme. Zinc polarizes the carbonyl group, Tyr-248 donates a proton and Glu-270 promotes nucleophilic attack by a molecule of water.

the peptide is also distorted, so carboxypeptidase exerts a strain on the substrate and is not proof of induced-fit catalysis.) Three interactions appear to account for the formation of the active Michaelis complex. First, the C-terminal side chain of a substrate fits into a pocket of relatively low specificity by displacing several molecules of water. Because the pocket is closed, carboxypeptidase has no endopeptidase activity. Second, the terminal carboxylate of the substrate is bound to the positively charged guanidino group of Arg-145. Finally, the carbonyl oxygen of the substrate's terminal peptide bond replaces the water ligand on the zinc. Two additional features seem important in catalysis. The phenolic group of Tyr-248 moves close to the peptide bond of the substrate, possibly forming a hydrogen bond with the amide of the peptide bond. This conformational change prevents the entry of water into the side-chain specificity pocket. The —OH of Tyr-248 may also bind to the amide of the penultimate peptide bond of the substrate. Models built with larger substrates suggest that the carboxyl group of Glu-270 is the nucleophilic catalyst of the enzyme. Figure 3 presents these interactions diagrammatically.

Examination of the model of carboxypeptidase shows that only Arg-145, the zinc atom, Glu-270 and Tyr-248 are close enough to the catalytic site to be involved in catalysis. The complexing of the carbonyl oxygen with the zinc probably withdraws electrons from the $-C=O$ group. The $-OH$ of Tyr-248 is suitably located to donate a proton to the amide nitrogen. The carboxyl group of Glu-270 could form a rapidly hydrolyzed anhydride intermediate as suggested by Vallee, or it could act as a general base with an intervening molecule of water as the nucleophile (as shown in Fig. 3). The conclusions can be drawn that in carboxypeptidase the zinc atom is an immobile acidic prosthetic group, and that it most likely behaves in a similar fashion in some other zinc metalloenzymes.

Breslow and Wernick (1976) have succeeded in explaining the only evidence favouring an anhydride of Glu-270 as an intermediate. Previous workers had demonstrated that carboxypeptidase A catalyzes ^{18}O incorporation from $H_2^{18}O$ into the carboxyl of N-acyl amino acids. Breslow and Wernick reasoned that an acyl–enzyme intermediate might not be involved. Instead, the exchange might be due to synthesis of a peptide if another amino acid were present. They prepared N-benzoylglycine with ^{18}O in the terminal carboxylate:

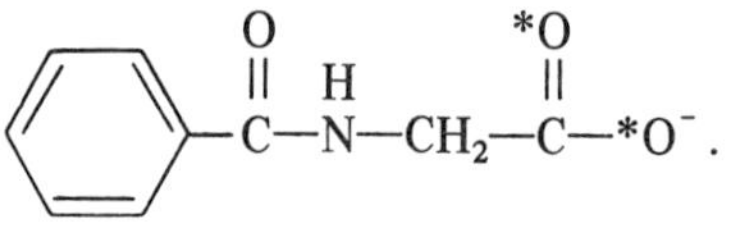

Carboxypeptidase catalyzes exchange of this labelled oxygen, but only in the presence of added phenylalanine or leucine. This result is to be expected if the exchange mechanism proceeds by resynthesis of a substrate. Therefore, the carboxyl of Glu-270 is a general base which delivers nucleophilic water to the carbonyl of the scissile peptide bond (Fig. 3). No intermediate in the hydrolysis has been observed yet.

Nitrogen Fixation (Mo)

Biological and chemical fixation of dinitrogen, i.e. reduction of the inert N_2 to NH_3, is an important reaction in agricultural economy.

The growing of large food crops requires enrichment of soil with usable nitrogen, in the form of ammonium sulphate fertilizer. The ammonium ion, NH_4^+, is needed by plants for biosynthesis of N-containing compounds, particularly proteins. Industrial chemical synthesis of ammonia by the Haber–Bosch process, using $N_2 + 3 H_2$, high pressure and temperature, and a heterogeneous catalyst, was developed early in this century. It is a relatively simple and cheap process, but nitrogen fertilizers are increasing in price rapidly because of the higher prices and dwindling supplies of oil and natural gas used for energy sources in the production of the H_2.

Legumes are often grown as pioneer plants, to prepare infertile soil. We now know that the legumes colonize the soil with nitrogen-fixing bacteria, thus supplying some fixed nitrogen. Infection of the root nodules of some legumes (e.g. peas, alfalfa, beans) by the bacteroid *Rhizobium* allows symbiotic nitrogen fixation. Some non-leguminous plants such as the alder tree also can carry out symbiotic N_2-fixation, in co-operation with unidentified bacteria. Recent experiments indicate that the plant partners of the nitrogen-fixing bacteria supply two carbon sources (a pentose and succinate) and a small amount of fixed nitrogen, explaining the role of the symbiotic host. These symbiotic systems and some photosynthetic organisms, such as the blue-green algae and photosynthetic bacteria, provide most of the NH_3 formed biologically. If the proportion of NH_3 nutrient supplied by the biological systems could be increased, then mankind would benefit greatly. Two approaches are being explored. Nitrogen-fixing bacteria (mutants or genetically altered strains) might be induced to grow symbiotically with major non-leguminous crops—especially grasses and grains—which now require synthetic fertilizer. Alternatively, the cereals might be infected with the genes from nitrogen-fixing bacteria, so that they could synthesize their own NH_3 from N_2. Sustained and multidisciplinary research will be needed, but the goal of increased protein production must be attained.

In the past 15 years, our understanding of biological N_2-fixation has progressed from ill-defined bacterial whole-cell systems to the availability in pure form of the enzyme responsible, nitrogenase. Information on this enzyme has been obtained most readily from the non-photosynthetic bacteria *Azotobacter* and *Clostridium* (summarized by Hardy *et al.*, 1971; Dalton and Mortenson, 1972), but nitrogenase has now been purified from 16 organisms including anaerobic, aerobic, facultative anaerobic and photosynthetic bacteria, bacteroids from root nodules, and blue-green algae. The methods of assay and purification of the proteins required for N_2-fixation are described in the last eight articles in "Methods in Enzymology", Volume 24.

Research with cell-free preparations of N_2-fixing bacteria established a number of fundamental properties of biological nitrogen fixation. The reduction of N_2 is an anaerobic process. Crude extracts of C. *pasteurianum* were separated into a nitrogen-activating system (crude nitrogenase) and a hydrogen-donating system. The hydrogen-donating system was found to be pyruvate synthase, and its fractionation led to the discovery of the protein coenzyme ferredoxin (p. 274). The pyruvate synthase-catalyzed reaction provides both reduced ferredoxin and acetyl CoA which is converted to ATP and acetate via acetyl phosphate. The requirements for any nitrogenase-catalyzed reaction system are now known to be ATP, Mg^{2+} and a source of reducing power. In non-photosynthetic bacteria, ferredoxin or flavodoxin supplies the hydrogen, and acetyl phosphate the ATP. In photosynthetic bacteria, the ATP is supplied by phosphorylation coupled to electron transport.

Enzymatic reduction of N_2 occurs without the accumulation of any detectable intermediates. The postulated diimide and hydrazine intermediates are probably enzyme-bound:

$$N{\equiv}N \longrightarrow HN{=}NH \longrightarrow H_2N{-}NH_2 \longrightarrow 2\,NH_3 \ .$$

| Dinitrogen | Diimide | Hydrazine | Ammonia |

Nitrogenase can catalyze the reduction of compounds other than N_2. Hydrazine can be reduced by the nitrogenase system (Bulen, quoted in Zumft and Mortenson, 1975), so the three-step reduction has its first direct experimental support. Non-physiological substrates include acetylene (HC${\equiv}$CH), azide (N${=}$N${=}$N$^-$), hydrogen cyanide (HCN) and methyl isonitrile (H$_3$C${-}$N${\equiv}$C). Reduction of each of these alternative substrates requires an even number of electrons and protons, though a variety of products is formed from some substrates (Table III). The requirement for an even number of electrons in all these reactions provides strong evidence for reductions catalyzed by nitrogenase to be in 2-electron steps. Each of these steps might be by

Table III. Products of reduction of alternative substrates by the nitrogenase system of *Azotobacter vinelandii*. From Hardy *et al.* (1971).

Substrate	Products	Electrons consumed
N_2	$2\ NH_3$	6
C_2H_2	C_2H_4	2
N_3^-	$NH_3 + N_2$	2
HCN	$CH_4 + NH_3$ (major)	6
,,	CH_3NH_2 (minor)	4
CH_3NC	$CH_3NH_2 + CH_4$ (major)	6
,,	C_2H_4 ⎫	8
,,	C_2H_6 ⎬ (minor)	10
,,	C_3H_6 ⎪	12
,,	C_3H_8 ⎭	14

either 1- or 2-electron increments. Use of acetylene as a substrate allows a rapid and sensitive assay, based on measurement by gas chromatography of the formation of ethylene. Reduction of all of these alternative substrates requires the same cofactors (ATP, Mg^{2+} and reduced ferredoxin) as does reduction of N_2.

Nitrogenase has been isolated from many micro-organisms. Because these enzymes are similar in molecular weight, metal content and reactions catalyzed, I will describe only the nitrogenase from *C. pasteurianum* purified by Mortenson and his colleagues (cf. review by Zumft and Mortenson, 1975). Nitrogenase, the nitrogen-activating complex of the bacterial extracts, contains two easily separable subunits that are required for reduction of N_2. These subunit proteins are called the Mo–Fe protein (or molybdoferredoxin by Mortenson) and the Fe protein (or azoferredoxin). They are both oxygen sensitive, so must be purified under anaerobic conditions. For example, azoferredoxin has a half-life of 20 seconds when exposed to air. The Mo–Fe protein is selectively precipitated with protamine sulphate, then redissolved and purified by anaerobic column chromatography. The Fe protein, not precipitated by the protamine sulphate treatment, is cold-labile and sensitive to high concentrations of salt. It too is purified by column chromatographic procedures. Purified Fe protein is stored by freezing into pellets in liquid nitrogen.

Table IV. Composition of nitrogenase from *Clostridium pasteurianum.*

	Mo–Fe Protein[a]	Fe Protein[b]
Molecular weight	220 000	55 000
Subunit structure	(2 × 60 000) + (2 × 50 000)	2 × 27 500
Mo per oligomer	2	0
Fe per oligomer	24 (approx.)	4
S^{2-} per oligomer	24 (approx.)	4

[a] Huang *et al.* (1973)
[b] Nakos and Mortenson (1971)

The Mo–Fe protein is an oligomeric protein with a molecular weight of 220 000. It is a tetrameric complex with two types of subunits, two of 60 000 and two of 50 000 daltons (Huang *et al.*, 1973). It contains 2 moles of molybdenum per 220 000 dalton oligomer, and iron and acid-labile sulphide as shown in Table IV. The Fe protein (Nakos and Mortenson, 1971) is a much smaller protein. It is dimeric, with each dimer containing 4 moles of iron and 4 moles of S^{2-} (Table IV). Both proteins are required for activity, measured by an acetylene assay (Dalton *et al.*, 1971) with dithionite as the reductant in place of ferredoxin:

$$HC\equiv CH + 2\,H_2O + Na_2S_2O_4 \xrightarrow[\text{Both proteins}]{Mg^{2+},\,ATP} H_2C{=}CH_2 + 2\,NaHSO_3.$$

ATP drives the electron transport of nitrogenase, but its function is not understood. Some recent experiments have lent support to the postulate that ATP is related to a change in conformation of a subunit of nitrogenase. Gel filtration experiments indicated that only the Fe protein binds significant amounts of the Mg–ATP complex. When the Fe protein is titrated with Mg–ATP (measured by low temperature e.p.r. spectroscopy), two molecules of ATP are bound per dimer of Fe protein (Zumft *et al.*, 1973). The ATP-induced change in the e.p.r. spectrum was attributed to a change in conformation of the protein. This conformation change leads to exposure of the Fe–S centre, as demonstrated by the availability of the iron to form a stable, red complex with α,α'-bipyridine in the presence of

Mg–ATP (Walker and Mortenson, 1974). Their results show that the Fe protein has at least three interconvertible states—an oxidized state, a reduced state and a reduced state which has ATP bound to it. Further e.p.r. experiments with the nitrogenase from *C. pasteurianum* (Mortenson *et al.*, 1973) indicate the following sequence of reactions. Reduced Fe protein rapidly complexes with Mg–ATP, and this complex then combines with reduced Mo–Fe protein to form a larger complex capable of reducing substrates. In the presence of ATP, the redox potential of the Fe protein shifts from -0.29 volts to -0.40 volts (Zumft *et al.*, 1974). A single electron per redox step

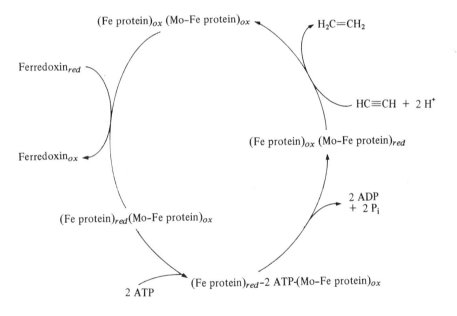

Fig. 4. A minimal reaction sequence for the clostridial nitrogenase reaction. The postulated oxidation levels of the metalloproteins are indicated by the subscripts *ox* (oxidized or partially oxidized form) and *red* (reduced form). The oligomeric structure has been assumed to be a dimer of each subunit (i.e. total molecular weight = 175 000 approx.) so that two electrons may be accommodated by the enzyme for reduction of substrate. Subunits may dissociate during the cycle, and ATP may be required for sites other than that shown. Here, acetylene has been shown as the substrate. In the reduction of one mole of N_2, three turns of the cycle, consuming 3 moles of reduced ferredoxin (6 electrons) and at least 6 moles of ATP, would be required for the formation of 2 moles of NH_3.

is transferred by the nitrogenase proteins. Therefore, consecutive multi-electron transfer or involvement of two subunits (or sites) must account for the even number of electrons donated to substrates. The rate limiting reaction in N_2 reduction may be an ATP-mediated electron flow from Fe protein to Mo-Fe protein. It has not yet been proven that substrates bind to the Mo-Fe protein, but most mechanisms incorporate the assumption that they bind to or near the molybdenum atom. A tentative abbreviated scheme for reduction of acetylene in the presence of nitrogenase is presented in Fig. 4. Ferredoxin donates electrons to oxidized Fe protein. Then MgATP binds to reduced Fe protein, and the reduced Fe protein-MgATP complex attaches to an oxidized form of Mo-Fe protein (a form of intermediate redox state). Fully reduced Mo-Fe protein in the active (Fe protein)-(Mo-Fe protein)-ATP complex transfers electrons to the substrate, and ATP is hydrolyzed. The acetylene may bind to the semioxidized Mo-Fe protein before it is incorporated into the active complex. If N_2 is the substrate, it should remain bound to the Mo-Fe protein until its reduction to ammonia is completed. Much more data are needed for a detailed mechanism to be worked out (e.g. knowledge of the stoichiometry of ATP hydrolysis, and even the determination of the ratio of Fe protein to Mo-Fe protein in nitrogenase).

Hardy et al. (1971) have proposed a mechanistic hypothesis for the reduction of both N_2 and $HC \equiv CH$. This mechanism, called the two-site hypothesis, suggests that nitrogen is bound to iron and is reduced by molybdenum, whereas acetylene is complexed to and reduced by the molybdenum. This proposal is in agreement with the formation of stable end-on complexes of N_2 ($N \equiv N$—Metal) and with inhibition experiments using nitrogenase. The reduction of N_2 would occur in three 2-proton, 2-electron steps while the nitrogen is complexed to an iron atom of the Mo-Fe protein. Steifel (1973) has extended this hypothetical mechanism to explain why Mo is the metal most suitable for use in nitrogenase. In all molybdenum-containing enzymes (nitrogenase, xanthine oxidase, nitrate reductase, sulphite oxidase and aldehyde oxidase), the molybdenum is believed to interact directly with substrate. He proposes that Mo behaves as both a redox and an acid-base reactant. The ligands of Mo^{4+} complexes are more highly protonated (more basic) than similar ligands bound to Mo^{6+}. Therefore, as electrons flow from Mo^{4+} to N_2, protons may also transfer to the N_2 from the ligands of the Mo. When the oxidized Mo–Fe protein is reduced by the Fe protein, its ligands incorporate two protons from water. The molybdenum of the Mo–Fe protein possibly binds to the N_2 by a side-on (π) bond, and in the first reductive step the N_2 is reduced to Fe-co-ordinated diimide by a coupled electron-proton transfer:

$$
\begin{array}{ccc}
\text{N}\equiv\text{N}\overset{\displaystyle\diagup\text{Fe}\text{---}}{} & & \text{HN}=\text{N}\overset{\displaystyle\diagup\text{Fe}\text{---}}{} \\
| & \longrightarrow & \backslash \\
\text{H}\ \ \ \ \ \ \ \text{H} & & \text{H} \\
\backslash\ \ \ \ \ \ /
\end{array}
$$

Steifel concludes that Mo alone has redox potentials and polarizing effects on its ligands suited to this type of reaction. Although the reaction scheme in Fig. 4 and the two-site hypothesis provide an appealing mechanism for reduction of N_2, much remains to be discovered about both biological and chemical nitrogen fixation. In reviewing molybdenum enzymes, Bray and Swann (1972) have suggested that reductive substrates react with xanthine oxidase by binding to Mo^{6+}, reducing it to the e.p.r.-detectable Mo^{5+}, and finally to Mo^{4+}. The oxidation states of molybdenum in nitrogenase are not yet known, but if similar to xanthine oxidase they would fit Steifel's postulate. A π-bound molybdenum nitrile chemical complex has been isolated (Thomas, 1975), but some chemical models of the nitrogenase system are based on end-on binding of substrate to molybdenum. Also a $Mo^{3+} \rightleftharpoons Mo^{5+}$ redox cycle may be involved in N_2 reduction.

The need for so many Fe–S centres per monomer in the Mo–Fe protein, the role for ATP and the mode of binding of N_2 to nitrogenase are just a few of the many aspects still not explained by experimental evidence. In fact, few of the 43 problems listed by Hardy *et al.* in 1971 have been answered to date. Regarding molybdenum, the focal point of this discussion, I can summarize best by quoting Zumft and Mortenson (1975, on their p. 47):

Finally, nothing is known about the role of molybdenum!

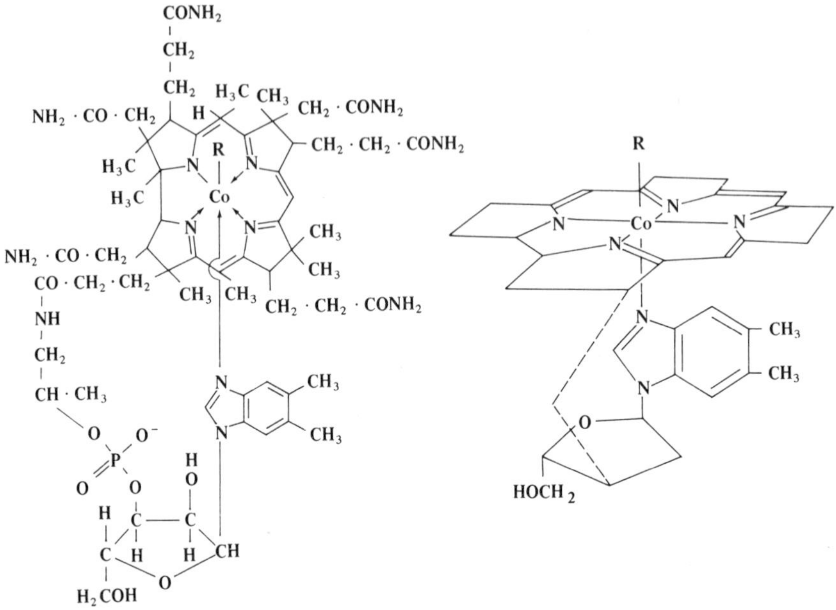

Fig. 5. Structure of vitamin B_{12} and its coenzyme forms. The complete structure is on the left, and an abbreviated spatial diagram is on the right. Although shown here as being planar, the corrin ring deviates slightly from planarity.

Regulation of nitrogen fixation is an elaborate process. The ATP/ADP ratio is a potential regulator of nitrogenase activity. Also, there must be some mechanism for protection from oxygen. Biosynthesis of nitrogenase is remarkably complicated, with high NH_4^+ levels causing repression. The ammonia acts indirectly, possibly by stimulating the adenylylation of glutamine synthetase (p. 452), with only the adenylylated form of the synthetase acting as the nitrogenase repressor. The future of nitrogenase looks bright but busy for collaborative research between inorganic chemists, biochemists, geneticists and botanists.

Vitamin B_{12} (Co)

The two alkyl coenzyme forms of vitamin B_{12} are the largest and most complicated of the low molecular weight organic coenzymes. The active centre of vitamin B_{12} is a Co^{3+} atom co-ordinated to the four nitrogen atoms of a corrin ring (Fig. 5). The porphyrin-like corrin ring consists of four reduced pyrrole rings, joined by three methylidyne radicals and by one direct C—C bond. The other two ligands of the cobalt atom of vitamin B_{12} are the nitrogen of 5,6-dimethylbenzimidazole, which is on a long side chain extending from the corrin ring, and cyanide. As shown on the right side of Fig. 5, the benzimidazole is depicted as binding below the planar corrin ring or in the α-position. Variation of the upper or β ligand (R) causes the differences between forms of vitamin B_{12}. The vitamin is isolated as an artifactual stable cyanide derivative, cyanocobalamin ($R=CN^-$). *In vivo*, vitamin B_{12} reacts as aquacobalamin, or B_{12a}, in which the R-group is H_2O. There are two main categories of B_{12}-mediated reactions, those which are methyl group transfers and those which involve intramolecular migration reactions. In the former group, methylcobalamin ($R = -CH_3$) is formed as an intermediate. The latter reactions utilize adenosylcobalamin, in which the β ligand is a 5′-deoxy-5′-adenosyl group. For simplicity, cobalamin derivatives can be abbreviated so that the nitrogen atoms of the corrin ring are represented by the corners of a parallelogram and the α ligand by a nitrogen atom. Vitamin B_{12} and its two isolated coenzyme forms, then, are:

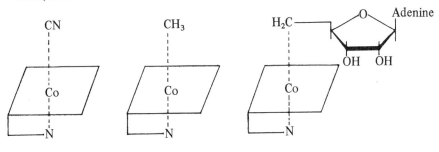

Cyanocobalamin   Methylcobalamin   Adenosylcobalamin

Before examining B_{12}-mediated enzymic reactions, it is necessary to consider some relevant chemical properties of vitamin B_{12} and its coenzymes (Huennekens, 1968; Wood and Brown, 1972). The Co^{3+} of vitamin B_{12a} (also called cob(III)alamin) can be reduced in 1-electron steps to compounds easily detected by their absorption spectra. The Co^{2+} species, vitamin B_{12r} (cob (II)alamin), is a free radical which gives the e.s.r. spectrum expected for a tetragonal low-spin complex. Vitamin B_{12r} is stable in anaerobic solutions. The fully reduced Co^+

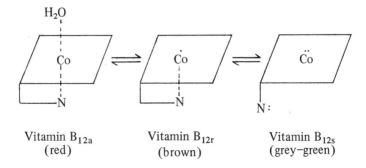

| Vitamin B_{12a} | Vitamin B_{12r} | Vitamin B_{12s} |
| (red) | (brown) | (grey–green) |

form, vitamin B_{12s} or cob(I)alamin, is a square planar complex lacking both the α and β ligands (Brodie and Poe, 1971). Vitamin B_{12s} is an extremely unstable and very nucleophilic reagent. It is oxidized rapidly to vitamin B_{12r} by water itself. It can react with methylating agents such as methyl iodide or S-adenosyl methionine to form methylcobalamin, or with adenosine-5′-tosylate to produce adenosylcobalamin. Biosynthesis of adenosylcobalamin by $C.$ $tetanomorphum$, and presumably other species, follows a pathway similar to the chemical synthesis. There is a stepwise reduction of vitamin B_{12a} to vitamin B_{12r} and then to vitamin B_{12s}, catalyzed by two separate flavoproteins, followed by adenosylation of vitamin B_{12s} by the adenosine group of ATP (Walker et $al.$, 1969). The methylation of enzyme-bound vitamin B_{12s} to produce methylcobalamin, discussed below, is also similar to the chemical synthetic route. The alkylcobalt vitamin B_{12} compounds are unstable in visible light, a property which has been used to detect their occurrence.

Methylcobalamin functions as an enzyme-bound intermediate in the de $novo$ biosynthesis of the methyl group of methionine (p. 245). Using serine as the C-1 source, the methylation of homocysteine catalyzed by extracts of $E.$ $coli$ requires three protein fractions and seven cofactors (for example, cf. Hatch et $al.$, 1961):

$$\text{Serine–3-}^{14}\text{C} + \text{RSH (Homocysteine)} \rightarrow \text{RS-}^{14}\text{CH}_3 \text{ (Methionine)} + \text{Glycine}$$

$$R = -(CH_2)_2-\underset{\underset{NH_3^+}{|}}{CH}-COO^-$$

Cofactors $=$ PLP, tetrahydrofolate, Mg^{2+}, ATP, NADH (or NADPH), vitamin B_{12} and FAD .

Systematic characterization of the biosynthesis of methionine has resolved the roles of the many cofactors and proteins. One protein is serine transhydroxymethylase (p. 249), which accounts for the requirements for PLP and tetrahydrofolate. The second protein is methylenetetrahydrofolate reductase, which catalyzes the reduction of methylenetetrahydrofolate to 5-methyltetrahydrofolate:

$$\text{N———N} + \text{NADH} + \text{H}^+ \xrightarrow{\text{FAD}} \text{N}^5\text{———N} + \text{NAD}^+ .$$

$$\underset{\text{CH}_2}{\diagdown\diagup} \qquad\qquad \underset{\text{CH}_3}{|} \;\; \underset{\text{H}}{|}$$

This reaction is thought not to be reversible under physiological conditions. When methyltetrahydrofolate and homocysteine are mixed with the third enzyme fraction, there is still a requirement for NADH and FAD. The third fraction contains protein-bound vitamin B_{12} (Takayama *et al.*, 1961), associated with the enzyme called either 5-methyltetrahydrofolate-homocysteine methyltransferase or methionine synthetase. The terminal step in methionine formation,

$$\underset{\substack{|\\ \text{CH}_3}}{\text{N}^5}\text{———}\underset{\substack{|\\ \text{H}}}{\text{N}} + \text{RSH} \xrightarrow[\text{(Vitamin } B_{12})]{\text{Methyltransferase}} \underset{\substack{|\\ \text{H}}}{\text{N}}\text{———}\underset{\substack{|\\ \text{H}}}{\text{N}} + \text{RSCH}_3 ,$$

requires the presence of S-adenosylmethionine (explaining the need for ATP, Mg^{2+}, and endogenous methionine) and $FADH_2$ or $FMNH_2$ which can replace NADH and a flavoprotein. Fujii and Huennekens (1974) have isolated from *E. coli* two flavoproteins that, acting sequentially, are the most efficient catalysts found to date for activation of methionine synthetase. In their reducing system, NADPH is the more efficient initial reductant:

NADPH → component R—(FAD) → component M—(FMN) →

                                                        methionine synthetase.

The preparation and testing of added methylcobalamin as a substrate in place of methyltetrahydrofolate (Guest *et al.*, 1962) strongly suggested that methylcobalamin was an intermediate in this reaction. However, it took seven years to establish that *enzyme-bound* methylcobalamin actually is the intermediate.

Proof of the involvement of methylcobalamin was obtained primarily with the methyltransferase partially purified from *E. coli* (reviewed by Taylor and Weissbach, 1973). This preparation contains vitamin B_{12} which is firmly but non-covalently bound. Several lines of evidence clearly indicate that, during the transmethylation reaction, the vitamin B_{12} molecule is methylated. First, the enzyme is inhibited by propyl iodide, but the inhibition can be reversed by a short exposure to light. This inhibition occurs by reductive alkylation of the vitamin B_{12} of the methyltransferase to form the light-sensitive propylcobalamin derivative:

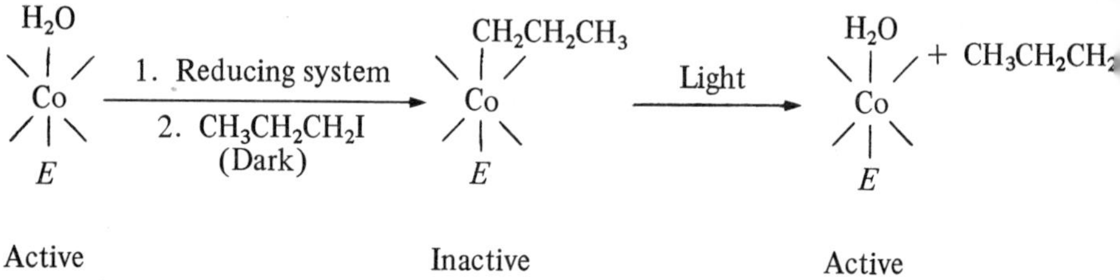

$$
\underset{\text{Active}}{\overset{\displaystyle H_2O}{\underset{\displaystyle E}{\overset{\diagdown\,|\,\diagup}{\underset{\diagup\,|\,\diagdown}{Co}}}}}
\xrightarrow[\substack{\text{2. } CH_3CH_2CH_2I\\(\text{Dark})}]{\text{1. Reducing system}}
\underset{\text{Inactive}}{\overset{\displaystyle CH_2CH_2CH_3}{\underset{\displaystyle E}{\overset{\diagdown\,|\,\diagup}{\underset{\diagup\,|\,\diagdown}{Co}}}}}
\xrightarrow{\text{Light}}
\underset{\text{Active}}{\overset{\displaystyle H_2O}{\underset{\displaystyle E}{\overset{\diagdown\,|\,\diagup}{\underset{\diagup\,|\,\diagdown}{Co}}}}} + CH_3CH_2CH_2
$$

Second, when methyl-labelled ^{14}C-methyltetrahydrofolate is incubated with the enzyme in the presence of adenosylmethionine and an $FMNH_2$-dithiol reducing system, the methyltransferase is methylated. The radioactive methylcobalamin form of the enzyme has been isolated and shown to be capable of transferring its methyl group to homocysteine. Also, labelled methylcobalamin has been extracted from the radioactive enzyme. The pathway for the methyltransferase-catalyzed reaction, therefore, is one in which the enzyme cycles between the vitamin B_{12s} and methylcobalamin forms:

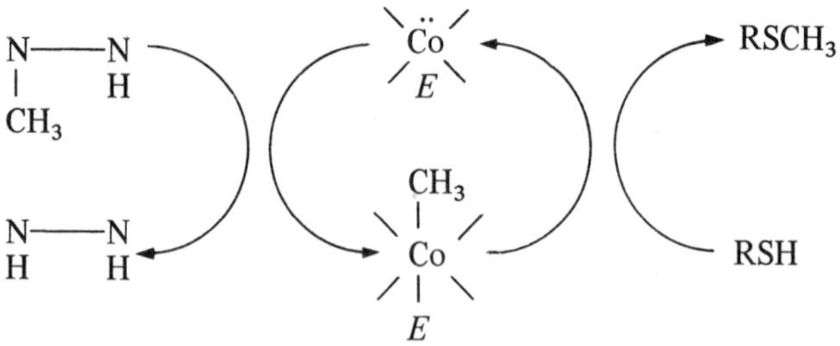

A flavoprotein reducing system and S-adenosylmethionine are necessary to convert inactive oxidized-cobalt enzyme into an enzymically active methylated species. A reaction sequence based on kinetic experiments with mammalian methyltransferase (Burke et al., 1971) is consistent with a priming role for adenosylmethionine in methionine synthesis. This scheme predicts two forms of reduced enzyme, one which can only react with adenosylmethionine (possibly a vitamin B_{12r} form) and one which can react with the less reactive methylating agent, methyltetrahydrofolate (the vitamin B_{12s} form).

Let us now turn our attention from methylcobalamin to adenosylcobalamin. Adenosylcobalamin is a cofactor for about ten known biochemical reactions (Babior, 1975), most of which occur in bacterial cells. With one exception (the reduction of nucleoside triphosphates), these are all isomerization or rearrangement reactions involving migration of an alkyl or acyl group or $-OH$ or $-NH_2$ to an adjacent carbon, and of a hydrogen atom from that adjacent carbon:

$$
\begin{array}{ccc}
\overset{H}{\underset{|}{}}\ \overset{X}{\underset{|}{}} & & \overset{X}{\underset{|}{}}\ \overset{H}{\underset{|}{}} \\
-C^a-C^b- & \longrightarrow & -C^a-C^b- \quad . \\
\ |\ \ \ \ |\ & & \ |\ \ \ \ |\ \\
\end{array}
$$

Adenosylcobalamin appears to act as the intermediate acceptor of the hydrogen that is transferred. These reactions include dehydrations (Abeles, 1971), rearrangements of carbon skeletons (Barker, 1972), amino group migrations (Stadtman, 1972) and reduction of nucleoside triphosphates to 2'-deoxynucleoside triphosphates.

Compared to the reactions of methylcobalamin, the elucidation of the mechanisms of adenosylcobalamin-requiring reactions has been an even more difficult task. Adenosylated cobalamins were discovered in 1958 by Barker et al., and shown by X-ray crystallography to bind the adenosine moiety by the unique Co—C bond (Lenhert and Hodgkin, 1961). Despite the belief that the cobalt must be involved in catalysis, it is difficult to envisage how the Co—C bond breaks, allowing exposure of the active cobalt, and then reforms. Suggestions based on non-enzymatic model systems have been made to explain the mechanism of adenosylcobalamin action, but it is not yet clear if any of these model systems are analogous to the enzymic reactions. Our present knowledge of the function of adenosylcobalamin is based almost entirely on studies of the enzyme systems themselves.

Much of the fundamental work on the mechanism of action of adenosylcobalamin has been performed using the enzyme dioldehydrase isolated from *Aerobacter aerogenes* (Abeles, 1971). Dioldehydrase catalyzes the conversion of 1,2-propanediol to propionaldehyde. Isotopic experiments showed that in this reaction two group transfers occur: the C^2 hydroxyl moves to C^1 and a hydrogen atom from C^1 moves to C^2. The first product is a 1,1-gem-diol which dehydrates stereospecifically:

$$
\underset{\underset{\text{1,2-Propanediol}}{}}{\overset{\overset{\text{OH}}{|}}{H_3C-CH-CHD}} \quad \to \quad \underset{\underset{\text{Gem-diol}}{}}{H_3C-CHD-\overset{|}{\underset{^{18}OH}{C}HOH}} \quad \xrightarrow{-H_2O} \quad \underset{\underset{\text{Propionaldehyde}}{}}{H_3C-CHD-CH=^{18}O}\ .
$$

with ^{18}OH under the first CH.

Note that the transfer of the C^1 hydrogen occurs without equilibration with the solvent. Abeles and Zagalak (1966), using a mixture of 1-^{3}H-1,2-propanediol and the alternative substrate ethylene glycol, showed that isotopic label was transferred to the acetaldehyde formed:

$$
\left.\begin{array}{c} CH_3-CHOH-C^*H_2OH \\ + \\ CH_2OH-CH_2OH \end{array}\right\} \xrightarrow[\text{Adenosylcobalamin}]{\text{Dioldehydrase}} \left\{\begin{array}{c} CH_3-C^*H_2-C^*HO \\ + \\ C^*H_3-CHO\ . \end{array}\right.
$$

This experiment suggested that hydrogen transfer was not necessarily intramolecular, but that adenosylcobalamin might act as an intermediate hydrogen carrier. This reasoning was followed by the demonstration (Frey et al., 1967) that labelled coenzyme could be isolated when adenosylcobalamin was incubated with enzyme and tritium-labelled substrate, and that this label resided at the $C^{5'}$-position of the adenosine group. Essenberg et al. (1971) examined the kinetics of the transfer of tritium from tritiated coenzyme to product and from tritiated substrate to coenzyme. Their results indicate that a scheme in which adenosylcobalamin accepts a hydrogen from the substrate to form an obligatory intermediate containing three equivalent hydrogens is kinetically permissible. 5'-Deoxyadenosine, or a compound readily converted to it during isolation, has been suggested as the intermediate hydrogen carrier. The most definitive evidence in favour of this intermediate was obtained first with another adenosylcobalamin-requiring enzyme, ethanolamine ammonia-lyase.

The enzyme ethanolamine ammonia-lyase, purified from a *Clos-*

tridium, catalyzes a reaction similar to the dioldehydrase reaction:

$$^+H_3N-CH_2-CD_2OH \longrightarrow \left[\begin{array}{c} OH \\ / \\ CH_2D-CD \\ \backslash \\ NH_3^+ \end{array} \right] \longrightarrow CH_2D-CDO + NH_4^+ .$$

Ethanolamine   1-Aminoethanol   Acetaldehyde

In 1970, Babior isolated small quantities of 5′-deoxyadenosine formed from adenosylcobalamin using enzymatically functioning solutions of the lyase. When L-2-aminopropanol (a more slowly reacting substrate) is deaminated by the lyase, the Co—C bond is rapidly cleaved and 5′-deoxyadenosine accumulates, accounting for up to 80% of the original adenosylcobalamin (Babior *et al.*, 1974a). When ethanolamine is added, the 5′-deoxyadenosine is converted back to adenosylcobalamin. This indicates that the formation of deoxyadenosine is reversible. Cleavage of the Co—C bond produces vitamin B_{12r} as another intermediate in the reaction. When the lyase–coenzyme complex is frozen in liquid nitrogen during catalysis, two peaks are observed in its e.s.r. spectrum (Babior *et al.*, 1972). One peak has been assigned to vitamin B_{12r}, and the other to an organic free radical. Transient formation of vitamin B_{12r} has also been detected by its light absorption with another dioldehydrase-like enzyme, glycerol-dehydrase (Cockle *et al.*, 1972), and with the lyase (Babior *et al.*, 1974a). The organic radical formed when 2-aminopropanol is added to the lyase-adenosylcobalamin complex has been identified (Babior *et al.*, 1974b) as the 2-aminopropanol-1-yl radical:

$$H_3C-CHNH_2-\overset{\cdot}{C}HOH .$$

The same homolytic mechanism operates with other adenosylcobalamin-dependent enzymes.

In summary, many adenosylcobalamin-mediated reactions proceed with a homolytic cleavage of the Co—C bond of the coenzyme and formation of 5′-deoxyadenosine as the intermediate hydrogen carrier. A partial mechanism incorporating these features is shown below. The first step is homolysis of the Co—C bond of the coenzyme. Binding of the substrate to the holoenzyme causes this homolysis, presumably by inducing a change in protein conformation. The second step is the abstraction of hydrogen by the 5′-deoxyadenosyl radical from the substrate. After rearrangement of substrate radical to product radical, the adenosylcobalamin is reformed. The mechanism of the migration of the other group (—X), including the mode of binding of the substrate to the cobalt of vitamin B_{12}, is not yet known, nor are the functional groups of the apoenzymes.

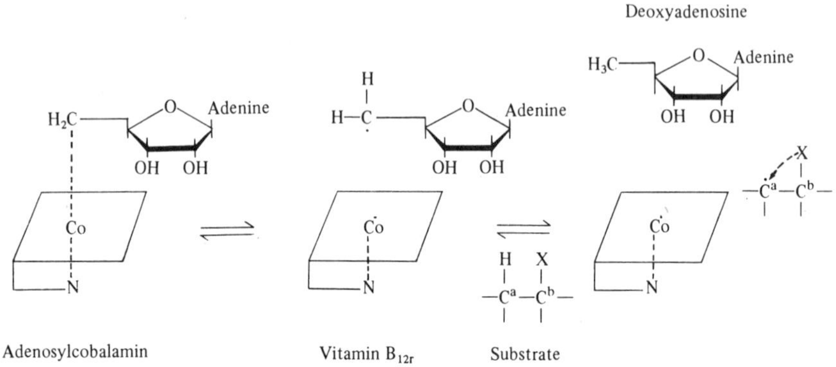

Adenosylcobalamin　　　　　Vitamin B_{12r}　　　Substrate

There are two health-related problems whose solutions depend on an understanding of vitamin B_{12}. The only metabolic reactions in mammals known to require vitamin B_{12} coenzymes are methionine biosynthesis (methylcobalamin) and the methylmalonyl CoA mutase-catalyzed reaction, which involves the isomerization of methylmalonyl CoA to succinyl CoA, this being a step in the catabolism of branch-chained amino acids (adenosylcobalamin). One of the symptoms often encountered in vitamin B_{12} deficiency states, such as pernicious anemia, is a lowering of the level of tetrahydrofolate in serum. This appears to be a result of interference with the activity of methyltetrahydrofolate homocysteine methyltransferase. Despite the presence of a vitamin B_{12}-dependent methyltransferase in mammalian liver, there is insufficient activity to supply the methionine needed for growth, making methionine an amino acid essential in the diet. The role of the mammalian methyltransferase is to release free tetrahydrofolate, needed for thymidylate and purine synthesis, from methyltetrahydrofolate. It cannot fulfil this role in the absence of vitamin B_{12}. The reason for neural degeneration in vitamin B_{12} deficiency, probably expressed through lack of adenosylcobalamin for the methylmalonyl CoA mutase, is not clear. Subsequent increases in the concentration of methylmalonyl CoA in neural tissue may cause biosynthesis of branch-chained fatty acids, or accumulation of a further metabolite, propionyl CoA, which could lead to odd-numbered fatty acids.

In bacteria, methylcobalamin is an intermediate in several reactions. These include the formation of methane by anaerobic microorganisms growing in river or lake sediments. These organisms can detoxify mercury from industrial waste by transferring a methyl carbanion from methylcobalamin to Hg^{2+}:

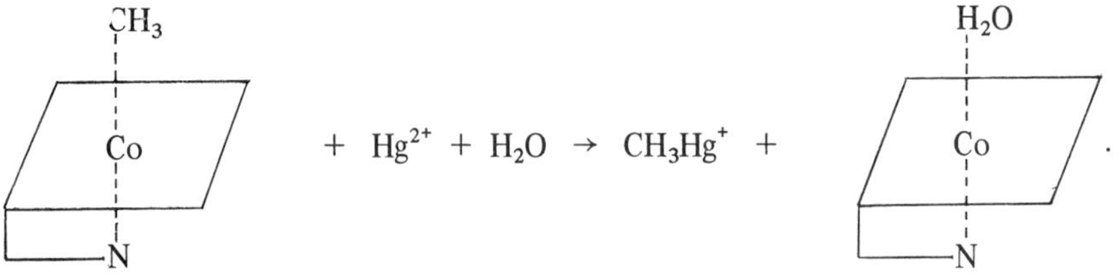

The methylmercury formed in this reaction is a potent neurotoxin. It is concentrated in the fish that ingest methylmercury, and causes a severe neurological disease (Minamata disease) in humans who consume mercury-contaminated fish. Methylcobalamin in anaerobic bacteria also can be the methyl donor in formation of poisonous methylarsenic compounds from arsenite. The reactions of methylcobalamin in the metabolism of potential pollutants are another active aspect of current research on vitamin B_{12} (Wood and Brown, 1972).

Hemoglobin (Fe)

Hemoproteins, the iron–porphyrin metalloproteins, participate both in redox reactions (e.g. cytochromes) and in oxygen transport. Hemoglobin (Hb) is the heme-containing dioxygen carrier protein in the red blood cells of vertebrates. It binds oxygen to the iron of the heme at the lungs, and delivers the oxygen to tissues. Hemoglobin is not an enzyme, but it shares many properties with enzymes. The hemes resemble coenzymes and the globin or protein portion corresponds to an apoenzyme. Hemoglobin which has full activity can be reconstituted when heme and globin are mixed properly. The O_2 ligand is like a substrate because its structure is altered while it is bound reversibly to the "active site" iron. No overall reaction is catalyzed by hemoglobin, though. Hemoglobin is composed of sub-units which undergo conformational changes in the presence of an allosteric effector (p. 404), 2,3-diphosphoglycerate or upon binding of O_2. Much more is known about the structure of hemoglobin and the alterations of its structure produced by ligand binding than for any subunit-containing enzyme.

Each of the hemes of deoxygenated hemoglobin (deoxyHb) is a five-co-ordinate complex which is prone to further substitution with O_2, CN^- or CO to form an octahedral complex. The ligands in the deoxy form are the four nitrogens of the porphyrin ring and the nitrogen of an imidazole group of a histidine residue:

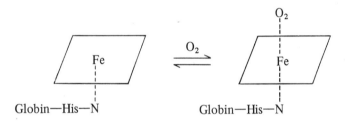

In the five-co-ordinate complexes, the iron atom is displaced out of the porphyrin ring toward the histidine ligand, but in the octahedral complexes the nitrogens and the iron are coplanar. In deoxyHb, the iron is in the high-spin ferrous state, but until recently the electronic configuration of the iron in the oxygenated form (oxyHb) was not known. The O_2-bearing hemoglobin was known to contain low-spin iron complexes, however. Either an $[Fe^{2+}-O_2]$ or an $[Fe^{3+}-O_2^-]$ structure could account for the diamagnetism of oxyHb.

Cobalt has been used as a probe in place of iron to distinguish between these two possible electronic structures. Basolo and his coworkers (Crumbliss and Basolo, 1970; Hoffman et al., 1970) succeeded in preparing a monomeric and reversible 1:1 oxygen model adduct. They crystallized and characterized the O_2 adduct of N,N'-ethylenebis(acetylacetoniminato)cobalt(II), or Co(acacen). The success of their synthesis depended on use of low temperatures and a non-aqueous solvent, and addition of a Lewis base (B = pyridine or dimethylformamide). The Co(acacen)–pyridine complex has an e.s.r. spectrum comparable to that of vitamin B_{12r}. E.s.r. studies with its O_2 adduct suggest that the unpaired electron is associated with the dioxygen molecule, not the cobalt, so the oxygen becomes a super-oxide anion and the cobalt diamagnetic Co^{3+} (as in vitamin B_{12a}). These experiments indicate that the most probable geometry of the Co(acacen)–Base–O_2 adduct is:

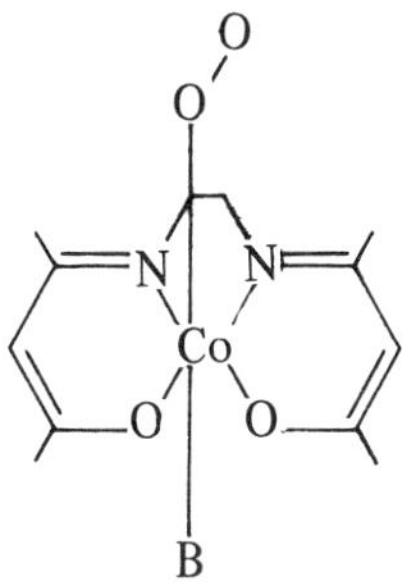

X-ray crystallographic studies of Fe^{2+}—O_2 model complexes (e.g. Collman *et al.*, 1974) have confirmed that dioxygen can co-ordinate with the metal "end-on," with a bent Fe—O—O bond. Model studies also suggest that the apoprotein of oxygen-carrier hemoproteins protects the heme from oxidation rather than assisting in the binding of the dioxygen.

The model compound work was extended to hemoglobin by Hoffman and Petering (1970), who formed the functional cobalt analogue of hemoglobin, coboglobin, by reconstitution from globin plus cobalt protoporphyrin-IX. The e.s.r. spectrum shows that deoxy-coboglobin has one unpaired electron, as expected, in a low-spin Co^{2+} five-co-ordinate complex. Upon oxygenation, the e.s.r. spectrum changes to one similar to that of Co(acacen)–Base–O_2. Thus, oxy-coboglobin is a protein containing Co^{3+}-superoxide complexes, and from this information the conclusion has been reached that oxyHb also contains (ferric iron)–(bound superoxide) diradical complexes. OxyHb is diamagnetic because of the interaction of the two unpaired spins on adjacent atoms. Investigators have provided additional evidence in support of the end-on binding of O_2 to Fe^{2+} or Co^{2+} in hemoglobin. For example, using $^{17}O_2$ and e.s.r. spectroscopy, Gupta *et al.* (1975) have established the asymmetric bonding of the oxygen molecule to the metal atom of coboglobin, with almost complete transfer of the unpaired electron from the Co^{2+} to the O_2.

Mammalian hemoglobins have a molecular weight of 64 500 dal-tons. A hemoglobin molecule contains two subunits of each of two readily distinguishable types, called α and β polypeptide chains. The α chains each have 141 amino acid residues and the β chains 146 residues. Each subunit bears one molecule of heme. The amino acid sequence of the α and β chains are homologous, and the tertiary structures of both subunit types are much alike. However, the α and β subunits are not exactly equivalent, neither in structure nor in reactivity.

Three-dimensional structures, based on sequences and 2.8 Å X-ray crystallographic data, have been proposed for both the oxyHb (Perutz *et al.*, 1968) and deoxyHb (Bolton and Perutz, 1970) from horse erythrocytes. Comparison of these two structures shows that significant structural changes occur when O_2 is bound to deoxyHb. Because oxyHb slowly autoxidizes to the ferric form, called methemoglobin (Fe^{3+}—Hb), the atomic model of oxyHb is based on the structure determined for the isomorphous methemoglobin. Perutz believes that small differences in structure exist between oxyHb and metHb, but the large changes due to binding of the O_2 ligands to deoxyHb are not obscured. The position of the O_2 molecule cannot be determined, though. Perhaps the structure of a liganded form of Hb will be determined (e.g. carbonmonoxyHb) for additional support of the oxyHb model. The four subunits of hemoglobin are arranged tetrahedrally, as shown in this diagram:

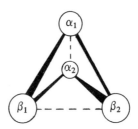

Upon dissociation, α—β dimers are formed. Each subunit of hemoglobin is made up of helical segments (designated by letters A to H), similar to those found in myoglobin, the monomeric O_2-binding hemoprotein in muscle. The heme groups of hemoglobin are located in nonpolar pockets of the globin chains. In deoxyHb, the ligand sites in the β subunits are blocked but in the α subunits the ligand sites are free. There are few bonds between like subunits, i.e. α_1 to α_2 or β_1 to β_2. Conversely, there are many non-covalent bonds, the majority being hydrophobic, between unlike chains. There is more

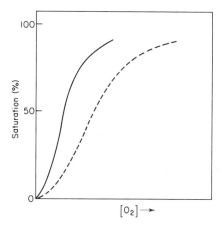

Fig. 6. Sigmoid curve for the binding of O_2 to hemoglobin (solid line). The effect of a slight decrease in pH is shown by the dashed line. Addition of 2,3-diphosphoglycerate would give results similar to the decrease in pH.

extensive contact between α_1 and β_1 or α_2 and β_2 than there is between α_1 and β_2 or α_2 and β_1. The $\alpha_1-\beta_1$ and $\alpha_2-\beta_2$ bonds undergo little change in the interconversion of deoxyHb and oxyHb, but there are large changes in the $\alpha_1-\beta_2$ and $\alpha_2-\beta_1$ contacts during oxygenation or deoxygenation. Having completed an elementary description of the structures of the deoxy and oxy forms of hemoglobin, I shall now describe the physiological role and a possible molecular mechanism for the interaction between the four subunits of hemoglobin.

The adaptation of hemoglobin to its physiological role is demonstrated by several of its thermodynamic properties. Hemoglobin avidly binds O_2 at the oxygen source, the lungs, but releases the ligand at the tissues which require the oxygen. In addition, its affinity for O_2 is lowered as the pH decreases (the Bohr effect), supplying more O_2 during times of lactic acid accumulation in muscle. The curve (Fig. 6) for the association of O_2 with hemoglobin (% saturation with ligand v. concentration of O_2) is S-shaped or sigmoid, not hyperbolic as in the binding curve for myoglobin. Adair (1925) suggested that the $Hb_4(O_2)_4$ molecule is formed in four stages, with a different equilibrium constant for each stage. As early as 1935, Pauling recognized that interaction between the heme groups could explain the shape of this curve. He estimated that binding of O_2 to deoxyHb is dependent on $[O_2]^n$ with n being about 2.6. Binding of O_2 to one heme group facilitates binding to the second heme, and so on, with the equilibrium constants increasing at each step. This behaviour is termed *positive co-operativity*. Co-operativity between the four subunits, expressed in the sigmoid binding curve, gives the increased tendency for remaining hemes to bind O_2 once one or two hemes are liganded, or for oxygenated hemoglobin to lose all four O_2 molecules under conditions where one O_2 is lost. This near all-or-none phenomenon has been interpreted (Perutz, 1970) by the biblical analogy in the parable of communication of knowledge:

> For the man who has will be given more, and the man who has not will forfeit even what he has (Mark, Chapter 4, verse 25).

2,3-Diphosphoglycerate (DPG) is present in high concentrations in red blood cells, and facilitates the unloading of O_2 from oxyHb (Benesch and Benesch, 1969). Hemoglobin freed from DPG by chromatography on Sephadex has a much greater affinity for O_2 than does hemoglobin from whole blood, which contains about 0.5 moles of DPG per mole of Hb. Arnone (1972) examined the crystal structure of the DPG–deoxyHb complex. He found that one molecule of DPG was bound to each hemoglobin molecule. Each of the two phosphate groups binds to a cationic site on a different β subunit. After the conformation changes which accompany the deoxyHb to oxyHb transition, the binding site for DPG no longer exists. Since the deoxy and oxy forms are in equilibrium, DPG shifts this equilibrium by stabilizing deoxyHb, the form which has the lower affinity for O_2. In this manner, diphosphoglycerate changes the affinity of hemoglobin for oxygen by binding at a site distinct from the O_2 binding site.

Based on the comparisons of the atomic models of deoxyHb, oxyHb and a modified hemoglobin held in the quaternary structure of oxyHb, Perutz (1970) proposed a stereochemical explanation of the co-operative binding of oxygen to hemoglobin. In considering this mechanism, remember that the binding affinity of deoxyHb is low compared to that of either isolated subunits or $Hb_4(O_2)_3$. In the atomic model of deoxyHb, only the hemes of the α chains are accessible to O_2. Perutz's basic supposition is that there is a step by step release of constraints on the unreactive quaternary structure, until, when both α hemes possess O_2 ligands, the quaternary structure clicks into the high affinity or oxyHb shape. In Monod's parlance

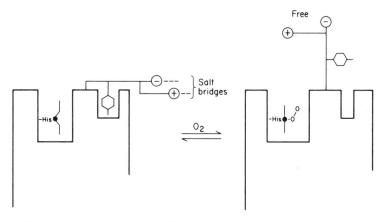

Fig. 7. Diagram showing how a small change in the shape of the iron complex of a hemoglobin subunit is amplified into a change in the tertiary structure of that subunit (Perutz, 1970). This mechanism is supported by structural studies of model heme compounds (Hoard and Scheidt, 1973). The heme becomes planar upon ligation, which pulls the porphyrin ring and the globin attached to it toward the proximal histidine ligand. This moves helix F (bearing the histidine) and squeezes the tyrosine next to the C-terminal residue from a pocket between the F and H helices. Expulsion of the penultimate tyrosine from its pocket leads to breakage of the salt bridges formed by the C-terminal amino acid.

(p. 421), this is a change from the T (tense, low-affinity, or un-reactive) state to the R (relaxed, high-affinity, or reactive) state.

The initiating step in the T to R conversion is envisaged as a change in the stereochemistry of the iron atoms in the α hemes, from high-spin five-co-ordination to low-spin six-co-ordination. Shrinking of the iron atoms and addition of the O_2 ligands moves the porphyrin ring and consequently affects the tertiary structure of the α poly-peptides. The proposed sequence of events in binding of one O_2 molecule is described in Fig. 7 and the overall changes are sum-marized in Table V. Eight salt bridges (two per C-terminal residue) of deoxyHb are absent in oxyHb. Each C-terminal residue is anchored by one (β) or two (α) salt bridges to neighbouring subunits in the T state. Movement of the α porphyrins produces conform-ational changes which break four salt bridges and release the α C-termini. It is thought that the salt bridges formed by the C-termini are the constraining crosslinks holding hemoglobin in the T or deoxy state. When the first four bonds are broken, the other four are weakened and the T to R transition can occur. Once the inter-subunit

Table V. Summary of the proposed steps in binding of oxygen to hemoglobin.

	DeoxyHb	OxyHb
1. Reactions of α subunits	α Hemes accessible to O_2	
	β Hemes inaccessible to O_2	$Hb_4(O_2)$ and $Hb_4(O_2)_2$ formed, leading to:
	High-spin iron, with larger atomic radius	Low-spin iron, with smaller atomic radius
	Fe-complex domed	Fe-complex planar
	His on helix F bound to iron	Iron moves toward plane of porphyrin ring, pulling porphyrin toward helix F
	Penultimate Tyr held by helix F	Penultimate Tyr expelled from its pocket
	Terminal carboxyls in salt linkages	C-termini freed (four salt linkages broken)
2. Quaternary structure change	*T state:* α_1 and α_2 deoxygenated	*R state:* α_1 and α_2 oxygenated
	Four salt bridges remain	No salt bridges remain
	DPG can bind, providing additional weak bonds	DPG cannot bind
		β Hemes become accessible by Fe, His, Tyr, and C-termini changes
		β_1 and β_2 oxygenated

constraints are removed, the β subunits are more flexible and their hemes become available for O_2 binding.

Perutz's postulated mechanism also accounts for the DPG and Bohr effects. DPG can bind to the T state in the $T \rightleftharpoons R$ equilibrium, and act as a constraining force. By preferentially binding to deoxyHb, it leads to the release of O_2. The Bohr effect is observed either as a release of protons upon oxygenation of hemes or as a decrease in O_2 affinity with a decrease in pH. Breakage of salt bridges of the C-termini releases protons. Protons can therefore influence the tertiary structure of the β subunits and the quaternary structure of the tetramer by opposing the rupture of the C-terminal salt bridges, and again stabilizing the low-affinity deoxy state of hemoglobin.

Elucidation of the crystal structures of the unliganded and fully liganded forms of hemoglobin has supplied information about the forces which stabilize these low- and high-affinity forms, respectively. Because the forms intermediate between Hb_4 and $Hb_4(O_2)_4$ are not available for crystallographic study, spectroscopic experiments must be performed to learn about the intermediate stages. Several experiments have shown that a single transition between low- and high-affinity forms can occur. Valency hybrids of hemoglobin, with one type of subunit in the ferric state and complexed with cyanide and the other type in the normal ferrous state, $(\alpha^{III}CN\beta^{II})_2$ or $(\alpha^{II}\beta^{III}CN)_2$, have been prepared. N.m.r. experiments by Ogawa and Shulman (1972) have shown that $(\alpha^{III}CN\beta^{II})_2$ has a spectrum similar to $(\alpha^{III}CN\beta^{II}O_2)_2$, but has a different spectrum when DPG is added. Ogawa and Shulman have attributed these different spectra to two quaternary forms of hemoglobin, with the same degree of ligation (i.e. when two subunits are ligated). The switch between these forms could be responsible for co-operativity. Using n.m.r. spectroscopy of normal and mutant hemoglobins, Ho et al. (1973) have found that the α hemes and the β hemes are structurally nonequivalent (i.e. have different spectra). Also, when hemoglobin is kept in the T form by the addition of high concentrations of DPG, O_2 is preferentially bound to the hemes of the α subunits. In agreement with Perutz's mechanism, these workers suggest that tertiary structures in the heme pockets affect the $\alpha_1-\beta_2$ subunit contacts and facilitate the T to R quaternary structural transition. Raftery and Huestis (1973) have monitored tertiary and quaternary structural changes as a function of amount of CO bound using a ^{19}F-n.m.r. probe. The ^{19}F-trifluoroacetonyl (TFA) probe is specifically bound to Cys-93 of the β chains, which is near His-92 (the heme ligand) and at the important $\alpha_1-\beta_2$ interface. The principal structural change detected by the TFA group is movement of the C-terminus from its bound to free state. Four tertiary structural forms have been detected in these experiments: Hb_4, $Hb_4(CO)_2$, $Hb_4(CO)_3$ and $Hb_4(CO)_4$. By comparison with the spectrum of the TFA-modified $(\alpha^{III}CN\beta^{II})_2$ hybrid, $Hb_4(CO)_2$ was identified as $(\alpha CO\beta)_2$. Raftery and Huestis describe the binding sequence as α, α, β and β. Binding of the third ligand produces large changes in the tertiary and quaternary structure, so that in $Hb_4(CO)_3$ the C-terminus is in the free state. ^{31}P-n.m.r. experiments, using ^{31}P-DPG and partially liganded hemoglobin solutions (Huestis and Raftery, 1975), indicate that in the presence of DPG the $T \rightleftharpoons R$ transition occurs in the three-liganded species.

To summarize, the accumulated results show that as ligands are bound to deoxyHb, both tertiary and quaternary structural changes take place, with a large quaternary transition occurring upon addition of either the second or third ligand. Although hemoglobin is not an enzyme and its subunit "active sites" are not equivalent, the process of its ligand binding gives insight into the binding of allosteric effectors, discussed in the next chapter. Because the properties of hemoglobin are not exactly consistent with either a simple con- certed (p. 420) or simple sequential (p. 424) allosteric mechanism, I have refrained from commenting on application of these theories to hemoglobin. The present results, extensive but far from complete, indicate that the co-operativity of O_2 binding and the allosteric effect of DPG can be explained in physical terms. The protein conformation changes associated with these phenomena occur at both the tertiary and quaternary structure levels.

Hamilton (1973) has summarized the functions of metal ions in enzyme reactions as

1. binding of substrates to enzymes,

2. activating the enzyme by binding to remote sites, and

3. acting as an acid catalyst to polarize bonds.

Other reactions, unique to redox metalloenzymes, are

4. electron transfer, and

5. complexing with O_2 or H_2O_2 so that it can be a more reactive metal-complexed species.

In this chapter, I have presented examples of each of these effects. I have also correlated the type of function to the strength of co- ordination bonds that a metal forms. The wide variety of naturally available metals, their versatile chemistry, and their ability to rapidly exchange ligands provides the basis for the use of specific metal ions in individual enzymically-catalyzed reactions.

REFERENCES

Abeles, R. H. (1971). *In* "The Enzymes" (P. D. Boyer, ed.), Third edition, Vol. 5, pp. 481-497. Academic Press, New York.

Abeles, R. H. and Zagalak, B. (1966). *J. biol. Chem.* **241**, 1245-1246.

Adair, G. S. (1925). *J. biol. Chem.* **63**, 529-545.

Arnone, A. (1972). *Nature, Lond.* **237**, 146-149.

Babior, B. M. (1970). *J. biol. Chem.* **245**, 6125-6133.

Babior, B. M. (1975). *Accts Chem. Res.* **8**, 376-384.

Babior, B. M., Moss, T. H. and Gould, D. C. (1972). *J. biol. Chem.* **247**, 4389-4392.

Babior, B. M., Carty, T. J. and Abeles, R. H. (1974a). *J. biol. Chem.* **249**, 1689-1695.

Babior, B. M., Moss, T. H., Orme-Johnson, W. H. and Beinert, H. (1974b). *J. biol. Chem.* **249**, 4537-4544.

Barker, H. A. (1972). *In* "The Enzymes" (P. D. Boyer, ed.), Third edition, Vol. 6, pp. 509-537. Academic Press, New York.

Barker, H. A., Weissbach, H. and Smyth, R. D. (1958). *Proc. natn. Acad. Sci. U.S.A.* **44**, 1093-1097.

Basolo, F. and Johnson, R. (1964). "Coordination Chemistry". W. A. Benjamin, New York.

Beem, K. M., Rich, W. E. and Rajagopalan, K. V. (1974). *J. biol. Chem.* **249**, 7298-7305.

Benesch, R. and Benesch, R. E. (1969). *Nature, Lond.* **221**, 618-622.

Bolton, W. and Perutz, M. (1970). *Nature, Lond.* **228**, 551-552.

Bradshaw, R. A., Ericsson, L. H., Walsh, K. A. and Neurath, H. (1969). *Proc. natn. Acad. Sci. U.S.A.* **63**, 1389-1394.

Bray, R. C. and Swann, J. C. (1972). *Structure and Bonding* **11**, 107-144.

Breslow, R. and Wernick, D. (1976). *J. Am. Chem. Soc.* **98**, 259-261.

Brodie, J. D. and Poe, M. (1971). *J. Am. Chem. Soc.* **10**, 914-922.

Burke, G. T., Mangum, J. H. and Brodie, J. D. (1971). *Biochemistry* **10**, 3079-3085.

Cockle, S. A., Hill, H. A. O., Williams, R. J. P., Davies, S. P. and Foster, M. A. (1972). *J. Am. Chem. Soc.* **94**, 275-277.

Cohn, M. (1963). *Biochemistry* **2**, 623-629.

Cohn, M. (1970). *Q. Rev. Biophys.* **3**, 61-89.

Cohn, M., Leigh, J. S., Jr. and Reed, G. H. (1971). *Cold Spring Harb. Symp. quant. Biol.* **36**, 533-540.

Coleman, J. E. (1971). *Prog. bioorg. Chem.* **1**, 159-344.

Collman, J. P., Gagne, R. R., Reed, C. A., Robinson, W. T. and Rodley, G. A. (1974). *Proc. natn. Acad. Sci. U.S.A.* **71**, 1326-1329.

Crumbliss, A. L. and Basolo, F. (1970). *J. Am. Chem. Soc.* **92**, 55-60.

Dalton, H. and Mortenson, L. E. (1972). *Bact. Rev.* **36**, 231-260.

Dalton, H., Morris, J. A., Ward, M. A. and Mortenson, L. E. (1971). *Biochemistry* **10**, 2066-2072.

Dinovo, E. C. and Boyer, P. D. (1971). *J. biol. Chem.* **246**, 4586-4593.

Dyrkacz, G. R., Libby, R. D. and Hamilton, G. A. (1976). *J. Am. Chem. Soc.* **98**, 626-628.

Essenberg, M. K., Frey, P. A. and Abeles, R. H. (1971). *Biochemistry* **93**, 1242-1251.

Evans, H. J. and Sorger, G. J. (1966). *A. Rev. Pl. Physiol.* **17**, 47-76.

Frey, P. A., Essenberg, M. K. and Abeles, R. H. (1967). *J. biol. Chem.* **242**, 5369-5377.

Fridovich, I. (1972). *Accts Chem. Res.* **5**, 321-326.

Fridovich, I. (1975). *A. Rev. Biochem.* **44**, 147-159.

Fujii, K. and Huennekens, F. M. (1974). *J. Biol. Chem.* **249**, 6745-6753.

Guest, J. R., Friedman, S., Woods, D. D. and Smith, E. L. (1962). *Nature, Lond.* **195**, 340-342.

Gupta, R. K., Mildvan, A. S., Yonetani, T. and Srivastava, T. S. (1975). *Biochem. biophys. Res. Commun.* **67**, 1005-1012.

Hamilton, G. A. (1973). *In* "Catalysis. Progress in Research." (Basolo, F. and Burwell, R. L., Jr., eds), pp. 47-50. Plenum Press, London.

Hardy, R. W. F., Burns, R. C. and Parshall, G. W. (1971). *In* "Bioinorganic Chemistry", Advances in Chemistry, Vol. 100, pp. 219-247. American Chemical Society, Washington.

Harmony, J. A. K., Shaffer, P. J. and Himes, R. H. (1974). *J. biol. Chem.* **249**, 394-401.

Hartsuck, J. A. and Lipscomb, W. N. (1971). *In* "The Enzymes" (P. D. Boyer, ed.), Third edition, Vol. 3, pp. 1-56. Academic Press, New York.

Hatch, F. T., Larrabee, A. R., Cathou, R. E. and Buchanan, J. M. (1961). *J. biol. Chem.* **236**, 1095-1101.

Ho, C., Lindstrom, T. R., Baldassare, J. J. and Breen, J. J. (1973). *Ann. N.Y. Acad. Sci.* **222**, 21-39.

Hoard, J. L. and Scheidt, W. R. (1973). *Proc. natn. Acad. Sci. U.S.A.* **70**, 3919-3922. (cf. Correction in **71** (1974), 1578.)

Hoffman, B. M. and Petering, D. H. (1970). *Proc. natn. Acad. Sci. U.S.A.* **67**, 637-643.

Hoffman, B. M., Diemente, D. L. and Basolo, F. (1970). *J. Am. Chem. Soc.* **92**, 61-65.

Huang, T. C., Zumft, W. G. and Mortenson, L. E. (1973). *J. Bact.* **113**, 884-890.

Huennekens, F. M. (1968). *In* "Biological Oxidations" (T. P. Singer, ed.), pp. 439-513. Interscience, New York.

Huestis, W. H. and Raftery, M. A. (1975). *Biochemistry* **14**, 1886-1892.

Kachmar, J. F. and Boyer, P. D. (1953). *J. biol. Chem.* **200**, 669-682.

Kayne, F. J. (1973). *In* "The Enzymes" (P. D. Boyer, ed.), Third edition, Vol. 8, pp. 353-382. Academic Press, New York.

Klug-Roth, D., Fridovich, I. and Rabani, J. (1973). *J. Am. Chem. Soc.* **95**, 2786-2790.

Knowles, P. F., Gibson, J. F., Pick, F. M. and Bray, R. C. (1969). *Biochem. J.* **111**, 53-58.

Lenhert, P. G. and Hodgkin, D. C. (1961). *Nature, Lond.* **192**, 937-938.

Lipscomb, W. N. (1974). *Tetrahedron* **30**, 1725-1732.

McCord, J. M. and Fridovich, I. (1969). *Fedn Proc.* **28**, 346.

Mildvan, A. S. (1970). *In* "The Enzymes" (P. D. Boyer, ed.), Third edition, Vol. 2, pp. 445-536. Academic Press, New York.

Morrison, J. F. and Cleland, W. W. (1966). *J. biol. Chem.* **241**, 673-683.

Morrison, J. F. and James, E. (1965). *Biochem. J.* **97**, 37-52.

Mortenson, L. E., Zumft, W. G. and Palmer, G. (1973). *Biochim. biophys. Acta.* **292**, 422-435.

Nakos, G. and Mortenson, L. (1971). *Biochemistry* **10**, 455-458.

Nowak, T. (1973). *J. biol. Chem.* **248**, 7191-7196.

Nowak, T. (1976). *J. biol. Chem.* **251**, 73-78.

Nowak, T. and Mildvan, A. S. (1972). *Biochemistry* **11**, 2819-2828.

Nowak, T., Mildvan, A. S. and Kenyon, G. L. (1973). *Biochemistry* **12**, 1690-1701.

Ogawa, S. and Shulman, R. G. (1972). *J. molec. Biol.* **70**, 315-336.

Pauling, L. (1935). *Proc. natn. Acad. Sci. U.S.A.* **21**, 186-191.

Perutz, M. F. (1970). *Nature, Lond.* **228**, 726-739.

Perutz, M. F., Muirhead, H., Cox, J. M. and Goaman, L. C. G. (1968). *Nature, Lond.* **219**, 131-139.

Quiocho, F. A. and Lipscomb, W. N. (1971). *Adv. Protein Chem.* **25**, 1-78.

Raftery, M. A. and Huestis, W. H. (1973). *Ann. N.Y. Acad. Sci.* **222**, 40-55.

Reuben, J. and Kayne, F. J. (1971). *J. biol. Chem.* **246**, 6227-6234.

Richardson, J. S., Thomas, K. A., Rubin, B. H. and Richardson, D. C. (1975). *Proc. natn. Acad. Sci. U.S.A.* **72**, 1349-1353.

Rotilio, G., Bray, R. C. and Fielden, E. M. (1972a). *Biochim. biophys. Acta.* **268**, 605-609.

Rotilio, G., Calabrese, L., Bossa, F., Barra, D., Agrò, A. F. and Mondovi, B. (1972b). *Biochemistry* **11**, 2182-2187.

Smith, E. L. and Hanson, H. T. (1949). *J. biol. Chem.* **179**, 803-813.

Stadtman, T. C. (1972). *In* "The Enzymes" (P. D. Boyer, ed.), Third edition, Vol. 6, pp. 539-563. Academic Press, New York.

Steifel, E. I. (1973). *Proc. natn. Acad. Sci. U.S.A.* **70**, 988-992.

Steinman, H. M. and Hill, R. L. (1973). *Proc. natn. Acad. Sci. U.S.A.* **70**, 3725-3729.

Steinman, H. M., Vishweshwar, R. N., Abernethy, J. L. and Hill, R. L. (1974). *J. biol. Chem.* **249**, 7326-7338.

Suelter, C. H. (1970). *Science, N.Y.* **168**, 789-795.

Takayama, S., Hatch, F. T. and Buchanan, J. M. (1961). *J. biol. Chem.* **236**, 1102-1108.

Taylor, R. T. and Weissbach, H. (1973). *In* "The Enzymes" (P. D. Boyer, ed.), Third edition, Vol. 9, pp. 121-165. Academic Press, New York.

Thomas, J. L. (1975). *J. Am. Chem. Soc.* **97**, 5943-5944.

Vallee, B. L. (1964). *Fedn Proc.* **23**, 8-17.

Vallee, B. L. and Neurath, H. (1954). *J. Am. Chem. Soc.* **76**, 5006-5007.

Vallee, B. L., Stein, E. A., Summerwell, W. N. and Fischer, E. H. (1959). *J. biol. Chem.* **234**, 2901-2905.

Walker, G. A. and Mortenson, L. E. (1974). *Biochemistry* 13, 2382-2388.

Walker, Glenn A., Murphy, S. and Huennekens, F. M. (1969). *Archs Biochem. Biophys.* 134, 95-102.

Watts, D. C. (1973). *In* "The Enzymes" (P. D. Boyer, ed.), Third edition, Vol. 8, pp. 383-455. Academic Press, New York.

Weisiger, R. A. and Fridovich, I. (1973). *J. biol. Chem.* 248, 4793-4796.

Williams, R. J. P. (1970). *Q. Rev., Lond.* 24, 331-365.

Wold, F. (1971). *In* "The Enzymes" (P. D. Boyer, ed.), Third edition, Vol. 5, pp. 499-538. Academic Press, New York.

Wood, J. M. and Brown, D. G. (1972). *Structure and Bonding* 11, 47-105.

Wu, T. W. and Scrimgeour, K. G. (1973). *Can. J. Biochem.* 51, 1391-1398.

Zumft, W. G. and Mortenson, L. E. (1975). *Biochim. biophys. Acta* 416, 1-52.

Zumft, W. G., Palmer, G. and Mortenson, L. E. (1973). *Biochim. biophys. Acta* 292, 413-421.

Zumft, W. G., Mortenson, L. E. and Palmer, G. (1974). *Eur. J. Biochem.* 46, 525-535.

Part III

Control

12 Quaternary Structure and Allosteric Control

Enzymes possess both fantastic catalytic capabilities and rigid chemical specificity for their substrates. In addition, some enzymes are not just passive catalysts but are regulated by chemical signals originating in or entering the cell. By this regulation, they provide co-ordination and control of the many interlocking pathways of metabolism. In this section of the book, I shall discuss how enzymes are controlled by their metabolic signals, and how their control leads to biological regulation.

The study of the control of enzyme catalysis is one of the newest aspects of enzymology. It started with the recognition of *feedback inhibition* in several biosynthetic pathways. Negative feedback control of a biosynthetic pathway is observed when an accumulation of the end-product of the pathway causes a decrease in the rate of its own biosynthesis. A few examples of this pattern of regulation in cells or in cell extracts had been reported in the 1940s and early 1950s, but the beginning of the molecular understanding of feedback inhibition was the realization by Umbarger (1956) and Yates and Pardee (1956) that the end-products were inhibitors of the enzymes catalyzing the first step of individual biosynthetic pathways. Umbarger reported that in extracts of *Escherichia coli* L-isoleucine is a potent and specific competitive inhibitor of the deamination of L-threonine, the first step in formation of isoleucine. Inhibition of this irreversible deamination allows isoleucine biosynthesis to proceed only when its cellular concentration is reduced to a low level. Isoleucine not only acts as a feedback inhibitor but also represses the synthesis of threonine deaminase (Umbarger and Brown, 1958). Both of these processes, feedback inhibition (rapid control) and repression of protein biosynthesis (slower control), are important in the economy of cellular energy in micro-organisms. Yates and Pardee, independently but simultaneously, reached the same conclusions about the control of another biosynthetic pathway, that of pyrimidines. They observed that low levels of uracil or cytosine blocked the first step in pyrimidine biosynthesis *in vivo* in *E. coli*. *In vitro* experiments showed that pyrimidine nucleotides are the actual inhibitors of the enzyme aspartate transcarbamylase. Yates and Pardee commented that inhibition of the first step of a biosynthetic pathway is a favourable response; its blockage prevents build-up of unneeded intermediates. Research on aspartate transcarbamylase progressed until, in 1962, Gerhart and Pardee proposed that there was a *second site* on the enzyme capable of binding the inhibitor metabolite. They suggested that a conformational change caused by binding of the inhibitor lowered the binding of the substrate to the active site. The evidence upon which they based the conclusion that the regulatory and catalytic sites are physically distinct is presented in the next chapter. These data provided the basis of the formulation of allosterism by Monod and his colleagues the following year. Many examples of the regulation of enzymic activity have been recognized since 1956. The large number of regulatory enzymes described in the review papers of Stadtman (1966, 1970) and Sanwal *et al.* (1971) and the book by Cohen (1968) prove the importance of allosteric modulation of metabolism.

Interaction of most modulators (signal compounds) with the regulatory enzymes is via non-covalent bond formation, but covalent binding also can occur. Modulator inhibition or activation normally proceeds with the binding of a low-molecular weight ligand to a ligand-specific regulatory site of the enzyme. This binding is a rapid and reversible process, and involves the formation of non-covalent bonds between the modulator ligand and the enzyme, much like the formation of an enzyme–substrate complex. Proteins can sometimes act as modulators of enzyme activity by forming non-covalent, or even covalent, complexes with enzymes. Several examples of protein-protein modulation of catalytic activity will be described in the following chapters. Covalent binding of small molecules also can regulate the activity of some enzymes. Covalent modifications usually require the cleavage of ATP, with transfer of either a phosphoryl group (with release of ADP) or of an adenylyl group (with release of PP_i) to the enzyme (Segal, 1973). Enzymes that can exist in active and inactive forms differing by the presence or absence of the covalent modifying group may also be subject to additional regulation by non-covalent ligand binding. Covalent modification reactions are reversible, but interconversion of the active and inactive forms requires a pair of enzymes (e.g. a protein kinase and a protein phosphatase). The enzymes that catalyze the modification reactions may in turn be under modulator control, so it is not difficult to see how very intricate the control mechanisms of some metabolic pathways must be.

Reversible binding of a small ligand to the control or allosteric site of an enzyme is presumed to induce a change in the conformation of the enzyme. A regulatory enzyme has two (or possibly more) conformations, each with a differing catalytic activity. The equilibrium between these conformations is governed by the law of mass action, i.e. by the binding constants for and the concentrations of the various inhibitors and activators. Covalent modification and protein modulation of enzyme activity probably also express their effects on catalytic rate by altering the structures of active sites of regulatory enzymes. There is not yet any structural information on the nature of ligand-induced conformational transitions of control enzymes, because of the complicated structures of these enzymes. The only protein–ligand system which has been examined in three-dimensional detail is the hemoglobin–O_2–diphosphoglycerate system (p. 390).

The final section of this book, on control of enzymic reactions, starts with a description of the inter-relationship of quaternary structure and regulation, and a presentation of the terminology and current theories of allosterism. The next chapter describes the properties of some well-established regulatory enzymes. Following these examples, there is a chapter on the organization of enzymes either by binding to membranes or to each other. The book concludes with chapters on the use of drugs as artificial enzyme-control agents and some extremely complex integrated control systems.

ALLOSTERIC CONTROL

In Chapter 2, while introducing quaternary structure of proteins, I purposely left the term "allosteric enzyme" undefined. Many enzymologists wish that this phrase had never been coined, as it has been overused by indiscriminate application to phenomena to which it does not apply. I agree with Stadtman (1970), though, that proper usage of the word "allosteric" should be retained, because the allosteric theory is a fundamental concept of enzyme regulation.

In 1963, Monod and his colleagues formulated a series of generalizations concerning the properties of regulatory enzymes. They themselves claimed not to be proposing a new theory, but their paper is usually cited as the first statement of the allosteric concept of enzyme control. As I have commented above, their proposals are based on the experimental findings of Umbarger and Pardee. Sufficient examples of feedback inhibition and ligand activation were available to Monod, Changeux and Jacob for them to be able to realize that the substrates and the modulating ligands were chemically and structurally unrelated; hence the term "*allo*steric," the antonym of "*iso*steric," occupying the same space. They applied this adjective to the second site at which the control ligand binds, and thus defined the *allosteric site* as being distinct from the active site. The ligand bound at this site they called the *allosteric effector*. The effector may be either an inhibitor or an activator. The effector does not participate in the reaction, and can be recovered unchanged after completion of the reaction. Binding of an allosteric effector to an enzyme causes a reversible alteration in the conformation of the protein—the *allosteric transition*—which changes the kinetic properties of the enzyme. I shall apply the term allosteric only to those enzymes that have been demonstrated to have second sites for ligand binding. This eliminates such enzymes as glyceraldehyde-3-phosphate dehydrogenase, which provides a great deal of information about the binding of substrates but has neither distinct allosteric sites nor a vital control function *in vivo*. Regulatory proteins other than enzymes may also be classed as allosteric, if their effector-binding and functional sites are spatially distinct. The binding of diphosphoglycerate to deoxyhemoglobin is an elementary example of non-enzymic allosterism. Chemoreceptors (Chapter 16) probably are or contain non-enzymic allosteric proteins.

Monod *et al.* (1963) summarized several properties of allosteric enzymes, and suggested that future work might prove these properties to be common to all regulatory enzymes. They are characteristic features, but, as data become available on more allosteric enzymes, more exceptions are found. One common feature of regulatory enzymes is the lability of their sensitivity to metabolic inhibitors. This loss of sensitivity to the inhibitory effector without loss of catalytic activity is called *desensitization*. The demonstration of an allosteric site by this type of selective denaturation often is the most definitive obtainable evidence of its existence. Chemical (e.g. mercurial reagents, urea or limited proteolysis) and physical (e.g. heating or ageing) treatments have been used to produce desensitization. Monod stated that a likely explanation of desensitization is the uncoupling of the interaction between the allosteric and active sites, not simply destruction of the effector-binding site. Another general property observed by Monod *et al.* is co-operativity of ligand binding. I have already described positive co-operativity in the binding of sequential O_2 molecules to hemoglobin (p. 390). Sigmoid curves, either of amount of ligand bound v. [ligand] or of v_i v. [S], are characteristic of positive co-operativity. However, sigmoid kinetic curves are not seen with all regulatory enzymes, and under some conditions non-regulatory single-chain enzymes can exhibit sigmoid kinetics. Co-operativity of ligand binding would be expected only with enzymes containing more than one subunit, because of the need for interaction between identical ligand-binding sites. In fact, only two allosteric enzymes, ribonucleoside triphosphate reductase (Panagou *et al.*, 1972) and pyruvate-UDP-*N*-acetylglucosamine transferase (Zemell and Anwar, 1975), are known to have only one polypeptide chain each.

The almost universal presence of quaternary structure in allosteric enzymes and the frequent discovery of their co-operative binding behaviour implicate subunit interactions in enzymic regulation. Therefore, it is necessary to study the occurrence and nature of quaternary structure. We must always keep in mind, though, that many more enzymes are known to have quaternary structure than are allosteric.

THE PHYSICAL BASIS OF QUATERNARY STRUCTURE

Klotz *et al.* (1975) have compiled a list of the subunit compositions of 300 proteins having quaternary structure. Their table lists only proteins having subunits held together by non-covalent bonds. Table I

Table I. The quaternary structure of enzymes, compiled from the assembled data of Klotz *et al.* (1975, their Table II). Column A lists those enzymes in which the subunits are all of the same molecular weight. Column B lists enzymes in which the subunits are of differing molecular weight (i.e. there are at least two types of protomers).

Number of Subunits	Number of Enzymes	
	Column A	Column B
2	82	6
3	3	0
4	74	13
5	0	1
6	21	0
8	13	0
10	1	2
12	5	1
14	0	1
16	2	1

is the data on subunits of the enzymes in their list. The enzymes are divided into two classes: those having seemingly identical protomeric units (column A), and those having protomeric units of more than one type (column B). Most of the oligomeric enzymes are either dimers or tetramers, and have subunits of identical molecular weight (and probably structure). Examination of data of this type has led to the conclusion that the quaternary structure in almost all proteins is arranged in definite geometric patterns. This is because ordered structures, like crystals, usually are more stable than disordered structures. The common occurrence of spatial symmetry in multimeric enzymes has been verified by electron microscopy and X-ray crystallography. The most likely geometry for a protein is that having the most areas of contact between subunits. For example, the tetrahedral form of a tetramer should be more stable than the square planar form because it has six rather than four inter-subunit contacts. Two theories have been proposed to explain ligand-induced allosteric transitions. As we shall see, the main difference between these theories is in the extent to which symmetry is retained during the steps of the transition.

An oligomeric protein is in equilibrium with smaller polymeric species and the protomers. A number of different reagents can affect this association–dissociation equilibrium (Klotz *et al.*, 1975). Equilibrium can be shifted towards either the oligomer or the smaller forms of some enzymes by low concentrations of specific ligands. Dissociation can be brought about by higher concentrations of nonspecific small molecules such as salts, sulphydryl reagents, or denaturants. Usually, lower concentrations of denaturants are needed to effect dissociation than to cause denaturation of the globular protomers. Subunit composition is determined by measuring the molecular weights of both the oligomeric species of the enzyme and the protomeric species. A method using gel electrophoresis was described in Chapter 3 (p. 71). With a few enzymes, dissociation can occur just by dilution, so that their apparent molecular weights measured by sedimentation are a function of their concentrations.

The non-covalent forces which form the inter-subunit links in oligomeric enzymes are in all probability similar to those which create the tertiary structure of globular proteins. In general, the number of such bonds forming a contact area between any two subunits must be fewer than the non-covalent bonds in a single-chain protein. One reason for this is that contact between two globulins is severely limited by their roughly spherical shape. Evidence for the lack of extensive inter-subunit contact is the ease of dissociation of oligomers compared to denaturation of protomers. X-ray studies are starting to provide information about subunit interactions. Insulin normally crystallizes as a hexamer containing two zinc atoms. This hexamer may play a physiological role in hormone storage in the pancreas. Analysis of the atomic contacts in the structure of the hexamer (Blundell *et al.*, 1972) has suggested the types of bonds involved in the sequential monomer–dimer–tetramer–hexamer association. In the monomer, most nonpolar residues are exposed to water. The more stable dimer is held together largely by contacts between side chains of nonpolar residues, but there is also a series of hydrogen bonds forming a β-pleated sheet between the monomeric units. Association of two dimers requires formation of nonpolar interactions, and possibly some hydrogen bonds. Finally, after the progressive shielding of side chains of nonpolar residues, the hexamer is formed. The hexamer has a polar surface. These workers concluded that the nonpolar groups exposed in the subunits are used in formation of the final quaternary structure. Because of the small size of insulin, more inter-subunit contacts are possible than in larger oligomeric proteins. We have already seen that most of the bonds between the subunits of hemoglobin are hydrophobic, but that there are some important electrostatic bonds (p. 388). Concanavalin A and the few dehydrogenases that have been examined by X-ray crystallography retain their quaternary structure by formation of appreciable interchain hydrogen bonds in β-structures. Chothia and Janin (1975), after analyzing the protein–protein interactions in the published structures of several known oligomers, have reached the conclusions that hydrophobicity is the major factor in stabilizing association, and that polar interactions contribute little to stability but are necessary for recognition between protomers. Since only three structures were examined, their conclusions—although quite reasonable—must be regarded as tentative, not general.

Binding of ligands to allosteric enzymes causes changes in both the tertiary and quaternary structures. Because these catalytically profound alterations involve small free energy changes (approximately several kcal per mole of subunit), they are probably initiated by rupture and reforming of only a few non-covalent bonds. X-ray crystallographic and n.m.r. examination of allosteric enzymes will be necessary for the elucidation of the types of bonds which stabilize the conformations of different catalytic activity.

An unusual property of a few multimeric enzymes is *cold lability*. These enzymes are inactivated when they are cooled from room temperature to $0–5°$ C. Cold lability is caused by dissociation of the enzyme into inactive protomeric units. For example, pyruvate carboxylase from chicken liver undergoes a slow dissociation of its active tetramer to monomers when it is cooled. Under appropriate warming conditions, both its activity and its tetrameric structure are regained (Irias *et al.*, 1969). An inactive tetramer appears to be an intermediate in the association of the protomer, since the reactivation is a first-order reaction. Hydrophobic bond formation is endothermic at low temperatures, and the strength of hydrophobic bonds increases with temperature (Scheraga *et al.*, 1962). Due to the fact that hydrophobic bonds are the only protein-stabilizing forces to become weaker with cooling, enzymes which exhibit cold sensitivity probably have their protomers held together by hydrophobic interactions. Cold lability might also be caused by formation at lowered temperatures of a conformation of the protomer which does not favour association. Lack of cold lability by most multimeric enzymes can be explained either by retention of catalytic activity by their monomers, or by inter-subunit binding by a significant number of bonds that are not hydrophobic.

An obvious but as yet unanswered question is: why do so many enzymes possess quaternary structure? There must be advantages to subunit structure, but what are they? For enzymes having more than one type of protomer, the need for subunit structure is clear. For example, the regulatory site and the catalytic site may reside on different protomer molecules, as in aspartate transcarbamylase and cyclic AMP-dependent protein kinase. Both types of subunits are essential for the complete function of these enzymes. Some enzymes exist as large complexes, with different protomers containing active sites for different chemical reactions. Their quaternary structure seems to be needed for subcellular organization of the metabolic reactions. However, I have already said that most oligomeric enzymes contain identical subunits and are not allosteric enzymes. Probably certain conformations of enzymes are only possible if their quaternary structure is intact. Those enzymes which are cold labile exemplify another type of enzyme which requires intact quaternary structure, because their protomers have no enzymic activity. Formyltetrahydrofolate synthetase must be in the tetrameric form before it can bind the substrate tetrahydrofolate (cf. p. 248). Quaternary structure may also confer stability to enzymes, as observed in the experiments with aldolase described below. Fersht (1975) has proposed that there can be a catalytic advantage to oligomeric structure. In an enzyme having two (or more) identical binding sites, binding of products to one site might cause a conformational change which accelerates the reaction of substrates at another site. Thus, there are several possible advantages of quaternary structure for enzymes, but it may be many years before the widespread occurrence of non-regulated oligomeric enzymes is fully understood.

Some indirect techniques have been used for the study of the activity of different polymeric forms of enzymes. Many enzymes undergo ligand-mediated association or dissociation, but the state of aggregation may not have a physiological regulatory role. Few of these enzymes have been tested directly for activity of protomers and oligomers, as this is a difficult task. Duncan *et al.* (1972) reported that the dimeric enzyme deoxycytidylate deaminase has second-order kinetics with respect to enzyme concentration. Assuming that the dimeric form was needed for activity, they tested the effect of the activator dCTP on the aggregation state of the deaminase by gel filtration. In the presence of dCTP, the association–dissociation equilibrium was shifted far towards the dimer species. Some oligomeric enzymes, called *hysteretic enzymes* by Frieden (1970), do not react instantaneously to rapid changes in ligand concentrations. Instead, they show lags up to the one minute range in progress curves (ΔP v. time). These lags can be attributed to slow changes in conformation (either tertiary or quaternary structure) or to slow displacement of other ligands. Many of the hysteretic enzymes are regulatory oligomers, so the time-dependent changes in kinetic properties could be caused by slow allosteric transitions.

Direct testing of the activity of different polymeric forms of oligomeric enzymes is technically difficult. Only a few methods are available. Fractions having different molecular weights, obtained either by gel filtration or by density gradient centrifugation, can be tested for activity but there is no assurance that the quaternary structure of the enzyme in the fractions has not changed upon mixing with assay reagents. Subunits can be covalently linked and then tested for activity. Quinonoid dihydropterin reductase (p. 266) retains full activity when restricted in its dimeric form by treatment with dimethylsuberimidate (Cheema *et al.*, 1973). This result tells nothing about the enzymic activity of the monomer, though. Chan has devised a method for comparing the activities of monomeric and oligomeric forms. He has covalently bound the tetrameric enzyme aldolase to Sepharose beads, at a concentration of enzyme that allows only one subunit to be attached to the matrix (Chan and Mawer, 1972). The matrix-bound aldolase is converted by mild denaturation to matrix-bound monomer and free aldolase subunits that can be removed. Both the monomeric and tetrameric forms of bound aldolase have the same catalytic properties. However, the matrix-bound subunit is less stable than the matrix-bound tetramer. The most conclusive method for testing the aggregation state of catalytically active forms of enzymes is sedimentation through a solution of substrate (Kemper and Everse, 1973). This method can only be applied to enzymes having assays with suitable changes in absorbance, so that the extent of reaction can be measured in an analytical ultracentrifuge. Using this method, Taylor *et al.* (1972) found that the predominant active form of pyruvate carboxylase is the tetramer for the carboxylase from yeast and liver, but the dimer for the carboxylase from bacteria. Haberland *et al.* (1972) also used sedimentation through substrates to demonstrate that all three forms—the protomer, the dimer and the tetramer—of phosphoenolpyruvate carboxytransphosphorylase are catalytically active. Sedimentation through an assay solution is the only test of activity of different quaternary structures that can be used on unmodified enzymes, but its use to date has been quite restricted.

OLIGOMERIC ISOENZYMES

Enzymes, even those from one organism or tissue, are not always a single chemical entity. Multiple forms of enzymes include genetic variants, catalytically similar enzymes in different cellular compartments, modified forms of enzymes and isoenzymes. *Isoenzymes,* also called isozymes, are different proteins from one biological species that catalyze the same chemical reaction (Wilkinson, 1970). They are detected by separation of active fractions, usually by either electrophoresis or column chromatography. Isoenzymes have different primary structures, so must be formed under the direction of different genes. Not all isoenzymes are oligomeric proteins, but many are. Oligomeric isoenzymes, even though they are chemically distinct, can be similar enough in the structure of their subunits so that subunits of the parent types can form hybrid quaternary structures. These isoenzymes and their hybrids are of both academic and medical interest.

Lactate dehydrogenase (LDH; p. 214) was the first enzyme in which two isoenzyme forms were discovered (reviewed in Chapter 6 of Wilkinson, 1970, and by Everse and Kaplan, 1973). Crystalline LDH from heart muscle was found to have one major and one minor protein fraction upon electrophoresis, with both fractions having the same molecular weight (Meister, 1950). In 1952, Neilands showed that both these fractions had enzymic activity. Subsequently, the test used for detection of isoenzymes of LDH has been solid-support electrophoresis of tissue extracts, followed by staining of the electropherogram for LDH activity. Up to five protein fractions having lactate dehydrogenase activity could be found in most tissues (Wieland and Pfleiderer, 1957). The lactate dehydrogenases of animal origin have a molecular weight of about 140 000 and consist of four protomers. Two types of LDH having four identical subunits have been isolated. They have been called the heart (*H*-type) and the muscle (*M*-type) isoenzymes because of their purification as major LDH components from these two tissues. The *H*- and *M*-forms of LDH have similar molecular weights, but different physical and kinetic properties. The occurrence of five isoenzymic forms in animal tissues arises from hybridization of subunits from the *H* and *M* parent isoenzyme types. If the abbreviations H_4 and M_4 are used to represent the parent tetramers, then the hybrids are H_3M, H_2M_2 and HM_3. The relative amounts of the five forms, the two parent types and the three hybrids, vary in different tissues from the same species. The *H*-type of LDH predominates in tissues having mainly aerobic metabolism, and the *M*-type in tissues relying on anaerobic metabolism.

The non-covalent bonds which maintain the quaternary structure of the LDH tetramers are not yet known (Everse and Kaplan, 1973). From the data available, it appears that tetramer formation involves both hydrophobic and electrostatic bonds. Chilson *et al.* (1965) showed that formation of subunits from purified H_4 and M_4 lactate dehydrogenases and their random recombination requires freezing and thawing, and proceeds most rapidly in the presence of sodium phosphate and chaotropic anions. Their experimental results favour the existence of some hydrophobic inter-subunit bonds. Dissociation of LDH can be accomplished by treatment with guanidine, urea or sodium dodecyl sulphate, so hydrogen bonds may also be involved in the subunit–subunit interactions. X-ray crystallography of the dogfish M_4 LDH revealed that the subunits are globular, with tails of about 20 amino acid residues at the N-terminal. Inter-subunit contact involves both tail-to-tail interactions and bonding (probably hydrogen bonds) between helical and sheet regions of the globular domains. Since the amino acid sequence of the LDH from dogfish muscle is not yet completely solved, the relative contributions of hydrophobic bonding and electrostatic forces cannot be fully assessed. The existence of hybrids of the two LDH types indicates that the structure needed for tetramer formation has been conserved throughout the evolution of the enzymes. LDH shows no co-operativity in its reactions. Therefore, the oligomeric form probably exists because it is more stable than the protomers.

A detailed examination of the hybridization properties of mammalian fructose diphosphate aldolase isoenzyme subunits was used to determine the quaternary structure of this enzyme (Penhoet *et al.*, 1967). Three parent aldolases, A from muscle, B from liver and C from brain, had been isolated. Despite a great deal of research, by 1967 the number of subunits in an aldolase oligomer had still not been determined accurately. Some results indicated that the enzyme was a trimer, and others could be explained only if aldolase were a tetramer. Rutter and his colleagues used electrophoretic techniques to show that five-membered hybrid sets are formed from reversible dissociation of binary mixtures of parent aldolase isoenzymes. Their results proved that aldolase contains four subunits. They mixed aldolase A from rabbit muscle with aldolase C from rabbit brain, hybridized the subunits by acidification and reneutralization at $0°$ C, and subjected the resulting protein solution to electrophoresis. Five activity bands were resolved. They also hybridized ^{3}H-leucyl-aldolase A with the C-form and isolated the A-C hybrid set. As expected for a four-subunit enzyme, five isoenzyme bands could be separated on DEAE-Sephadex. The relative specific radioactivities of the isoenzymes were close to the ratios of $1 (A_4) : 0.75 (A_3C) : 0.5 (A_2C_2) : 0.25 (AC_3) : 0 (C_4)$. Penhoet *et al.* concluded that each subunit comprises one-quarter of the total enzyme. Their hybridization experiments have been verified by electron microscopy and X-ray crystallography, both of which show aldolase to have a tetrahedral arrangement of subunits.

One of the most active areas of research in clinical biochemistry is the examination of the levels of isoenzymes in tissues and blood. The isoenzyme levels released into blood can be an indication of the tissue of origin of the enzyme. Most clinical assays for isoenzymes use kinetic measurements, but some laboratories use electrophoresis to measure isoenzyme levels. Human heart contains mostly H_4 and H_3M LDH. Lactate dehydrogenase is released into the blood stream after a heart attack (myocardial infarction) because the heart tissue undergoes partial breakdown. The level of the dehydrogenase (or of the H_4, H_3M, and H_2M_2 forms) typically reaches a maximum about 2 days after a heart attack, then slowly decreases to normal. The principal LDH isoenzyme in human liver is M_4. In cirrhosis or hepatitis, more of the muscle isoenzymes can be detected in serum. At present, analysis of alkaline phosphatase isoenzymes in serum is most commonly used, but the theoretical and experimental principles are the same as those for LDH. The availability of accurate assays for more isoenzymes, when combined with other medical procedures, will greatly aid and simplify diagnosis of diseased states.

Precautions must be taken to ensure that separable fractions of enzyme activity are actually polypeptides that differ in structure *in vivo*. Artefacts can be observed due to limited proteolysis or other degradation occurring during either tissue extraction or the separation. Isoenzymes may also differ in their stability upon storage, so that stability in extracts and in partially purified form should always be tested. Multiple forms of enzymes, resembling isoenzymes, can be observed for other reasons. Alkaline phosphatase from *Escherichia coli* (Reid and Wilson, 1971) forms multiple bands in electrophoresis which have been called isoenzymes. This enzyme is a dimer, and three bands are commonly observed. However, all of the forms of alkaline phosphatase are coded for by the same gene. It appears that conjugation with carbohydrate material to differing extents causes the multiple forms of this bacterial enzyme. Dihydrofolate reductase from *Lactobacillus casei* has two separable forms, the apoenzyme and the enzyme–NADPH complex (p. 241).

Many enzymes exist as oligomeric isoenzymes. For many of these, there is no proven allosteric control function. The multiple catalytic forms may have more subtle control functions. For example, Kaplan believes that *M*-type LDH is used for reduction of pyruvate and *H*-type LDH for oxidation of lactate. Such control functions are difficult to prove, but attempts are being made to find out why there is a need for isoenzymes.

BINDING OF LIGANDS TO OLIGOMERS

For the examination of allosteric control of enzymes, one approach is to characterize the interaction of oligomeric proteins with low molecular weight ligands, including both substrates and effectors (Koshland, 1970; Janin, 1973; Gutfreund, 1974; Klotz, 1974). Simple kinetic studies led to the prediction of the existence of *ES* complexes and to an elementary explanation of enzymic catalysis.

In a similar way, studies of ligand–oligomer complex formation have provided data about enzymic control at a level of sophistication equivalent to the Michaelis–Menten catalysis theory. The use of more complicated and specialized techniques (Gutfreund, 1974) will provide the data needed to advance binding experiments to the current level of accomplishments of the science of enzyme kinetics. Here, as in Chapter 5 with kinetics, I will describe the experimental techniques available to most laboratories and explain how results obtained can be applied to studies on the mechanism of enzymatic control.

The first material needed for any extensive research on ligand binding is a source of highly purified enzyme. The molecular weight and quaternary structure of the enzyme must be accurately determined. With many enzymes, the measurement of the effects of concentrations of substrates, inhibitors and activators on the rates of enzyme-catalyzed reactions can be used as a measure of ligand binding (cf. p. 368 in Koshland, 1970, for limitations). The earliest experiments done with allosteric enzymes were kinetic in nature. However, true thermodynamic experiments are more desirable. The equilibrium techniques employed are either methods for the determination of both bound and free ligand concentrations, or optical methods such as spectroscopy or fluorescence. A system in which a spectral or fluorescence change occurs upon binding of a chromophoric ligand is preferable because the data is obtained more directly (Gutfreund, 1972). For example, the binding of reduced pyridine nucleotides to dehydrogenases can be measured by fluorescence enhancement or suppression. If possible, several methods should be used for verification of results.

Dialysis is a common method for equilibrium analysis of ligand binding to oligomeric enzymes. An equilibrium dialysis cell consists of two cavities separated by a dialysis membrane (Myer and Schellman, 1962). The two compartments are tightened together with set-screws, and filled with solutions or sampled by syringe and needle through capillary holes normally closed with screws. One half of the cell is filled with a solution of buffered protein, and the other half with a solution of ligand (usually radioactive ligand, to achieve maximal sensitivity). Attainment of equilibrium takes from 4 to 24 hours, depending on conditions such as the porosity of the membrane, the temperature and whether or not the dialysis cell is shaken or rotated. After equilibrium is established, the concentrations of free and enzyme-bound ligand are determined. The enzyme compartment contains free ligand plus bound ligand, and the other compartment only free ligand, so the difference between the two measured concentrations equals the concentration of bound ligand. Due to the length of time needed for equilibrium to be reached, the stability of both proteins and ligands must be checked under dialysis conditions.

Womack and Colowick (1973) have developed a steady-state dialysis technique for rapid determination of ligand binding to enzymes. Their method employs a stirred two-chamber apparatus with the upper chamber containing the enzyme solution and the lower chamber having a constant flow of buffer through it, feeding into a fraction collector. Successive additions of radioactive ligand are made to the upper chamber. The fraction of free ligand in the enzyme-containing chamber is determined from the radioactivity per ml of the effluent solution compared with the radioactivity per ml in the effluent when enzyme is omitted in the upper chamber or when the enzyme is saturated with ligand. The rapidity of a steady-state dialysis experiment (about 20 minutes) allows more labile reactants to be used.

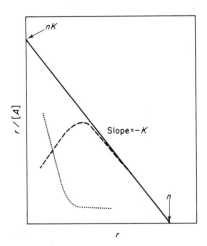

Fig. 1. Typical Scatchard plots for ligand binding to oligomers. A Scatchard plot is similar to the Eadie–Hofstee plot used in enzyme kinetics, and is therefore more indicative of curvature than some graphical methods. The solid line shows the results obtained if the binding sites are independent and equal. The dashed line shows a typical example for positive co-operativity, and the dotted line that for negative co-operativity.

Gel filtration can also be used to examine the binding of ligands to enzymes (Hummel and Dreyer, 1962; Ackers, 1973). The only equipment needed is a column of gel filtration beads which will cleanly separate the enzyme and the ligand and a device to monitor the concentration of ligand (usually a spectrophotometer). The column is equilibrated with the ligand at a known concentration (the concentration of free ligand). Then a small sample of enzyme and ligand is added, and the enzyme is eluted with a solution containing the same concentration of ligand used to equilibrate the column. The elution profile exhibits a peak and then a trough of ligand concentration. The peak is the excess ligand bound to the enzyme, and the trough—at the normal elution position of the ligand—is the depletion resulting from ligand binding to the enzyme. The amount of ligand bound to the known amount of enzyme can be calculated from the size of the trough in the elution profile.

Data from equilibrium binding experiments are usually interpreted by the graphical method of Scatchard (1949), which gives the stoichiometry, and under some conditions the affinity, of the binding of ligand to protein.* If the binding sites on the oligomer are independent and have equal affinities for the ligand, A, then:

$$\frac{r}{[A]} = K(n-r)$$

where r = moles of A bound per mole of oligomer
    $[A]$ = molar concentration of free ligand A
     K = binding constant of enzyme and A (i.e. reciprocal of the dissociation constant)
and n = number of binding sites per oligomer.

When $r/[A]$ is plotted against r, the intercept on the abscissa is n, and the slope of the straight line is $-K$ (Fig. 1). Curvature of the line in a Scatchard plot indicates that there is more than one class of sites (i.e. different binding constants for ligand-binding sites). Positive co-operativity produces a bell-shaped curve, and negative co-operativity a curve upward from the abscissa (Fig. 1). Klotz (1974) has com-

* Eisenthal and Cornish-Bowden (1974) have suggested that the direct linear plot (cf. p. 117) can be applied successfully to binding experiments.

mented that, because saturation is rarely achieved and it is often necessary to extrapolate $[A]$ to infinity, it is preferable to perform experiments over a wide range of ligand concentrations (e.g. 1000-fold) and to use a logarithmic scale for $[A]$. If $\log r$ v. $\log [A]$ is plotted, then at high values of $[A]$, $\log r$ asymptotically approaches $\log n$, making n readily obtainable (Thompson and Klotz, 1971). Individual site binding constants can be deduced graphically for co-operative cases, but their solutions are difficult and require very precise analytical data.

In cases where co-operativity of binding is observed, the extent of co-operativity can be estimated from a Hill plot (Hill, 1910). This approach was first used to interpret the binding of O_2 to hemoglobin (p. 389), giving a value of $n = 2.8$. The Hill coefficient n is the slope of either the kinetic plot:

$$\log \frac{v}{V-v} \quad \text{v.} \quad \log [S]$$

or the equilibrium binding plot:

$$\log \frac{Y}{1-Y} \quad \text{v.} \quad \log [A],$$

where Y = the fraction of sites occupied.

At both very low and very high ligand concentrations, n should approach unity. However, the intermediate slope indicates the strength of interaction between binding sites. The closer n is to the number of sites per oligomer (i.e. to the number of protomers or to n from a Scatchard plot), the greater is the co-operative interaction. A value of 1.0 for the Hill coefficient indicates that the binding sites do not interact. If n is less than unity, there is negative co-operativity (p. 425). Cornish-Bowden and Koshland (1975) have shown how Hill plots can be used to obtain additional information about co-operativity of ligand binding. If data are available over a wide range of saturation (e.g. $Y = 0.05-0.95$), then K_1 and K_n can be obtained from intercepts of the asymptotes. Also, visual inspection, and comparison to their published plots, may allow discrimination between different models of co-operativity.

CO-OPERATIVITY OF LIGAND BINDING

The observation of sigmoid rather than hyperbolic v v. $[S]$ curves for many allosteric enzymes and the presence of subunits suggested that there is co-operativity of subunit function. Co-operativity in an oligomeric enzyme is defined as the influence of one subunit on the conformation and activity of other subunits. For example, the effect of positive co-operativity is binding enhancement; the binding of substrate to the first protomer increases the affinity of other protomers for subsequent substrate molecules. High co-operativity, as exhibited through a steep sigmoid kinetic curve, would have profound metabolic function. One should recall that in cells enzymes tend to operate with concentrations of reactants near half-saturation* concentrations. A small increase in the substrate concentration near the half-saturation value should cause a large increase in the catalytic activity of the enzyme. This increase is greater if the saturation curve is sigmoid. If the enzyme is regulated by allosteric effectors, the presence of an activator will shift the half-saturation concentration to lower concentrations of substrate. Addition of an allosteric inhibitor *in vitro* or its build-up in the cell raises the concentration of substrate needed for any given catalytic rate (Fig. 2). Non-hyperbolic kinetics most often arise from co-operative effects, but several kinetic mechanisms unrelated to subunit interactions can also produce a sigmoid curve. Co-operativity should always be verified by equilibrium binding experiments which can detect deviations from independence of sites.

Two theories proposed to explain the co-operativity observed in binding of some ligands by oligomeric enzymes have gained general recognition. Both attempt to describe co-operative transitions in relatively simple quantitative terms. Often, these theories have been incorrectly described as allosteric theories rather than theories describing co-operative effects. Fortunately, some allosteric phenomena *can* be represented by these models. The two theories are the concerted or symmetry-driven theory and the sequential or ligand-induced co-operativity theory. Both theories have some intuitively pleasing

---

* Strictly, the Michaelis constant, K_m, can be reported only for enzymes having rectangular hyperbolic saturation curves. For enzymes having non-hyperbolic saturation kinetics, the terms "ligand concentration at half-saturation" or "apparent K_m" are preferred.

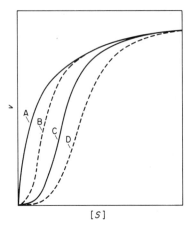

[S]

Fig. 2. Hyperbolic (curve A) and sigmoid (curve C) kinetic saturation patterns for an enzyme. Notice that curve C provides a greater change in v for a small increment or decrease in [S] than would a hyperbolic curve at its half-saturation point. Curve B shows the response expected in the presence of an allosteric activator, and curve D that in the presence of an allosteric inhibitor, with the enzyme control represented by curve C.

exclusive features. Neither has been proven by physical evidence from enzymes. The mechanisms of allosteric control will undoubtedly be different for different enzymes, so no simple theory will explain the behaviour of all regulatory enzymes. Some enzymes favour the concerted mechanism and others the sequential theory. The properties of some enzymes seem to fit a combination of the two theories. In this chapter, I will summarize the basic tenets of these two theories of co-operativity. In the next chapter, when looking at examples of controlled enzymes, there will be an opportunity to see how individual reactions are controlled. It will require much more complete knowledge of the chemical and physical properties of any given enzyme before its complete mechanism of regulation can be described. Hopefully, some general features should be evident, as expected from present evidence.

The Concerted Theory

In a paper entitled "On the Nature of Allosteric Transitions: A Plausible Model", Monod, Wyman and Changeux (1965) presented a relatively simple theory designed to explain co-operative ligand-binding effects. This theory is usually referred to as the concerted transition or concerted theory, or the MWC theory. Their use of the word "allosteric", or rather their redefinition of the term from the original meaning (p. 404), has limited the application of the theory and introduced unnecessary confusion. As previously stated, to prevent misunderstanding, I will reserve "allosteric" for enzymes having control ligand second-site effects. The concerted theory redefines an allosteric effector as a specific ligand endowed with different affinities toward two conformational states of an oligomeric enzyme. Two other phrases are introduced in this paper. *Homotropic effects* are interactions between identical ligands (i.e. binding of an effector or substrate to one protomer alters the affinity of other protomers for the same ligand). *Heterotropic effects* are interactions between different ligands (i.e. one ligand affects the affinity of the enzyme for another type of ligand).

The concerted model was devised to explain homotropic effects in a series of six statements. All of these assumptions are needed to invoke the symmetry retention proposed by Monod for the oligomeric conformational transition:

1. the protomers of an oligomeric protein all occupy equivalent geometric positions,
2. there is only one stereospecific binding site per protomer for each ligand,
3. the conformation of each protomer is constrained by its association with the other protomers,
4. two states (or at least two states) are reversibly accessible to an oligomeric protein (i.e. the R, or high-affinity, state and the T, or low-affinity, state are in equilibrium),
5. when the conformational transition occurs, the affinity of ligand-binding sites also changes, and
6. when the oligomeric protein undergoes a conformational transition, its molecular symmetry is conserved.

The concerted theory was then expressed mathematically. For complete retention of symmetry, it must be assumed that a ligand is bound to each protomer with the same affinity, in any given conformational state. Thus, the dissociation constant (K_R) for a ligand–protomer complex is the same for all protomers when the oligomer is in the R state, and similarly K_T is the same for all protomers when the enzyme is in the T state. L is the equilibrium constant for the $R \rightleftharpoons T$ equilibrium, and c the ratio of dissociation constants for a ligand between the R and T states, K_R/K_T. Equations were derived to show that homotropic binding is dependent on both L and c. Co-operativity is more marked when L is large (i.e. the T state is favoured) and when c is small (i.e. when the ligand is bound much more tightly to the R state). If $c = 1$ and L is negligible, then Michaelis–Menten hyperbolic kinetic curves are observed.

Next, the concerted theory was extended to cover heterotropic effects, which more closely describe allosteric enzymes. For simplicity, the assumptions were made that activators and substrates bind only to the R conformation, and that inhibitors bind only to the T state. (More complicated calculations result if, as in actual cases, ligands can bind to both states.) Heterotropic effects are due exclusively to displacement of the $R \rightleftharpoons T$ equilibrium. An inhibitor, when bound to the oligomer, usually should increase the co-operativity of substrate binding and raise the half-saturation point for substrate. The presence of activator molecules should decrease the co-operativity of binding of substrate and lower the apparent K_m of the substrate. Monod recognized that some enzymes (called V systems) existed in which the substrate had the same affinity in both the reactive (R) and unreactive (T) conformations.

The concerted theory is based on a fundamental principle of protein structure, the common occurrence of spatial symmetry (p. 406). It proposes a simple relationship between function and symmetry retention, with unequal binding of ligands to different conformational states causing the co-operative transition between the states. The sequential theory differs from the all-or-none MWC theory not by totally eliminating symmetry but by permitting hybrid conformations in the partially liganded states.

By kinetic experiments, phosphofructokinase (PFK) from $E.$ $coli$ shows both homotropic and heterotropic co-operative interactions that can be fully accounted for by the concerted theory. To ensure that the kinetic results represented co-operative binding, as should be done when sufficient pure enzyme is available, equilibrium dialysis experiments were later performed. A brief look at these kinetic and equilibrium experiments shows the type of problem facing enzymologists wishing to determine control mechanisms.

Phosphofructokinase is a tetrameric enzyme catalyzing a step in glycolysis that is under allosteric regulation:

$$\text{Fructose-6-phosphate} + \text{ATP} \xrightarrow{\text{Mg}^{2+}} \text{Fructose-1,6-diphosphate} + \text{ADP}.$$

Citrate and high concentrations of ATP are feedback inhibitors of mammalian PFK, but in *E. coli* PFK is inhibited by phosphoenolpyruvate and activated by nucleoside diphosphates such as GDP. Blangy *et al.* (1968), using PFK partially purified from *E. coli*, measured the kinetic parameters of the two substrates and the two types of effectors. The saturation curves for ATP are hyperbolic, with the K_m value being 6×10^{-5} M at all concentrations of fructose-6-phosphate (F-6-P). In the absence of effectors, the binding of F-6-P is co-operative, with a Hill coefficient of 3.8. Under these conditions, the co-operativity is high because 3.8 is close to the maximal n of four protomers. Addition of the inhibitor phosphoenolpyruvate (PEP) causes the Hill coefficient for F-6-P to approach 1 and the apparent K_m for F-6-P to exceed 4×10^{-2} M. The presence of sufficient activator, GDP, has a similar effect on the Hill coefficient, but the apparent K_m for F-6-P becomes 1.2×10^{-5} M. The authors concluded that high concentrations of PEP or GDP completely convert PFK to the T or R conformation, respectively. In either of these extreme states, hyperbolic kinetics for F-6-P would be in accordance with the concerted theory. Between the two frozen conformational states, when both R and T tetramers exist, co-operativity for F-6-P binding should be observed.

When the enzyme was purified and the binding of ligands measured by equilibrium dialysis, it was discovered that the allosteric transition is more complicated than originally envisaged. Blangy (1971) confirmed that GDP and PEP are antagonistic allosteric effectors and that ATP does not participate in the allosteric transition equilibrium. However, binding of F-6-P to the free enzyme could not be detected, possibly because of the existence of additional conformational species. In support of this suggestion, Blangy found that the allosteric equilibrium constant L determined by kinetics differs greatly from that determined in the equilibrium binding experiments.

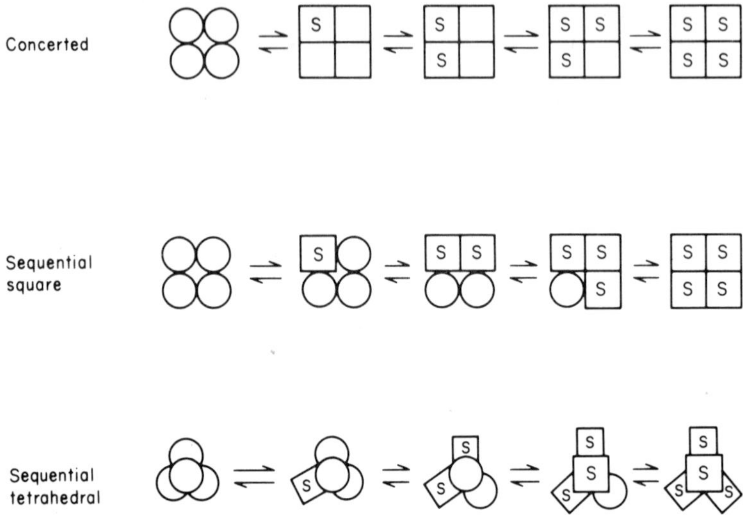

Fig. 3. Comparison of the binding of a ligand S to a tetrameric protein by the concerted, sequential square and sequential tetrahedral mechanisms. Individual protomers in the T (low affinity) state are designated by circles, and those in the R (high affinity) state by squares. Adapted from Koshland *et al.* (1966).

The Sequential Theory

The year after Monod *et al.* proposed the concerted theory, Koshland, Némethy and Filmer suggested an alternative and more general theory to interpret ligand saturation curves (Koshland *et al.*, 1966). Their postulate is called the sequential or the KNF theory. It is based on the induced-fit principle that a ligand may induce changes in the tertiary structure of the protomer to which it binds (p. 31), and this protomer–ligand complex may change the conformational stability of neighbouring subunits or the quaternary structure to varying extents. The mathematics of the sequential theory were designed to explain the binding of only one type of ligand to a tetrameric enzyme. Koshland *et al.*, like Monod and his colleagues, assumed that individual protomers can exist in two conformations and that only one confor-mation binds the ligand. The difference in the sequential theory is the possible inclusion of hybrid oligomers in which both high- and low-affinity protomers can exist at fractional saturation. The diagrams of Fig. 3 show three of the possible mechanisms. In the square mechanism, later favoured by Koshland (1970) as the simplest

Table II. Association constants predicted for binding of ligands to a tetrameric protein. K_1, K_2, K_3 and K_4 represent the intrinsic constants for the binding of the first through fourth molecules of ligand, respectively, to a tetrameric protein. Adapted from Cornish-Bowden and Koshland (1970).

Theoretical Model	Relationship of Binding Constants
Michaelis–Menten	$K_1 = K_2 = K_3 = K_4$
Concerted Theory	$K_1 < K_2 \leqslant K_3 \leqslant K_4$
Sequential Square	$1/3 \leqslant \dfrac{K_2}{K_1} = \dfrac{K_4}{K_3} \geqslant \dfrac{K_3}{K_2}$
Simple Positive Co-operativity	$K_1 < K_2 < K_3 < K_4$
Simple Negative Co-operativity	$K_1 > K_2 > K_3 > K_4$

sequential model, a protomer can interact with only two adjacent protomers. The tetrahedral mechanism assumes that each protomer can interact with all three other protomers. Cornish-Bowden and Koshland (1970) have suggested a curve-fitting programme that can give estimates of ligand binding constants. As shown in Table II, the relative values of these constants could identify the conformational model most likely to explain observed co-operative effects. However, the present experimental limits of kinetic and binding measurements make differentiation of many mechanisms impossible.

The sequential theory allows the co-operative binding system the flexibility of protein structure now assumed to be almost universal in globular proteins. It permits but does not insist on oligomeric symmetry and identical binding constants of all subunits. The concerted theory did not make any provision for *negative co-operativity,* the decrease in affinity as subsequent ligand molecules are bound to the oligomer. The more general equations of sequential co-operativity can accommodate this phenomenon.

Negative Co-operativity and Half-Site Reactivity

Negative co-operativity was first noted for the binding of NAD^+ to the tetrameric enzyme glyceraldehyde-3-phosphate dehydrogenase

(p. 326) from rabbit muscle. For many years, the number of NAD^+ sites in this dehydrogenase was in doubt, because classical optical titrations of the apoenzyme gave confusing results. Conway and Koshland (1968) identified the source of this anomaly as negative co-operativity in the binding of NAD^+. They found, by equilibrium dialysis of the stable $E(NAD^+)_3$ complex with NAD^+, that the fourth molecule of NAD^+ is loosely bound in comparison with the other three molecules. They estimated the values of the dissociation constants K_1 through to K_4 for NAD^+ to be $<10^{-11}$ M, $<10^{-9}$ M, 3×10^{-7} M and 2.6×10^{-5} M, respectively. Using freshly prepared apoenzyme of high specific activity, Bell and Dalziel (1975) have used fluorimetry to directly determine the dissociation constants. Their results provided the values $K_1 = 1 \times 10^{-8}$ M, $K_2 = 9 \times 10^{-8}$ M, $K_3 = 4 \times 10^{-6}$ M and $K_4 = 3.6 \times 10^{-5}$ M. Thus, there are four NAD^+ binding sites per tetramer, differing in their affinity but falling into two distinct classes: two of high affinity and two of low affinity.

The crystal structure of lobster glyceraldehyde-3-phosphate dehydrogenase bearing a full complement of four molecules of NAD^+ has been determined at 2.9 Å resolution in Rossmann's laboratory (Moras et al., 1975), where further structural studies are being performed. Lactate dehydrogenase, which shows no co-operativity of NAD^+ binding, has the NAD^+ sites completely within each subunit and on the outside of the molecule. In glyceraldehyde-3-phosphate dehydrogenase, though, the NAD^+ binding site of each protomer is close to the interface with an adjacent protomer. There are differences in conformation between subunits in the crystals, with two types of subunits being observed. The largest differences between the two types of protomers occur in the vicinity of the active site. One type of subunit has a structure compatible with firmer binding of NAD^+, in agreement with the presence of two high-affinity sites per tetramer. The small but significant asymmetry of the tetramer can be described best as a pair of related dimers. Indeed, in some modification reactions, the enzyme behaves as two independent dimers. Elucidation of the three-dimensional structures of the apoenzyme, partially-saturated enzymes and the fully-saturated enzyme may be required to completely explain the mechanism of negative co-operativity of glyceraldehyde-3-phosphate dehydrogenase.

Some homogeneous oligomeric enzymes exhibit a mixture of negative and positive co-operativity in their saturation curves. For example, glyceraldehyde-3-phosphate dehydrogenase from yeast, in a Scatchard plot of NAD$^+$ binding, shows deviations from linearity indicative of positive co-operativity. In the Hill plot, there are two slopes, with the break between these occurring at about 50% saturation. Cook and Koshland (1970) concluded that their data fit best with a relation of association constants of $K_1 < K_2 > K_3 > K_4$. Teipel and Koshland (1969) have analyzed kinetic and binding saturation curves for several enzymes which show intermediate plateau regions. They proposed that curves with two inflection points, bumpy saturation curves, require more than two ligand binding sites and are caused by negative co-operativity and then positive co-operativity of binding. The kinetic saturation curve of IMP dehydrogenase for its substrate IMP, in the presence of the allosteric inhibitor GMP, has *three* inflection points under conditions that give maximal departure from classical competitive inhibition (pH 7.3 in Fig. 4). Unfortunately, not enough pure enzyme is available yet to do the equilibrium binding experiments needed to prove a mixture of positive, negative and finally positive co-operativity during the saturation of the IMP dehydrogenase tetramer.

Conway and Koshland (1968) suggested that the physiological role of negative co-operativity is to make enzymes exhibiting this phenomenon less sensitive to fluctuations in environmental substrate

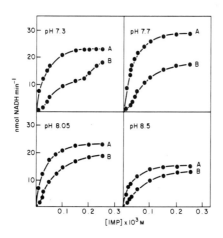

Fig. 4. Inhibition of IMP dehydrogenase by GMP. Curve A is in the absence of GMP, and curve B in the presence of 2×10^{-4} M GMP. From Wu and Scrimgeour (1973), with permission.

concentrations. Calculations by Cornish-Bowden (1975), using the Hill equation, indicate that the more negatively co-operative an oligomer is, the less sensitive it is to changes in concentration of ligand at all ligand concentrations. However, whether or not negative or mixed co-operativity has any physiological significance is not yet known.

A phenomenon related to negative co-operativity is the *half-site reactivity* of some oligomeric enzymes (reviewed by Seydoux *et al.*, 1974). Half-site reactivity is usually detected either by using inhibitors which cause irreversible inactivation or by using transient kinetics. A number of reactions have been observed in which only half of the oligomer's active sites participate in a reaction; i.e. one-half are inhibited by an excess of inhibitor or the burst-size is half of that expected, and the other half of the active sites do not react at all. Seydoux *et al.* point out that half-site reactivity may be considered an extreme example of negative co-operativity. As an explanation of the two classes of binding sites, they favour a model in which the tetramer exists as two asymmetric dimers, in each of which binding of one ligand transmits a negative co-operative effect across a single inter-protomer binding domain.

The validity and meaning of half-site reactivity is still in question. Most cases are still under active examination. As an example, succinyl CoA synthetase from *E. coli*—an $\alpha_2\beta_2$ tetramer (p. 331)—appeared to bind ^{32}P from $\gamma\text{-}^{32}P\text{-ATP}$ to only one of its two active-site histidine residues. Repetition of experiments with undenatured enzyme showed full reactivity in phosphohistidine formation (Bowman and Nishimura, 1975). Rabbit muscle glyceraldehyde-3-phosphate dehydrogenase shows half-site reactivity with some reagents which alkylate the active-site thiol, but full reactivity with other alkylating agents. Levitzki (1974) has suggested that the compounds in the group of alkylating molecules which produce all-of-the-sites reactivity are too small or not properly oriented to sterically hinder the active site —SH of the neighbouring protomer in the dimer of dimers. A few examples of half-site reactivity with natural substrates have been uncovered, and probably the most useful information will come from their study.

Let us now move from theory to experimental examples. Undoubtedly, allosteric control is exerted through conformational alteration in the presence of the responsible ligand or under the conformation-freezing influence of covalent modification. Study of the properties of some allosteric enzymes will be more useful than any further pursuit at present of theoretical explanations of co-operativity of ligand binding.

REFERENCES

Ackers, G. K. (1973). In "Methods in Enzymology" (Hirs, C. H. W. and Tima-sheff, S. N., eds), Vol. 27, pp. 441-455. Academic Press, New York.

Bell, J. E. and Dalziel, K. (1975). Biochim. biophys. Acta 391, 249-258.

Blangy, D. (1971). Biochimie 53, 135-149.

Blangy, D., Buc, H. and Monod, J. (1968). J. molec. Biol. 31, 13-35.

Blundell, T., Dodson, G., Hodgkin, D. and Mercola, D. (1972). Adv. Protein Chem. 26, 279-402.

Bowman, C. M. and Nishimura, J. S. (1975). J. biol. Chem. 250, 5609-5613.

Chan, W. W.-C. and Mawer, H. M. (1972). Archs Biochem. Biophys. 149, 136-145.

Cheema, S., Soldin, S. J., Knapp, A., Hofmann, T. and Scrimgeour, K. G. (1973). Can. J. Biochem. 51, 1229-1239.

Chilson, O. P., Costello, L. A. and Kaplan, N. O. (1965). Biochemistry 4, 271-281.

Chothia, C. and Janin, J. (1975). Nature, Lond. 256, 705-708.

Cohen, G. N. (1968). "The Regulation of Cell Metabolism." Holt, Rinehart and Winston, New York.

Conway, A. and Koshland, D. E., Jr. (1968). Biochemistry 7, 4011-4023.

Cook, R. A. and Koshland, D. E., Jr. (1970). Biochemistry 9, 3337-3342.

Cornish-Bowden, A. (1975). J. theor. Biol. 51, 233-235.

Cornish-Bowden, A. and Koshland, D. E., Jr. (1970). Biochemistry 9, 3325-3336.

Cornish-Bowden, A. and Koshland, D. E., Jr. (1975). J. molec. Biol. 95, 201-212.

Duncan, B. K., Diamond, G. R. and Bessman, M. J. (1972). J. biol. Chem. 247, 8136-8138.

Eisenthal, R. and Cornish-Bowden, A. (1974). Biochem. J. 139, 715-720.

Everse, J. and Kaplan, N. O. (1973). Adv. Enzymol. 37, 61-133.

Fersht, A. R. (1975). Biochemistry 14, 5-12.

Frieden, C. (1970). J. biol. Chem. 245, 5788-5799.

Gerhart, J. C. and Pardee, A. B. (1962). J. biol. Chem. 237, 891-896.

Gutfreund, H. (1972). "Enzymes: Physical Principles", pp. 68-72. Wiley-Interscience, London.

Gutfreund, H. (1974). In "Chemistry of Macromolecules" (H. Gutfreund, ed.), pp. 261-286. Butterworths, London.

Haberland, M. E., Willard, J. M. and Wood, H. G. (1972). Biochemistry 11, 712-722.

Hill, A. V. (1910). J. Physiol., Lond. 40, IV-VIII.

Hummel, J. P. and Dreyer, W. J. (1962). *Biochim. biophys. Acta* **63**, 530-532.
Irias, J. J., Olmsted, M. R. and Utter, M. F. (1969). *Biochemistry* **8**, 5136-5148.
Janin, J. (1973). *Prog. Biophys. molec. Biol.* **27**, 77-120.
Kemper, D. L. and Everse, J. (1973). *In* "Methods in Enzymology" (Hirs, C. H. W. and Timasheff, S. N., eds), Vol. 27, pp. 67-82. Academic Press, New York.
Klotz, I. M. (1974). *Accts Chem. Res.* **7**, 162-168.
Klotz, I. M., Darnall, D. W. and Langerman, N. R. (1975). *In* "The Proteins" (Neurath, H. and Hill, R. L., eds), Third edition, Vol. 1, pp. 293-411. Academic Press, New York.
Koshland, D. E., Jr. (1970). *In* "The Enzymes" (P. D. Boyer, ed.), Third edition, Vol. 1, pp. 341-396. Academic Press, New York.
Koshland, D. E., Jr., Némethy, G. and Filmer, D. (1966). *Biochemistry* **5**, 365-385.
Levitzki, A. (1974). *J. molec. Biol.* **90**, 451-458.
Meister, A. (1950). *J. biol. Chem.* **184**, 117-129.
Monod, J., Changeux, J.-P. and Jacob, F. (1963). *J. molec. Biol.* **6**, 306-329.
Monod, J., Wyman, J. and Changeux, J.-P. (1965). *J. molec. Biol.* **12**, 88-118.
Moras, D., Olsen, K. W., Sabesan, M. N., Buehner, M., Ford, G. C. and Rossmann, M. G. (1975). *Proc. natn. Acad. Sci. U.S.A.* **250**, 9137-9162.
Myer, Y. P. and Schellman, J. A. (1962). *Biochim. biophys. Acta* **55**, 361-373.
Neilands, J. B. (1952). *J. biol. Chem.* **199**, 373-381.
Panagou, D., Orr, M. D., Dunstone, J. R. and Blakley, R. L. (1972). *Biochemistry* **11**, 2378-2388.
Penhoet, E., Kochman, M., Valentine, R. and Rutter, W. J. (1967). *Biochemistry* **6**, 2940-2949.
Reid, T. W. and Wilson, I. B. (1971). *In* "The Enzymes" (P. D. Boyer, ed.), Third edition, Vol. 4, pp. 384-387. Academic Press, New York.
Sanwal, B. D., Kapoor, M. and Duckworth, H. W. (1971). *Curr. Topics cell. Reguln* **3**, 1-115.
Scatchard, G. (1949). *Ann. N.Y. Acad. Sci.* **51**, 660-672.
Scheraga, H. A., Némethy, G. and Steinberg, I. Z. (1962). *J. biol. Chem.* **237**, 2506-2508.
Segal, H. L. (1973). *Science, N.Y.* **180**, 25-32.
Seydoux, F., Malhotra, O. P. and Bernhard, S. A. (1974). *Crit. Rev. Biochem.* **2**, 227-257.
Stadtman, E. R. (1966). *Adv. Enzymol.* **28**, 41-154.
Stadtman, E. R. (1970). *In* "The Enzymes" (P. D. Boyer, ed.), Third edition, Vol. 1, pp. 397-459. Academic Press, New York.
Taylor, B. L., Barden, R. E. and Utter, M. F. (1972). *J. biol. Chem.* **247**, 7383-7390.
Teipel, J. and Koshland, D. E., Jr. (1969). *Biochemistry* **8**, 4656-4663.
Thompson, C. J. and Klotz, I. M. (1971). *Archs Biochem. Biophys.* **147**, 178-185.
Umbarger, H. E. (1956). *Science, N.Y.* **123**, 848.
Umbarger, H. E. and Brown, B. (1958). *J. biol. Chem.* **233**, 415-420.

Wieland, T. and Pfleiderer, G. (1957). *Biochem. Z.* **329**, 112-116.

Wilkinson, J. H. (1970). "Isoenzymes", Second edition. Chapman and Hall, London.

Womack, F. C. and Colowick, S. P. (1973). *In* "Methods in Enzymology" (Hirs, C. H. W. and Timasheff, S. N., eds), Vol. 27, pp. 464-471. Academic Press, New York.

Wu, T. W. and Scrimgeour, K. G. (1973). *Can. J. Biochem.* **51**, 1391-1398.

Yates, R. A. and Pardee, A. B. (1956). *J. biol. Chem.* **221**, 757-770.

Zemell, R. I. and Anwar, R. A. (1975). *J. biol. Chem.* **250**, 3185-3192.

13 Regulated Enzyme Reactions

INTRODUCTION

The regulation of metabolic reactions is an essential function of living organisms. Without co-ordination of anabolism, catabolism and cell division, there would be so much chemical chaos that life could not be sustained. Allosteric phenomena play a major role in the imposition of orderliness on cellular chemistry. The number of recognized allosteric enzymes is now great, as is the variety of their control mechanisms. Despite the multitude of allosteric controls, all those enzyme reactions proven to be under allosteric regulation occur at rational metabolic locations, although not only at the first committed step of a pathway. Extension of results from *in vitro* experiments to whole organisms often is a tremendous task, but it must be achieved for a full understanding of metabolic regulation. Much of the early research was performed on biosynthetic pathways in unicellular organisms that lack membranous organelles. With these relatively simple cells, the correlation of studies of isolated enzymes with metabolism in whole cells has been easier. Examination of enzymic regulation in higher organisms has demonstrated many similar, but also some more complex, regulation modes.

The allosteric control of metabolic pathways can be relatively simple if the pathway is unbranched, or can be quite complicated for a branched pathway (reviewed by Stadtman, 1966; Umbarger, 1969). If a linear pathway A to E (below) is considered, the product E may

$$A \rightarrow B \rightarrow C \rightarrow D \rightarrow E$$

inhibit the enzyme which catalyzes the conversion of A to B. Inhibition of the first committed step of the pathway is most logical if the cell desires to prevent accumulation of intermediates or of too much of the ultimate end-product. As already noted, this is a common occurrence, but not universal. Intermediates may inhibit the first step, or activation of the pathway at a strategic site may occur. Many biosynthetic pathways are branched, so the rate of production of all the end-products must be co-ordinated. In the following pathway, if an excess of E inhibited the initial step, the formation of

$$A \rightarrow B \rightarrow C \rightarrow D \begin{array}{c} \nearrow E \\ \\ \searrow F \end{array}$$

product F would also be limited. Under many growth conditions, this would be harmful to a cell. Therefore, cells have developed two major mechanisms for rapidly controlling multi-functional pathways —enzyme multiplicity and concerted feedback inhibition. Enzyme multiplicity in higher organisms is sometimes accompanied by compartmentalization of enzyme activity. Several names have been given to inhibition by combinations of end-products, depending on the quantitative nature of the effects observed. However, I shall refer to a greater than additive inhibition by two end-products only as *concerted feedback inhibition.* For example, in the branched pathway drawn above, a combination of E and F might be more effective than either E or F alone in inhibiting the formation of B from A.

The control of the branched biosynthetic pathway leading from aspartate to four amino acids in bacteria has been thoroughly studied (Cohen, 1968). In micro-organisms, a multi-branched pathway emanates from the four-carbon chain of aspartate ultimately to lysine, methionine, threonine and isoleucine (Fig. 1). After the first two steps of the pathway, which are common to the synthesis of all four amino acids, there are three main branches: to lysine, to methionine and to threonine (and isoleucine). The first step, the formation of aspartyl phosphate catalyzed by aspartokinase, is one site of cellular regulation. Different bacteria have different methods for controlling this initial reaction.

The control exerted at the aspartokinase-catalyzed step is quite simple in some bacteria. For example, *Rhodopseudomonas spheroides* regulates the activity of its aspartokinase by feedback inhibition by the intermediate aspartate β-semialdehyde (Datta and Prakash, 1966). However, in most bacteria the level of aspartokinase activity is controlled by the concentrations of the ultimate amino acid end-products, not that of a key intermediate. In *R. capsulata* and in the genus

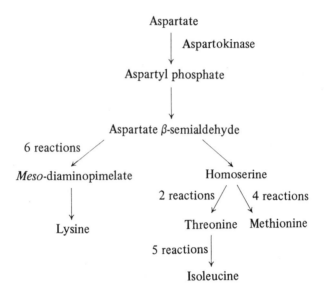

Fig. 1. Outline of the biosynthetic pathways leading from L-aspartate in bacteria.

Bacillus, aspartokinase activity is regulated by concerted feedback inhibition by two amino acids, threonine and lysine. The asparto-kinase of *B. polymyxa* is a constitutive enzyme synthesized at a rate just sufficient to supply the aspartyl phosphate needed to make the four amino acid end-products. It is inhibited by a concerted feedback mechanism. Paulus and Gray (1967) studied the inhibition of asparto-kinase partially purified from *B. polymyxa* in detail. Combinations of threonine and lysine at concentrations below 1 mM strongly inhibit the enzyme (e.g. if both amino acids are present at 0.6 mM, there is about 70% inhibition of aspartokinase activity). Threonine or lysine alone can inhibit, but only at much higher concentrations. A concentration in excess of 10 mM is required for 50% inhibition by either amino acid. This aspartokinase is an $\alpha_2\beta_2$ oligomer, composed of four polypeptide chains, two each of 43 000 and 17 000 daltons. The larger chains carry both the active site and the inhibitor binding sites. The smaller subunits seem to function in the folding or maturation of the enzyme (Biswas and Paulus, 1973). Control of the phosphoryl-ation of aspartate in *Bacillus* is imperfect, since methionine has no controlling influence. This metabolic flaw has been corrected in *Pseudomonas fluorescens*, though. The aspartokinase from this organism also is inhibited by a combination of threonine plus lysine, but in addition methionine plus threonine can inhibit the enzyme (Dungan and Datta, 1973).

The control of the aspartokinase reaction in *B. subtilis* appears to be more complex than that of *B. polymyxa*, which is grown under non-sporulating conditions. Rosner and Paulus (1971) have shown that *B. subtilis* contains two aspartokinases. One is inhibited by a combination of threonine and lysine, and presumably supplies the precursors of amino acids needed for protein synthesis. The other aspartokinase is inhibited by *meso*-diaminopimelate. This second aspartokinase may be required for the formation of dipicolinate for spores, or diaminopimelate for cell wall biosynthesis.

In *Escherichia coli*, several control mechanisms are used to produce an effective regulation of the biosynthesis of all four amino acids from aspartate. Repression and negative feedback inhibition controls are exerted at many locations. In addition, *E. coli* has multiple aspartokinases. The existence of these isoenzymes, each regulated by different metabolites, confers an evolutionary advantage to the cells by allowing independent control of the first step in the branched pathway by the four amino acids produced. Three different aspartokinases have been detected and isolated from *E. coli* (Patte *et al.*, 1967). Aspartokinase I is inhibited by threonine and by homoserine, and is repressed by threonine plus isoleucine. Aspartokinase II is not controlled by feedback inhibition, but is repressed by the accumulation of methionine. Aspartokinase III, which accounts for about 80% of the total aspartokinase activity, is both inhibited and repressed by lysine. The flow of common intermediates from aspartate is regulated by the existence and separate control of these three aspartokinases in *E. coli*. Control at the initial reaction is reinforced by regulation at other reaction sites. For example, there are feedback inhibitions by lysine, methionine, threonine and isoleucine at the first divergent steps of their own branches. A sequential feedback is also exerted in *E. coli* in the aspartate pathway. An accumulation of isoleucine causes inhibition of threonine deaminase and an accumulation of threonine. Excess threonine can then inhibit homoserine kinase with a resultant accumulation of homoserine. The high levels of homoserine inhibit the activity of aspartokinase I.

As examples of well-characterized regulatory enzymes, I have chosen aspartate transcarbamylase, ribonucleotide reductase, glutamine synthetase and lactose synthetase. Each of these enzymes performs a clearly defined metabolic function whose regulation can now be described in molecular terms. It is not possible to describe the precise conformational changes that must occur in their allosteric transitions; a future writer will have the pleasure of reviewing those discoveries. These four allosteric enzyme systems could not be called typical or average control enzymes. There may not be any such enzyme! Each of the four demonstrates a different control mechanism, and each has a different quaternary structure. Aspartate transcarbamylase provided the original evidence for separation of regulatory and catalytic binding sites. Ribonucleotide reductase is an enzyme having its substrate specifity controlled by allosteric conformational changes. Glutamine synthetase introduces covalent modification and cascade control effects. Lactose synthetase provides an illustration of alteration of enzyme activity by non-covalent binding of a modifier protein. Many of the regulatory principles presented in these four examples will crop up again in the concluding chapters of the book.

ASPARTATE TRANSCARBAMYLASE

Research on aspartate transcarbamylase (ATCase) isolated from *E. coli* (reviewed by Jacobson and Stark, 1973) provided the evidence for the second site or allosteric concept (cf. p. 402). My description of allosteric enzymes starts with the *E. coli* transcarbamylase because of its historical prominence and wealth of illustrative data. ATCases from some other organisms are described in Chapter 14 (p. 479).

The study of the ATCase of *E. coli* has proceeded in several stages. First, in metabolic experiments with whole cells, the reaction catalyzed by the ATCase was established as the site of negative feedback in pyrimidine biosynthesis. This reaction, the formation of carbamyl-L-aspartate from carbamyl phosphate and aspartate (Fig. 2), is the first step unique to pyrimidine nucleotide synthesis. Next, the enzyme was purified and kinetic experiments were performed on activation and inhibition by allosteric effectors. Experiments designed to determine the structure of the transcarbamylase were then initiated, and are still in progress. Finally, physicochemical solution techniques are being applied to the enzyme to answer some of the questions about the interrelation of enzymic structure and function.

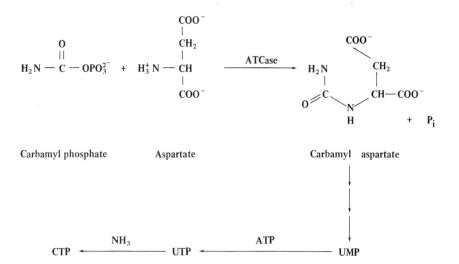

Fig. 2. Abbreviated pathway for biosynthesis of UTP and CTP. For the complete scheme (to UMP), see Fig. 3, Chapter 14.

When Yates and Pardee (1956) published their now classical paper on control of pyrimidine biosynthesis, it was already known that addition of pyrimidines to growth media resulted in preferential incorporation of the preformed bases into nucleic acids. It had also been demonstrated that the presence of uracil inhibited the formation of intermediates preceding orotic acid in the biosynthetic pathway (see Fig. 3 in Chapter 14). To see if a self-regulatory cellular mechanism exists to permit synthesis of only sufficient pyrimidines for growth, Yates and Pardee examined the control of the biosynthetic pathway both *in vivo* and *in vitro*. They observed that when *E. coli* cells are grown in the presence of traces of uracil, no intermediates (carbamyl aspartate, dihydro-orotate and orotate) are accumulated until the added uracil is used up. If RNA and DNA synthesis are blocked by ultraviolet irradiation of bacterial cells, causing an accumulation of mononucleotides, the rate of synthesis of dihydro-orotate rapidly decreases. These *in vivo* experiments showed that the site of inhibition is at or before the formation of carbamyl aspartate (called ureidosuccinate at that time). With cell-free extracts, they discovered that cytidine and CMP inhibited the formation of carbamylaspartate from carbamyl phosphate and aspartate. Later experiments showed that in *E. coli* CTP is the most potent inhibitor of the reaction catalyzed by ATCase. Yates and Pardee concluded that *de novo* synthesis of pyrimidines is regulated by the concentrations of pyrimidine nucleotides, with the site of control being aspartate transcarbamylase.

To prepare ATCase for detailed kinetic studies, a procedure for the isolation of homogeneous preparations of the enzyme was devised. Normally, synthesis of ATCase is repressed by the presence of uracil or one of its metabolites (UDP or UTP is probably the corepressor). Shepherdson and Pardee (1960) used a pyrimidine-requiring mutant strain of *E. coli*, grown in media containing a limiting amount of uracil. When the uracil was depleted, dihydro-orotate was added to permit slow growth. Release of repression in this manner caused ATCase to rise to levels as high as 7% of the bacterial protein. The enzyme was isolated by sonication of the cells, heating of the crude extract to 55° C, fractionation with ammonium sulphate and chromatography on DEAE-cellulose. Seven years later, Gerhart and Holoubek selected a special strain of *E. coli* (cf. p. 44) and published a scheme for purification of ATCase of slightly higher activity, in greater yield. Their procedure provided access to large quantities of ATCase needed for physicochemical experiments.

Using purified aspartate transcarbamylase, Gerhart and Pardee (1962) examined by kinetic techniques the mechanistic basis of feedback control. Cytosine derivatives inhibit highly purified transcarb-

Table I. Inhibition of ATCase by cytosine derivatives. The pyrimidines or derivatives were present at 2×10^{-3} M. Data taken from Gerhart and Pardee (1962).

Compound	% Inhibition
Cytosine	0
Cytidine	24
CMP	38
CDP	68
CTP	86
UTP	8
*d*TTP	23

amylase (Table I), whereas cytosine itself does not inhibit. The derivatives CMP, CDP, UTP and dTTP are less effective than CTP. Therefore, the base, the sugar and three phosphate groups must all contribute to the binding of the inhibitor. CTP appeared to be an inhibitor of ATCase which competes with aspartate, because high concentrations of aspartate reverse the inhibition. The inhibition by CTP alters only the concentration of substrate needed for half-saturation, not the maximal velocity. This raised the fundamental question: how does CTP—structurally so dissimilar to aspartate—compete with the substrate? The kinetic saturation curve for aspartate and ATCase is sigmoidal, with CTP making the curve more sigmoidal. Gerhart and Pardee found that ATP activates ATCase, and changes the aspartate saturation curve to a less sigmoid shape. Their results are quite similar to the theoretical curves in Fig. 2 of Chapter 12. (Positive co-operativity in the binding of succinate, a non-reactive analogue of aspartate, was confirmed when other workers used equilibrium dialysis several years later.) Gerhart and Pardee succeeded in desensitizing ATCase by heat treatment, or by selective inactivation with urea or heavy metal reagents. The selective destruction of inhibition, which provided evidence for spatial separation of inhibitor and substrate binding sites, was accompanied by conversion of the aspartate saturation curve to a hyperbolic shape and an increase in the maximal velocity. Gerhart and Pardee favoured a two-site mechanism in which binding of CTP to the inhibitor site increases the apparent K_m for aspartate at the active site, probably by a conformational change which distorts the active site. From their data and conclusions, the idea of allosteric regulation developed, with the suggestion that inhibitors and substrates compete for different conformational states of the enzyme.

Gerhart and Schachman (1965) directly demonstrated that ATCase binds CTP and its substrates at distinct sites. They dissociated the native transcarbamylase into two types of subunits under desensitizing conditions, with p-mercuribenzoate treatment. They detected separation of the native enzyme (11.3 S or 310 000 daltons) into subunits of 5.8 S (about 96 000 daltons) and 2.8 S (about 30 000 daltons) by analytical ultracentrifugation. Small quantities of each of the subunits were prepared by sucrose-density sedimentation. The larger subunit, called the catalytic subunit, possesses all the catalytic activity, but is not inhibited by CTP. The smaller subunit (the regulatory subunit), but not the catalytic subunit, can bind CTP. Native ATCase activity can be reconstituted by mixing the two subunits under appropriate conditions. Larger quantities of both subunits have since been prepared by desensitization, followed by either column chromatography or selective precipitation. The finding of separate catalytic and regulatory subunits with ATCase proves the existence of second sites. However, it does not mean that the regulatory site is in a different polypeptide chain from the catalytic site in all allosteric enzymes. On the contrary, the existence of identical subunits in allosteric oligomers is more common.

Although Gerhart and Schachman suggested that native ATCase was composed of four regulatory and four catalytic subunits, this turned out to be an incorrect assignment. Weber (1968), using SDS polyacrylamide gel electrophoresis, determined the molecular weights of the isolated subunits as 33 000 for the catalytic subunit and 17 000 for the regulatory subunit. He proposed that the values of 33 000 and 17 000 were for individual polypeptide chains, C and R, of the larger catalytic and regulatory subunits, respectively. Weber also determined the amino acid sequence of the R polypeptide. Its primary structure was in agreement with a molecular weight of 17 000 daltons. For clarity, the subunit composition of ATCase is summarized in Table II. The enzyme is described as a dimer of catalytic subunits and a trimer of regulatory subunits, in the arrangement $(C_3)_2(R_2)_3$.

Results from electrophoretic hybridization experiments (cf. p. 413) have confirmed the $(C_3)_2(R_2)_3$ quaternary structure of ATCase. Meighen *et al.* (1970) isolated catalytic subunits from ATCase, and

Table II. Quaternary structure of ATCase.

	Molecular weight	Number of chains	Composition
Catalytic polypeptide chain	33 000	1	C
Catalytic subunit	100 000	3	C_3
Regulatory polypeptide chain	17 000	1	R
Regulatory subunit	34 000	2	R_2
Native enzyme	310 000	12	$(C_3)_2(R_2)_3$

treated a solution of these subunits with succinic anhydride to obtain succinylated catalytic subunits. This modification made the modified protein more negatively charged than the native catalytic subunit protein, so the two forms could be separated by electrophoresis on cellulose acetate strips. Let us use the abbreviations C_n for a native catalytic polypeptide chain (33 000 molecular weight) and C_s for a succinylated polypeptide. When native and succinylated catalytic subunits were mixed with an excess of regulatory subunits and the reconstituted ATCase tested by electrophoresis, three protein bands were observed: $(C_nC_nC_n)_2(R_2)_3$, $(C_nC_nC_n)(C_sC_sC_s)(R_2)_3$, and $(C_sC_sC_s)_2$ $(R_2)_3$. This result confirmed that there are two catalytic subunits per molecule of ATCase, and that the catalytic subunits are stable oligomers. To prove that each catalytic subunit is a trimer, Meighen *et al.* denatured a mixture of native and modified catalytic subunits with guanidinium hydrochloride, renatured them by dialysis, and tested the hybrid mixture of catalytic subunits. Four protein bands were found: $(C_n)_3$, $(C_nC_nC_s)$, $(C_nC_sC_s)$ and $(C_s)_3$. Nagel and Schachman (1975) have done a similar series of hybridization experiments to prove the stoichiometry of the regulatory polypeptide chains. Using succinylated and native regulatory subunits, they observed a four-membered hybrid set of reconstituted ATCase-like complexes.

Use of electron microscopy and X-ray crystallography elucidated the arrangement of the catalytic and regulatory subunits in the ATCase structure. Cohlberg *et al.* (1972) proposed a model, based largely on electron microscopy data (Richards and Williams, 1972), in which the disc-like C chains of the $(C)_3$ subunits are in a triangular arrangement. The catalytic trimers are layered above each other in an eclipsed configuration but do not touch. They proposed that the regulatory dimers interconnect the upper and lower catalytic trimers, possibly by linking C chains 120° apart rather than those closest to each other. From X-ray diffraction studies to 5.5 Å resolution, Warren *et al.* (1973) proposed a model quite similar to that based on electron microscopy. They found that the two catalytic subunits are separated, but joined by the three regulatory dimers which project out from the equatorial region of the ATCase, leaving a central aqueous cavity. The three-dimensional structure, based on diffraction data to 5.9 Å resolution, of a CTP–ATCase complex has been reported by the same laboratory (Edwards *et al.*, 1974). An inhibitor binding site is located within each of the R chains. There is no large quaternary conformation difference between the enzyme–CTP complex and the free enzyme, but there are small changes, mostly in tertiary structure. X-ray crystallographic data at higher resolution is required for a more detailed description of the structure of ATCase.

Successful reconstitution of ATCase from isolated subunits requires the presence of a divalent metal ion in the regulatory polypeptide chain (Nelbach *et al.*, 1972; Rosenbusch and Weber, 1971). Analyses of intact ATCase indicated that there were six atoms of Zn^{2+} per mole of enzyme. When ATCase was treated with *p*-mercuribenzoate to obtain individual subunits, zinc could not be detected in either the regulatory or the catalytic subunits. However, one atom of Hg^{2+}, derived from the degradation of *p*-mercuribenzoate, had replaced the zinc originally in each R chain. After the workers realized that there is zinc in the enzyme, methods were devised for preparation of both zinc-containing and metal-free regulatory subunits. Since the catalytic subunit contains no metal, zinc cannot be required for catalytic activity. The metal-free regulatory subunit can bind the allosteric inhibitor CTP but cannot recombine with the catalytic trimers. Therefore, the zinc must be required for stabilization of the conformation of the (R_2) subunit that can recombine with the (C_3) subunits. The zinc ion probably is bound to the R chain by interaction with the four thiol groups of each R polypeptide. Desensitization by treatment with mercurials can thus be explained as release of bound zinc by reaction of 24 —SH groups (four per R chain) with the mercurial reagent, with ensuing weakening of the inter-subunit non-covalent bonds. (There is one —SH group in each C chain, but the catalytic subunit thiols are relatively unreactive.)

The mode of regulation of ATCase activity has been partially explained by the results of kinetic experiments using ATCase-like complexes which lack one molecule of either the catalytic or the regulatory subunit. By adding excess (R_2) to dilute solutions of (C_3), Mort and Chan (1975) prepared a relatively unstable ATCase complex deficient in one catalytic subunit, i.e. a $(C_3)(R_2)_3$ complex. This complex has the kinetic properties expected for the R state of the enzyme. It is unaffected by ATP or CTP, and has a hyperbolic saturation curve for aspartate with a lower V and a lower K_m than the isolated catalytic subunit. Chan has concluded that the lack of a second (C_3) subunit prevents the complex from existing in the T state. A more stable $(C_3)_2(R_2)_2$ complex, a species lacking a single regulatory dimer, has been obtained as an intermediate in both the reassembly of isolated subunits (Yang *et al.*, 1974) and the mercurial-induced dissociation of native ATCase (Evans *et al.*, 1975). This complex has the same specific activity as native ATCase but shows reduced homotropic and heterotropic interactions. For example, the saturation curve for aspartate is less sigmoidal under any given set of experimental conditions than that for aspartate and native ATCase. Both ATP and CTP are less effective as allosteric modifiers with the $(C_3)_2(R_2)_2$ complex than they are with the native enzyme. From these experiments, it can be concluded that ATCase must be in its native conformation, with three regulatory dimers and two catalytic trimers, for full allosteric control. The results suggest that, as described below, ATCase can best be represented as a trimer of $C(R_2)C$ units.

The binding of ligands to ATCase, now extensively studied, shows a complicated pattern of reactivity. Using the method of continuous variation and spectral techniques, Hammes *et al.* (1970) found that, in the presence of the unreactive aspartate analogue succinate, one mole of ATCase binds six moles of carbamyl phosphate, three molecules going to each catalytic trimer. ATCase also binds six moles of a CTP analogue, one molecule to each R chain. Rosenbusch and Griffin (1973) reported that ATCase binds only three moles of carbamyl phosphate if succinate is not present. In their experiments, the concentration of carbamyl phosphate was 0.24 μM or less. They confirmed that isolated catalytic subunits each bind three molecules of carbamyl phosphate in the absence of succinate. They concluded that the binding of succinate (and hence of aspartate, also) to native ATCase causes an unmasking of the remaining half of the carbamyl phosphate sites. If carbamyl phosphate binds before aspartate, as seems probable from kinetic studies (Wedler and Gasser, 1974), then an aspartate-induced increase in the number of effective carbamyl phosphate binding sites would explain, at least partially, the positive co-operativity in aspartate binding and kinetic reactivity. Jacobson and Stark (1975) have reported that dicarboxylic acids which are competitive inhibitors of aspartate (e.g. succinate) can activate ATCase, even when the enzyme is saturated with carbamyl phosphate (5 mM). Therefore, dicarboxylate activation must involve an increase in the affinity for aspartate. Thus, occupancy of the aspartate site, too, could contribute to the homotropic co-operativity. The binding of CTP also is more complicated than first envisaged. For example,

Winlund and Chamberlin (1970) provided evidence by equilibrium dialysis that there are two discrete classes of CTP binding sites on each ATCase molecule: three sites of high affinity and three of low affinity. One likely explanation of these and similar results is negative co-operativity of CTP binding within regulatory dimers of the native ATCase.

Tetraiodofluorescein is a dye that mimics the adenosine portion of the ATCase activator, ATP (Jacobsberg *et al.*, 1975). It activates the enzyme at very low concentrations. The authors have suggested that binding of the dye to only one regulatory site activates all six catalytic sites, making it a more potent effector than any metabolite. Both ATP and CTP compete with tetraiodofluorescein at the regulatory site, indicating that both of the natural antagonistic effectors exert their effects in the same region of the (R_2) subunit.

In summary, the properties of ATCase rule out a simple two-state model for allosteric control, although an individual step in the overall

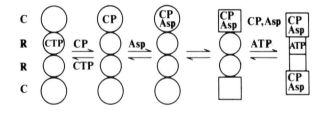

Extreme
T state

Extreme
R state

Fig. 3. Diagrammatic representation of some of the possible conformational states of the $C(R_2)C$ primary allosteric unit of ATCase. The regulatory dimer is depicted as showing full conversion to the T state when only one mole of CTP is bound per dimer, and to the R state when only one mole of ATP is bound per dimer. I have omitted possible intermediates in which both the nucleotide ligand and the substrates (CP = carbamyl phosphate and Asp = aspartate) are bound, and also unit—$(CP)_2$ intermediates. The kinetic mechanism favoured at present is:

regulatory mechanism may follow a concerted type of transition. Changes in the conformation of ATCase or its subunits upon addition of ligands have been observed by the use of several physical techniques, but definitive interpretations of all of the results are not yet possible. Intermediate conformational states can exist, but concerted transitions appear in other experiments. Kerbiriou and Hervé (1973) have isolated a modified ATCase from a mutant of *E. coli*. This modified enzyme lacks co-operativity in its binding of aspartate but retains feedback control by CTP. They concluded that the homotropic and heterotropic interactions in the wild-type enzyme are distinct but related phenomena. At present, the simplest concept of regulation of ATCase is to consider it a trimer of individual $C(R_2)C$ enzyme units (reviewed by Chan, 1975). The positive co-operative response is transmitted from one C chain to its mate in the other catalytic subunit through the regulatory dimer in the $C(R_2)C$ allosteric unit (Fig. 3). Similarly, the regulatory dimer transmits the heterotropic effect from the allosteric site on the R chain to both C chains in the primary allosteric unit. Many models have been proposed for the regulatory mechanism, but none have yet been proven. The structures of Fig. 3 are not meant to suggest a mechanism, but to summarize some of the pertinent experiments. It appears that both substrates and allosteric ligands induce conformational changes at the tertiary structure level, and in so doing cause quaternary structural changes. The negative co-operativity of CTP binding and the half-site reactivity of carbamyl phosphate support a sequential mechanism for co-operativity in some steps. At this point, it appears that a combination of Monod's and Koshland's theories applies to ATCase and that a complete answer to the question of how CTP competes with aspartate will only be obtained after high resolution X-ray diffraction experiments have been completed.

RIBONUCLEOTIDE REDUCTASE

The supply of deoxyribonucleoside triphosphates for biosynthesis of DNA is controlled by the activity of the enzyme ribonucleotide reductase (also called ribonucleoside diphosphate reductase). The discovery and subsequent isolation of this enzyme required the use of radioactive substrates of high specific activity, and rapidly dividing tissues or cells as a source of enzyme. Reichard and his collaborators have succeeded in characterizing the mechanisms of action and regulation of the ribonucleotide reductase system from *E. coli* (reviewed by Reichard, 1967, 1968). The *E. coli* reductase is representative of that of most organisms. However, some micro-organisms (e.g. *Lactobacillus leichmannii*) use an alternative pathway for producing deoxyribonucleotides–reduction of ribonucleoside triphosphates, catalyzed by an enzyme containing adenosyl cobalamin. Ribonucleotide reductase catalyzes the reduction of all four commonly occurring nucleoside diphosphates: GDP, ADP, CDP and UDP. The overall reduction of ribonucleotides by NADPH is catalyzed not by a single protein but by a system of four separable proteins. These proteins are the dithiol protein coenzyme thioredoxin (cf. pp. 287–290), thioredoxin reductase and the ribonucleotide reductase subunits B1 and B2. Thioredoxin reductase catalyzes the reduction of the $-S-S-$ group of thioredoxin, producing the reducing agent for the reduction of the ribose moieties:

$$\text{Thioredoxin}-S_2 + \text{NADPH} + H^+ \to \text{Thioredoxin}-(SH)_2 + \text{NADP}^+.$$

The reactions catalyzed by the ribonucleotide reductase components (B1 + B2) are shown by the first step in this pathway:

$$
\left.
\begin{array}{c}
\text{GDP} \\
\text{or} \\
\text{ADP} \\
\text{or} \\
\text{CDP} \\
\text{or} \\
\text{UDP}
\end{array}
\right\}
\xrightarrow[\text{B1, B2}]{\text{Thioredoxin}-(SH)_2}
\left\{
\begin{array}{ccc}
d\text{GDP} & \longrightarrow & d\text{GTP} \\
\text{or} & & + \\
d\text{ADP} & \longrightarrow & d\text{ATP} \\
\text{or} & & + \\
d\text{CDP} & \longrightarrow & d\text{CTP} \\
\text{or} & & + \\
d\text{UDP} & \to \to \to & d\text{TTP}
\end{array}
\right\}
\longrightarrow \text{DNA}.
$$

The most interesting aspect of the control of ribonucleotide reductase activity is the alteration of substrate specificity by allosteric interactions. Four effectors are known to alter the activity of the enzyme. Early work demonstrated that ATP is required, but not consumed, during the reduction of CDP to dCDP. ATP has since been shown to be the effector which stimulates the reduction of pyrimidine ribonucleotides. The other effectors are dGTP, which stimulates the reduction of purine ribonucleotides; dTTP, which is an activator for all the substrates; and dATP, which at high concentrations is an inhibitor of the reductase. Let us first examine the composition of ribonucleotide reductase from *E. coli*, and then the control of its activity.

Ribonucleotide reductase has been purified from extracts of a thymine-requiring mutant of *E. coli*, after growth of the organism under derepressing conditions (Brown *et al.*, 1969a). The purification procedure is reproducible but must be executed carefully because the enzyme can easily denature. As an example of the technical difficulties encountered, DEAE-cellulose which is available commercially could not be used effectively, so the ion exchanger had to be synthesized by the authors. The subunits B1 and B2 copurify for several steps in the procedure, and then are separated by column chromatography on hydroxyapatite. Each subunit is then further purified by separate procedures. More recently, Thelander (1973) has used affinity chromatography on a column of *d*ATP-Sepharose for a simpler and more reproducible preparation of B1. He has reported that the B1 subunit (molecular weight 160 000) is composed of two polypeptide chains, both of similar or identical size. Both chains of B1 have isoleucine as their C-terminal residues, but one has glutamate and the other aspartate at the N-terminal position. The molecular weight of protein B2 is 78 000 daltons, and it is composed of two identical polypeptide chains. Ribonucleotide reductase of *E. coli* therefore is a 1:1 complex of B1 and B2, with a composition that can be described as $\alpha\alpha'\beta_2$.

That ribonucleotide reductase is composed of two non-identical subunits, B1 and B2, suggested that it might be analogous to aspartate transcarbamylase by having separate catalytic and regulatory subunits. Separately, each subunit is catalytically inactive. Protein B2 contains two atoms of iron per 78 000 dalton unit (Brown *et al.*, 1969b). Removal of the iron from B2 inactivates the reductase. Reichard and his coworkers proposed that the iron participates in the reduction of the $>$CHOH group of the ribose, so it appeared that B2 contains all or part of the active site. Atkin *et al.* (1973) have suggested that the function of the iron in B2 is to generate an organic free radical which participates in the reduction of ribonucleotides catalyzed by the B1–B2 complex. Their more recent research shows that only protein B1 binds both nucleoside diphosphate substrates and nucleoside triphosphate effectors. The substrates and effectors are bound at different sites on the B1 subunit. Whether thioredoxin—$(SH)_2$ binds to B1 or B2, or to both, is not yet known. Mixing of equimolar amounts of B1 (7.8 S) and B2 (5.5 S) in the presence of Mg^{2+} generates a 1:1 complex, detectable by ultracentrifugation (9.7–10.1 S) and possessing enzymatic activity. The association between B1 and B2 in the presence of Mg^{2+} is quite loose, though. Using gentle lysis procedures, Eriksson (1975) has prepared extracts of *E. coli* that have much higher activity than extracts prepared by harsh mechanical disruption of the cells. Eriksson's extracts showed little or no stimulation when either purified B1 or B2 was added, suggesting that the ribonucleotide reductase B1–B2 complex was intact in his crude preparations. His results indicated that an unknown factor—possibly an additional protein, association with membrane structure or existence of a multi-enzyme complex—prevented the dissociation of the B1–B2 complex *in vivo*. In summary, unlike aspartate transcarbamylase, the active site of ribonucleotide reductase is formed only in the presence of both subunits and Mg^{2+}, with each subunit contributing a portion of the active site. Availability of the three-dimensional structure of ribonucleotide reductase may be needed to clarify the exact roles of the individual subunits.

The effect of nucleoside triphosphate allosteric effectors on ribonucleotide reductase is a systematic but complex phenomenon. Initially, the reduction of CDP to *d*CDP was used as an assay for the purification of the enzyme. Larsson and Reichard (1966) demonstrated that reduction of ^{3}H-CDP to ^{3}H-*d*CDP was inhibited competitively by UDP. Similarly, GDP and ADP were shown to compete with CDP for the enzyme. In addition, the ability to reduce all four ribonucleotides was purified in parallel during the isolation of B1 and B2. Therefore, the same site on the reductase probably catalyzes the reduction of all the ribonucleotides. Despite an earlier tentative conclusion that different enzyme components were involved in the reductions of CDP and ADP, Larsson and Reichard recognized that the difference in requirements for reduction of purine and pyrimidine ribonucleotides was associated with allosteric control of substrate specificity. The reduction of each nucleotide requires B1, B2, Mg^{2+} and an allosteric activator (summarized in Table III). ATP stimulates the reduction of CDP and UDP only, whereas in the presence of *d*TTP, the reduction of all the ribonucleotides is possible. Binding of *d*GTP converts the reductase into a form which catalyzes the reduction of ADP and GDP, while interaction of *d*ATP with the reductase results in inactivation of the enzyme. Reichard has proposed that the

Table III. Control of the activity and substrate specificity of ribonucleotide reductase by allosteric effectors. Adapted from Larsson and Reichard (1966).

Effector added	Activity state
None	Relatively inactive
dATP	Inactive
ATP	CDP and UDP reduced
dTTP	ADP, CDP, GDP and UDP reduced
dGTP	ADP and GDP reduced

four catalytic states shown in Table III have the following temporal sequence and physiological implications. The presence of high concentrations of ATP initiates the formation of the pyrimidine deoxyribonucleotides, CDP and UDP. In a second phase, dUDP is converted to dTTP by reactions catalyzed by other enzymes, and the formation of dTTP stimulates the formation of purine nucleotides, including dGTP. The synthesis of purine deoxyribonucleotides catalyzed by the reductase is accelerated by dGTP, until sufficient dATP accumulates and the reductase activity is turned off by a general feedback inhibition in the final phase. While DNA synthesis is proceeding, dATP does not accumulate and the reductase continues to catalyze the production of the necessary deoxyribonucleotides. Feedback inhibition by dATP prevents overproduction of deoxyribonucleotides, but this inhibition by dATP can be reversed by high concentrations of ATP, which starts the catalytic cycle again.

The regulatory effects observed by Larsson and Reichard with the partially purified ribonucleotide reductase have been verified with homogeneous enzyme material (Brown *et al.*, 1969a). The one exception was the observation that dATP has a stimulatory effect on CDP reduction at low concentrations (1μM) of the effector. At higher concentrations of dATP, feedback inhibition was observed as previously reported. A more complete description of the control of ribonucleotide reductase is presented in Table IV. These data include the effects of certain combinations of ligands, and the concentrations of ligands that produce maximal effects. The physical basis of inactivation of the reductase by dATP seems to be a polymerization

Table IV. Influence of ligands on the activity of ribonucleotide reductase. The concentrations of the ligands are those at which their allosteric effects are fully developed. Adapted from p. 52 of Brown and Reichard (1969).

Effector(s)	Activity
ATP $(2 \times 10^{-3}$ M$)$	Stimulation of pyrimidine nucleotide reduction
dATP $(10^{-6}$ M$)$	Stimulation of pyrimidine nucleotide reduction
dATP $(10^{-4}$ M$)$	Inhibition
dGTP $(10^{-5}$ M$)$	Stimulation of purine nucleotide reduction
dTTP $(10^{-5}$ M$)$	Stimulation of both purine and pyrimidine nucleotide reduction
dATP $(10^{-4}$ M$)$ + ATP $(2 \times 10^{-3}$ M$)$	Stimulation of pyrimidine nucleotide reduction
dATP $(10^{-6}$ M$)$ + dTTP $(10^{-4}$ M$)$	Inhibition
dATP $(10^{-6}$ M$)$ + dGTP $(10^{-4}$ M$)$	Inhibition
ATP $(10^{-4}$ M$)$ + dTTP $(5 \times 10^{-4}$ M$)$	Inhibition

of the protein (Brown and Reichard, 1969), but the size of the inhibited complex is unknown (Thelander, 1973). As described above, only B1 has the capacity to bind allosteric effectors. Brown and Reichard (1969) performed binding studies using the four allosteric effectors: dATP, ATP, dGTP and dTTP. Experiments were done with B1 alone and with the B1–B2 complex, with virtually the same results. Scatchard plots of dATP binding indicate that there are four non-equivalent binding sites for this nucleotide. Two molecules of dATP bind with high affinity, and two with a much lower affinity. dGTP and dTTP are bound only to the high affinity sites, with their n values being 2. The binding of ATP is more complicated, occurring at both the high and low affinity sites ($n = 4$). This is reasonable, since ATP has two functions; it stimulates reduction of pyrimidine ribonucleotides and reverses inhibition by dATP. ATP also inhibits the reductase, but only in the presence of dTTP (Table IV). Competition experiments showed that dGTP and dTTP compete with dATP for the high affinity site. All of Brown's and Reichard's data can be explained by the presence of four binding sites per B1 subunit (160 000 daltons, or two polypeptide chains). There are two high affinity sites, which can bind any of the four effectors and, by this binding, control the substrate specificity of ribonucleotide reductase. The two low affinity sites can bind only dATP and ATP. Binding of either of these two effectors at the low affinity sites regulates the overall catalytic activity of the reductase. The activators, the inhibitors and the substrates bind at different sites, so the enzyme meets the requirements for description as an allosteric enzyme. No homotropic co-operative effects have been observed with ribonucleotide reductase. The data described above predict the existence of many different conformational states of the reductase, so a simple two-state MWC model cannot explain the properties.

Characterization of the ribonucleoside diphosphate reductase from mammalian tissues has been even more difficult than that of the $E.$ $coli$ reductase. Nonetheless, the available data provide evidence that the enzyme from animals is similar to that from $E.$ $coli$ in reaction stoichiometry and allosteric regulation, and that it has a major role in control of DNA biosynthesis in animals. In 1969, Larsson found that the activity of the reductase could not be detected in crude extracts of rat liver, but activity was present in crude extracts of regenerating liver. After partial purification of material suspected to contain reductase, the enzyme could be detected in normal liver—but at much lower levels than in the regenerating tissue. Elford et $al.$ (1970) compared the activities of the reductase in normal rat liver, a series of rat liver tumours of differing growth rate, and regenerating and foetal livers. They observed an amazingly close correlation between the rate of cell growth and the specific activity of ribonucleotide reductase, with the fastest growing tissue having at least a 3000-fold greater specific activity than normal liver. Differences in specific activity up to 200-fold between the slowest and fastest growing tumours were reported! The tissue having the greatest activity, foetal liver, was three times more active than the fastest growing tumour, the Novikoff hepatoma, while, as Larsson had observed, the normal liver had no observable activity at all in the crude extracts. The regenerating liver proved to be comparable to tumours of medium growth rate.

Ribonucleotide reductase activity cannot be detected in unfertilized sea urchin eggs, but catalytic activity appears soon after fertilization (Noronha *et al.*, 1972). This is the first known example of an enzyme synthesized *de novo* in an egg in response to fertilization. It appears, therefore, that cells adapt to their requirement for DNA by synthesizing more ribonucleotide reductase (or preventing its degradation). An additional link between DNA synthesis and reduction of ribonucleotides has been discovered by Elford (1974). In the Novikoff hepatoma and in regenerating rat liver, over 90% of the ribonucleotide reductase activity is associated with a membrane fraction derived from the postmicrosomal supernatant fraction. This membrane fraction also contains a DNA polymerase, the activity of which is related to cell proliferation. Possibly a multi-enzyme complex exists, to catalyze the synthesis of both deoxyribonucleotides and DNA. The source of this membrane fraction and its enzymic composition must be determined before its biological significance can be evaluated. One component of the ribonucleotide reductase system, thioredoxin, has been found to be directly involved in DNA biosynthesis in virus-infected bacteria. DNA polymerase in T7-infected *E. coli* is composed of two protein components. Mark and Modrich (1975) identified the two protein components as a protein specified by the viral gene and *E. coli* thioredoxin. The relevance of this finding to the interrelation of ribonucleotide reductase and DNA polymerase in other cells is not known.

THE GLUTAMINE SYNTHETASE OF *ESCHERICHIA COLI*

The glutamine synthetase of *E. coli*, and also of other Gram-negative bacteria, is part of a highly organized regulatory enzyme system. As well as being regulated by negative feedback control, this glutamine synthetase undergoes reversible covalent modification—adenylylation —catalyzed by enzymes which are themselves under regulation by metabolites. Much of the research on the *E. coli* glutamine synthetase has been performed in the laboratory of Earl Stadtman, and has been recently reviewed (Stadtman and Ginsburg, 1974). Each year for more than a decade, new data have been published on the mechanism of regulation of glutamine synthetase. It will probably take another decade to fully explain at the molecular level the complexities of this controlled system.

Glutamine synthetase catalyzes the energy-requiring amination of the γ-carboxylate of glutamate:

$$\text{Glutamate} + \text{ATP} + \text{NH}_3 \xrightleftharpoons{\text{Me}^{2+}} \text{Glutamine} + \text{ADP} + \text{P}_i.$$

The unmodified form of the synthetase requires Mg^{2+} as the divalent metal ion, and the adenylylated form requires Mn^{2+}. The biosynthesis of glutamine via this reaction is necessary not only for the formation of proteins but also for the supply of many nitrogen-containing compounds. It is not surprising, therefore, that the activity of the synthetase is modulated by the presence of some of these metabolic products. The control of glutamine synthetase activity by feedback inhibition and by covalent modification will be the principal subject of this discussion.

Glutamine synthetase is an oligomeric protein with a molecular weight of 600 000 daltons. It is composed of 12 identical subunits. The conclusion that the protomeric units are identical is based upon the finding of only one N-terminal (serine) and one C-terminal (valine) residue, and the predicted number of peptide fragments after either trypsin digestion or cyanogen bromide cleavage. Electron microscopic experiments (Valentine et al., 1968) have suggested that the 12 subunits are arranged in two hexagonal rings, in which the two layers of hexagons are eclipsed. The dimensions of these aggregates agree with the molecular weight estimates obtained by ultracentrifugation and gel electrophoresis.

The feedback inhibition of the synthetase was largely studied before the existence of covalent modification was known. Woolfolk and Stadtman (1967) tested many compounds and found that purified glutamine synthetase is inhibited specifically by nine products of glutamine metabolism: alanine, glycine, serine, histidine, tryptophan, CTP, AMP, carbamyl phosphate and glucosamine. Because each of these compounds requires the amide group of glutamine for its biosynthesis, inhibition by these metabolites must be more than coincidental. At physiologically reasonable concentrations, one of these metabolites alone cannot fully inhibit the synthetase. Almost complete inhibition can be achieved if moderate concentrations of all of the metabolites are added. Stadtman has termed this type of regulation *cumulative feedback inhibition.* Cumulative inhibition is an effective means of controlling this one step common to many pathways. Activity of glutamine synthetase is progressively decreased by accumulation of each end-product. Later experiments (quoted in Stadtman and Ginsburg, 1974) have established that under appropriate conditions glutamine synthetase can be fully inhibited by saturation with any one inhibitor, but under *in vivo* concentration conditions each inhibitor causes only partial inhibition. The cumulative inhibition pattern is most easily explained by assuming that each inhibitor acts at an independent regulatory site on each subunit, but this appears not to be the case. The inhibitors fall into three categories: those that compete with glutamate (glycine, CTP, tryptophan and alanine), those that compete with NH_3 (histidine and glucosamine-6-phosphate) and those that are noncompetitive with either substrate (AMP and carbamyl phosphate). Kinetic experiments using pairs of inhibitors suggest that most of the inhibitors act at individual regulatory sites. For example, by the criterion of mutual exclusion, there appear to be separate binding sites for at least alanine, tryptophan, histidine, AMP and CTP. On the contrary, glycine, serine and alanine act as though they bind at the same site. The existence of different binding sites for tryptophan and AMP has been demonstrated by Ross and Ginsburg (1969), who used a microcalorimetry technique. Successful use of calorimetry for testing the

Table V. Possible subsites of the active site for binding of feedback inhibitors of glutamine synthetase. Adapted from Dahlquist and Purich (1975).

Metabolic Inhibitor	Region of Active Site
Alanine Glycine Histidine Tryptophan	$-CH-COO^-$ $\quad\mid$ NH_3^+ portion of glutamate
AMP CTP	Nucleoside-phosphate of ATP
Glucosamine-6-phosphate Carbamyl phosphate	γ-Phosphate of ATP Glutamate, NH_3 and ATP sites

binding of other pairs of ligands has not been reported. In contrast to the assertion that there are six or seven independent regulatory sites, all of an allosteric nature, Dahlquist and Purich (1975) have proposed that each of the feedback effectors binds to the active-site region of the synthetase. They have provided evidence, from the effect of Mn^{2+} on the proton magnetic resonance spectra of ligands, that alanine and the substrate glutamate bind at the same site. The ATP site may also bind AMP. They have suggested that glutamine synthetase has an active site capable of both specific catalysis and control. Possible binding regions for feedback inhibitors are listed in Table V. In favour of feedback inhibition using the binding specificity of the active site is the difficulty in imagining how 15 or 16 independent binding sites could exist on the surface of a polypeptide of only 50 000 daltons. Independent sites had been postulated for six or seven feedback inhibitors, three substrates and several divalent metal ions; an exposed region must be available for covalent modification, and interfaces with three neighbouring subunits must exist. The lack of a report of clearcut desensitization of the synthetase may also reflect a close relationship between the regulatory and catalytic sites.

Reversible covalent modification of glutamine synthetase is a control superimposed upon the negative feedback inhibitions. The first indications of the existence of two forms of glutamine synthetase came from independent observations in the laboratories of Holzer and Stadtman. Mecke *et al.* (1966) partially purified a factor, called "glutamine synthetase inactivating enzyme," that rapidly inactivates purified synthetase in the presence of ATP, Mg^{2+} and glutamine. They considered acylation, amination or phosphorylation of glutamine synthetase as possible mechanisms of inactivation, analogous to the modulation of glycogen phosphorylase activity by phosphorylation (p. 558). In 1967, Shapiro *et al.* reported that glutamine synthetase isolated from *E. coli* cells grown under conditions differing in nitrogen nutrition contain different amounts of an ultraviolet-absorbing compound. They identified the absorbing material as adenosine-5'-phosphate. Up to one residue of AMP could be incorporated covalently per subunit of the synthetase (i.e. 12 AMP per molecule of oligomer). The adenylylation of glutamine synthetase is catalyzed by an enzyme which Stadtman calls adenylyltransferase (ATase). Holzer *et al.* (1968) showed that the inactivating enzyme and adenylyltransferase are identical. ATase catalyzes the reaction:

$$\text{Glutamine synthetase} + \text{ATP} \xrightarrow[\text{Glutamine}]{\text{ATase}} \text{Adenylylated glutamine synthetase.}$$

(Active)   (Inactive)

Adenylylation does not cause complete inactivation, but the 5'-adenylylated and the unadenylylated synthetases have vastly different kinetic properties. Glutamine, formed from excess NH_4^+ in the culture media, ultimately inhibits its own formation by stimulating ATase. Extension of these initial experiments has uncovered a regulation system requiring a series of proteins and which is controlled in a systematic manner by metabolites. These proteins and metabolites manipulate the state of adenylylation and thus the activity of the *E. coli* glutamine synthetase.

Adenylylation of glutamine synthetase changes it from a Mg^{2+}-to a Mn^{2+}-requiring enzyme. This effect is probably the major physiological consequence of adenylylation, since insufficient Mn^{2+} is present in culture media to support glutamine biosynthesis catalyzed by the adenylylated enzyme. Adenylylation also increases the susceptibility of the synthetase to inhibition by histidine, tryptophan, CTP and AMP, and yet it decreases the extent of inhibition by alanine and glycine.[*] The full significance of adenylylation on feedback inhibition has not yet been ascertained.

Unadenylylated synthetase is isolated from *E. coli* cells grown on limiting amounts of NH_4^+. *Escherichia coli* grown in a medium containing excess glutamate produce adenylylated enzyme. The extent of adenylylation is quite variable, and synthetase molecules at almost any stage of covalent modification [from 0 to 12 mol AMP (mol synthetase)$^{-1}$] can be obtained under appropriate culture conditions. Adenylylation and deadenylylation can also be effected by *in vitro* procedures, using the ATase or snake venom phosphodiesterase as catalysts. Heinrikson and Kingdon (1971) prepared radioactively labelled adenylylated synthetase by incubating deadenylylated enzyme with ^{14}C-ATP and ATase. From tryptic digestion of the labelled synthetase, they obtained a single labelled peptide, 21 residues long, with the adenylyl group bound to the hydroxyl group

---

[*] Stadtman and Ginsburg (1974) have estimated that the synthetase used by Woolfolk in the cumulative feedback experiments contained about six adenylylated subunits per molecule.

of a tyrosine residue. Their observation of only one labelled peptide provides further evidence for the identity of the 12 protomers of glutamine synthetase. The physical properties of the deadenylylated and adenylylated synthetases are identical. The two forms are indistinguishable by electron microscopy, sedimentation analysis or immunochemical properties. Therefore, the changes in catalytic capability are not the result of large changes in either the tertiary or quaternary structure.

The control mechanism now being elucidated is the cascade control of the reversible adenylylation–deadenylylation of glutamine synthetase. At first, it appeared that two enzymes—the ATase and a deadenylylating enzyme—catalyzed the covalent modification:

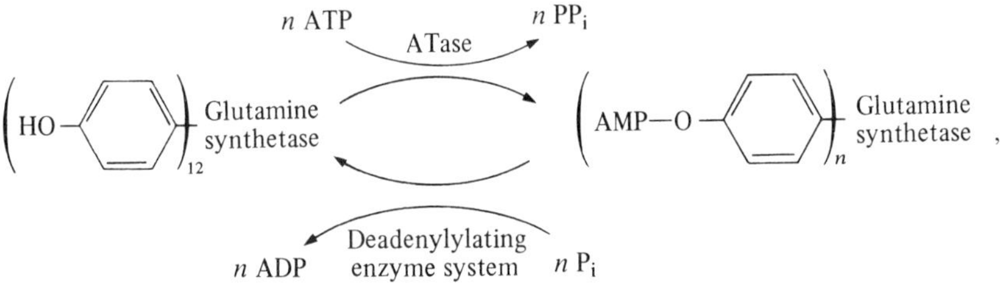

where n can be as high as 12.

However, the adenylylation–deadenylylation cycle requires at least three protein components and the substrate, glutamine synthetase. Shapiro (1969) resolved the deadenylylating enzyme into two protein fractions, P_I and P_{II}. As well as both these protein fractions, the deadenylylation of glutamine synthetase in Shapiro's assays required α-ketoglutarate, UTP and ATP. P_{II} alone has no known catalytic activity, but highly purified preparations of the P_I protein component possess not only the activity needed for deadenylylation but also ATase activity (Anderson et al., 1970). Adenylyltransferase, therefore, is a bifunctional enzyme, catalyzing both the covalent modification and the reactivation of glutamine synthetase. Indirect evidence suggests that there are different sites on ATase for catalysis of the two reactions. ATase binds α-ketoglutarate as an inhibitor of adenylylation and an activator of deadenylylation, and glutamine as an inhibitor of deadenylylation and an activator of adenylylation. Using partially purified P_I and P_{II} fractions, Brown et al. (1971) found that the P_{II} protein exists in two interconvertible forms. The form originally observed, P_{IID}, stimulates deadenylylation (hence, the subscript D) in the presence of ATase. It has little effect on adenylylation. The other form, P_{IIA} (with the subscript A because it promotes adenylylation), stimulates the adenylylation reaction catalyzed by ATase but does not greatly affect the deadenylylation reaction. Brown *et al.* reported preliminary evidence for covalent attachment of UMP to P_{IID}, explaining the requirement of UTP for deadenylylation. P_{II} is a regulatory protein which specifies the activity of ATase. It may be considered as a regulatory subunit of an (ATase–P_{II}) complex which, depending on the absence or presence of UMP in P_{II}, catalyzes either:

$$\text{Glutamine synthetase} + \text{ATP} \xrightarrow{\text{ATase-}P_{IIA}}$$

$$\text{Glutamine synthetase}-(\text{AMP}) + \text{PP}_i$$

or

$$\text{Glutamine synthetase}-(\text{AMP}) + \text{P}_i \xrightarrow{\text{ATase-}P_{IID}}$$

$$\text{Glutamine synthetase} + \text{ADP}.$$

Control of these two reactions by P_{II} is necessary so that a futile adenylylation–deadenylylation cycle is not established, leading to aimless formation of $\text{ADP} + \text{PP}_i$ from $\text{ATP} + \text{P}_i$ (the sum of the above two reactions).

Mangum *et al.* (1973) showed that there are two additional enzyme activities present in *E. coli*, primarily in the P_I fraction, that catalyze the attachment and detachment of UMP. One activity, called uridylyltransferase (UTase), catalyzes the modification of P_{IIA} to P_{IID}. By isotope techniques, they verified that the UMP moiety of UTP is covalently bound to P_{IID}. α-Ketoglutarate and ATP appear to activate UTase by an allosteric mechanism, just as they do the (ATase–P_{IID})-catalyzed deadenylylation reaction:

$$P_{IIA} + x\,\text{UTP} \xrightarrow[\text{ATP, }\alpha\text{-KG, Me}^{2+}]{\text{UTase}} P_{IID} + x\,\text{PP}_i.$$

Mangum *et al.* also showed that the conversion of P_{IID} back to P_{IIA}, the unmodified form, is catalyzed by a uridylyl removing enzyme (UR-enzyme):

$$P_{IID} \xrightarrow[\text{Mn}^{2+}]{\text{UR-enzyme}} P_{IIA} + x\,\text{UMP}.$$

P_{II}, now purified to homogeneity, is a tetrameric protein of 44 000 daltons. It can bind UMP to one tyrosine in each of its four identical subunits (Adler *et al.*, 1975). The UR-enzyme and UTase activities copurify through many steps, so it may be a single enzyme or an enzyme complex.

Now that the components and reactions of the covalent modification control system have been established, an integrated regulatory pathway can be written. The control of glutamine synthetase activity through variation in the level of adenylylation is regulated by the concentrations of metabolites such as glutamine, α-ketoglutarate, ATP and UTP, and of the divalent metal ions, Mg^{2+} and Mn^{2+}. The effects of ligands on adenylylation and deadenylylation, presumably expressed by allosteric activations or inhibitions, are summarized in Table VI. Excess glutamine favours adenylylation, resulting in lowered catalytic potential for glutamine synthetase. It activates the ATase–P_{IIA} complex, and inhibits the activity of the ATase–P_{IID} complex and of UTase. The action of α-ketoglutarate opposes that of glutamine. It promotes deadenylylation by activation of AMP removal from the synthetase and inhibition of adenylylation. For adenylylation to occur, UTP must be available for uridylylation of P_{IIA} to form the modified form of P_{II}. ATP at low concentrations activates deadenylylation, but at high concentrations favours adenylylation

Table VI. Effector ligands for the four covalent modifications. Me^{2+} indicates that either Mg^{2+} or Mn^{2+} is required.

Activity	Effectors	
	Activators	Inhibitors
Adenylylation ($ATase$-P_{IIA})	Glutamine + Me^{2+}	α-Ketoglutarate
Deadenylylation ($ATase$-P_{IID})	α-Ketoglutarate + ATP + Me^{2+}	Glutamine
Uridylylation (UTase)	α-Ketoglutarate + ATP + Me^{2+}	Glutamine
Deuridylylation (UR-enzyme)	Mn^{2+} (or Mg^{2+} + α-KG + ATP)	–

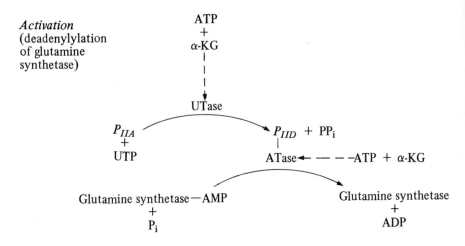

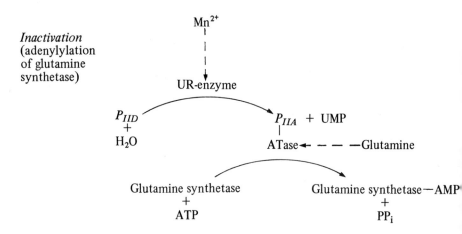

Fig. 4. The two cascade control mechanisms involved in the regulation of the activity of glutamine synthetase. The diagram shows only the metabolic activators, not the inhibitors, of the modifying enzymes.

because of its role as substrate. The contrary effects of α-ketoglutarate and glutamine on the synthetase activity through covalent modification are depicted in Fig. 4 by two opposing cascade control pathways. α-Ketoglutarate and ATP stimulate not only the catalyst of uridylylation but also the next catalyst in the cascade, the deadenylylation activity of ATase. Glutamine has no effect on UR-enzyme activity, but inhibits UTase and activates the adenylylation reaction. As expected in a cascade or amplification process, the proteins initiating regulation—the uridylylation–deuridylylation enzyme(s)—are present in the lowest concentrations in the bacterial cell. P_{II} and ATase are present at intermediate concentrations, and glutamine synthetase at a much higher concentration.

Research is now being directed towards obtaining a quantitative understanding of how adenylylation alters the rate of glutamine biosynthesis. The Mg^{2+}-dependent biosynthetic activity of glutamine synthetase does not vary in a linear manner with the extent of adenylylation. The largest decrease in activity occurs after the binding of the first three AMP residues per oligomer. This indicates that there must be interaction between adenylylated and unadenylylated subunits in single hybrid glutamine synthetase molecules. Using a reconstituted system which allows both adenylylation and deadenylylation to proceed simultaneously, Segal et al. (1974) have concluded that both the attachment and the release of the 5′-AMP groups occur until a steady state is reached, producing a constant number of adenylylated subunits. When the rates of adenylylation and deadenylylation are equal, the extent of adenylylation depends on the relative concentrations of the regulatory metabolites. Stadtman has predicted that the maintenance of a dynamic steady state utilizes cellular energy, but the cost of imposing this strict metabolic control is less than 0.1% of the energy used in the synthesis of glutamine. Thus, by governing the relative rates of adenylylation and deadenylylation, the key metabolites of the E. coli cell effectively control the biosynthesis of the central chemical in nitrogen anabolism, glutamine. Many experiments remain to be performed, especially with the less easily purified protein components such as P_{II}, ATase and the uridylylation enzyme activities, and also in assessing the precise roles of Mg^{2+} and Mn^{2+} concentrations.

LACTOSE SYNTHETASE

The disaccharide lactose is the carbohydrate in highest concentration in the milk of most mammals. Both the galactosyl and glucosyl portions of lactose are derived metabolically from blood glucose. The

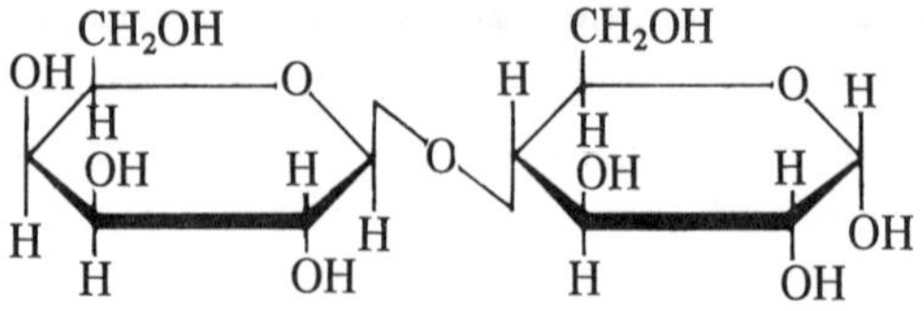

Lactose, or β–D–galactosyl–1,4–α–D–glucose

regulation of the final step in the biosynthesis of this nutrient sugar is a fascinating process, introducing the principle of modification of enzyme activity by the presence of a non-catalytic protein. The *modifier protein* is α-lactalbumin, a protein characterized many years ago but whose biological function was discerned only about ten years ago (reviewed by Ebner, 1970, 1973; Hill and Brew, 1975).

Milk contains two principal classes of proteins: caseins, a mixture of phosphoproteins that precipitate when milk is adjusted to pH 4.5, and whey proteins which are not precipitated at this pH value. β-Lactoglobulin and α-lactalbumin are the most common proteins of whey. Crystalline α-lactalbumin has a molecular weight of 15 500 daltons. Because no specific biological role for lactalbumin was known, biochemists had assumed that it served a nutritional function in milk. This evaluation of its role was dramatically altered when α-lactalbumin was discovered to be a component of the lactose synthetase enzyme system.

Crude preparations of mammary tissue extracts can catalyze the formation of lactose from uridine diphosphate-D-galactose (UDPGal) and D-glucose:

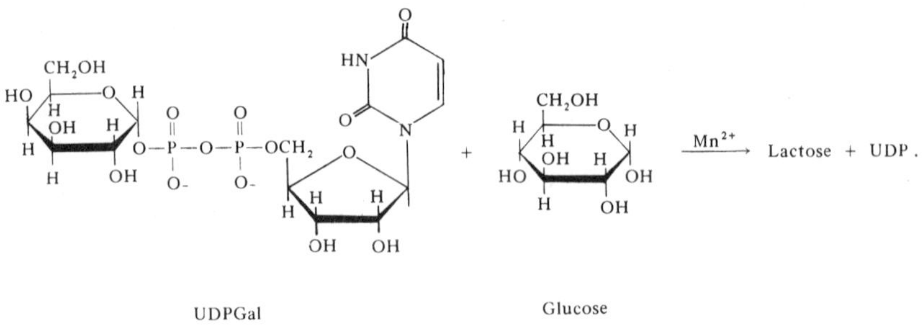

UDPGal   Glucose

In this reaction, there is the formation of a β-1,4-bond between the sugar donated by the UDPGal and the acceptor monosaccharide, glucose. The enzyme which catalyzes this reaction is present only in the microsomal fraction of mammary tissues. As with many membrane-bound proteins, its purification is difficult, with loss of much activity occurring after solubilization. Babad and Hassid (1966) found that lactose synthetase is *soluble* in milk, so they used cow's milk as a reproducible source for partial (approximately 50-fold) purification of the synthetase. If they attempted further purification, they observed drastic losses in activity. Because of its non-particulate state in milks, lactose synthetase has been purified only from milk. The lactose synthetase system is secreted into milk and also into serum, then the enzyme is solubilized—probably by limited proteolysis. Two proteins from whey, called A and B, can be separated by gel filtration. Because both the A and B proteins are required for lactose synthetase activity, I shall refer to them as non-identical subunits of the synthetase. Brodbeck *et al.* (1967) purified the lower molecular weight B subunit of lactose synthetase from milk and from mammary tissue. They showed that the B protein is identical to the whey protein α-lactalbumin not only by complete interchangeability in enzymatic reactions but also by a variety of chemical and physical criteria. Brew *et al.* (1968) reported that α-lactalbumin is devoid of enzymic activity, and demonstrated that the higher molecular weight or A subunit is a potent galactosyltransferase. For example, it catalyzes the formation of N-acetyl-D-glucosamine from UDPGal and the galactosyl acceptor N-acetyl-D-glucosamine (GlcNAc):

$$\text{UDPGal} + \text{GlcNAc} \rightarrow \textit{N}\text{-Acetyllactosamine} + \text{UDP}.$$

Brew *et al.* concluded that the larger protein of lactose synthetase, the galactosyltransferase subunit, is similar or identical to the galactosyltransferase occurring in the microsomal fraction of many tissues. Galactosyltransferase is one of a series of glycosyltransferases catalyzing the addition of specific carbohydrates to the growing oligosaccharides of glycoproteins. These membrane-bound glycosyltransferases are located in the Golgi apparatus (Schachter *et al.*, 1970), and form a multi-enzyme system responsible for the completion and eventual secretion of glycoproteins. Subunit A of lactose synthetase from milk has been obtained free from contaminating proteins. It is a glycoprotein itself, with a molecular weight of about 45 000 daltons.

Several active forms, possibly resulting from limited proteolysis and/or differing carbohydrate content, are detected by gel filtration or SDS electrophoresis testing.

During lactation, the mammary gland modifies the activity of the pre-existing galactosyltransferase to allow the biosynthesis of lactose. The synthesis of α-lactalbumin, under hormonal control, starts at childbirth. α-Lactalbumin activates the synthesis of lactose in the presence of glucose and galactosyltransferase, and inhibits the galactosylation of N-acetylglucosamine (or glycoproteins). Because the mammary glands are the only tissue able to synthesize α-lactalbumin, they are unique in possessing high lactose synthetase activity. In mammary tissue, both galactosyltransferase reactions—synthesis of glycoprotein and of lactose—can be maintained during lactation. I will next discuss the manner of the activation of lactose synthesis by the modifier subunit.

Initial-velocity kinetic experiments have greatly aided in the elucidation of the mechanism of the reaction catalyzed by lactose synthetase. However, the kinetic results are very involved, and because the reaction is not reversible, product inhibition studies are not possible. The kinetic mechanism can be described, but with the limitation that the assigned order of binding of α-lactalbumin to the transferase is only tentative. Morrison and Ebner (1971; verified by Geren et al., 1975) concluded that the transfer of the galactosyl group from UDPGal to monosaccharides follows an ordered mechanism. The reactants add in the order Mn^{2+}, UDPGal and galactosyl acceptor (monosaccharide or MS), and the products are released in the order disaccharide (DS) and UDP. Mn^{2+} need not dissociate from the enzyme after each catalytic cycle. The simplest productive mechanism consistent with their data is:

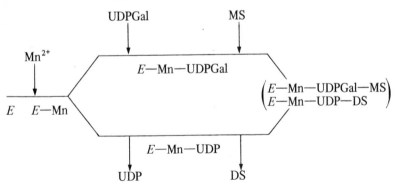

When Morrison and Ebner examined the effect of α-lactalbumin on the kinetics of the lactose synthetase reaction, they observed that the most marked effect is a decrease of the apparent K_m for glucose from a value well above 1 M in the absence of α-lactalbumin to a value in the millimolar range in the presence of lactalbumin. These authors concluded that α-lactalbumin acts kinetically like a substrate, adding to and dissociating from the enzyme at each turn of the catalytic cycle. They favour the following kinetic mechanism for lactose synthesis in the presence of α-lactalbumin:

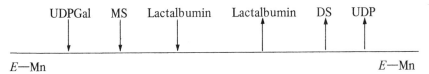

In view of the fact that α-lactalbumin causes no change in the basic mechanism and because lactalbumin–galactosyltransferase complexes are unstable, they reasoned that α-lactalbumin adds immediately before the formation of the central complex. Its dissociation before the disaccharide and UDP is in accordance with the feature of ordered reactions that the product of the last substrate to add is the first to dissociate. They ascribed the activating effect of the lactalbumin modifier on lactose synthesis to its ability, when combining with the ternary complex, to displace already established equilibria. In other words, α-lactalbumin suppresses the dissociation of glucose from the central complex to reform the E–Mn–UDPGal complex. Morrison and Ebner suggest that lactalbumin differs from small allosteric effectors by exerting its influence only after Mn^{2+} and both substrates are already bound to the active site. Whether α-lactalbumin adds immediately before or immediately after glucose (or if their addition order is random) has not been proven by the kinetic experiments, though. In fact, physical chemical studies indicate that glucose need not be present for the modifier subunit to add to the galactosyltransferase subunit.

Direct interaction between the two subunits of lactose synthetase has been difficult to demonstrate. Complex formation between these two proteins is stable only in the presence of substrates of the galactosyltransferase. This interaction has been used as a step in purifying the transferase subunit. It is retained on a Sepharose-α-lactalbumin affinity column when the elution buffer contains a substrate such as

N-acetylglucosamine. After inert proteins are removed, the transferase is eluted from the affinity column by omission of the substrate from the buffer solution. Klee and Klee (1972) showed the protein–protein interaction more directly by sedimentation analysis. They found that a complex containing one molecule each of transferase and lactalbumin was formed, but they could only detect the complex in the presence of Mn^{2+} and either UDPGal or N-acetylglucosamine. Osborne and Steiner (1974) have used fluorescent conjugates of α-lactalbumin to detect complex formation by fluorescence polarization. A lactalbumin–galactosyltransferase complex is formed in the presence of either UDPGal or N-acetylglucosamine alone, and Mn^{2+} further stabilizes the complex. Complex formation, therefore, does not require prior binding to the galactosyltransferase of both substrates. Neither of these studies proves that the observed complexes are catalytically active. Indeed, they may be dead-end complexes (i.e. lie on non-productive side pathways).

Brew *et al.* (1975) have employed dimethylpimelimidate (cf. p. 71) to crosslink potential complexes. Starting with mixtures containing a ten-fold excess of lactalbumin over transferase, they examined the effects of combinations of substrates on crosslinking. As an assay of 1 : 1 complex formation, they measured the appearance of protein of appropriate mobility on SDS gels. They found that the formation of crosslinked complexes is dependent on the presence of substrates. No complex was formed in the absence of N-acetylglucosamine, while presence of Mn^{2+} and UDPGal in addition produced maximal crosslinking. The crosslinked complex was isolated and tested for catalytic activity. While it possessed only about 1% of the specific activity of unmodified lactose synthetase, it did not require the addition of α-lactalbumin to catalyze the synthesis of lactose at low concentrations of glucose. Thus, the formation of an enzymically active covalent complex had been achieved, but its conformation must differ from that of the natural non-covalent complex. Stabilization of the lactalbumin–transferase complex by covalent crosslinking may allow structural studies for elucidation of the mechanism by which α-lactalbumin redirects galactosyltransferase to the synthesis of lactose.

Of the four regulatory enzymes described in this chapter, aspartate transcarbamylase and ribonucleotide reductase are clearly allosteric enzymes. The non-covalent feedback control of glutamine synthetase may or may not be allosteric, but its covalent modification certainly is. Whether lactose synthetase can be called "allosteric" in the true sense remains undecided. Above all, the terminology is overshadowed by the potent physiological controls achieved by these catalytic macromolecules.

REFERENCES

Adler, S. P., Purich, D. and Stadtman, E. R. (1975). *J. biol. Chem.* **250**, 6264-6272.

Anderson, W. B., Hennig, S. B., Ginsburg, A. and Stadtman, E. R. (1970). *Proc. natn. Acad. Sci. U.S.A.* **67**, 1417-1424.

Atkin, C. L., Thelander, L., Reichard, P. and Lang, G. (1973). *J. biol. Chem.* **248**, 7464-7472.

Babad, H. and Hassid, W. Z. (1966). *J. biol. Chem.* **241**, 2672-2678.

Biswas, C. and Paulus, H. (1973). *J. biol. Chem.* **248**, 2894-2900.

Brew, K., Vanaman, T. C. and Hill, R. L. (1968). *Proc. natn. Acad. Sci. U.S.A.* **59**, 491-497.

Brew, K. Shaper, J. H., Olsen, K. W., Trayer, I. P. and Hill, R. L. (1975). *J. biol. Chem.* **250**, 1434-1444.

Brodbeck, U., Denton, W. L., Tanahashi, N. and Ebner, K. E. (1967). *J. biol. Chem.* **242**, 1391-1397.

Brown, M. S., Segal, A. and Stadtman, E. R. (1971). *Proc. natn. Acad. Sci. U.S.A.* **68**, 2949-2953.

Brown, N. C. and Reichard, P. (1969). *J. molec. Biol.* **46**, 25-55.

Brown, N. C., Canellakis, Z. N., Lundin, B., Reichard, P. and Thelander, L. (1969a). *Eur. J. Biochem.* **9**, 561-573.

Brown, N. C., Eliasson, R., Reichard, P. and Thelander, L. (1969b). *Eur. J. Biochem.* **9**, 512-518.

Chan, W. W.-C. (1975). *J. biol. Chem.* **250**, 668-674.

Cohen, G. N. (1968). "The Regulation of Cell Metabolism", pp. 101-127. Holt, Rinehart and Winston, New York.

Cohlberg, J. A., Pigiet, V. P., Jr. and Schachman, H. K. (1972). *Biochemistry* **11**, 3396-3411.

Dahlquist, F. W. and Purich, D. L. (1975). *Biochemistry* **14**, 1980-1989.

Datta, P. and Prakash, L. (1966). *J. biol. Chem.* **241**, 5827-5835.

Dungan, S. M. and Datta, P. (1973). *J. biol. Chem.* **248**, 8534-8546.

Ebner, K. E. (1970). *Accts Chem. Res.* **3**, 41-47.

Ebner, K. E. (1973). *In* "The Enzymes" (P. D. Boyer, ed.), Third edition, Vol. 9, pp. 363-377. Academic Press, New York.

Edwards, B. F. P., Evans, D. R., Warren, S. G., Monaco, H. L., Landfear, S. M., Eisele, G., Crawford, J. L., Wiley, D. C. and Lipscomb, W. N. (1974). *Proc. natn. Acad. Sci. U.S.A.* **71**, 4437-4441.

Elford, H. L. (1974). *Archs Biochem. Biophys.* **163**, 537-543.
Elford, H. L., Freese, M., Passamani, E. and Morris, H. P. (1970). *J. biol. Chem.* **245**, 5228-5233.
Eriksson, S. (1975). *Eur. J. Biochem.* **56**, 289-294.
Evans, D. R., Pastra-Landis, S. C. and Lipscomb, W. N. (1975). *J. biol. Chem.* **250**, 3571-3583.
Geren, C. R., Geren, L. M. and Ebner, K. E. (1975). *Biochem. biophys. Res. Commun.* **66**, 139-143.
Gerhart, J. C. and Pardee, A. B. (1962). *J. biol. Chem.* **237**, 891-896.
Gerhart, J. C. and Schachman, H. K. (1965). *Biochemistry* **4**, 1054-1062.
Hammes, G. G., Porter, R. W. and Wu, C.-W. (1970). *Biochemistry* **9**, 2992-2994.
Heinrikson, R. L. and Kingdon, H. S. (1971). *J. biol. Chem.* **246**, 1099-1106.
Hill, R. L. and Brew, K. (1975). *Adv. Enzymol.* **43**, 411-490.
Holzer, H., Schutt, H., Mašek, Z. and Mecke, D. (1968). *Proc. natn. Acad. Sci. U.S.A.* **60**, 721-724.
Jacobsberg, L. B., Kantrowitz, E. R. and Lipscomb, W. N. (1975). *J. biol. Chem.* **250**, 9238-9249.
Jacobson, G. R. and Stark, G. R. (1973). In "The Enzymes" (P. D. Boyer, ed.), Third edition, Vol. 9, pp. 225-308. Academic Press, New York.
Jacobson, G. R. and Stark, G. R. (1975). *J. biol. Chem.* **250**, 6852-6860.
Kerbiriou, D. and Hervé, G. (1973). *J. molec. Biol.* **78**, 687-702.
Klee, W. E. and Klee, C. B. (1972). *J. biol. Chem.* **247**, 2336-2344.
Larsson, A. (1969). *Eur. J. Biochem.* **11**, 113-121.
Larsson, A. and Reichard, P. (1966). *J. biol. Chem.* **241**, 2533-2549.
Mangum, J. H., Magni, G. and Stadtman, E. R. (1973). *Archs Biochem. Biophys.* **158**, 514-525.
Mark, D. and Modrich, P. (1975). *Fedn Proc.* **34**, 639.
Mecke, D., Wulff, K., Liess, K. and Holzer, H. (1966). *Biochem. biophys. Res. Commun.* **24**, 452-458.
Meighen, E. A., Pigiet, V. and Schachman, H. K. (1970). *Proc. natn. Acad. Sci. U.S.A.* **65**, 234-241.
Morrison, J. F. and Ebner, K. E. (1971). *J. biol. Chem.* **246**, 3977-3998.
Mort, J. S. and Chan, W. W.-C. (1975). *J. biol. Chem.* **250**, 653-660.
Nagel, G. M. and Schachman, H. K. (1975). *Biochemistry* **14**, 3195-3203.
Nelbach, M. E., Pigiet, V. P., Jr., Gerhart, J. C. and Schachman, H. K. (1972). *Biochemistry* **11**, 315-327.
Noronha, J. M., Sheys, G. H. and Buchanan, J. M. (1972). *Proc. natn. Acad. Sci. U.S.A.* **69**, 2006-2010.
Osborne, J. C. and Steiner, R. F. (1974). *Archs Biochem. Biophys.* **165**, 615-627.
Patte, J.-C., LeBras, G. and Cohen, G. N. (1967). *Biochim. biophys. Acta* **136**, 245-257.
Paulus, H. and Gray, E. (1967). *J. biol. Chem.* **242**, 4980-4986.
Reichard, P. (1967). "The Biosynthesis of Deoxyribose." John Wiley and Sons, New York.

Reichard, P. (1968). *Eur. J. Biochem.* **3**, 259-266.
Richards, K. E. and Williams, R. C. (1972). *Biochemistry* **11**, 3393-3395.
Rosenbusch, J. P. and Griffin, J. H. (1973). *J. biol. Chem.* **248**, 5063-5066.
Rosenbusch, J. P. and Weber, K. (1971). *Proc. natn. Acad. Sci. U.S.A.* **68**, 1019-1023.
Rosner, A. and Paulus, H. (1971). *J. biol. Chem.* **246**, 2965-2971.
Ross, P. D. and Ginsburg, A. (1969). *Biochemistry* **8**, 4690-4695.
Schachter, H., Jabbal, I., Hudgin, R. L., Pinteric, L., McGuire, E. J. and Roseman, S. (1970). *J. biol. Chem.* **245**, 1090-1100.
Segal, A., Brown, M. S. and Stadtman, E. R. (1974). *Archs Biochem. Biophys.* **161**, 319-327.
Shapiro, B. M. (1969). *Biochemistry* **8**, 659-670.
Shapiro, B. M., Kingdon, H. S. and Stadtman, E. R. (1967). *Proc. natn. Acad. Sci. U.S.A.* **58**, 642-649.
Shepherdson, M. and Pardee, A. B. (1960). *J. biol. Chem.* **235**, 3233-3237.
Stadtman, E. R. (1966). *Adv. Enzymol.* **28**, 41-154.
Stadtman, E. R. and Ginsburg, A. (1974). *In* "The Enzymes" (P. D. Boyer, ed.), Third edition, Vol. 10, pp. 755-807. Academic Press, New York.
Thelander, L. (1973). *J. biol. Chem.* **248**, 4591-4601.
Umbarger, H. E. (1969). *A. Rev. Biochem.* **38**, 323-370.
Valentine, R. C., Shapiro, B. M. and Stadtman, E. R. (1968). *Biochemistry* **7**, 2143-2152.
Warren, S. G., Edwards, B. F. P., Evans, D. R., Wiley, D. C. and Lipscomb, W. N. (1973). *Proc. natn. Acad. Sci. U.S.A.* **70**, 1117-1121.
Weber, K. (1968). *Nature, Lond.* **218**, 1116-1119.
Wedler, F. C. and Gasser, F. J. (1974). *Archs Biochem. Biophys.* **163**, 57-68.
Winlund, C. C. and Chamberlin, M. J. (1970). *Biochem. biophys. Res. Commun.* **40**, 43-49.
Woolfolk, C. A. and Stadtman, E. R. (1967). *Archs Biochem. Biophys.* **118**, 736-755.
Yang, Y. R., Syvanen, J. M., Nagel, G. M. and Schachman, H. K. (1974). *Proc. natn. Acad. Sci. U.S.A.* **71**, 918-922.
Yates, R. A. and Pardee, A. B. (1956). *J. biol. Chem.* **221**, 757-770.

14 Physical Organization of Enzymes

Many enzymes do not exist as freely soluble and diffusible molecules, but are localized in organized systems. Their organization may take the form of binding to related enzymes, binding to a cellular membrane, enclosure in an organelle, or a combination of some of these spatial compartmentations. One distinct advantage of physical organization is an easier access to substrates which may also be partitioned in the cell. In addition, some enzymes, as integral parts of membranes, exhibit a vectorial influence by catalyzing reactions with substrates and products on opposite sides of the membranes. While proximity to higher concentrations of substrate or to enzymes catalyzing subsequent reactions could only be considered as an efficiency mechanism and not as a regulatory mechanism, in some cases the physical coupling of reaction systems does impose a control of reaction rates. The control of these systems is completely lost when their physical integrity is destroyed. I consider even the simplest assembly of enzyme activities a form of regulation. Enzyme assemblies extend from oligomers having two types of protomers to subcellular organizations of tremendous complexity. I therefore feel it is appropriate to discuss the organization of enzymes in the context of control of enzymic activity.

MULTI-ENZYME COMPLEXES

Multi-enzyme complexes are soluble aggregates of enzymes, held together by non-covalent bonds, catalyzing two or more sequential reactions in a metabolic pathway (Ginsburg and Stadtman, 1970; Reed, 1974). The existence of enzyme aggregates of this sort is a specialized expression of quaternary structure. As with other proteins having quaternary structure, multi-enzyme complexes have definite structural and symmetrical composition. The organization and function of multi-enzyme complexes is simpler than that of the membrane-bound enzyme arrays, but the collection of several enzymes into a soluble complex may serve the same purpose as the concentration of several enzymes in a membrane matrix. The efficiency of a series of metabolically related enzymes could be increased if the enzymes were not randomly scattered throughout the cell, but were in enzyme aggregates. Theoretically, the greatest efficiency would be achieved if the product of the first enzyme, when released, was close to the active site of the enzyme catalyzing the second reaction, and so on. Channelling of metabolites in this manner occurs with some multi-enzyme complexes. Feedback inhibition and covalent modification can act as control mechanisms for some complexes.

The detection of multi-enzyme complexes usually is an accidental process. Most complexes have been found when several enzymes have copurified as a result of the purification of one of them. Naturally, the most easily isolated complexes are those in which the enzymes are most firmly bound to each other. Because of weak quaternary interactions, many complexes may have gone undetected to date. For example, the enzymes catalyzing the reactions of glycolysis are generally considered soluble and freely diffusible proteins. There have been suggestions that these enzymes might be physically as well as metabolically adjacent, but until recently there was no direct evidence for a complex. In 1973, however, Clarke and Masters reported that sedimentation of a cytoplasmic fraction of rat muscle showed two boundaries each for aldolase, lactate dehydrogenase and pyruvate kinase activities. One boundary was that of the free enzyme, and the other was a common 23 S boundary, presumed to be caused by the association of the glycolytic enzymes. A complex of glycolytic enzymes has not yet been isolated, though. MacKenzie (1973) found that three tetrahydrofolate-requiring enzymes from pig liver, all previously assumed to be separate proteins, copurified through four isolation steps, suggesting that they exist as an enzyme complex. Not all cytoplasmic enzymes are in multi-enzyme aggregates, but more examples of weakly bound complexes will be discovered. Let us now look at some well characterized enzyme complexes, chosen to exhibit the principles of physical and functional organization. Then let us examine the organizational properties of some membrane-bound enzyme systems.

Fatty Acid Biosynthesis

The biosynthesis of fatty acids occurs in two phases, the carboxylation of acetyl CoA to form malonyl CoA, and the condensation and reduction of acyl groups to form longer-chained fatty acid derivatives. In many organisms, these two sets of reactions are catalyzed by two soluble high molecular weight enzyme systems, acetyl CoA carboxylase and fatty acid synthetase (reviewed by Volpe and Vagelos, 1973).

Since the acetyl CoA carboxylase-catalyzed reaction is the first unique step in fatty acid biosynthesis, the irreversible carboxylation of acetyl CoA to form malonyl CoA, it is, as expected, a site of physiological control. The carboxylase is composed of three protein subunits: biotin carboxylase, carboxyltransferase and biotin carboxyl carrier protein (cf. p. 294). All acetyl CoA carboxylases of animal origin are activated by citrate (Lane and Moss, 1971). The stimulation of carboxylase activity by citrate is associated with a protomer-polymer transition which produces a filamentous carboxylase oligomer of very high molecular weight. In animals, fatty acids are used as a reserve of energy, so feedforward control of their synthesis by cytoplasmic citrate—indicating an accumulation of acetyl CoA, NADPH and ATP, all needed for fatty acid synthesis—seems logical.* In *Escherichia coli*, which makes fatty acids principally for phospholipid formation rather than for energy storage, citrate has no effect on the carboxylase. Fatty acid biosynthesis in *E. coli* can be controlled by the levels of guanosine-5′-diphosphate-3′-diphosphate (ppGpp), which inhibits the carboxyltransferase component of acetyl CoA carboxylase at physiologically significant concentrations (Polakis *et al.*, 1973).

Both acetyl CoA and malonyl CoA are utilized in the formation of long-chain fatty acids. In yeast and animals, the enzymes which catalyze the synthesis of fatty acids from these two precursors are aggregated in stable multi-enzyme complexes, to which the simple

* Recent evidence indicates that both acetyl CoA carboxylase and fatty acid synthetase in animal livers are inactivated by phosphorylation, catalyzed by ATP-dependent kinases, and the phosphoenzymes are activated by Mg^{2+}-dependent phosphatases (e.g. Qureshi *et al.*, 1975).

name fatty acid synthetase has been assigned. In plants and most bacteria, the same reactions are catalyzed by easily dissociable complexes or individual non-aggregating enzymes. The availability of the separate enzymes from *E. coli* enabled easier study of the intermediate steps, especially the role of acyl carrier protein (cf. p. 298). Because the earliest work on the fatty acid synthetase complexes was performed with the yeast complex (Lynen, 1967), I shall detail only its properties. In most respects, the other fatty acid synthetase complexes and the non-associated synthetases catalyze comparable reactions.

Acetyl CoA is the primer of the synthesis of fatty acids. Two-carbon units derived from malonyl CoA are added to the acetyl residue during the synthetic sequence, so that in the final product only the two carbons farthest from the carboxylate are derived from acetyl CoA. Lynen and his colleagues isolated from yeast a homogeneous protein fraction, of molecular weight 2 300 000 daltons, which catalyzes the formation of a mixture of palmityl CoA and stearyl CoA:

Acetyl CoA $+ n$ malonyl CoA $+ 2n$ NADPH $+ 2n$ H$^+$ $\rightarrow$

CH$_3$(CH$_2$CH$_2$)$_n$C($=$O)S$-$CoA $+ n$ CO$_2$ $+ 2n$ NADP$^+$ $+ n$ H$_2$O $+ n$ CoASH,

where $n = 7$ or 8.

Low molecular weight intermediates in this reaction could not be detected, leading to Lynen's suggestion that all of the intermediates are bound to a thiol group of the enzyme. This has been proved, with the demonstration of intermediates bound to acyl carrier protein (ACP) in the *E. coli* system. The yeast multi-enzyme complex also contains acyl carrier protein as one of its associated components (Willecke *et al.*, 1969). The ACP, called the central thiol group by Lynen, carries the covalently bound acyl intermediates from one active subunit to the next, until fatty acid synthesis is completed. Lynen has proposed that the carrier is located in the centre of the complex, providing a flexible arm which can move from one active site to the next, around the organized complex. Direct evidence for this location has not yet been obtained, though. The inability to isolate active subunits of the yeast fatty acid synthetase complex indicates that the proteins are stabilized by their aggregation.

The sequence of reactions in the condensation and reduction of

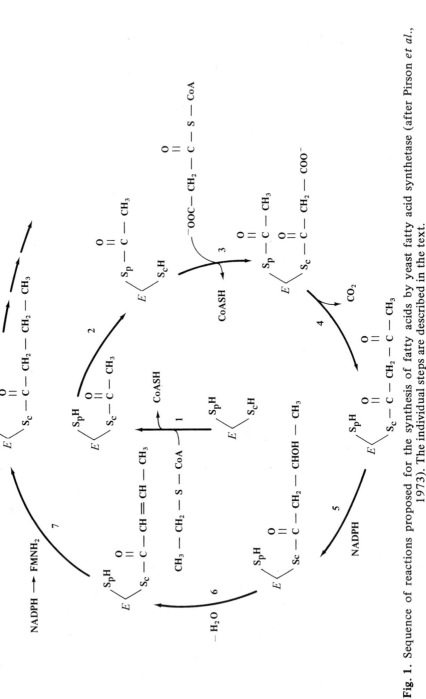

Fig. 1. Sequence of reactions proposed for the synthesis of fatty acids by yeast fatty acid synthetase (after Pirson *et al.*, 1973). The individual steps are described in the text.

the fatty acid precursors is shown in Fig. 1. Step 1 is the binding of the acetyl group to the central thiol ($-S_cH$), the 4'-phosphopantetheine moiety of the acyl carrier protein. In step 2, this acetyl group is transferred to the peripheral thiol ($-S_pH$), a cysteine residue of the condensing enzyme. Next, the malonyl group is bound to the central thiol, and then in step 4 condensation to form acetoacetyl-enzyme occurs, with the acyl group bound for the remainder of the cycle to the ACP (i.e. central thiol). The decarboxylation in the condensation step provides the energy to overcome what would be an unfavourable equilibrium, if two molecules of acetyl-ACP were to condense. Steps 5–7 are the first reduction, the dehydration, and the second reduction needed to form butyryl-enzyme from acetoacetyl-enzyme. The C_4-enzyme then proceeds, by a series of reactions identical to those of the acetyl-enzyme, to form either C_{16^-} or C_{18^-} enzyme as the ultimate enzyme-bound acyl products. The terminal reaction in yeast is transfer of the acyl group from the central thiol to CoA, to form either palmityl or stearyl CoA. Although enzymes catalyzing individual steps have not been isolated, the partial reactions have all been demonstrated with model substrates (acyl derivatives of pantetheine or N-acetylcysteamine).

Characterization of individual polypeptides of the yeast multi-enzyme complex will be a difficult task. To date, only the ACP has been isolated. It is known that the acetyl and malonyl groups form covalent intermediates with different serine residues (so also presumably with different transferases) before they are transferred to ACP (Ziegenhorn et al., 1972). If single polypeptide chains were responsible for the catalysis of each step, by analogy with the E. coli fatty acid synthetase enzymes yeast would contain seven enzymes and the protein coenzyme ACP. Electron microscopy of the synthetase complex indicates that it is an oval complex, with proteins arranged on the surface. ACP may be located in the hollow central space. Aggregation of these proteins into a multi-enzyme complex does confer stability, perhaps of the type seen in quaternary structural effects with identical subunits (cf. p. 410). Adjacent proteins may also catalyze metabolically adjacent reactions. Types of control, such as determination of product size, still must be ascertained. Quaternary structure probably is maintained by both hydrophobic and electrostatic interactions, as it is in the pigeon liver fatty acid synthetase (Kumar et al., 1972).

Mycobacterium smegmatis (formerly called *M. phlei*) is one of the few bacteria from which the fatty acid synthetase complex has been purified (Vance *et al.*, 1973). Sedimentation velocity experiments indicate that the complex has a molecular weight of 1 400 000 in 0.5 M phosphate buffer, but in solutions of low ionic strength it dissociates into subunits of 250 000 daltons. Some activity can be obtained by incubation of subunits in 0.5 M phosphate, suggesting that the antichaotropic properties of high concentrations of phosphate (p. 488) restore or maintain hydrophobic bonds between otherwise inactive subunits. The *M. smegmatis* complex has an unusually high K_m for acetyl CoA, but addition of mycobacterial polysaccharides stimulates fatty acid synthesis by dramatically lowering this K_m value. Palmityl CoA forms stable complexes with the polysaccharides. It also inhibits fatty acid synthesis. At physiologically observable concentrations (0.5 mM) in 0.5 M phosphate, palmityl CoA causes complete dissociation of the complex into 250 000 dalton subunits (Flick and Bloch, 1975). The palmityl CoA is bound to the protein, but when it is removed, the fatty acid synthetase complex can reform. Therefore, Bloch and his associates have concluded that the stimulation of fatty acid biosynthesis by naturally occurring polysaccharides is a relief of inhibition by sequestering the product or feedback inhibitor, palmityl CoA. Experiments with fatty acid synthetases from animal tissues indicate that they too may be regulated by dissociation and association; dissociation (inactivation) occurs in the presence of fatty acyl CoA, and association (activation) in the presence of NADPH.

Pyruvate Dehydrogenase Complex

High molecular weight complexes which catalyze the oxidative decarboxylation of the α-keto acids pyruvate and α-ketoglutarate have been isolated from both prokaryotic and eukaryotic cells (Reed and Cox, 1970; Reed, 1974). Each complex is composed of three enzymes: a thiamine pyrophosphate (TPP)-containing dehydrogenase (E_1), a lipoic acid (LipS$_2$)-containing transacylase (E_2) and a dihydrolipoyl (Lip(SH)$_2$) flavoprotein dehydrogenase. The number of polypeptide chains in three of the α-keto acid dehydrogenases is shown in Table I. In each of these complexes, the lipoate-containing

Table I. Subunit composition of α-keto acid dehydrogenase complexes. Data from Reed (1974).

Enzyme	Molecular Weight	Subunits per Molecule of Complex		
		Dehydrogenase (E_1)	Transacylase (E_2)	Flavoprotein (E_3)
Pyruvate dehydrogenase complex from E. coli	4.6×10^6	24	24	12
Pyruvate dehydrogenase complex from beef kidney[a]	7×10^6	40	60	10
α-Ketoglutarate dehydrogenase complex from E. coli	2.5×10^6	12	24	12

[a] The beef pyruvate dehydrogenase complex also contains a kinase and a phosphatase (approx. 5 molecules of each per molecule of complex).

enzyme is the structural core to which the other two enzymes are non-covalently bound. Electron micrographs of the purified complexes show that they are molecules organized in symmetrical arrays, and Reed and his collaborators have built relatively simple models of the quaternary structure of the complexes (Reed and Cox, 1970). For example, the pyruvate dehydrogenase complex of E. coli seems to be assembled from a cube of transacetylase molecules, with 12 pyruvate dehydrogenase dimers on the edges and six dihydrolipoyl dehydrogenases in the faces.

Pyruvate dehydrogenase complex catalyzes the reaction:

Pyruvate + NAD^+ + CoASH → Acetyl CoA + NADH + CO_2.

As shown in Fig. 2, the enzyme-bound lipoate undergoes a series of reactions: reductive acetylation by α-hydroxyethyl-TPP, acetyl trans-

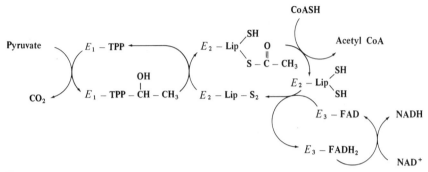

Fig. 2. Pathway for the oxidative decarboxylation of pyruvate. E_1 is pyruvate dehydrogenase, E_2 is dihydrolipoyl transacetylase and E_3 is dihydrolipoyl dehydrogenase.

fer to CoA, and oxidation by the FAD of dihydrolipoyl dehydrogenase. The lipoid acid prosthetic group is bound by its carboxyl group in an amide linkage to the ϵ-amino group of a lysine residue of the transacetylase (E_2). Reed has proposed that the lipoyllysine group acts as a flexible arm, rotating among the catalytic sites of the three enzymes. Both initial-velocity and product inhibition kinetic studies have been performed with the pyruvate dehydrogenase complex from beef kidney mitochondrial matrix (Tsai *et al.*, 1973). One should recall that some biotin-containing enzymes have two-site substituted enzyme mechanisms, in which the biotin accepts a carboxyl group at one site and releases it at a second site (pp. 295–298). With pyruvate dehydrogenase, there is also a possibility of a multi-site ping-pong mechanism (Cleland, 1973). The results of Tsai *et al.* (1973) show that a ping-pong mechanism occurs at two steps of the pyruvate dehydrogenase-catalyzed reaction:

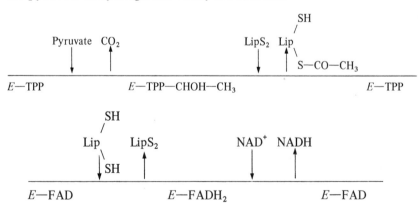

Two anomalous product inhibition results were observed, but these were explained as resulting from protein–protein interactions between the subunits of the dehydrogenase complex, E_1 with E_2 and E_2 with E_3. The transacetylation reaction, catalyzed by either the intact complex or uncomplexed transacetylase, proceeds by a random rapid equilibrium mechanism (Butterworth *et al.*, 1975). The transacetylase appears to have two adjacent binding sites, one to accommodate CoA and acetyl CoA and the other to bind oxidized lipoate and *S*-acetyl-dihydrolipoate. These kinetic results all support Reed's suggestion that the lipoyl prosthetic group rotates to all three of the active sites.

Pyruvate dehydrogenase complex is a regulated enzyme system. In *E. coli*, the dehydrogenase is stimulated by nucleoside monophosphates and inhibited by nucleoside triphosphates by feedback regulation. The mitochondrial pyruvate dehydrogenase complexes are regulated by covalent modification (Reed, 1974). A kinase catalyzes phosphorylation of a serine residue of the dehydrogenase component (E_1), to inactivate the complex. A phosphatase in the complex, which requires Ca^{2+} and Mg^{2+}, catalyzes the dephosphorylation of the phosphodehydrogenase. It seems likely that the ATP/ADP ratio inside mitochondria regulates the activity of the phosphatase, and hence the entire complex. High levels of ATP would chelate Mg^{2+} more efficiently than would ADP. ADP is also an inhibitor of the pyruvate dehydrogenase kinase. In addition, acetyl CoA/CoA and NADH/NAD$^+$ ratios modulate the phosphorylation of the dehydrogenase (Pettit *et al.*, 1975). The pyruvate dehydrogenase kinase is stimulated by acetyl CoA and NAD$^+$; the dehydrogenase phosphatase is inhibited by NADH. Although insulin can affect the level of phosphorylation of the complex in tissues, this hormone acts by an indirect route because cyclic nucleotides have no effect on the pyruvate dehydrogenase kinase or phosphatase. The structural integration of the three enzyme activities of the pyruvate dehydrogenase complex provides efficiency of catalysis, and the inclusion of the phosphatase and kinase in the mammalian complex adds a mechanism for control of an enzyme central in cellular metabolism.

Pyrimidine Nucleotide Biosynthesis

Six enzymes are required to catalyze the formation of uridine-5'-monophosphate (UMP) from bicarbonate, ATP and glutamine (Fig.

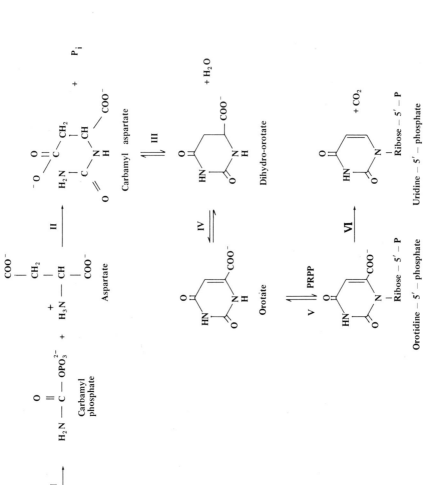

Fig. 3. Pathway for the biosynthesis of UMP. The six enzymes catalyzing this pathway are carbamyl phosphate synthetase (I), aspartate transcarbamylase (II), dihydro-orotase (III), dihydro-orotate dehydrogenase (IV), orotidylate phosphoribosyl transferase (V) and orotidylate decarboxylase (VI).

3). The positions of feedback control and the physical organization of the enzymes of this pathway are different in different organisms (Hartman, 1970). In many organisms, some of these enzymes occur in the pathway (cf. p. 436). The aspartate transcarbamylase from pathway, starting with bacteria and concluding with higher organisms.

Carbamyl phosphate is a metabolite in two pathways, the biosynthesis of pyrimidine nucleotides and the biosynthesis of arginine. Most cells synthesize carbamyl phosphate by a reaction catalyzed by a glutamine-dependent carbamyl phosphate synthetase:

$$\text{Glutamine} + \text{HCO}_3^- + 2\,\text{ATP} + \text{H}_2\text{O} \xrightarrow{\text{Carbamyl phosphate synthetase}}$$

$$\text{Carbamyl phosphate} + \text{glutamate} + 2\,\text{ADP} + \text{P}_i.$$

In mammalian tissues, this enzyme is located in the soluble supernatant fraction (Hager and Jones, 1967). In the livers of ureotelic animals, a second carbamyl phosphate synthetase, using ammonia as its physiological substrate, is located in mitochondria. The latter enzyme provides the carbamyl phosphate substrate for citrulline (and eventually urea) synthesis via arginine. In urea-excreting animals, then, the mitochondrial pool of carbamyl phosphate is intended for synthesis of arginine and the cytoplasmic pool for pyrimidine synthesis.

In *E. coli*, there is only one carbamyl phosphate synthetase. It supplies carbamyl phosphate for the biosynthesis of both pyrimidines and arginine. This enzyme shows a combination of allosteric effects suited to its dual role. It is inhibited by UMP, but stimulated by ornithine, a precursor of arginine.

In *E. coli*, the formation of carbamyl aspartate (step 2 in Fig. 1) is the first step totally committed to pyrimidine biosynthesis. Thus, aspartate transcarbamylase is the primary site of feedback inhibition in the pathway (cf. p. 436). The aspartate transcarbamylase from *E. coli* is not, however, typical of the majority of aspartate transcarbamylases. In bacteria, there are three classes of aspartate transcarbamylases, separable by their kinetic properties and molecular weights (Bethell and Jones, 1969). Class A transcarbamylases, exemplified by that purified from *Pseudomonas fluorescens* (Adair and Jones, 1972), are potently inhibited by UTP, CTP and ATP, and are the largest molecules of the three bacterial types. The *P.*

fluorescens enzyme has a molecular weight of 360 000. Only a single band of 180 000 daltons is obtained when it is subjected to SDS gel electrophoresis. The *E. coli* transcarbamylase, with its activation by ATP, its inhibition by CTP and its 12 polypeptide chains (C_6R_6), is typical of class B aspartate transcarbamylases. The transcarbamylase from *Streptococcus faecalis* (Chang and Jones, 1974) is an example of class C. It is insensitive to 10^{-3} M pyrimidine nucleotides, and has a lower molecular weight than the other two enzymes (four identical subunits of 32 500 daltons). There is no evidence for aggregation of any bacterial aspartate transcarbamylase with carbamyl phosphate synthetase.

In yeast, the synthesis of both carbamyl phosphate synthetase and aspartate transcarbamylase is controlled by the same gene, the *ura-2* locus. Lue and Kaplan (1971) purified these two enzymes from baker's yeast. Most of the activity of the purified enzymes co-sediments in a peak of molecular weight 800 000 daltons. Both activities are strongly inhibited by UTP. Under some conditions, a fraction containing only carbamyl phosphate synthetase activity can be separated from the complex. This fraction, still inhibited by UTP, has a molecular weight of 250 000. Heating of the 800 000 dalton aggregate, followed by chromatography on Sepharose, permitted isolation of an aspartate transcarbamylase subunit of molecular weight 140 000, which is no longer inhibited by UTP (Aitken *et al.*, 1973). The molecular composition of the multi-enzyme complex has been estimated as two molecules of carbamyl phosphate synthetase ($2 \times 250\,000$) and two molecules of aspartate transcarbamylase ($2 \times 140\,000$). Two evolutionary advantages of the aggregation of these two enzymes in yeast have been noted by Kaplan. One is the sensitivity of the aspartate transcarbamylase to feedback inhibition only in the aggregated form. The other is the channelling of carbamyl phosphate from the synthetase to the transcarbamylase. Carbamyl phosphate produced by the carbamyl phosphate synthetase of the complex is preferentially used to synthesize carbamyl phosphate, not citrulline.

In the mould *Neurospora crassa*, the pyrimidine-specific carbamyl phosphate synthetase and aspartate transcarbamylase exist as an enzyme complex, too. Although the synthetase is inhibited by UTP, no feedback inhibition of the transcarbamylase has been observed. In *Neurospora*, pyrimidine biosynthesis occurs in the nuclei and arginine biosynthesis in mitochondria. Efficient *in vivo* channelling of carbamyl phosphate is achieved by compartmentation of both the metabolite and the synthetases (Davis, 1972).

In Ehrlich ascites tumour cells, the first three enzymes of the pyrimidine biosynthetic pathway are a multi-enzyme complex, as demonstrated by their co-purification and co-sedimentation. The fifth and sixth enzymes form a second complex (Shoaf and Jones, 1973). All five of these enzyme activities are isolated in the cytoplasmic fraction. The fourth enzyme, dihydro-orotate dehydrogenase, occurs in the mitochondria (Miller *et al.*, 1968; Kennedy, 1973) of other mammalian tissues. Jones has suggested that the two enzyme complexes (enzymes I to III, and V and VI) may be associated with the dehydrogenase to form a larger complex containing all six enzymes. As she has pointed out, such a master complex is improbable if the dihydro-orotate dehydrogenase is in the interior of the mitochondrion. Mammalian aspartate transcarbamylase appears similar to the *Neurospora* enzyme in its occurrence in a high molecular weight complex, with the pyrimidine-biosynthetic carbamyl phosphate synthetase, not the transcarbamylase, being the target for feedback inhibition by UTP.

Examination of the different ways organisms have evolved the enzymes of pyrimidine nucleotide biosynthesis is an example of how several types of adaptation can occur. Similar metabolic control has been achieved by three distinct methods: allosteric control of aspartate transcarbamylase, physical organization into multi-enzyme complexes and intracellular compartmentation.

MEMBRANE-BOUND ENZYMES

One major organizational level for enzymes is their binding on or in biological membranes. In bacteria, the cellular membrane can be the site of fixation, while in higher plant or animal cells enzymes are attached not only to the plasma membrane but also to all the other ubiquitous membranes such as those of the endoplasmic reticulum and the mitochondria. Some enzymes also are bound to membranes of more specialized cells or organelles, such as erythrocytes and chloroplasts. As I have mentioned earlier (p. 48), enzyme activity measurements can be used as markers for subcellular particles.

In order to retain their catalytic activity, membrane-bound enzymes often must be treated by different techniques from those used for cytoplasmic enzymes. In general, the extent of difference between a soluble and a membrane-bound enzyme is a reflection of the amount of interaction between the enzyme and its membrane matrix. Membrane-bound enzymes are globulins with hydrophobic cores, but the limited evidence available predicts that as a rule they differ from soluble enzymes by having extensive hydrophobic regions on their exterior surfaces. It is these external hydrophobic domains that allow incorporation of the enzyme proteins into the lipid phase of the membrane. It is also these hydrophobic domains that cause technical difficulties, such as self-aggregation of membrane proteins in aqueous solution after their isolation.

The Fluid Mosaic Model of Membranes

Only a few years ago, the chemical structure of biological membranes was obscure. Although present descriptions of membranes have not been proven, they are well documented and certainly more than idle speculation. First, the chemical composition of most membranes has been accurately determined. Second, the accumulation of chemical and physical data on membrane structure has brought successively more reasonable explanations of membrane structure and function. The concept that is most consistent with analytical data, and is most widely accepted, is the *fluid mosaic model* of membrane structure (Singer and Nicholson, 1972).

Most membranes are 50% or less lipid by weight, but the basic properties of membranes are those of a fluid bilayer of lipid material (Tanford, 1973). Membranes from different sources have differing lipid compositions, but the majority of membrane lipids are amphiphilic molecules with two hydrocarbon chains. An *amphiphile* is a molecule made up of a hydrocarbon portion (the tail) and an ionic group (the polar head). Phosphoglycerides are the most prevalent amphiphilic lipids in membranes:

$$
\begin{array}{c}
\quad\ \ O \\
\quad\ \ \| \\
O \quad CH_2-O-C-R_1 \\
\| \quad | \\
R_2-C-C \qquad\ \ O \\
| \qquad\ \ \| \\
CH_2-O-P-O-X \\
| \\
O^-
\end{array}
$$

The fatty acids containing the hydrocarbon chains, R_1 and R_2, are usually 16-24 carbons long. There is a tendency for R_1—COOH to be a saturated fatty acid and for R_2—COOH to be unsaturated. In higher organisms, the head group —O—P(=O)O⁻—OX is either anionic or neutral. Phospholipids, because of their two hydrocarbon tails, form large planar bilayers in aqueous suspension, not simple ellipsoidal micelles (cf. the description of micelles of monohydrocarbon detergents, p. 154). Their hydrophilic head groups are on the two external surfaces of the bilayer, with the tail groups forming a central layer of fluid-like hydrocarbon. The bilayers can fold into spherical sacs or vesicles, which are filled with the aqueous solvent.

Proteins are the other major component of membranes. By weight, they range from 50 to 80% of the total membrane material. Because of their prevalence, proteins must be both important structural and functional participants in the membranes. Membranes, then, are essentially protein–lipid complexes. The greatest uncertainty about membrane structure is the nature of the non-covalent protein–protein and protein–lipid interactions, and the location of the proteins in the membrane. The advance of the fluid mosaic model over previous membrane theories is to place most of the proteins *in* the membrane, not on the surfaces. In doing so, the differences between pure lipid bilayers and actual membranes can be explained. Many kinds of proteins are associated with most membranes. The heterogeneity of membranous protein composition became apparent when sodium dodecyl sulphate gel electrophoresis was developed and applied to membranes. It is now necessary to isolate and characterize membrane proteins in order to fully understand membrane function.

Singer and Nicholson (1972) have divided membrane-bound proteins into two general classes, depending on their location in membranes. The *peripheral* (or extrinsic) proteins are on or near the surface of the membrane, and can be removed easily from the membrane by mild treatments that do not disrupt the lipid matrix of the membrane. Suitable mild treatments include changing the ionic strength or the pH, or addition of a chelating agent. When dissociated, peripheral proteins tend to behave like cytoplasmic proteins in their freedom from lipid and their solubility in aqueous media. Singer (1974) has proposed that peripheral proteins are bound relatively weakly by polar bonds to specific integral proteins in the membrane, not to the lipids. There also could be electrostatic interaction of

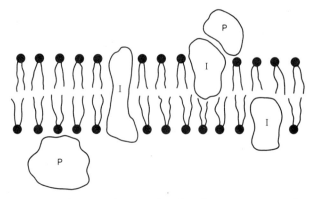

Fig. 4. Cross-section of a membrane, showing features representing the fluid mosaic model. The solid spheres are the polar head groups of phospholipid molecules, in contact with water. The two wavy lines of each phospholipid are the hydrophobic tails. I and P indicate integral and peripheral proteins, respectively.

positively charged regions of the protein surface to the phosphate of phospholipids. The *integral* (or intrinsic) proteins, which include most membrane-bound proteins, are more firmly attached to the membrane, by interaction with the membrane lipids. Integral proteins may be deeply embedded in the bilayer, and in some cases probably completely span the membrane. They can only be dissociated from the membranes by more drastic treatments, with reagents such as detergents or organic solvents. They may retain bound lipid after dissociation, and are relatively insoluble in neutral aqueous media.

Membrane proteins appear to be globular in shape. It has been suggested that they have hydrophilic regions exposed to the aqueous phase and hydrophobic surface regions which allow hydrophobic binding to the lipid of the membrane. Some membranes also contain oligosaccharides (as glycoproteins or glycolipids), probably located at the polar surfaces of the membrane.

The fluid mosaic theory considers that membranes are composed of a lipid bilayer matrix in which globular proteins or protein aggregates are dispersed (Fig. 4). The integral proteins form the mosaic, or two dimensional solution, of proteins in the lipid. The experimentally observed fluid nature of membranes is support for suspension of proteins in a fluid lipid matrix, rather than lipid dispersion in a more rigid protein matrix. With the dynamic character of the protein mosaic, integral proteins are allowed translational movement. However, each protein molecule should always be facing in only one direction, because its polar domain orients it towards one membrane surface. With this insight into the properties of integral proteins, we can now see why the isolation of membrane-bound enzymes requires different methods from those applied to water-soluble enzymes.

Solubilization of Integral Proteins

Isolation of a protein from the membrane to which it is firmly bound *in vivo* is called *solubilization*. Seldom is the solubilization of an integral protein the same as dissolving a protein in aqueous buffer. Usually, it is the separation of the protein—and some very closely associated lipid—from the bilayer it had penetrated, with the formation of a metastable suspension of the protein or lipoprotein in an aqueous medium. The suspended enzyme, however, can be a homogeneous material. Quantitative definitions of solubilization differ from laboratory to laboratory, with criteria for achievement of solubilization ranging from lack of sedimentation of the protein under a force of 100 000 X g or greater applied for several hours, to the lack of microscopically observable membrane vesicles in an active supernatant fraction. Solubilization of membrane-bound enzymes is essential for purification and determination of molecular parameters, and desirable for both mechanism studies and reconstitution experiments. The ultimate aim is to effect a solubilization that provides a homogeneous preparation of the desired enzyme, as the smallest active protein unit, with the preservation of as many of the natural properties as possible.

Solubilization procedures are essentially disruptions of the lipid–protein complexes of the membrane (cf. Coleman, 1973). Extraction of the lipids, for example with alcohols (*n*-butanol or ethanol) or acetone, inactivates or drastically modifies many enzymes. Milder treatments are needed for most enzymes. Treatment of membranes with phospholipase is a rapid and often convenient procedure. The lysophospholipids produced from membrane phospholipid by the action of the phospholipase can act as detergents with solubilizing properties. Limited proteolysis has been used to release some enzymes from membranes, but the resulting proteins are partially degraded. Chaotropic ions (e.g. tribromoacetate, trichloroacetate, SCN⁻, guanidinium and ClO_4^-) are ions that can disorder water molecules, making them more lipophilic and thereby enhancing the solubility of nonelectrolytes.* Hatefi and Hanstein (1969) suggested that the presence of chaotropic anion denaturants could increase the solubility in water of particulate proteins. High concentrations of these ions (approx. 0.5–2 M) should weaken the hydrophobic bonds holding some integral proteins in lipid bilayers. Subsequently, Hatefi's group has used chaotropes for the resolution of some of the components of the mitochondrial electron transfer system (Hatefi and Hanstein, 1974). Addition of antichaotropic anions (SO_4^{2-}, HPO_4^{2-}, and F^-) or use of D_2O rather than H_2O as solvent can strengthen hydrophobic interactions, i.e. cause a reversal of chaotropic destabilization. Therefore, antichaotropic salts or D_2O can sometimes be used to stabilize a solubilized integral protein or to assist in reconstitution experiments.

The most generally useful solubilizing agents are detergents (Helenius and Simons, 1975; Tzagoloff and Penefsky, 1971). The detergent amphiphiles must be present at a concentration high enough to both disrupt the membrane by solubilizing the lipids and saturate the detergent-binding hydrophobic loci of the membrane-bound proteins. Because the monomeric species of the detergent, not the micellar form, binds to both the protein and the lipid, the concentrations of detergent needed are not high. The exact amount of detergent required varies with the amount of membrane, because much of the added detergent is bound to the membrane being lysed. As expected, high concentrations of detergents result in the loss of activity by denaturation of enzymes. Detergents solubilize proteins by exchanging their apolar tails for the apolar regions of the membrane lipids bound to the hydrocarbon-like domains of the proteins. Therefore, continued metastable solution of these proteins usually requires the presence of a dispersing agent.

Detergents vary in the harshness of their effects on proteins. Sodium dodecyl sulphate is one of the most potent solubilizing detergents, but it is so potent that it usually causes denaturation of the dislodged enzymes. However, it has great utility in the determination of the protein composition of membranes and in the estimation of molecular weights of these proteins (cf. p. 69). The bile salts sodium cholate and sodium deoxycholate have been used to solubilize

---

* The chaotropic series is the opposite of the Hofmeister series that ranks anions by their effectiveness in salting-out proteins.

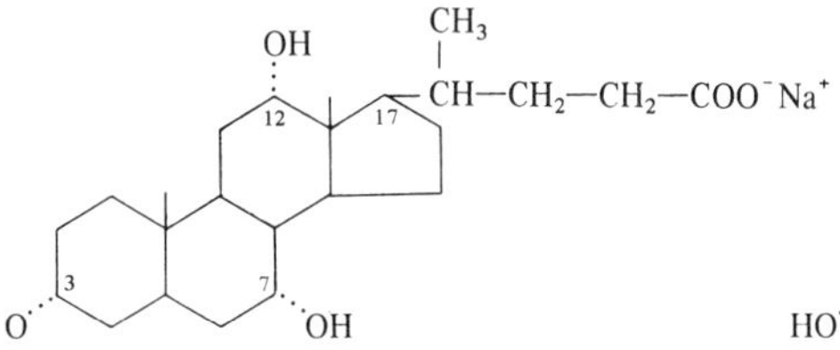

Sodium cholate

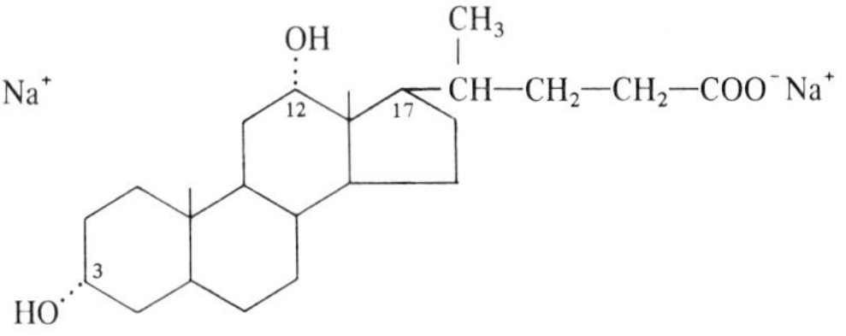

Sodium deoxycholate

many proteins with the retention of their catalytic activity. Bile salts do not have just one well-defined polar head group, but have several distributed polar groups. The carboxylate is on the aliphatic chain, and the hydroxyl groups are all on one side of the ring structure. Thus, there is an apolar face of cholate or deoxycholate to bind to the protein or lipid, and a polar face which is oriented toward the water. Bile acids must be used at pH values of 7 or higher; otherwise, they precipitate. The solubilization capabilities of bile salts are often enhanced in the presence of $0.2-1.0$ M NaCl or KCl. Of the synthetic surfactants, the most useful for solution of enzymes are the non-ionic detergents Triton X-100 and Lubrol WX. Both are derivatives of polyoxyethyleneglycol (PEG). Their effects are also enhanced by NaCl or KCl. These non-ionic detergents are heterogeneous, either in

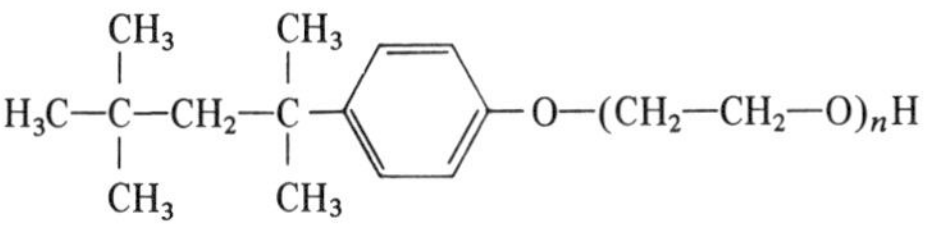

$n = 9$ or 10

Triton X-100

$R = CH_3(CH_2)_{15}—$ to $CH_3(CH_2)_{17}—$

Lubrol WX

the polyoxyethylene head groups or the apolar tails. Generally, they are the most advantageous solubilization agents because they tend to allow retention of native conformation by integral proteins.

During the purification of membrane-bound enzymes, isolation of the organelle or particulate material gives a partial purification before solubilization is attempted. After solubilization, further purification may be hampered by the presence of the detergent. Solubilized enzymes can be delipidated by separation of lipid-detergent mixed micelles and the protein–detergent complexes by gel filtration or ion-exchange chromatography. Ammonium sulphate fractionation seldom can be applied successfully in the presence of non-ionic detergents. Detergent removal, still an area of limited technical experience, can be effected by dialysis or column chromatography. It leads to exposure of the hydrophobic domains of the proteins, and thus often to aggregation of the proteins into amorphous precipitates. Gradual replacement of detergent by phospholipid, and thus reconstitution into synthetic membrane systems, appears to be the most desirable method for examining the catalytic properties of those integral proteins which aggregate when the dispersing agent is removed.

Treatment of membranes by detergents almost always alters the observed catalytic activity of their enzymes. There is usually a stimulation of activity. Activation could be due either to a direct effect on the enzyme or to a change in its environment (e.g. an increase in accessibility of substrate to an enzyme previously on the inside surface of a membrane vesicle). A typical example of activation upon solubilization is IMP dehydrogenase from *Bacillus subtilis*, which exists primarily bound to a cellular membrane (Wu and Scrimgeour, 1973). The dehydrogenase is released rapidly from the membrane by treatment with either phospholipase A or detergents, with a doubling of enzymatic activity but no changes in kinetic parameters. The solubilized IMP dehydrogenase is much less stable when stored than is the membrane-bound enzyme, but the soluble enzyme can be stabilized by either membrane material or phosphate buffer. Release from membrane may cause alterations in the kinetic properties of enzymes. Purified D-lactate dehydrogenase from *E. coli*, solubilized from a membrane suspension using detergents, exhibits both a different pH optimum (8–9 v. 7) and a different K_m for lactate (6×10^{-4} M v. 2.2×10^{-3} M) than the activity in crude membranes (Futai, 1973). L-Glycerol 3-phosphate dehydrogenase is located on the inner membrane surface in *E. coli* (Weiner and Heppel, 1972). Its K_m for glycerophosphate shows a large decrease, from 23 mM to 0.8 mM, when the dehydrogenase is released from membrane and purified. In some cases, detergents can inhibit the activity of a solubilized enzyme. Baugh and King (1972) found that NADH dehydrogenase solubilized from heart mitochondria is strongly inhibited by Triton X-100, which must be removed by exchange with cholate and by binding of remaining traces of the detergent to serum albumin. Coleman (1973) has listed two dozen examples of delipidated enzymes that show reactivation correlated with rebinding of lipid.

Their successful reconstitution requires mixing with the right type of lipid under correct conditions. An example of these lipid-requiring enzymes is D-β-hydroxybutyrate dehydrogenase.

D-β-Hydroxybutyrate is a ketone body, formed from acetoacetate during times of fasting or in diabetes. Its formation is catalyzed by D-β-hydroxybutyrate dehydrogenase:

$$H_3C-\overset{\overset{\displaystyle O}{\|}}{C}-CH_2-COO^- + NADH + H^+ \xrightleftharpoons[\text{dehydrogenase}]{\beta\text{-Hydroxybutyrate}}$$

$$H_3C-CHOH-CH_2-COO^- + NAD^+.$$

This dehydrogenase is firmly bound to the mitochondrial inner membrane, and can be used as a marker for the identification of mitochondrial preparations. A homogeneous, lipid-free and water-soluble preparation of the dehydrogenase has been obtained from beef heart mitochondrial membranes by Bock and Fleischer (1975). The enzyme was solubilized by treatment with phospholipase A, aided by a second solubilization step using the chaotrope LiBr. The dehydrogenase is unstable after its release from the membrane, but is stabilized at pH 6.5 in the presence of LiBr and dithiothreitol. The major step in its purification is adsorption chromatography on a column of controlled pore glass beads, with elution being effected by 1 M LiBr. The purified enzyme is inactive, but it can be reactivated specifically by addition of lecithin, or phospholipids containing lecithin (Gazzotti *et al.*, 1975). Reactivation involves the formation of a phospholipid-dehydrogenase complex. The lipid-free enzyme is incapable of binding NADH. After formation of the phospholipid–enzyme complex, the enzyme can again bind the coenzyme. The absolute specificity for lecithin has been ascribed to its molecular fit with the enzyme. Precise structural information about the lipid–dehydrogenase–NADH holoenzyme would be invaluable in assessing the role of the lipid.

Cytochrome b_5

One of the most completely documented cases of protein–membrane interactions is the binding of the proteins involved in the electron transfer needed for the desaturation of stearyl CoA by mammalian liver microsomes. The two enzymes (cytochrome b_5 reductase and stearyl CoA desaturase) and the protein coenzyme (cytochrome b_5) involved in this system are all localized on the outer or cytoplasmic surface of the endoplasmic reticulum. They catalyze the mixed function oxidation reaction in which the $C^9=C^{10}$ double bond is introduced into stearyl CoA:

$$\text{Stearyl CoA} + H^+ + O_2 + \text{NADH} \xrightarrow{\text{3 proteins}} \text{Oleyl CoA} + 2H_2O + NAD^+.$$

The substrates and the products are in the cytoplasm, and the three enzyme components are integral proteins of the microsomal membrane. The electron transfer sequence is from the reductant (NADH) to the flavin of the reductase, to the heme of cytochrome b_5, then to the iron atom of the desaturase, and finally to the oxygen molecule:

$$\text{NADH} \rightarrow \text{Reductase—FAD} \rightarrow \text{Cytochrome } b_5 \rightarrow \text{Desaturase—Fe} \begin{array}{c} \text{Stearyl CoA} \searrow \text{Oleyl CoA} \\ \xrightarrow{\hspace{1cm}} O_2 \end{array}.$$

Work in Strittmatter's laboratory, elucidating the manner of protein binding to the membrane, has been based on the initial observation by Ito and Sato (p. 42) that detergent-solubilized cytochrome b_5 has a higher molecular weight than trypsin-solubilized cytochrome b_5. The non-hydrolyzed forms of both the cytochrome b_5 reductase (Spatz and Strittmatter, 1971) and cytochrome b_5 (Spatz and Strittmatter, 1973) have been solubilized using Triton X-100 and deoxycholate, purified and characterized. These proteins each contain a globular catalytic core and an additional hydrophobic membrane-binding domain (about 25% of the total molecular weight), with the two polypeptide segments being connected by short chains which are sensitive to mild proteolysis. The hydrophilic fragments of these two proteins cannot bind to the microsomal membrane, whereas intact cytochrome b_5 and cytochrome b_5 reductase can bind firmly. The desaturase also has been detergent-solubilized and purified, from the livers of rats induced for high activity by feeding on a fat-free test diet (Strittmatter *et al.*, 1974). The purified desaturase forms inactive aggregates in aqueous solvents, and is unstable in the presence of detergents. Therefore, it must be purified rapidly and stored at $-70°$ C in low detergent concentrations. For its assay, however, it must be dispersed with fairly high concentrations of

Table II. Properties of the three proteins involved in microsomal desaturation of stearyl CoA.

Property	Protein		
	Cytochrome b_5 reductase	Cytochrome b_5	Desaturase
Molecular weight	43 000 daltons	16 700 daltons	53 000 daltons
Hydrophilic segment	33 000	11 000	
Hydrophobic segment	10 000	4600	
% Hydrophobic amino acid residues[a]	58%	48%	62%
Hydrophilic segment	55%	49%	
Hydrophobic segment	65%	67%	
Reference	Spatz and Strittmatter, 1973	Spatz and Strittmatter, 1971	Strittmatter et al., 1974

[a] A typical cytoplasmic protein has a content of about 53% hydrophobic residues, and a typical membrane integral protein about 60%.

detergent. It requires phospholipid for full activity. The three proteins have been reconstituted by incorporation into synthetic liposomes. In this form, they exhibit the same kinetic properties observable in microsomal preparations. Some of the properties of the three protein components are summarized in Table II. All three consist of only one polypeptide chain each. The flavoprotein and the cytochrome both contain large globular segments that extend into the cytoplasm, but the desaturase, although also on the membrane's outer surface, appears to be deeply embedded in the membrane. The contents of nonpolar amino acid residues in the proteins and in their hydrophobic domains are in agreement with the proposed orientations and catalytic functions.

Of these three proteins, cytochrome b_5 is the simplest and most fully examined. Forty of its 141 amino acid residues are in the hydrophobic N-terminal appendage which binds to the microsomal membrane (Spatz and Strittmatter, 1971). When freed of detergent by chromatography on a Sephadex column, cytochrome b_5 readily undergoes aggregation. Dehlinger et al. (1974) have used spin-labelled lipids and e.s.r. measurements to show that

1. fatty acids bind to intact cytochrome b_5 but not to the polar fragment;
2. the hydrophobic peptide segment is the site of immobilization of fatty acids; and
3. the lipid-binding regions of cytochrome b_5 are more polar than the interior of a phospholipid bilayer.

Robinson and Tanford (1975) have studied the binding of detergents and phospholipid vesicles to isolated native cytochrome b_5 and its polar and hydrophobic fragments. Deoxycholate and Triton X-100 are bound only to the small hydrophobic domain or fragment, as expected. Binding of SDS to the polar fragment (0.7 g (g protein)$^{-1}$) is below the typical 1.4 g SDS (g protein)$^{-1}$, but its binding to the hydrophobic fragment is unusually great (approx. 3 g (g protein)$^{-1}$). These results are consistent with formation of detergent–cytochrome b_5 micelles, with the polar domain being exposed to solvent. Direct evidence for the incorporation of cytochrome b_5 into phospholipid vesicles was obtained by gel chromatography. The polar peptide is not incorporated into the vesicles. Robinson and Tanford concluded that the hydrophobic domain of cytochrome b_5 will insert into any available hydrophobic environment.

Strittmatter has found that when purified cytochrome b_5 is added to microsome vesicles, it is functionally indistinguishable from the hemoprotein originally present in the microsomes. All the cytochrome b_5 (the endogenous and the added cytochrome) can be reduced by NADH, so all the cytochrome molecules must be able to interact with the reductase. Rogers and Strittmatter (1974) have convincingly demonstrated that the extra cytochrome is reduced solely by the cytochrome b_5 reductase, not by electron transfer from endogenous cytochrome b_5 molecules. They also found that when

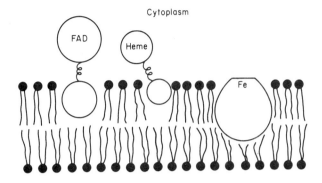

Fig. 5. Possible arrangement of the three integral proteins of the cytochrome b_5 electron transport system of the endoplasmic reticulum (ER). The proteins are considered to be distributed randomly and to undergo translational movement in the lipid bilayer. The reductase (left) and cytochrome b_5 both are composed of two domains, but the desaturase (right) is embedded in the membrane, on the cytoplasmic surface. The substrates—NADH, O_2 and stearyl CoA—are all in the cytoplasm.

90% of the cytochrome b_5 reductase was inactivated by reaction with N-ethylmaleimide all the cytochrome could still be reduced. This shows that any molecule of cytochrome b_5 can interact with any nearby molecule of reductase. Similar experiments have been performed using cytochrome b_5 and reductase bound to egg lecithin liposomes, artificial phospholipid-containing vesicles (Rogers and Strittmatter, 1975), and they substantiate Strittmatter's results with microsomal vesicles.

The data of Strittmatter and his coworkers are best accommodated by a model in which the three proteins are not bound to the phospholipid of the microsomal membrane in a fixed array but are randomly distributed over the membrane surface, as depicted in Fig. 5. The three microsomal proteins appear to be in a two-dimensional solution in the lipid and to interact with each other by relatively unrestricted translational movement (diffusion and random collision) in a fluid bilayer. The existence of the fluid bilayer which permits protein diffusion has been demonstrated indirectly by the effect of temperature on the enzymatic activity of dimyristoyl lecithin liposomes containing cytochrome b_5 reductase and cytochrome b_5 (Strittmatter and Rogers, 1975). There is an abrupt decrease in

R

catalysis at temperatures below $25°$ C, shown kinetically to be due to a decrease in electron transfer from reductase to cytochrome b_5. The phospholipid in the liposomes undergoes a transition from liquid crystal to the more rigid gel phase at about $24°$ C. Physicochemical studies on cytochrome b_5 or the hydrophobic fragment bound to deoxycholate micelles and on the polar fragment in aqueous solution indicate that the two domains are globular particles (Visser *et al.*, 1975). The interdomain link region, about 15 residues long, appears to be highly ordered, perhaps in a helical array. The reductase and the cytochrome molecules must be flexible enough between their polar and apolar segments to allow collision of the heme of the cytochrome with the FAD of the reductase and with the anchored iron of the desaturase. This model gives strong support to the basic principles of the fluid mosaic structure of membranes discussed above.

DePierre and Dallner (1975) have reviewed the structure of the membrane of the endoplasmic reticulum, with emphasis placed on the orientation of protein molecules bound to the membrane. The endoplasmic reticulum has a large surface area, and the enzymes bound to it catalyze a number of indispensable functions. Ribosomes are attached, and the NADH-driven cytochrome b_5 and the NADPH-driven cytochrome P-450 electron transfer systems are bound, to the cytoplasmic surface, while glucose-6-phosphatase and several other enzymes have been shown to be localized on the internal (or luminal) surface. One major task is to see if other integral proteins are bound in a manner resembling that of the three proteins I have described.

(Na$^+$ + K$^+$)-ATPase

Sodium and potassium ion-activated adenosine triphosphatase (abbreviated as (Na$^+$ + K$^+$)-ATPase) is an integral protein which spans the plasma membrane in most, perhaps all, animal cells. This enzyme is responsible for maintaining the high concentration of K$^+$ and the low concentration of Na$^+$ in cells, by an active transport process of these ions against a concentration gradient. The driving force for this ion pump is the coupling of the hydrolysis of the terminal phosphate group of ATP to ion transport. The stoichiometry of the transport reaction is the exit of three sodium ions and the entry of two potassium ions for each molecule of ATP hydrolyzed. (Na$^+$ + K$^+$)-ATPase activity has been studied in many tissues (e.g. axon, kidney and erythrocyte) and the properties of the ATPase are similar in all tissues (reviewed by Dahl and Hokin, 1974; Schwartz *et al.*, 1975). Because of its binding to membranes, isolation of the ATPase and determination of its mechanism of action has been hindered. Unlike the cytochrome b_5 system, in which the microsomal membrane is an anchoring and orienting agent, the plasma membrane and its intrinsic cation pump control or maintain many physiological processes such as absorption of sugars and amino acids, excitability of nerve and muscle, and osmotic pressure of cells. A thorough elucidation of the properties of $(Na^+ + K^+)$-ATPase will eventually explain not only the vectorial transport of monovalent cations but also a great deal about membrane function and about energy conservation.

The first recognition that an ATPase activity was associated with active extrusion of Na^+ from cells was the suggestion by Skou (1957) that the stimulation by monovalent cations of a Mg^{2+}-dependent ATPase activity in crab nerve membrane preparations was part of the ion translocation process. Simultaneous addition of Na^+ and K^+ gave maximal activation of the ATPase used in Skou's experiments. Many lines of evidence have since been used to support the direct role of the $(Na^+ + K^+)$-ATPase in ion transport. The enzyme is absolutely specific in its requirement for Na^+, but other ions can bind and activate at the K^+ site. These ions are the same, and have the same order of effectiveness, as the activators of IMP dehydrogenase (p. 354). Sodium and ATP bind to the inside surface of the membrane-bound ATPase, and potassium and the inhibitor ouabain bind to the external surface of the ATPase. This spatial orientation can be deduced from the effects of these four compounds on both ATPase activity and ion transport in erythrocytes.

The pathway of hydrolysis of ATP by the $(Na^+ + K^+)$-ATPase is known to be a multi-step sequence, although the exact mechanism is not known yet. A phosphoenzyme occurs as an intermediate in the reaction cycle. The pH hydrolysis profile of this phosphorylated intermediate indicates that the phosphoryl group is present as an acyl phosphate. Post and Kume (1973) labelled the ATPase by incubation with γ-^{32}P-ATP, Mg^{2+} and Na^+. After digestion of the phosphoenzyme with pronase, they isolated a tripeptide with the structure of either Thr-X(^{32}P)-Lys or Ser-X(^{32}P)-Lys, where X is either Asp or Glu. The chemical behaviour of this phosphorylated tripeptide is more similar to synthetic Pro-Asp(P)-Lys than to Pro-

Glu(P)-Lys, so they concluded that the phosphoenzyme contained a β-aspartyl phosphate. Nishigaki *et al.* (1974) have confirmed that it is an aspartate residue that is phosphorylated by reductive cleavage of the ^{32}P-labelled phosphoenzyme and isolation of homoserine derived from the aspartyl phosphate. The phosphorylated intermediate is formed in the presence of Na^+, and is hydrolyzed in the presence of K^+. I will discuss the role of the phosphoenzyme below, when considering the reversibility of the individual reaction steps catalyzed by the ATPase.

The rigorous extraction procedures needed for purification of the $(Na^+ + K^+)$-ATPase indicate how firmly the enzyme is attached to the plasma membrane. First, the plasma membrane is prepared as part of the microsome fraction. Most purification procedures then follow this general protocol. Extraneous proteins are removed by treatment with the chaotrope NaI or the detergent deoxycholate. The ATPase is then solubilized either with Lubrol WX or with deoxycholate-salt.

SDS gel electrophoresis of the ATPase purified from several different sources indicates that the enzyme is composed of two polypeptides, a larger protein (approx. 90 000 daltons) and a smaller glycoprotein (Kyte, 1972). The larger protein contains the site of phosphorylation by ATP (Uesugi *et al.*, 1971). Because of its apparent interaction with Na^+, ATP, K^+ and ouabain, it presumably traverses the membrane. The two polypeptides copurify, and must be close to each other in the purified enzyme because they can be covalently linked by treatment with dimethylsuberimidate (Kyte, 1972). While the minimal molecular weight of the ATPase seems to be about 140 000 daltons, there are indications that it is an $\alpha_2\beta_2$ oligomer. Based on the stoichiometry of ligand binding and the extent of phosphorylation, Jørgensen (1974) has suggested that the molecular weight is 250 000–280 000. Kyte had previously shown that the two chains are in a 1 : 1 molar ratio (1.7 : 1 mass ratio, large to small polypeptide chain) when isolated. In support of the $\alpha_2\beta_2$ quaternary structure, Kyte since has found conditions under which a covalent dimer of the larger chains can be formed (Kyte, 1975). Estimates made by other workers are similar. The values I have presented for the molecular weight and chain composition must be viewed with caution until more rigorous techniques can be applied to the ATPase.

The two proteins have been separated only by dissociation with SDS, followed by gel filtration, so successful reconstitution of the holoenzyme has not yet been achieved. Therefore, the essentiality of the glycoprotein has not yet been proved. Delipidation of partially purified ATPase preparations causes a loss of activity, which can be regained upon addition of phospholipids. The lipid compound which gives the greatest stimulation varies with the method used for lipid depletion. Where reported, the lipid content of purified ATPase is about 25% by weight. Purified $(Na^+ + K^+)$-ATPase cannot be spontaneously incorporated into phospholipid vesicles, but it has been incorporated into artificial membranes by dialysis of a phospholipid–enzyme–cholate mixture (Hilden *et al.*, 1974). The vesicles, formed after removal of the detergent by dialysis, pump Na^+ inwards but not optimally.

What can be learned about the mechanism of transport from the $(Na^+ + K^+)$-ATPase? We have seen that the enzyme is a complex of two polypeptides. The 90 000 molecular weight protein is large enough to span a lipid bilayer, even if it is spherical. The hydrophilic sugars of the glycoprotein make it unlikely that the smaller protein rotates about an axis parallel to the membrane surface (Kyte, 1972). Passage of a polar domain of a protein through the hydrocarbon interior of a membrane is thermodynamically unfavourable. In fact, transmembrane rotations of integral proteins have been looked for but not yet detected (Singer, 1974). Two hypotheses for transport of small hydrophilic molecules have been advanced (summarized on p. 826 of Singer, 1974). These are *carrier mechanisms,* in which the transport protein rotates across the membrane, and *fixed pore mechanisms,* in which a protein-lined pore undergoes a conformational change to translocate the binding site. Kyte (1974) prepared antibodies to both purified $(Na^+ + K^+)$-ATPase holoenzyme and to the lipid-free large protein component. All but a minor component of these specific antibodies cause no inhibition of $(Na^+ + K^+)$-ATPase activity. The anti-holoenzyme binds to both sides of plasma membrane vesicles. Anti-large chain γ-globulins are located exclusively on the inside surface of the cell membrane of intact cells. This location for the anti-large chain antigenic site agrees with the phosphorylation of the large protein by ATP, which approaches ATPase from the interior of the cell. Binding of antibodies with retention of activity suggests that active transport of Na^+ and K^+ (assuming that ATPase activity and Na^+,K^+ transport are related) cannot involve large movements of the ATPase across the membrane. If the channel or pore mechanism applies, then only minimal conformational rearrangements of the transmembranous protein complex can occur. The conformational change needed to pass Na^+ out of the cell would be driven by the phosphorylation of the enzyme, and K^+ would enter the cell when the enzyme returned to its initial conformation.

I will conclude this description of $(Na^+ + K^+)$-ATPase with a summary of the individual reaction steps of ATP hydrolysis and a consideration of the energetics of these reactions (cf. Dahl and Hokin, 1974). A simplified scheme can be written based upon three major observations:

1. the existence of a phosphoenzyme intermediate, its Na^+-dependent formation and K^+-dependent dephosphorylation, as described above;
2. the existence of different conformations of the enzyme, E_1 and E_2, and of the phosphoenzyme, $E_1{\sim}P$ and $E_2{-}P$ (E_1 accepts a phosphate group from ATP, and E_2 forms $E_2{-}P$ by reaction with P_i); and
3. the reversibility of the steps of the entire pump cycle.

The sum of four equations (the steps involving formation of ES complexes have been omitted) accounts for the sodium- and potassium-dependent hydrolysis of ATP. The stoichiometry of transport is 3 Na^+ leaving the cell and 2 K^+ entering the cell per mole of ATP hydrolyzed. The inorganic phosphate released from the $E_2{-}P$ intermediate in the presence of K^+ is retained inside the cell. Conformational changes, as required for validity of the fixed pore transport mechanism, occur in the second and fourth steps:

(1) $Na_3{-}E_1 + ATP \rightleftharpoons Na_3{-}E_1{\sim}P + ADP$

(2) $Na_3{-}E_1{\sim}P \rightleftharpoons E_2{-}P + 3\,Na^+ \text{(outside)}$

(3) $E_2{-}P + 2\,K^+ \text{(outside)} \rightleftharpoons K_2{-}E_2 + P_i$

(4) $K_2{-}E_2 + 3\,Na^+ \text{(inside)} \rightleftharpoons Na_3{-}E_1 + 2\,K^+ \text{(inside)}.$

Post has written the first two steps as:

$$\text{Na}{-}E_1 \xrightleftharpoons[\quad\quad]{\text{ATP} + \text{Mg}^{2+} \quad \text{ADP}} \text{Mg}{-}\text{Na}{-}E_1{\sim}P \xrightleftharpoons[\quad]{\text{Na}^+} \text{Mg}{-}E_2{-}P$$

and his laboratory has demonstrated the reversibility of these two steps in leaky-membrane preparations of kidney ($Na^+ + K^+$)-ATPase. Taniguchi and Post (1975) prepared $Mg-E_2-P$ by incubation of ATPase with Mg^{2+}, ^{32}P-phosphate and low concentrations of Na^+. The potassium-sensitive phosphoenzyme, E_2-P, in the presence of high concentrations of Na^+, could rapidly and quantitatively transfer its phosphoryl group to ADP to form ^{32}P-ATP. Phospholipids, but not intact membranes, are essential for the synthesis of ATP. These workers concluded that the binding of sodium was sufficient to cause ATP synthesis from functional phosphoenzyme and ADP, by converting the phosphoenzyme from a conformation equilibrating with P_i to one equilibrating with ATP. They estimated that the binding of 3 Na^+ ions causes a change in free energy of hydrolysis of the $E-P$ bond of about $10-11 \text{ kcal mol}^{-1}$. Their results are compatible with the conformational-coupling hypothesis proposed by Boyer to explain the conservation of energy from electron transport in mitochondria, discussed in the next section of this chapter.

It is too early for any definitive mechanism to be proposed for the ($Na^+ + K^+$)-ATPase. More information is needed about the structure of the enzyme, the roles of the two non-identical subunits, the nature of the conformational changes and the architecture of the active site. There are indications that the ATPase is only partially phosphorylated by ATP, so half-site reactivity may be involved in its mechanism. ($Na^+ + K^+$)-ATPase seems to be representative of a group of membrane transport enzymes, all of which are oligomers that undergo conformational changes during their transport and catalytic functions.

The Mitochondrial Electron Transport System

The most complicated of the membrane-bound enzyme systems are the mitochondrial electron transport and the chloroplast electron transport systems. Research on the former has finally furnished the identity of most of the oxidation–reduction components, and similar techniques are being applied to the less well advanced problem of chloroplast redox mechanism (cf. pp. 281–283). With few exceptions, the sequence of reactions is known for the passage of reducing equivalents from either NADH or succinate to molecular oxygen during mitochondrial respiration. Segments of the overall process are catalyzed by lipoprotein complex enzymes bound to the inner mitochondrial membrane. A short discussion of these enzyme complexes is appropriate because they exemplify the theme of this chapter—the physical organization of enzyme reactions and their control through organization.

The sequence of reactions in electron transfer by the mitochondrial respiratory chain has been determined in three ways. The components can be placed in an order consistent with their standard reduction potentials at pH 7 (E_0' values). The reduction potentials become more positive during the sequence, going from the E_0' of the primary substrate couple, $NAD^+/NADH$ at -0.32 volts, through those of the electron transferring agents in the membrane-bound enzyme complexes, to the E_0' of the $1/2$ O_2/H_2O couple, $+0.82$ volts. Measurements by optical methods of the state of reduction of individual electron carriers in intact mitochondria has been used to great advantage (Chance and Williams, 1956). Difference spectra between aerobic mitochondria and mitochondria reduced by interaction with tricarboxylic acid cycle intermediates can be measured. The light-absorbing cytochromes show, in the steady state, an extent of reduction related to their proximity to the reducing (NADH) end of the respiratory chain. Similar experiments can be performed by observing the difference spectra between the oxidized mitochondria and the reoxidation of fully reduced mitochondria, or observing the time sequence of oxidation or reduction of the electron carriers. If an inhibitor of the respiratory chain is added to a mitochondrial system, then redox carriers on the NADH side of the inhibition site, or crossover point, are reduced and those in the direction of O_2 are oxidized. More recently, e.p.r. spectroscopy has shown that many iron–sulphur proteins, perhaps more than there are cytochromes, are redox components of mitochondria. The final aid to the sequencing of the electron carriers has been the isolation and characterization of four submitochondrial particles, each of which catalyzes a section of the passage of electrons from NADH or succinate to O_2. Actually, five complexes have been isolated, but I shall leave discussion of the mitochondrial ATPase until last. *In vivo*, the enzyme complexes are somehow coupled, a process in which the membrane must be involved. Physical organization prevents the oxidation by O_2 of any but the final redox component.

Mitochondria have enzymes in four locations: the outer membrane, the inner membrane, between these two surfaces and in the matrix—the gel-like inner compartment. All the enzymes of the electron transport system are associated with the inner membrane. The inner membrane has many folds, or cristae, extending into the matrix. The cristae greatly increase the surface area of the inner membrane. The mitochondrial inner membrane contains about 75% protein (at least 60 different polypeptides) and 25% lipid. Most of the proteins are integral proteins, and therefore difficult to solubilize. This is true of all of the proteins of the electron transfer chain except the peripheral protein coenzyme cytochrome *c.* Although cytochrome *c* can be extracted with salt solutions, it readily binds to lipid vesicles and to other proteins, especially cytochrome *c* oxidase.

In the early 1960s, four discrete high molecular weight lipoprotein complexes were isolated at the Enzyme Institute in Wisconsin from beef heart mitochondria by mild bile acid–salt fractionation (reviewed by Hatefi, 1966). These multi-protein–lipid complexes, units of the respiratory chain, are each composed of smaller protein subunits. When supplemented with phospholipids, the four complexes can be associated into vesicular membranes which have all the apparent enzymic properties of the intact electron transport chain. The relative concentrations of the complexes in mitochondria vary with the tissue of mitochondrial origin. Binding of the complexes to the inner membrane may just orient the complexes, as observed in the microsomal cytochrome b_5 system (p. 491), but *both* sides of the inner membrane appear to be involved in mitochondria. In contrast to the apparent fluid environment of the complexes, the current view is that the protein subunits inside each complex are fixed in position. The only mobile elements of the respiratory chain are the coenzymes which link the complexes, coenzyme Q (also called ubiquinone) and

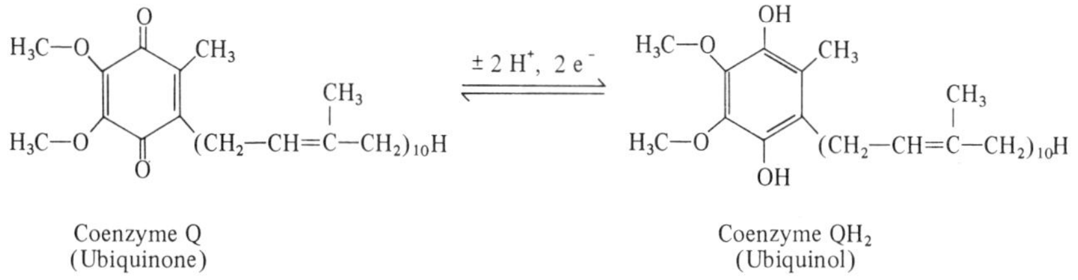

Coenzyme Q
(Ubiquinone)

$\pm 2\,H^+,\ 2\,e^-$

Coenzyme QH$_2$
(Ubiquinol)

cytochrome *c.* At least, these lipid-soluble coenzymes can be considered as the mobile factors in the functioning of the sequence of complexes (Fig. 6). The complexes are numbered I–IV, with names

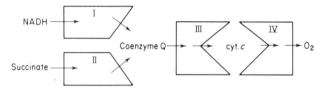

Fig. 6. The path of electron flow through the complexes of the electron transport system. Reduction of NAD^+ by compounds such as pyruvate, isocitrate,
α-ketoglutarate and malate is catalyzed by specific dehydrogenases located in
the mitochondrial matrix. Coenzyme Q and cytochrome c are the low molecular
weight coenzymes which shuttle electrons between specific donor and receptor
sites on the appropriate complexes. Both of the complex-joining coenzymes may
function as 1-electron carriers, as current evidence suggests that coenzyme Q
shuttles between the fully reduced and semiquinone forms in the mitochondrial
membrane.

assigned by the partial reaction catalyzed:

Complex I   = NADH–coenzyme Q reductase
Complex II   = Succinate–coenzyme Q reductase
Complex III = Coenzyme QH_2–cytochrome c reductase
Complex IV   = Cytochrome c oxidase.

The four complexes are not just artefacts of isolation; their organization and compositions are directly related to their physiological roles
in energy conservations. The reactions catalyzed by three of the
complexes supply energy which is coupled to either the synthesis of
ATP from ADP and P_i or to active transport.

The structure of the four complexes has been extensively studied
(Hatefi *et al.*, 1974a, 1975). All but a few electron acceptors have
been identified. Complexes I and II are both flavoproteins, metabolically connected to complex III *in vivo* by coenzyme Q. The flavin
of the NADH dehydrogenase subunit of complex I is FMN. NADH
dehydrogenase also contains one iron–sulphur (Fe–S) cluster. Three
other Fe–S proteins occur in complex I. Complex II, the succinate
dehydrogenase complex, has FAD as its flavin coenzyme and also
contains several (three) Fe–S proteins (Ohnishi *et al.*, 1974). It can
be separated into two subunits, one containing FAD and an Fe–S
protein and the other containing additional Fe–S components. Complex III is the least well-characterized of the four enzyme complexes.
It contains at least two cytochrome b components, cytochrome c_1
and an Fe–S centre (Orme-Johnson *et al.*, 1974). Cytochrome c
oxidase (complex IV) catalyzes the terminal step of respiration, the

I. NADH ——[FMN-$(Fe-S)_1$ → $(Fe-S)_3$; $(Fe-S)_4$ → $(Fe-S)_2$]—→ Coenzyme Q

  −0.32   −0.30   −0.24   −0.02   +0.09

II. Succinate ——[FAD-$(Fe-S)$ → $(Fe-S)$]—→ Coenzyme Q

  0.00   ?   ?   +0.09

III. Coenzyme QH_2 ——[Cytochromes b → $(Fe-S)$ → Cytochrome c_1]—→ Cytochrome c
                                                    (Fe^{3+} form)
  +0.09   −0.06 and   +0.03   +0.22   +0.26
               +0.04   to 0.00

IV. Cytochrome c ——[Cytochrome a → Cu^{2+} → Cytochrome a_3]—→ O_2
  (Fe^{2+} form)
  +0.26   +0.20 to   +0.28   +0.35 to   +0.82
               +0.22   +0.40

Fig. 7. Pathway proposed for electron flow through the four multi-protein–lipid complexes. The current estimates of the E'_0 values, in volts, are shown below each component. Those portions of the reaction occurring in a lipid-bound complex are shown by inclusion in square brackets.

reduction of O_2 to H_2O. It has a large number of proteins, and its electron carriers are cytochrome a, cytochrome a_3 and protein-bound copper (Nicholls and Chance, 1974). All of the complexes contain lipid. The apparent sequence of electron transfer reactions through these complexes is depicted in Fig. 7.

The standard free energy liberated when two electrons pass through the respiratory chain to O_2 via the complexes can easily be calculated (Hatefi, 1965). For the reactions catalyzed by complexes I, III and IV, the values for $\Delta G°$ are −19, −8 and −25 kcal mol^{-1}, respectively. If coupled through an isoenergetic chemical reaction, each of these exergonic oxidation steps could produce one mole of ATP per two electrons transferred. These calculations agree with the experimentally observed sites of oxidative phosphorylation in mitochondria: between NADH and coenzyme Q; between cytochrome b and cytochrome c; and between cytochrome a and O_2. The $\Delta G°$ for the reaction catalyzed by complex II is only −4 kcal per two electrons, insufficient for the formation of an ATP γ-phosphoryl bond, for which the $\Delta G°$ for hydrolysis is −7.3 kcal mol^{-1}.

Since oxidative phosphorylation has not yet been detected with enzyme preparations lacking an intact membrane, it appears that the membrane somehow integrates electron transfer with the phosphorylation of ADP. The integration of the oxidative and phosphorylative phases of energy production is illustrated by the measurement of respiration rates of carefully prepared tightly coupled mitochondria (Chance and Williams, 1956). When these mitochondria are suspended in a phosphate-containing buffer containing an oxidizable substrate, there is little oxygen consumption until ADP, the phosphate acceptor, is added. The ratio of the rates in the presence and absence of ADP, called the respiratory control index, is a measure of the integrity of the isolated mitochondria. Uncoupling agents, such as 2,4-dinitrophenol, can also relieve the inhibition of respiration in tightly-coupled mitochondria. At the same time, they give rise to the appearance of an ATPase activity in the uncoupled mitochondria. Ageing or physical damage of mitochondria also stimulates the latent ATPase activity. For many years, this ATPase has been considered to be the "ATP synthetase" of the intact mitochondria, normally catalyzing the reversal of the uncoupler-induced ATPase activity. Therefore, let us now examine the mitochondrial ATPase complex, the fifth multi-subunit complex.

Senior (1973), in a delightful historical review, has described the properties of the mitochondrial ATPase. In comparison with the already complicated $(Na^+ + K^+)$-ATPase, he describes an ATPase complex which is much more intricate. It is composed of at least ten subunits, arranged in a membrane-sector, a stalk section and the actual ATPase catalytic centre—a sphere connected by the stalk to the inner membrane. Racker and his colleagues (Penefsky et al., 1960) isolated the first soluble ATPase preparation, which they called coupling factor F_1, from beef heart mitochondria by mechanical agitation with glass beads and centrifugation from particulate matter. The particulate fraction alone catalyzed oxidation of substrates, but needed added F_1 for oxidative phosphorylation. The F_1 fragment of the ATPase is not inhibited by oligomycin, an antibiotic which prevents both stimulation of mitochondrial O_2 consumption by ADP and synthesis of ATP. Subsequently, the fifth mitochondrial complex—an oligomycin-sensitive ATPase containing F_1, other proteins and lipid—was obtained from mitochondria by detergent solubilization. One of the proteins absent from F_1 but present in the complex is a soluble protein called, because of its function, the oligomycin sensitivity conferring protein (MacLennan and Tzagoloff,

1968). This protein is also a coupling factor, because it stimulates energy-linked reactions. None of the oligomycin-sensitive ATPase preparations can synthesize ATP, but if carefully incorporated into phospholipid vesicles (e.g. Kagawa and Racker, 1971), they can catalyze a $^{32}P_i$-ATP exchange, one of the partial reactions of oxidative phosphorylation. Present data indicate that F_1 is the sphere, and the oligomycin sensitivity conferring protein is located in the stalk region of the ATPase complex.

Recently, Hatefi et al. (1974b) have isolated a soluble complex by deoxycholate-KCl treatment of mitochondria which they call complex V. It differs from the oligomycin-sensitive ATPase complex described above in some of its polypeptide composition and in its ability to catalyze ^{32}P-ATP exchange by just mixing with phospholipids, not by careful dialysis or sonication. Complex V contains F_1, the oligomycin sensitivity conferring protein and a protein which appears to bind uncoupling agents. It remains to be seen how this complex differs exactly from other ATPase complexes, and how either type of ATPase complex interacts with complexes I, III and IV for energy conservation.

The mechanistic question asked in the 1940s is still unanswered: how does the energy obtained from oxidation of substrates cause the reversal of the ATPase reaction? Three theories have been suggested to explain oxidative phosphorylation. The first is the postulate of energy-coupled generation of a high-energy phosphorylated intermediate which, in a subsequent step, isoenergetically transfers its phosphate to ADP. Despite intensive research, no high-energy intermediate has been found. A second theory is the chemiosmotic coupling theory (Mitchell, 1961, 1967; Lehninger et al., 1967). This theory is based on the requirement of an intact mitochondrial membrane for oxidative phosphorylation and the impermeability of the membrane to protons. It is supported by evidence that cation and proton translocations can be in reversible equilibrium with the hydrolysis or synthesis of ATP. Mitchell's theory proposes that protons are actively transported from mitochondria during oxidation, creating a high-energy state, not an activated chemical intermediate. Two protons should be ejected per pair of electrons at each energy-conserving site. A recent reappraisal of the H^+ per site ratio (Brand et al., 1976) indicates that previous experiments underestimated the number of protons ejected, and that the true value is at least 3.0 and possibly 4.0. Because phosphorylation is the removal of water from ADP and P_i, with the oxygen being lost from the P_i, Mitchell suggests that the mitochondrial ATPase allows removal of an OH^- group from P_i by the external H^+ excess and a removal of a proton from ADP by the OH^- excess in the matrix of the mitochondria. This theory requires both the ATPase and the electron-transfer molecules to be specifically oriented by the inner membrane. The action of un-coupling agents, most of which are lipid-soluble, is envisaged as a disruption of the membrane permitting protons to enter freely. The sequence of redox components necessary to fulfil the Mitchell hypothesis (i.e. placing of cytochrome b closer to NADH than coenzyme Q is) does not fit with accepted evidence. Also, the fact that the catalytic centre of the ATPase (F_1) is held away from the membrane by the stalk seems incompatible with direct contact of the ATPase with the outer surface of the inner membrane, as needed for catalysis of water formation. A third and simpler theory suggests that the energy of oxidation is used to change the conformation of the site on the ATPase which binds ATP, causing tightly but non-covalently bound ATP to be released (Boyer et al., 1973; Cross and Boyer, 1975). This theory suggests that the formation of tightly bound ATP may be catalyzed by the entropic effects of ADP and P_i collection at the active site and/or by reduction of the water concentration there. The ATP is formed spontaneously, but cannot be released from the ATPase until after the oxidation transfer and the resultant conformational change occur. This theory is supported by the finding of traces of tightly bound ATP, by the existence of possibly analogous systems such as $(Na^+ + K^+)$-ATPase (p. 500), and by indirect evidence that cytochrome c oxidase and the ATPase are coupled physically (Wilson and Fairs, 1974). The quaternary structure of the ATPase may change with electron transport, with dissociation of an inhibitor polypeptide (cf. discussion by Slater, 1974). Perhaps this reversible dissociation is driven by proton formation. However, the conformational coupling theory itself does not explain why an intact mitochondrial membrane is essential for respiration-linked phosphorylation. Williams (1975, and references cited therein) has suggested that the protonation of phosphate is effected by high local concentrations of protons, i.e. the protons formed during substrate oxidation are retained inside the membrane. He strongly opposes any mechanism in which the protons leave the membrane to be diluted in the cytoplasm. But they do, in many experiments, leave the membrane! The possibility that a membrane-requiring proton gradient causes the conformational change at the active site of the ATPase has been considered, but the validity of the suggestion is unresolved at present (Boyer, 1975a,b; Mitchell, 1975).

To list all the unanswered questions about mitochondrial oxidations would take as much space as my treatment of the topic. Certainly the prime question is how, under the physical control of the membrane, does one ATPase complex interact with three other enzymes to retain the substrate-generated energy as ATP molecules. Many facts have accumulated over the past 50 years, but it will take many decisive experiments to completely elucidate "mitochondriology". Each year brings us closer to that goal.

In this chapter, we have examined how relatively simple physical contact between proteins can aid in the control of enzymic catalysis, and then how binding to membranes can be an aid to catalysis. We have looked at examples of enzymes from each of the three major cellular membranes of animals—the endoplasmic reticulum, the plasma membrane and the mitochondria membrane. Let us now turn our attention from control by physical organization to artificial control of metabolism through chemotherapeutic regulation of enzyme activity.

REFERENCES

Adair, L. B. and Jones, M. E. (1972). *J. biol. Chem.* **247**, 2308-2315.
Aitken, D. M., Bhatti, A. R. and Kaplan, J. G. (1973). *Biochim. biophys. Acta* **309**, 50-57.
Baugh, R. F. and King, T. E. (1972). *Biochem. biophys. Res. Commun.* **49**, 1165-1173.
Bethell, M. R. and Jones, M. E. (1969). *Archs Biochem. Biophys.* **134**, 352-365.
Bock, H.-G. and Fleischer, S. (1975). *J. biol. Chem.* **250**, 5774-5781.
Boyer, P. D. (1975a). *FEBS Letters* **50**, 91-94.
Boyer, P. D. (1975b). *FEBS Letters* **58**, 1-6.
Boyer, P. D., Cross, R. L. and Momsen, W. (1973). *Proc. natn. Acad. Sci. U.S.A.* **70**, 2837-2839.
Brand, M. D., Reynafarje, B. and Lehninger, A. L. (1976). *Proc. natn. Acad. Sci. U.S.A.* **73**, 437-441.
Butterworth, P. J., Tsai, C. S., Eley, M. H., Roche, T. E. and Reed, L. J. (1975). *J. biol. Chem.* **250**, 1921-1925.

Chance, B. and Williams, G. R. (1956). *Adv. Enzymol.* **17**, 65–134.

Chang, T.-Y. and Jones, M. E. (1974). *Biochemistry* **13**, 629–638.

Clarke, F. M. and Masters, C. J. (1973). *Biochim. biophys. Acta* **327**, 223–226.

Cleland, W. W. (1973). *J. biol. Chem.* **248**, 8353–8355.

Coleman, R. (1973). *Biochim. biophys. Acta* **300**, 1–30.

Cross, R. L. and Boyer, P. D. (1975). *Biochemistry* **14**, 392–398.

Dahl, J. L. and Hokin, L. E. (1974). *A. Rev. Biochem.* **43**, 327–356.

Davis, R. H. (1972). *Science, N.Y.* **178**, 835–840.

Dehlinger, P. J., Jost, P. C. and Griffith, O. H. (1974). *Proc. natn. Acad. Sci. U.S.A.* **71**, 2280–2284.

DePierre, J. W. and Dallner, G. (1975). *Biochim. biophys. Acta* **415**, 411–472.

Flick, P. K. and Bloch, K. (1975). *J. biol. Chem.* **250**, 3348–3351.

Futai, M. (1973). *Biochemistry* **12**, 2468–2474.

Gazotti, P., Bock, H.-G. and Fleischer, S. (1975). *J. biol. Chem.* **250**, 5782–5790.

Ginsburg, A. and Stadtman, E. R. (1970). *A. Rev. Biochem.* **39**, 429–472.

Hager, S. E. and Jones, M. E. (1967). *J. biol. Chem.* **242**, 5674–5680.

Hartman, S. C. (1970). *In* "Metabolic Pathways" (D. M. Greenberg, ed.), Third edition, Vol. 4, pp. 39–68. Academic Press, New York.

Hatefi, Y. (1965). *Clin. Chem.* **11**, 198–212.

Hatefi, Y. (1966). *Comprehensive Biochem.* **14**, 199–231.

Hatefi, Y. and Hanstein, W. G. (1969). *Proc. natn. Acad. Sci. U.S.A.* **62**, 1129–1136.

Hatefi, Y. and Hanstein, W. G. (1974). *In* "Methods in Enzymology" (Fleischer, S. and Packer, L., eds), Vol. 31, pp. 770–790. Academic Press, New York.

Hatefi, Y., Hanstein, W. G., Davis, K. A. and You, K. S. (1974a). *Ann. N.Y. Acad. Sci.* **227**, 504–520.

Hatefi, Y., Stiggall, D. L., Galante, Y. and Hanstein, W. G. (1974b). *Biochem. biophys. Res. Commun.* **61**, 313–321.

Hatefi, Y., Hanstein, W. G., Galante, Y. and Stiggall, D. L. (1975). *Fedn Proc.* **34**, 1699–1706.

Helenius, A. and Simons, K. (1975). *Biochim. biophys. Acta* **415**, 29–79.

Hilden, S., Rhee, H. M. and Hokin, L. E. (1974). *J. biol. Chem.* **249**, 7432–7440.

Jørgensen, P. L. (1974). *Biochim. biophys. Acta* **356**, 53–67.

Kagawa, Y. and Racker, E. (1971). *J. biol. Chem.* **246**, 5477–5487.

Kennedy, J. (1973). *Archs Biochem. Biophys.* **157**, 369–373.

Kumar, S., Muesing, R. A. and Porter, J. W. (1972). *J. biol. Chem.* **247**, 4749–4762.

Kyte, J. (1972). *J. biol. Chem.* **247**, 7642–7649.

Kyte, J. (1974). *J. biol. Chem.* **249**, 3652–3660.

Kyte, J. (1975). *J. biol. Chem.* **250**, 7443–7449.

Lane, M. D. and Moss, J. (1971). *In* "Metabolic Pathways" (H. J. Vogel, ed.), Third edition, Vol. 5, pp. 34–46. Academic Press, New York.

Lehninger, A. L., Carafoli, E. and Rossi, C. S. (1967). *Adv. Enzymol.* **29**, 303–311.

Lue, P. F. and Kaplan, J. G. (1971). *Can. J. Biochem.* **49**, 403–411.

Lynen, F. (1967). *Biochem. J.* **102**, 381-400.
MacKenzie, R. E. (1973). *Biochem. biophys. Res. Commun.* **53**, 1088-1095.
MacLennan, D. H. and Tzagoloff, A. (1968). *Biochemistry* **7**, 1603-1610.
Miller, R. W., Kerr, C. T. and Curry, J. R. (1968). *Can. J. Biochem.* **46**, 1099-1106.
Mitchell, P. (1961). *Nature, Lond.* **191**, 144-148.
Mitchell, P. (1967). *Fedn Proc.* **26**, 1370-1379.
Mitchell, P. (1975). *FEBS Letters* **50**, 95-97.
Nicholls, P. and Chance, B. (1974). *In* "Molecular Mechanisms of Oxygen Activation" (O. Hayaishi, ed.), pp. 479-534. Academic Press, New York.
Nishigaki, I., Chen, F. T. and Hokin, L. E. (1974). *J. biol. Chem.* **249**, 4911-4916.
Ohnishi, T., Winter, D. B., Lim, J. and King, T. E. (1974). *Biochem. biophys. Res. Commun.* **61**, 1017-1025.
Orme-Johnson, N. R., Hansen, R. E. and Beinert, H. (1974). *J. biol. Chem.* **249**, 1928-1939.
Penefsky, H. S., Pullman, M. E., Datta, A. and Racker, E. (1960). *J. biol. Chem.* **235**, 3330-3336.
Pettit, F. H., Pelley, J. W. and Reed, L. J. (1975). *Biochem. biophys. Res. Commun.* **65**, 575-582.
Pirson, W., Schuhmann, L. and Lynen, F. (1973). *Eur. J. Biochem.* **36**, 16-24.
Polakis, S. E., Guchhait, R. B. and Lane, M. D. (1973). *J. biol. Chem.* **248**, 7957-7966.
Post, R. L. and Kume, S. (1973). *J. biol. Chem.* **248**, 6993-7000.
Qureshi, A. A., Jenik, R. A., Kim, M., Lornitzo, F. A. and Porter, J. W. (1975). *Biochem. biophys. Res. Commun.* **66**, 344-351.
Reed, L. J. (1974). *Accts Chem. Res.* **7**, 40-46.
Reed, L. J. and Cox, D. J. (1970). *In* "The Enzymes" (P. D. Boyer, ed.), Third edition, Vol. 1, pp. 214-225. Academic Press, New York.
Robinson, N. C. and Tanford, C. (1975). *Biochemistry* **14**, 369-378.
Rogers, M. J. and Strittmatter, P. (1974). *J. biol. Chem.* **249**, 895-900.
Rogers, M. J. and Strittmatter, P. (1975). *J. biol. Chem.* **250**, 5713-5718.
Schwartz, A., Lindenmayer, G. E. and Allen, J. C. (1975). **27**, 3-134.
Senior, A. E. (1973). *Biochim. biophys. Acta* **301**, 249-277.
Shoaf, W. T. and Jones, M. E. (1973). *Biochemistry* **12**, 4039-4051.
Singer, S. J. (1974). *A. Rev. Biochem.* **43**, 805-833.
Singer, S. J. and Nicholson, G. L. (1972). *Science, N.Y.* **175**, 720-731.
Skou, J. C. (1957). *Biochim. biophys. Acta* **23**, 394-401.
Slater, E. C. (1974). *Biochem. Soc. Trans.* **2**, 1149-1163.
Spatz, L. and Strittmatter, P. (1971). *Proc. natn. Acad. Sci. U.S.A.* **68**, 1042-1046.
Spatz, L. and Strittmatter, P. (1973). *J. biol. Chem.* **248**, 793-799.
Strittmatter, P. and Rogers, M. J. (1975). *Proc. natn. Acad. Sci. U.S.A.* **72**, 2658-2661.

Strittmatter, P., Spatz, L., Corcoran, D., Rogers, M. J., Setlow, B. and Redline, R. (1974). *Proc. natn. Acad. Sci. U.S.A.* **71**, 4565–4569.

Tanford, C. (1973). "The Hydrophobic Effect: Formation of Micelles and Biological Membranes." Wiley-Interscience, New York.

Taniguchi, K. and Post, R. L. (1975). *J. biol. Chem.* **250**, 3010–3018.

Tsai, C. S., Burgett, M. W. and Reed, L. J. (1973). *J. biol. Chem.* **248**, 8348–8352.

Tzagoloff, A. and Penefsky, H. S. (1971). *In* "Methods in Enzymology" (W. B. Jakoby, ed.), Vol. 22, pp. 219–230. Academic Press, New York.

Uesugi, S., Dulak, N. C., Dixon, J. F., Hexum, T. D., Dahl, J. L., Perdue, J. F. and Hokin, L. E. (1971). *J. biol. Chem.* **246**, 531–543.

Vance, D. E., Mitsuhashi, O. and Bloch, K. (1973). *J. biol. Chem.* **248**, 2303–2309.

Visser, L., Robinson, N. C. and Tanford, C. (1975). *Biochemistry* **14**, 1194–1199.

Volpe, J. J. and Vagelos, P. R. (1973). *A. Rev. Biochem.* **42**, 21–60.

Weiner, J. H. and Heppel, L. A. (1972). *Biochem. biophys. Res. Commun.* **47**, 1360–1365.

Willecke, K., Ritter, E. and Lynen, F. (1969). *Eur. J. Biochem.* **8**, 503–509.

Williams, R. J. P. (1975). *FEBS Letters* **53**, 123–125.

Wilson, D. F. and Fairs, K. (1974). *Archs Biochem. Biophys.* **163**, 491–497.

Wu, T. W. and Scrimgeour, K. G. (1973). *Can. J. Biochem.* **51**, 1380–1390.

Ziegenhorn, J., Niedermeier, R., Nüssler, C. and Lynen, F. (1972). *Eur. J. Biochem.* **30**, 285–300.

15 Chemotherapeutic Control of Enzyme Reactions

A living organism has many enzymic and other types of vital control mechanisms. Modern medicine is sometimes called upon to apply an entirely artificial type of control, chemotherapy. *Chemotherapy* is the use of chemical agents for the control or elimination of infectious diseases caused by parasitic cells or viruses. Successful implementation of chemotherapy relies on "selective toxicity" (Albert, 1973), the ability to kill or arrest the growth of an invasive cell with little or no adverse effect on the host organism. Because of the great practical importance of chemotherapeutic agents, or drugs, which are enzyme inhibitors, I shall discuss several examples of the control of enzyme reactions by selective inhibitors.

Artificial control of enzyme reactions is a delicate but necessary art. Side-effects, due to lack of absolute drug specificity or to different responses by individual patients, present a hazard which must be scientifically controlled. Selective toxicity is applied in both human and veterinary medicine, and in agriculture in the design of insecticides and herbicides. The cost of health research, such as the experiments which I shall describe, is high, but the rewards are higher—in better health, longer useful and comfortable life, and economic savings.

One of the longest continuing projects has been a search for exploitable metabolic differences between normal cells and cancerous cells. Due to the eradication of so many infectious diseases, cancer has become the second most common cause of death (heart disease is the number one cause) in Western society. Surgery or radiation therapy is used on localized neoplasms and for tumours detected at an early stage. However, successful drug treatment is desirable for a cancer that has spread through the body. A major approach in cancer research has been to seek drugs which can selectively destroy tumour cells, primarily through prevention of cell division (i.e. DNA synthesis) in the more rapidly growing neoplastic cells. Cancer is not one but many diseases, related by cellular growth characteristics but differing in tissue of origin and cause. Cancer research workers often have been disappointed when well planned attacks have given minimal or no useful results, for no apparent reason. However, as a result of the massive research completed to date, many neoplastic diseases can be at least temporarily arrested and cures are now available for some cancers. These successes have occurred when such diversified disciplines as synthetic organic chemistry, pharmacology, genetics and enzymology have been combined with clinical trials. As with other large research projects, cancer research has resulted in the development of fundamental knowledge, and in the synthesis of new drugs which are extremely useful for other medical applications. In this chapter, I will describe the theory of selective inhibition or artificial control of enzyme reactions, and the mechanisms of action of several of the more successful anti-tumour agents. These latter drugs are of added interest because of the differences in their modes of action.

Early development of cancer chemotherapy hinged upon the antimetabolite theory, which described useful drugs as inhibitors of *target enzymes* and thus controllers of key metabolic reactions. A classical *antimetabolite* is an enzyme inhibitor which has only small changes from the structure of the substrate of the enzyme which it inhibits. The antimetabolite approach arose from the discovery that sulphonamides are compounds antagonistic to and structurally related to the bacterial growth factor, p-aminobenzoic acid. Chemotherapy has been extended past the level of simple synthetic structural analogues, and now includes use of naturally occurring and modified antibiotics (Gale *et al.*, 1972), active site-directed covalent inhibitors, artificial negative feedback inhibitors and use of more than one drug at a time, as a double blockade of alternative or sequential pathways. A good general review of the pharmacological use of anti-neoplastic agents has been published recently (Chabner *et al.*, 1975). It outlines the current clinical uses of antimetabolites, and also of antibiotics and alkylating agents.

SULFA DRUGS

The sulfa drugs or sulphonamides ($R-SO_2-NHR'$) are important historically in chemotherapy, both as the first safe and effective anti-bacterial drugs and as the foundation of the antimetabolite theory. They are still used for some bacterial infections (e.g. streptococcal infection, some types of meningitis and enteritis) but have been widely replaced or supplemented by antibiotics. In the late 1930s, it was discovered that the anti-streptococcal azo dye prontosil was active only after cleavage to sulphanilamide. Woods (1940) found that an aromatic acid occurring in yeast extract and other sources prevented the anti-bacterial effects of sulphonamide. He identified

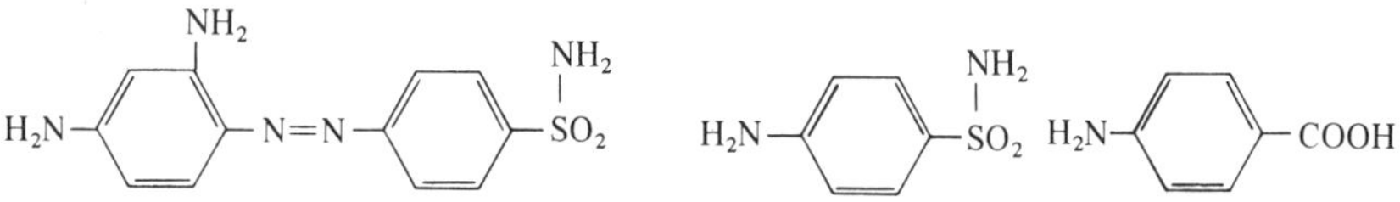

| Prontosil | Sulphanilamide | p-Aminobenzoic acid |

this acid as p-aminobenzoic acid. Woods hypothesized that the antagonism between sulphonamides and p-aminobenzoate is due to competitive inhibition of an enzyme reaction requiring p-amino-benzoate. The chemical and steric resemblance between these two types of compounds supported his suggestion for the mode of action of the sulfa drugs. It also led to a rational approach to chemotherapy, based on testing of structural analogues of substrates of target enzymes.

In 1940, it was known that p-aminobenzoate was required for the growth of many bacteria. It has since been shown to be essential for the formation of the coenzyme tetrahydrofolate, in the pathway of biosynthesis of 7,8-dihydrofolate (p. 239) from GTP, as shown on the next page. Sulphonamides inhibit the conversion of p-aminobenzoate to dihydropteroate, catalyzed by dihydropteroate synthetase (Brown, 1971). Although the drugs are inhibitors which compete with p-aminobenzoic acid in this reaction, they are also poor substrates for the enzyme and condense with the dihydropterin diphosphate to form a sulphanilic acid-containing analogue of dihydropteroate. Formation of this metabolically inactive analogue appears to aid in prevention of bacterial growth by using up the small supplies of the dihydro-pterin diphosphate precursor. Without dihydropteroate and the resulting tetrahydrofolate, the bacteria cannot grow or divide.

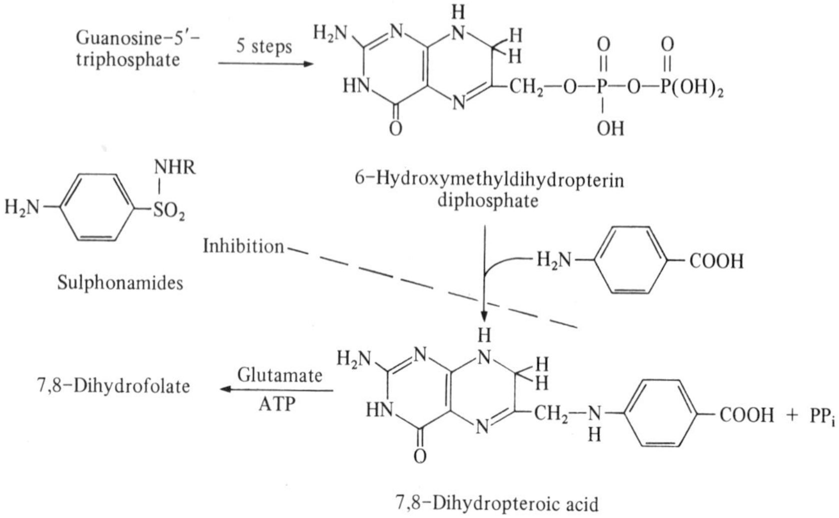

Three factors contribute to the selective toxicity of the sulfa drugs. First, the inhibited reaction does not occur in man, the host, who must rely on dietary sources of folate. Second, the continued synthesis of folate coenzymes is essential for bacterial growth. Finally, the bacteria cannot absorb preformed folate compounds from the host because they lack a transport system for folate. Therefore, the drugs are toxic primarily to the parasitic bacteria, although some patients receiving sulfa drugs exhibit toxic symptoms or side-effects such as allergies, renal toxicity or nausea.

INHIBITORS OF THYMIDYLATE BIOSYNTHESIS

The enzymic synthesis of thymidylic acid (TMP) is the target for a number of clinically useful chemotherapeutic agents (Friedkin, 1973). That TMP biosynthetic enzymes are the receptor sites of these drugs has been determined either by a search for the target of successful drugs, or by synthesis and testing of specific structural analogues. In the TMP synthesis cycle (p. 256), both dihydrofolate reductase and TMP synthetase are susceptible to chemotherapeutic inhibition:

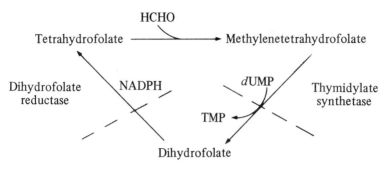

Inhibition of either or both of these enzymes should prevent cell division by causing a deficiency of TMP, the nucleotide which occurs in DNA but not RNA. In bacteria, a similar phenomenon called *thymineless death* can occur when DNA synthesis is blocked by thymine deprivation but synthesis of RNA and protein is allowed to continue (Cohen, 1971). Thymine starvation not only halts cell proliferation but also leads to irreversible cell damage. Degradation of DNA in thymine-starved bacteria is the apparent reason for thymineless death.* The rationale for chemotherapy by inhibitors of dihydrofolate reductase or TMP synthetase is either a selective inhibition of an enzyme in an infecting organism or a differential effect of drugs on neoplastic cells because of their higher rate of DNA synthesis. Cancer chemotherapy agents are difficult to employ because the drugs do not bind preferentially to the enzymes of neoplastic cells, and doses producing measurable toxicity to the host must be administered to achieve maximal response (Frei, 1967).

It is easy to see how inhibitors of TMP synthetase can directly inhibit the biosynthesis of thymidylate. However, it is less obvious how anti-folate drugs block the TMP synthetase cycle in a relatively

---

* Purine deficiency, caused by blocking of *de novo* purine nucleotide synthesis, also may contribute to the lethal effect of amethopterin in mammalian cells (Hryniuk, 1972). The probable reason for the sensitivity of purine biosynthesis to antifolates is the requirement for tetrahydrofolate coenzymes in two steps of the synthetic pathway (cf. p. 246). From studies with cells grown in tissue culture, Hryniuk (1975) has concluded that amethopterin toxicity occurs in three stages:
1. a shrinking of the purine pools,
2. expansion of these pools, possibly from breakdown of nucleic acids, and
3. eventual killing of cells by inhibition of thymidylate synthesis.

specific manner. These agents inhibit many tetrahydrofolate-dependent enzyme reactions at concentrations near 10^{-4} M, but only dihydrofolate reductase at physiologically tolerated concentrations. Still, the product of reduction of dihydrofolate, the coenzyme tetrahydrofolate, is required for many important metabolic reactions (Chapter 8, Fig. 8). The explanation of the specificity of anti-folates for prevention of TMP formation lies in the mechanism of the reaction catalyzed by TMP synthetase. This is the only reaction in which tetrahydrofolate is a reductant and therefore used in substrate rather than coenzyme amounts. When the reductase is inhibited, dihydrofolate accumulates and TMP synthesis effectively ceases, although sufficient tetrahydrofolate probably is generated in the presence of the anti-folate inhibitor to allow some RNA and protein synthesis. I will now describe the biochemistry of these two types of inhibitors of TMP biosynthesis in more detail.

Inhibition of Dihydrofolate Reductase

Soon after the structure of folic acid had been determined, the 2,4-diamino analogue of folic acid—aminopterin—was synthesized as a potential antimetabolite of folate. This compound, when tested as an anti-leukemia agent, gave temporary remission of leukemia in children (Farber et al., 1948). This was the first successful use of an antimetabolite against cancer in man. The 10-methyl derivative of aminopterin, amethopterin (Methotrexate or MTX), is the most significant anti-folate compound in treatment of cancers:

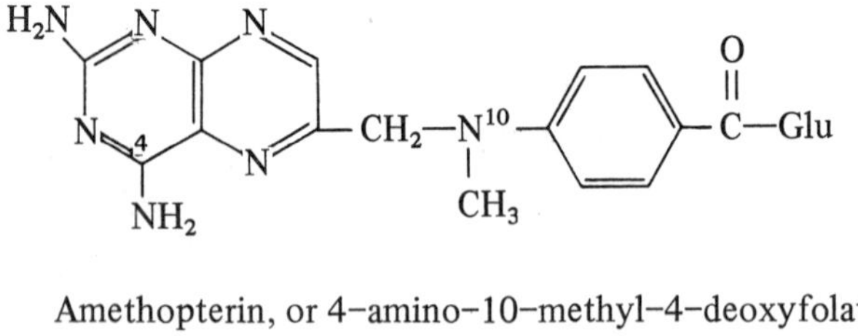

Amethopterin, or 4–amino–10–methyl–4–deoxyfolate .

Amethopterin, either alone or in combination with other drugs, can cure several types of cancer and is used in treating psoriasis and in suppression of antibody production. The activity of amethopterin is exerted with no prior metabolic transformation, in contrast to fluorouracil and mercaptopurine, which will be discussed below. When aminopterin and amethopterin were synthesized, the target of their action was not known. The site of action of these drugs was quickly narrowed down to the conversion of folate to tetrahydrofolate coenzymes, until 7,8-dihydrofolate reductase (p. 239) was demonstrated to be the target enzyme.

The inhibition of dihydrofolate reductase by aminopterin or amethopterin is competitive, but because of the extremely high affinity of the inhibitors for the enzyme, the kinetics of inhibition are not simple. From the structural similarities between the substrates, folate and dihydrofolate, and the inhibitors, competitive inhibition would be expected. The initial observations of Osborn *et al.* (1958) indicated that at pH 7.5 aminopterin and amethopterin are noncompetitive inhibitors of dihydrofolate reductase with K_i values near 10^{-9} M. In 1961, Werkheiser described the inhibition at pH 6.1 as competitive stoichiometric. *Stoichiometric inhibition* occurs when the inhibitor, at levels too low to saturate the enzyme, is so firmly bound to the enzyme that most or all of it is enzyme-bound. It results in biphasic v v. $[E]$ curves (Fig. 1, curve B). The binding of the anti-folate inhibitors is not covalent, as can be demonstrated by

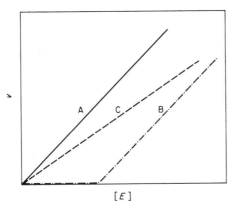

$[E]$

Fig. 1. Inhibition of dihydrofolate reductase by amethopterin. This is a composite figure, showing results from a number of reported experiments. Curve A represents the control, with no inhibitor. Curve B shows inhibition with amethopterin present at about 10^{-8} M at pH 6. In curve C, amethopterin is present at pH 7.6. A curve similar to curve B is obtained when the reductase is preincubated with NADPH and amethopterin at pH 7.2.

their release from the reductase in some experiments by dialysis or column chromatography. Under some assay conditions, the binding of amethopterin to dihydrofolate reductase is so strong that the enzyme activity can be titrated with the drug. These titrations indicate that one molecule of inhibitor is bound to each molecule of enzyme. Bertino (1963) found that inhibition was maximal and stoichiometric at pH 5.9, but that classical competitive inhibition could be observed at pH 7.6 (Fig. 1, curve C). The binding constant of the reductase for dihydrofolate or for amethopterin is decreased in the presence of the other substrate, NADPH. In fact, the ternary E-NADPH-amethopterin $1:1:1$ complex of the *Lactobacillus casei* reductase is so stable that it can be separated from both the apo-enzyme (form I, p. 241) and the holoenzyme (form II) by electrophoresis, and it even can be crystallized (Otting and Huennekens, 1972). Williams *et al.* (1973) have found that inhibition becomes stoichiometric with the *E. coli* reductase at pH 7.2 if the enzyme is preincubated with NADPH and amethopterin. The K_i values for the 2,4-diamino analogues of folate are dependent upon assay conditions, but are in the range of 10^{-10} to 10^{-9} M, below the K_m for dihydrofolate by a factor of at least 10^3.

Despite potent and selective binding to dihydrofolate reductase, amethopterin has several limitations in treatment of cancer. The drug does not penetrate the blood-brain barrier, so is not useful for brain tumours. Some neoplastic diseases, especially solid tumours, have a natural resistance to amethopterin. Rapidly dividing host cells, such as bone marrow and intestinal mucosa, can suffer toxicity during amethopterin therapy. The greatest limitation, though, is acquired resistance in patients with tumours that initially were sensitive to amethopterin. The lifespan of patients can be prolonged by variation of the schedule of drug administration or by combination chemotherapy using amethopterin with other drugs (e.g. fluorouracil, mercaptopurine, vincristine and/or prednisone), but longer remissions or eradication of tumours is eventually prevented by drug resistance. Development of resistance to amethopterin is accompanied by a large increase in the amount of dihydrofolate reductase in resistant cells (Bertino, 1963). Inhibition of this increased level of reductase would require higher drug concentrations than the patient could tolerate. Although resistance in bacteria is caused by mutations, the resistance in most mammalian cells appears to be analogous to substrate stabilization of enzymes (p. 50). The amethopterin–enzyme complex seems to accumulate in resistant cells because of its reduced rate of degradation. Elevated levels of reductase, even though most is inhibited, allow sufficient production of tetrahydrofolate to satisfy the metabolic needs of the neoplastic cells. Thus, amethopterin can have two counteracting effects on its target enzyme, inhibition and stabilization.

A promising development in treatment with amethopterin is the "rescue" approach. It has been found that superior therapeutic results can be obtained with some tumours when amethopterin is supplied in extremely high doses (as high as 100 times normal doses), and then a few hours later a tetrahydrofolate derivative—usually 5-formyltetrahydrofolate (leucovorin)–is administered repeatedly. The leucovorin protects host cells which normally would be killed by the excessive drug level. Two hypotheses for the success of high-dose rescue programmes are that very high concentrations of the antifolate drug will destroy cells by a mechanism other than that of conventional doses, and that the tetrahydrofolate-rescue coenzyme may selectively rescue normal cells (Frei et al., 1975). The latter postulate is favoured at present, but the mechanism of rescue is unknown, and may vary with cell type, drug level and duration of exposure before rescue. Several compounds other than formyltetrahydrofolate have been used for the rescue of experimental animal tumours. For example, Tattersall et al. (1975) have found that thymidine prevents amethopterin toxicity of normal cells but does not rescue tumour cells. 5-Methyltetrahydrofolate and amethopterin can be used in combination to selectively kill some malignant cells in tissue culture (Halpern et al., 1975). These tumour cells have lower than normal levels of methionine synthetase activity (p. 379), so there is a rational basis for the rescue with methyltetrahydrofolate. Studies still must be performed to establish how best to rescue only normal cells, and to establish the site or sites of lethality caused by amethopterin-induced tetrahydrofolate deficiency.

Baker and his associates attempted the preparation of clinically useful *active site-directed inhibitors* of dihydrofolate reductase. In a book describing his rational design of this type of inhibitor, Baker (1967) gives full details of the application of affinity-labelling (p. 34) to chemotherapy. Inhibition by an active site-directed inhibitor $(I-X)$ occurs in two kinetic steps, the reversible binding of the inhibitor to the enzyme and then covalent binding of I at a site near the catalytic centre:

$$E + I\!-\!X \; \underset{k_{-1}}{\overset{k_1}{\rightleftharpoons}} \; E\!\cdots\!I\!-\!X \; \overset{k_2}{\longrightarrow} \; E\!\cdots\!I \; + \; X \, .$$

Proper positioning of the leaving group X on the inhibitor is essential for step 2, the covalent binding to a side chain of the enzyme. Often, these inhibitors are *nonclassical antimetabolites,* having large but appropriate structural differences from the substrate. The use of the active site-directed irreversible inhibitors DFP and TPCK in the study of chymotrypsin has already been described (pp. 316 and 320). Baker's research on the reductase started with experiments designed to find the mode of binding of inhibitors to the enzyme. By synthesizing and testing many nonclassical reversible inhibitors, he concluded that the 2,4-diamino group of compounds like amethopterin was bound near the active site, the *p*-aminobenzoate ring was bound to a hydrophobic region and the glutamate moiety was bound to a polar site of the reductase. An irreversible inhibitor which fulfils the binding requirements and can penetrate cell membranes by passive diffusion is shown in Fig. 2. This compound has some specificity for the reductase of tumour cells of experimental animals because normal cells can inactivate the sulphonyl fluoride group. Surprisingly, related reversible inhibitors are just as effective as the irreversible inhibitor.

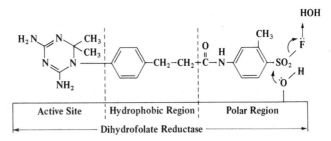

Fig. 2. A nonclassical irreversible inhibitor of dihydrofolate reductase (Baker, 1971). Four regions of the inhibitor bind to the reductase:
1. the substituted dihydrotriazine ring binds to the active site,
2. the phenylethyl group binds to a hydrophobic region,
3. the benzenesulphonyl fluoride binds to a polar region, and then
4. the sulphonyl fluoride group reacts with a nucleophilic group (presumed to be the —OH of a serine) in the irreversible step which anchors the inhibitor near the catalytic site.

One of these, triazinate (Baker, 1971), is currently undergoing pharmaceutical testing (Cashmore *et al.*, 1975):

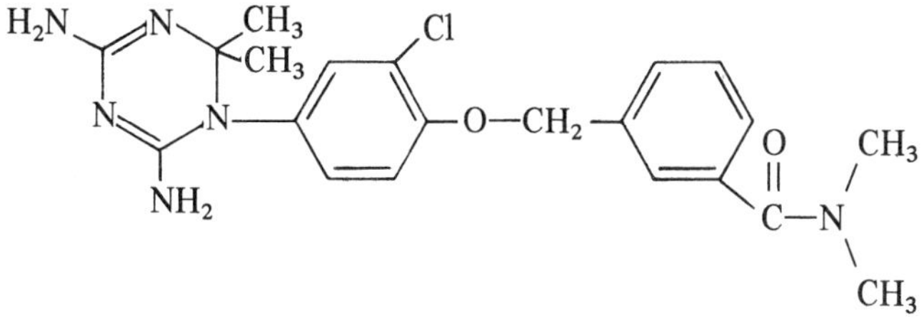

Triazinate

Triazinate is of potential clinical importance because, in contrast to amethopterin, it can enter the cerebrospinal fluid. Although no therapeutically useful irreversible inhibitors of dihydrofolate reductase are available yet, extension by other workers of Baker's *modus operandi* may produce a chemotherapeutic covalent inhibitor of the reductase.

Two synthetic 2,4-diaminopyrimidines, prepared for use as antimalarial agents, now are known to be highly selective inhibitors of microbial dihydrofolate reductases. These are the nonclassical antimetabolites pyrimethamine and trimethoprim:

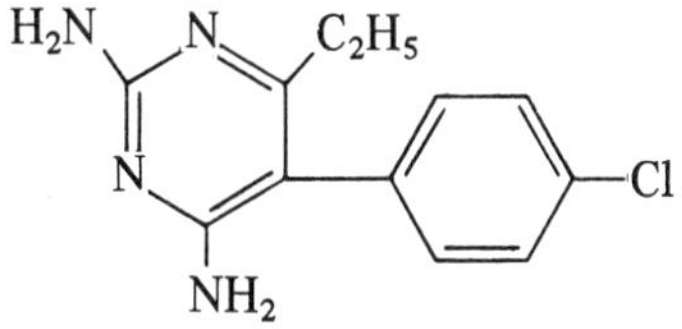

Pyrimethamine 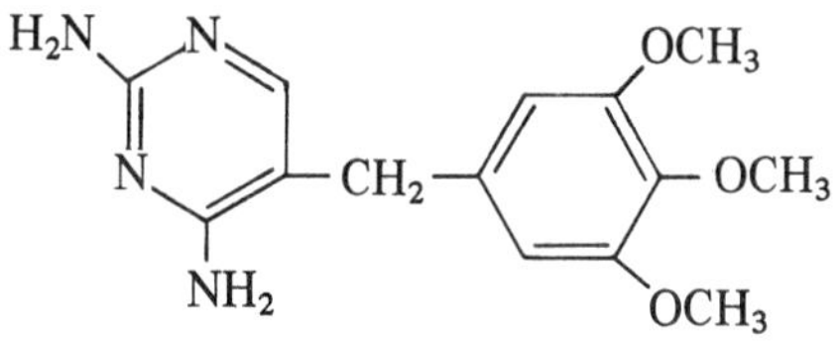   Trimethoprim

Pyrimethamine, in clinical use since 1951, can produce nearly complete inhibition of the reductase of the malarial parasite at inhibitor levels having negligible effects on the host enzyme. Trimethoprim, the most potent known inhibitor of bacterial reductases, is about 60 000 times more potent as an inhibitor of bacterial than human reductase. Both drugs are competitive stoichiometric inhibitors of microbial reductases. They both can pass through the cell walls of micro-organisms, in contrast to folate and amethopterin. Now that their target site is known, these drugs can be used in combination with sulphonamides (p. 514) to effect a double blockade of the pathway of biosynthesis of tetrahydrofolate (Hitchings, 1971).

Inhibition of Thymidylate Synthetase

No anti-folate drugs have been found which are as potent inhibitors of thymidylate synthetase as amethopterin is of dihydrofolate reductase. Several fluorinated pyrimidines, however, are converted metabolically into excellent synthetase inhibitors. At present, the best understood and most commonly used fluoropyrimidine drug is 5-fluorouracil (Heidelberger, 1965, 1970). Fluorouracil was synthesized in 1957 as an antimetabolite of uracil which might be effective in cancer chemotherapy. Heidelberger reasoned that 5-substitution of uracil would prevent its enzymic methylation. This compound was found to have anti-tumour activity against several mouse and rat tumours and is of clinical use in the treatment of some solid cancers (Heidelberger and Ansfield, 1963). The deoxyriboside of fluorouracil is a more effective drug, and is less toxic. Both compounds are converted *in vivo* into 5-fluoro-2′-deoxyuridylate (FdUMP) before exerting their inhibitory effects:

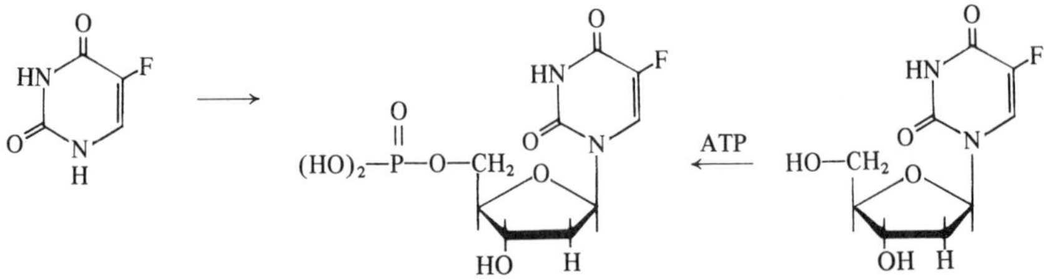

| Fluorouracil | 5–Fluoro–2′–deoxyuridylate | Fluorodeoxyuridine |

Because cells are considered to be impermeable to nucleotides,* FdUMP is formed from either fluorouracil or fluorodeoxyuridine by intracellular anabolism. The principal target of FdUMP is thymidylate synthetase (p. 254), blocking DNA synthesis and causing a condition

---

* Cohen (1975) has commented recently that, though penetration of nucleotides is slow, some of them do enter cells. This observation indicates that testing of nucleotides for therapy is warranted.

similar to thymineless death. In common with amethopterin, fluorinated pyrimidines cause only temporary regression or palliation of susceptible tumours, not cures, because of the onset of drug resistance. Acquired resistance to fluorouracil appears to be accompanied by several metabolic changes, including lowering of the activity of enzymes (phosphoribosyltransferase or ribonucleosidediphosphate reductase) necessary for the incorporation of fluorouracil into FdUMP. There is some selective toxicity of these agents in tumour cells because normal but not neoplastic tissues can detoxify fluorouracil.

Cohen and his coworkers (Cohen et al., 1958) examined the effects of fluorouracil and its derivatives on growing and virus-infected E. coli. They showed that the most potent agent, fluorodeoxyuridine, induced thymineless death in the bacteria. They prepared FdUMP from fluorodeoxyuridine by incubation either with whole cells or with ATP and thymidine kinase. The activity of partially purified E. coli thymidylate synthetase was irreversibly lost when it was preincubated with the FdUMP they prepared. In 1963, Mathews and Cohen reported that inhibition of the E. coli thymidylate synthetase by FdUMP was competitive with respect to dUMP if there was no preincubation, but stoichiometric if the enzyme was preincubated with the inhibitor. Initial kinetic results with the synthetase in extracts of Ehrlich ascites cells differed slightly from those with the bacterial enzyme. Reyes and Heidelberger (1965) showed that FdUMP is a competitive inhibitor, even when preincubated with the extracts. Their work suggested that methylenetetrahydrofolate reacted with the synthetase before either dUMP or FdUMP, in an ordered sequential mechanism. Inhibition by FdUMP could be reversed by dialysis, but this release of the inhibitor was retarded in the presence of methylenetetrahydrofolate. When these kinetic experiments were extended using purified mammalian reductase, Langenbach et al. (1972) found that FdUMP was bound covalently to the synthetase, but that this binding required methylenetetrahydrofolate. The latter substrate also is firmly bound to the enzyme. The laboratories of Heidelberger (Danenberg et al., 1974), Santi (Santi et al., 1974) and others are continuing the study of the mechanism of inhibition of FdUMP using the thymidylate synthetase purified from L. casei (p. 257). Both the above groups suggest that FdUMP is a quasisubstrate which undergoes only part of the dUMP to TMP conversion, but including the nucleophilic addition of an enzyme side chain to C^6 of the pyrimidine. The structure of the stable ternary E–cofactor–FdUMP complex (or complexes) has yet to be determined. Sommer and Santi (1974) have isolated and characterized a hexapeptide obtained from pronase digestion of inactivated methylenetetrahydrofolate–FdUMP–TMP synthetase ternary complex. Santi proposed that the peptide has this structure:

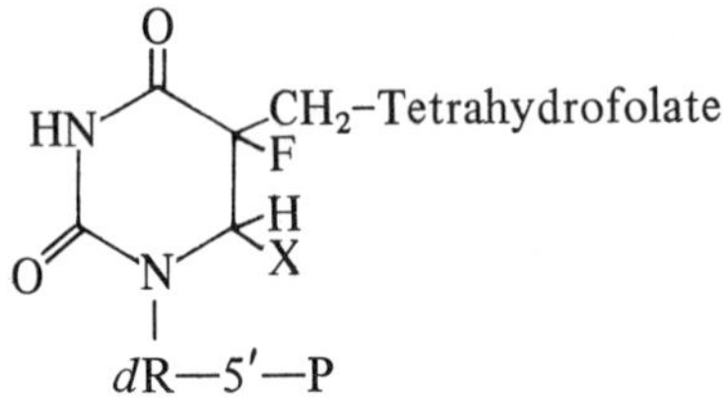

where X = (Thr, His, Ala, Leu, Pro$_2$)

This is a structure analogous to that of the enzyme-bound intermediate postulated to occur in the formation of TMP (cf. Chapter 8, Fig. 10). Because the C—F bond is stable, the covalent complex is stabilized and normal products cannot be formed. Santi cites the isolation of this cysteine-free hexapeptide as strong evidence that an —SH group is not the nucleophile on the enzyme which adds to position-6 of FdUMP (and dUMP in the normal reaction). He has concluded that the —OH group of threonine or the imidazole of histidine is the nucleophile. Other investigators believe that the hexapeptide is an artefact of isolation, with the FdUMP having migrated during the denaturation from a cysteine-SH to a nearby threonine-OH (or histidine imidazole group). The covalent bond(s) of the enzyme to FdUMP need not be permanent. For example, Dunlap's laboratory (Aull et al., 1974) have isolated and separated the 1:1:1 and 2:2:1 ternary complexes (i.e. those in which one subunit and two subunits, respectively, are substituted). Both of these complexes can revert to native enzyme upon prolonged storage. Folate compounds that cannot form bridge complexes with the pyrimidine also support tight binding of FdUMP to TMP synthetase (Galivan et al., 1975). Resolution of the controversy over the nature of FdUMP–synthetase complexes will provide not only a rationale for fluorouracil chemotherapy but also mechanistic information on a very complicated enzymic reaction. It is ironic that the long-sought irreversible inhibitor of thymidylate biosynthesis may be a metabolite of a compound, fluorouracil, supposed to be a classical antimetabolite. This is the sort of unplanned good fortune which occurs in science (cf. also the amusing story of the design of homofolic acid, Friedkin *et al.*, 1971).

Another biologically active fluorinated pyrimidine derivative is 5-trifluoromethyl-2'-deoxyuridylate (Heidelberger, 1970). This compound is both an inhibitor of thymidylate synthetase and incorporated into DNA. Its deoxyriboside is a promising anti-cancer drug, and is an extremely potent agent against DNA viruses such as in Herpes simplex infection of the cornea (a major source of blindness in man). All the fluorinated pyrimidines have anti-fungal activity, but only 5-fluorocytosine, which is the only one not toxic to mammals, can be used against fungal infections.

Pyrimidine nucleosides are amongst the few classes of compounds known to be useful in anti-viral chemotherapy (reviewed by Shugar, 1974). 5-Iodo-2'-deoxyuridine has been used most to date. Its

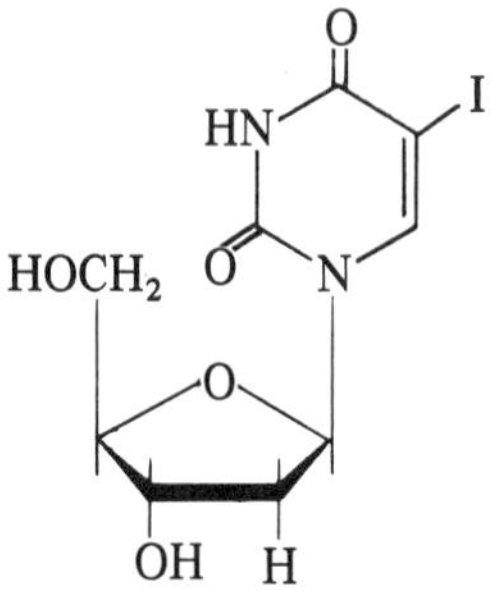

5–Iodo–2'–deoxyuridine

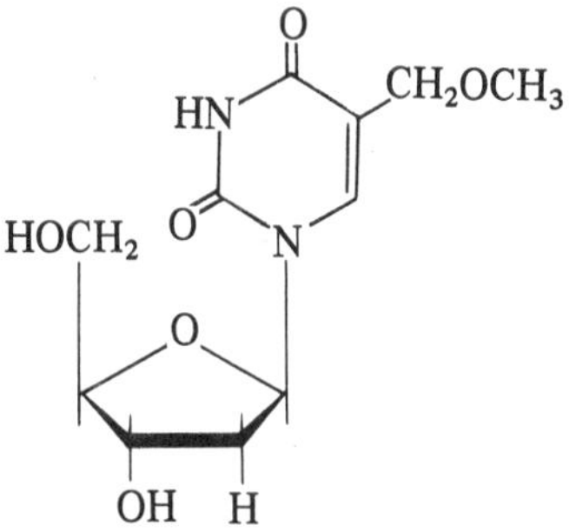

5–Methoxymethyl–2'–deoxyuridine

mechanism of action is not known; presumably it is converted into a nucleotide, and then incorporated into viral DNA. Iododeoxyuridine has beneficial effects against several viruses, but it is toxic and mutagenic. Babiuk *et al.* (1975) have found that another synthetic pyrimidine nucleoside, 5-methoxymethyl-2'-deoxyuridine, has anti-viral activity against Herpes simplex and appears to have a different mechanism of action (possibly interference with the biosynthesis of thymidylate from thymidine). Its extraordinary feature

S

is its very low mammalian toxicity, suggesting that it may prove useful in treatment of Herpes simplex virus infections in humans. Other pyrimidine nucleoside analogues have different target sites, providing a wide variety of compounds to be tested in combination for synergistic therapy.

6-MERCAPTOPURINE

The biochemistry and pharmacology of 6-mercaptopurine (Elion, 1967; Hitchings, 1969) is one of the most intriguing developments in cancer chemotherapy. Mercaptopurine, synthesized in 1951 as an antimetabolite of hypoxanthine, causes temporary remissions of acute leukemia in children. Unlike fluorouracil, mercaptopurine was

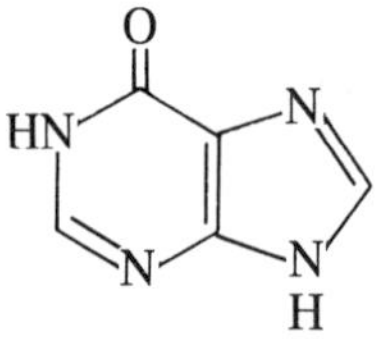

Hypoxanthine 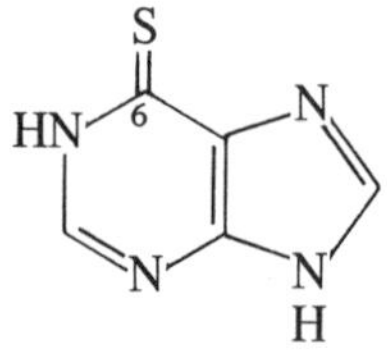   6−Mercaptopurine

not synthesized to inhibit a specific enzyme reaction. In fact, at least a half dozen enzymes are inhibited by metabolites of mercapto-purine, so there may be more than one primary target enzyme in malignant cells. 6-Mercaptopurine (or a derivative) substitutes for hypoxanthine, adenine and/or guanine (or their derivatives) as a substrate in many reactions.

Before it becomes an effective inhibitor, mercaptopurine must be anabolized to its ribonucleoside-5'-monophosphate, thioinosinic acid. This conversion is catalyzed by hypoxanthine-guanine phospho-ribosyltransferase:

$$\text{Mercaptopurine} + \text{PRPP} \rightarrow \text{Thioinosinate} + \text{PP}_i \,.$$

After thioinosinate is formed, it may be phosphorylated to the di- and triphosphate forms which, as analogues of ADP and ATP, can be used for biosynthesis of faulty coenzymes (e.g. NAD^+) or nucleic acids. The puzzle of how 6-mercaptopurine exerts its chemotherapeutic effect is a result of the broad substrate specificity of some of the enzymes catalyzing purine nucleotide metabolic reactions. Three of the possible target sites for inhibition by thioinosinate are two enzymes involved in purine nucleotide interconversion, IMP dehydrogenase and adenylosuccinate synthetase, and the first enzyme in the pathway of *de novo* purine ribonucleotide biosynthesis, phosphoribosyl pyrophosphate amidotransferase. In the former two instances, thioinosinate acts as a substrate analogue (a covalent inhibitor and a competitive inhibitor, respectively). Assuming that these three enzymes are the most sensitive to inhibition by thioinosinate, formation of AMP and GMP (and thus all nucleic acids and some coenzymes) would be slowed down by 6-mercaptopurine treatment by a combination of two sequential blocks.

Inhibition of the amidotransferase by thioinosinate apparently demonstrates a novel principle, *pseudofeedback inhibition,* negative feedback by an unnatural end-product. Accumulation of naturally occurring purine nucleotides is believed to exert negative feedback inhibition on the first specific reaction of purine biosynthesis:

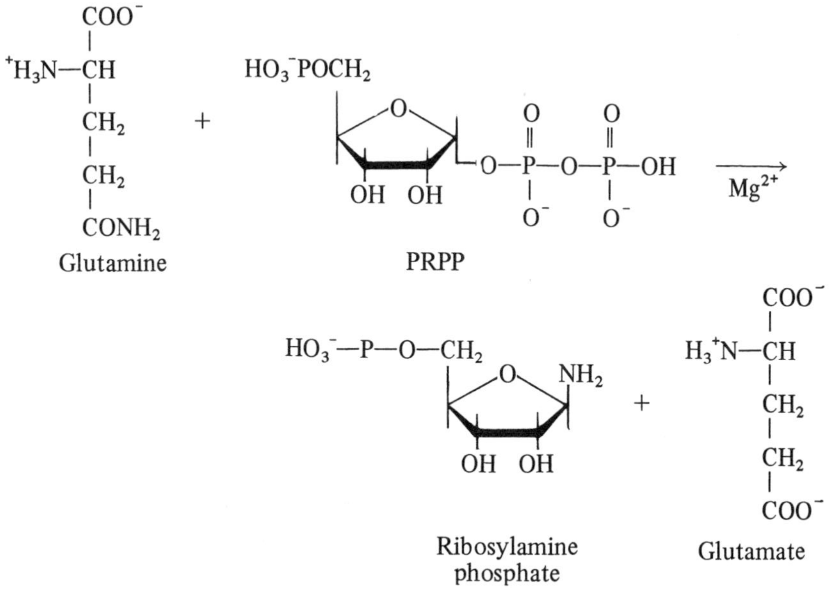

Thioinosinate also is an inhibitor of this reaction, thereby mimicking the action of natural ribonucleotides (McCollister *et al.,* 1964).

Because of the instability of the amidotransferase, leading to conflicting kinetic results, the mechanisms of the reaction and the end-product inhibition are unclear. Henderson (1972), in an excellent assessment of the role of the amidotransferase as a site of metabolic regulation, has pointed out the need for a thorough and carefully controlled study of the kinetics of amidotransferases from several sources.

When ^{14}C-6-mercaptopurine is injected into mice bearing tumours, radioactivity can be found in the isolated tumour DNA (Elion, 1967). The ^{14}C is found mainly in deoxythioguanosine, not deoxythioinosine. Until recently, this incorporation into DNA has been neglected as a target for 6-mercaptopurine cytotoxicity. Following up on this earlier observation and knowing of the incorporation of 6-thioguanine into DNA, Tidd and Paterson (1974) proposed a mechanism for 6-mercaptopurine inhibition involving its metabolism to 6-thioguanosine phosphates and incorporation into DNA as a thioguanine derivative. Their conclusions are based upon experiments in which ^{35}S-6-mercaptopurine was incubated with cultured cells, and the nucleic acids isolated and degraded. The cells were killed by 6-mercaptopurine treatment, but only after a delay of about 24 hours. Associated with this delayed cytotoxicity was the incorporation of radioactivity into DNA. The ^{35}S in the dying cells was in the form of thioguanine. The extent of incorporation of thioguanine into DNA correlated with the toxicity of the drug. Their work strongly suggests that the other sites of 6-mercaptopurine action are not related to cell death. Other workers have since published data obtained with other cellular systems verifying this conclusion.

One major mechanism of resistance to 6-mercaptopurine is related to its anabolism. Brockman (1963) has shown that acquired resistance in most nonhuman systems (micro-organisms and experimental tumours) is associated with loss of the capacity to form thioinosinate from mercaptopurine. Specifically, the activity of hypoxanthine-guanine phosphoribosyltransferase is either lowered or completely lost. Resistant cells, therefore, no longer catalyze a reaction which

$$\begin{matrix} \text{Guanine} \\ \text{or} \\ \text{Hypoxanthine} \\ \text{or} \\ \text{6-Mercaptopurine} \end{matrix} + \text{PRPP} \xrightarrow{\begin{subarray}{c}\text{Hypoxanthine-guanine}\\ \text{phosphoribosyltransferase}\end{subarray}} \begin{matrix} \text{GMP} \\ \text{or} \\ \text{IMP} \\ \text{or} \\ \text{Thioinosinate} \end{matrix} + \text{PP}_i.$$

would lead to their own death. Other possible mechanisms of resistance include loss of ability to transport 6-mercaptopurine into cells and increase in catabolism of 6-mercaptopurine. Several investigators reported that a decrease in hypoxanthine-guanine phosphoribosyltransferase does not always appear in cell-free extracts of human leukemic cells resistant to 6-mercaptopurine. However, Kessel and Hall (1969) have observed good correlation between the ability of isolated cells (human or mouse leukemia cells) to convert ^{14}C-mercaptopurine into a labelled non-diffusible form (i.e. thioinosinate) and the responsiveness of the cells to 6-mercaptopurine. This apparent conflict appears to be due to a large increase in alkaline phosphatase, catalyzing the hydrolysis of thioinosinate, in resistant cells (Wolpert *et al.*, 1971). Rosman *et al.* (1974) have expanded the previous studies by assaying both hypoxanthine-guanine phosphoribosyltransferase and phosphatase activities in leucocytes from 18 patients resistant to 6-thiopurines. One patient had a severe deficiency of the phosphoribosyltransferase, and six others showed moderate enzyme deficiency. Eight of the patients had elevated levels of particulate-bound alkaline phosphatase. In conclusion, resistance in human tumours often can be related to a decrease in the available levels of thioinosinate, caused in some cases by prevention of its formation and in others by increase in its catabolism.

The catabolism of mercaptopurine also is important in its effectiveness as an anti-cancer drug (Elion, 1967). Some administered drug is anabolized to thioinosinate, but most is excreted as catabolites. The major pathway of breakdown is the direct oxidation of 6-mercaptopurine catalyzed by xanthine oxidase:

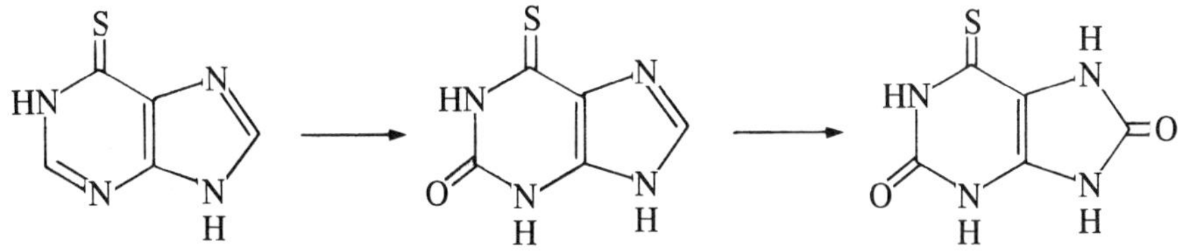

6-Mercaptopurine   6-Thioxanthine   Thiourate

In order to improve the efficiency of utilization of 6-mercaptopurine, Hitchings and Elion used two approaches, inhibition of xanthine oxidase and S-substitution of mercaptopurine to provide a slowly released or latent form of 6-mercaptopurine. Neither trick was particularly useful in cancer chemotherapy, but both led to major drugs for other uses!

Allopurinol is a structural analogue of hypoxanthine, and a potent inhibitor of xanthine oxidase. Although allopurinol exerts a sparing action on mercaptopurine upon concurrent administration, this effect is of no practical advantage. Another clinical application of allopurinol is in reduction of uric acid biosynthesis, especially in gout. Gout is a metabolic disorder in which an elevated level of serum uric acid leads to crystallization of uric acid in tissues, especially on cartilage and tendons, resulting in inflammation of the joints. Administration of allopurinol causes inhibition of xanthine oxidase, and accumulation and excretion of the more soluble purines hypoxanthine and xanthine. Allopurinol exerts its inhibitory effect on xanthine oxidase in an interesting fashion. It is a substrate of the enzyme, forming a product, alloxanthine or oxipurinol, which is a stoichiometric inhibitor (Massey *et al.*, 1970; Spector and Johns, 1970) which binds to the reduced oxidase:

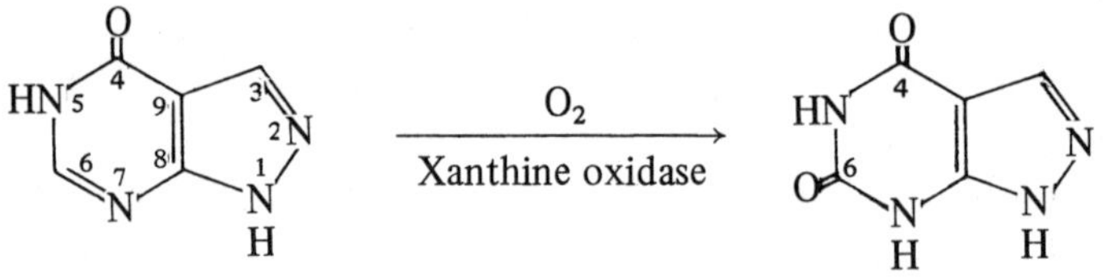

Allopurinol   Oxipurinol

Azathioprene (Imuran) is an S-substituted derivative of 6-mercaptopurine which was synthesized to circumvent the rapid catabolism of mercaptopurine. As administered, azathioprene is inactive, but it is cleaved to 6-mercaptopurine and then metabolized to thioinosinate, see diagram on next page. This latent form of mercaptopurine has been no more useful than the parent drug for treatment of cancer, but it is the most widely used immunosuppressive drug (Hitchings and Elion, 1969). It is routinely used as a continuing treatment to prevent tissue rejection after organ (e.g. kidney) transplantations, with no serious toxic effects. The exact mechanism of inhibition of antibody biosynthesis by the thioinosinate formed from azathioprene

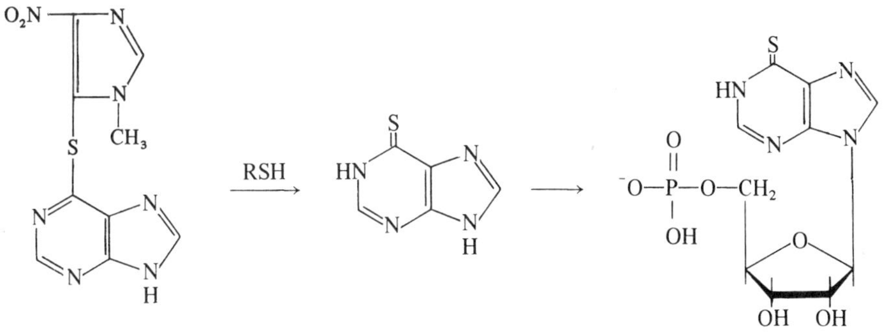

Azathioprene   Mercaptopurine   Thioinosinate

is not known, nor is the reason why azathioprene is a more effective immunosuppressive drug than 6-mercaptopurine.

ENZYMES AS DRUGS

Asparaginase

Therapy with the enzyme asparaginase (Crowther, 1971; Capizzi and Handschumacher, 1973) provides an unconventional approach to the selective death of neoplastic cells. It involves a new mechanism of drug action—an enzyme-induced nutritional deficiency. Although few types of human tumours (notably, acute lymphocytic leukemia) are inhibited by this enzyme drug, its unique mechanism permits treatment of patients resistant to conventional anti-leukemic drugs. Use of an enzyme as a drug presents novel clinical possibilities and difficulties, both well worth examining.

The possible use of asparaginase as a therapeutic agent arose from the observations that intraperitoneal injection of guinea pig serum inhibited growth of transplanted lymphomas in mice (Kidd, 1953) and that L-asparaginase was the active protein factor in this serum (Broome, 1961). The primary effect of asparaginase appears to be hydrolysis of plasma asparagine to aspartate and NH_4^+. For some tumour cells, asparagine is an essential nutrient. Asparaginase-induced removal of endogenous asparagine causes a rapid inhibition of protein synthesis in the sensitive tumour cells, but less drastic effects on normal cells. Asparaginase purified from *E. coli* is used for most clinical studies because it has a high affinity for its substrate, asparagine, and because it is retained in plasma for a relatively long time. In contrast to most drugs, asparaginase is effective at concentrations well below the drastically toxic range. In addition to inhibiting the growth of many animal tumours, asparaginase treatment has caused complete but short remissions in up to 60% of patients with acute lymphocytic leukemia (Capizzi and Handschumacher, 1973). Resistance, due to appearance of elevated levels of asparagine synthetase in the previously sensitive cells, appears in only a few months, so that the duration of remission is short (2–4 months). Patients relapsing during or after asparaginase treatment seldom respond to later treatment with the enzyme.

Introduction of a foreign enzyme as a drug requires a number of precautions and special considerations. The asparaginase must be highly purified, to be free of contaminating bacterial proteins which could cause toxic allergic reactions. Injection is the only effective method for administration of asparaginase. Usually, to achieve complete remission of leukemia, the drug is injected daily for about a month. Allergic responses to asparaginase have been observed in about 20% of patients. Higher doses of asparaginase seem to cause less allergic reaction by causing immunosuppression. However, toxicity to the liver, kidneys and other organs does occur. The use of asparaginase is limited by the small number of susceptible tumours, the toxicity and the early development of resistance.

Other purified enzymes are being tested as possible anti-cancer drugs. Bertino *et al.* (1971) suggested that a carboxypeptidase isolated from *Pseudomonas stutzeri* capable of catalyzing the cleavage of the terminal glutamate residues from folic acid might cause a folate deficiency which could selectively inhibit tumour growth. Similarly, Abell *et al.* (1973) have tested phenylalanine ammonia lyase from yeast as a method of depleting the level of phenylalanine in mouse tumour cells. When injected into tumour-bearing mice, the lyase caused cures in up to 43% of the animals. No successful clinical applications of enzymes other than asparaginase have been reported to date, however.

Chemotherapeutic treatment of cancer has been of value in gradually allowing extension of the lives of patients. By combination therapy with several drugs or by use of drugs plus either surgery or radiation, remissions of several cancers have been extended from a few months to several years and some cancers have been cured. While research concentrates on the isolation and characterization of tumour viruses, the chemotherapeutic approach will be in use for the foreseeable future, as long as it provides benefits. Resistance and toxicity remain the major problems. As the research continues, using the concepts outlined in this chapter and possible new approaches, the number of cancers for which there is a successful anti-tumour agent will surely increase. Last, and far from least, medicine has benefited greatly from cancer chemotherapy research with the development of drugs now used successfully in the treatment of many other diseases.

REFERENCES

Abell, C. W., Hodgins, D. S. and Stith, W. J. (1973). *Cancer Res.* **33**, 2529-2532.
Albert, A. (1973). "Selective Toxicity", Fifth edition. Chapman and Hall, London.
Aull, J. L., Lyon, J. A. and Dunlap, R. B. (1974). *Archs Biochem. Biophys.* **165**, 805-808.
Babiuk, L. A., Meldrum, B., Gupta, V. S. and Rouse, B. T. (1975). *Antimicrobial Agents and Chemotherapy* **8**, 643-650.
Baker, B. R. (1967). "Design of Active-Site-Directed Irreversible Enzyme Inhibitors". John Wiley and Sons, New York.
Baker, B. R. (1971). *Ann. N.Y. Acad. Sci.* **186**, 214-226.
Bertino, J. R. (1963). *Cancer Res.* **23**, 1286-1306.
Bertino, J. R., O'Brien, P. and McCullough, J. L. (1971). *Science, N.Y.* **172**, 161-162.
Brockman, R. W. (1963). *Adv. Cancer Res.* **7**, 155-160.
Broome, J. D. (1961). *Nature, Lond.* **191**, 1114-1115.
Brown, G. M. (1971). *Adv. Enzymol.* **35**, 35-77.
Capizzi, R. L. and Handschumacher, R. E. (1973). *In* "Cancer Medicine" (Holland, J. F. and Frei, E. III, eds), pp. 850-859. Lea and Febiger, Philadelphia.
Cashmore, A. R., Skeel, R. T., Makulu, D. R., Gralla, E. J. and Bertino, J. R. (1975). *Cancer Res.* **35**, 17-22.
Chabner, B. A., Myers, C. E., Coleman, C. N. and Johns, D. G. (1975). *New Engl. J. Med.* **292**, 1107-1113 and 1159-1168.
Cohen, S. S. (1971). *Ann. N.Y. Acad. Sci.* **186**, 292-301.
Cohen, S. S. (1975). *Biochem. Pharmac.* **24**, 1929-1932.
Cohen, S. S., Flaks, J. G., Barner, H. D., Loeb, M. R. and Lichtenstein, J. (1958). *Proc. natn. Acad. Sci. U.S.A.* **44**, 1004-1012.
Crowther, D. (1971). *Nature, Lond.* **229**, 168-171.
Danenberg, P. V., Langenbach, R. J. and Heidelberger, C. (1974). *Biochemistry* **13**, 926-933.

Elion, G. B. (1967). *Fedn Proc.* **26**, 898-903.
Farber, S., Diamond, L. K., Mercer, R. D., Sylvester, R. F., Jr. and Wolff, J. A. (1948). *New Engl. J. Med.* **238**, 787-793.
Frei, E. III. (1967). *Fedn Proc.* **26**, 918-922.
Frei, E. III, Jaffe, N., Tattersall, M. H. N., Pitman, S. and Parker, L. (1975). *New Engl. J. Med.* **292**, 846-851.
Friedkin, M. (1973). *Adv. Enzymol.* **38**, 235-292.
Friedkin, M., Crawford, E. J. and Plante, L. T. (1971). *Ann. N.Y. Acad. Sci.* **186**, 209-213.
Gale, E. F., Cundliffe, E., Reynolds, P. E., Richmond, M. H. and Waring, M. J. (1972). "The Molecular Basis of Antibiotic Action". John Wiley and Sons, London.
Galivan, J. H., Maley, G. F. and Maley, F. (1975). *Biochemistry* **14**, 3338-3344.
Halpern, R. M., Halpern, B. C., Clark, B. R., Ashe, H., Hardy, D. N., Jenkinson, P. Y., Chou, S.-C. and Smith, R. A. (1975). *Proc. natn. Acad. Sci. U.S.A.* **72**, 4018-4022.
Heidelberger, C. (1965). *Prog. Nucleic Acid Res. molec. Biol.* **4**, 1-50.
Heidelberger, C. (1970). *Cancer Res.* **30**, 1549-1569.
Heidelberger, C. and Ansfield, F. J. (1963). *Cancer Res.* **23**, 1226-1243.
Henderson, J. F. (1972). "Regulation of Purine Biosynthesis", pp. 174-188. American Chemical Society, Washington.
Hitchings, G. H. (1969). *Cancer Res.* **29**, 1895-1903.
Hitchings, G. H. (1971). *Ann. N.Y. Acad. Sci.* **186**, 444-451.
Hitchings, G. H. and Elion, G. B. (1969). *Accts Chem. Res.* **2**, 202-209.
Hryniuk, W. M. (1972). *Cancer Res.* **32**, 1506-1511.
Hryniuk, W. M. (1975). *Cancer Res.* **35**, 1085-1092.
Kessel, D. and Hall, T. C. (1969). *Cancer Res.* **29**, 2116-2119.
Kidd, J. G. (1953). *J. exp. Med.* **98**, 565-606.
Langenbach, R. J., Danenberg, P. V. and Heidelberger, C. (1972). *Biochem. biophys. Res. Commun.* **48**, 1565-1571.
Massey, V., Komai, H., Palmer, G. and Elion, G. (1970). *J. biol. Chem.* **245**, 2837-2844.
Mathews, C. K. and Cohen, S. S. (1963). *J. biol. Chem.* **238**, 367-370.
McCollister, R. J., Gilbert, W. R., Jr., Ashton, D. M. and Wyngaarden, J. B. (1964). *J. biol. Chem.* **239**, 1560-1563.
Osborn, M. J., Freeman, M. and Huennekens, F. M. (1958). *Proc. Soc. exp. Biol. Med.* **97**, 429-431.
Otting, F. and Huennekens, F. M. (1972). *Archs Biochem. Biophys.* **152**, 429-431.
Reyes, P. and Heidelberger, C. (1965). *Molec. Pharmac.* **1**, 14-30.
Rosman, M., Lee, M. H., Creasey, W. A. and Sartorelli, A. C. (1974). *Cancer Res.* **34**, 1952-1956.
Santi, D. V., McHenry, C. S. and Sommer, H. (1974). *Biochemistry* **13**, 471-481.
Shugar, D. (1974). *FEBS Letters* **40**, S48-S62.
Sommer, H. and Santi, D. V. (1974). *Biochem. biophys. Res. Commun.* **57**, 689-695.

Spector, T. and Johns, D. G. (1970). *J. biol. Chem.* **245**, 5079-5085.
Tattersall, M. H. N., Brown, B. and Frei, E. III. (1975). *Nature, Lond.* **253**, 198-200.
Tidd, D. M. and Paterson, A. R. P. (1974). *Cancer Res.* **34**, 738-746.
Werkheiser, W. C. (1961). *J. biol. Chem.* **236**, 888-893.
Williams, M. N., Poe, M., Greenfield, N. J., Hirshfield, J. M. and Hoogsteen, K. (1973). *J. biol. Chem.* **248**, 6375-6379.
Wolpert, M. K., Damle, S. P., Brown, J. E., Sznycer, K. C., Agrawal, K. C. and Sartorelli, A. C. (1971). *Cancer Res.* **31**, 1620-1626.
Woods, D. D. (1940). *Br. J. exp. Path.* **21**, 74-90.

16 Complex Allosteric Control Systems

In the final chapter devoted to control of enzyme activity, I will describe several complex but physiologically important systems which embody a number of the control principles previously discussed. I shall emphasize just the enzyme-related features of these processes not only because of their relevance to this text but also for concise treatment of these subjects. References to fuller discussions of them are given and should be consulted for a more complete understanding. The topics I have chosen are the mechanism of hormone expression by the adenylate cyclase- and protein kinase-catalyzed reactions, the repression of enzyme synthesis in bacteria and the stereochemical theory of odour.

CYCLIC AMP AND HORMONE ACTION

In mammalian cells, the effects of many hormones are elicited through a series of enzymes which metabolize the nucleotide adenosine-3', 5'-monophosphate, cyclic AMP. Some of these enzymes are mem-

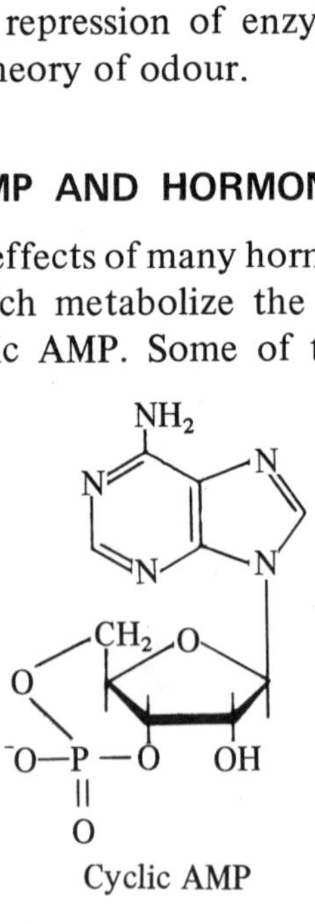

Cyclic AMP

brane-bound and others cytoplasmic. They exhibit a number of control characteristics including allosteric activations and inactivations, some by covalent modification.

Cyclic AMP was discovered by Sutherland and his colleagues while examining the stimulatory effects of the hormones epinephrine (adrenaline) and glucagon on the catabolism of glycogen in liver cells and cell homogenates. They found that the particulate fraction of cell homogenates catalyzes the synthesis of a heat-stable and dialyzable factor which stimulates glycogen phosphorylase. The discovery, chemistry, wide distribution in nature and many of the metabolic effects of this factor, cyclic AMP, have been reviewed in a comprehensive monograph (Robison et al., 1971). Cyclic AMP is maintained at a concentration of about 1 μM or less in mammalian cells, but changes in concentration under certain conditions. The continuous presence of cyclic AMP in cells implies that it is continuously synthesized. Many hormones stimulate the activity of adenylate cyclase, the enzyme which catalyzes the synthesis of cyclic AMP from ATP. The range of hormones able to stimulate adenylate cyclase in any given tissue is narrow. Hormone-stimulated adenylate cyclase therefore consists of a hormone receptor coupled in some fashion to a catalytic subunit which is responsible for the increased rate of cyclic AMP synthesis during hormone stimulation. Cyclic AMP has been described as a *second messenger*, generated on the inner surface of the plasma membrane after a signal is received from the first messenger, the hormone. The second messenger, being able to diffuse throughout the cell, brings about the metabolic response to the hormone.

Hormones are regulator molecules secreted by the internal secretion organs, the endocrine glands. When they are released by the glands, hormones circulate through the bloodstream to their distant target organs or tissues, and there exert their physiological effects. Hormones are capable of acting at extremely low concentrations (e.g. below 10^{-8} M). The specific cellular component which recognizes and binds a hormone is called a *receptor*. There are two types of hormones, those such as the steroid hormones which give a slow but long lasting response and those which produce a rapid response of short duration. The slow acting hormones enter their target cells and bind to cytoplasmic receptors. The receptor–ligand complexes then translocate to the nuclei, where they bind to chromosomes and stimulate the synthesis of specific protein molecules (reviewed by Buller and O'Malley, 1976). The hormones that produce rapid responses bind to receptors which are localized in cell membranes and have hormone-binding sites on the external surfaces of the cells. It is primarily the latter hormones that use cyclic AMP as their second messenger, to affect enzymic activities in target cells. Thanks to the large amount of research on cyclic AMP-related phenomena, some aspects of hormone action now can be described in molecular terms.

Hormones that change the cellular concentration of cyclic AMP include the catecholamines, glucagon and insulin. The former two groups of hormones usually raise the level of cyclic AMP, while insulin lowers its concentration. The two naturally occurring catecholamines are epinephrine and norepinephrine. Epinephrine is synthesized and secreted by the adrenal medulla, and norepinephrine both by other cells in the adrenal medulla and by nerve synapses. There are two classes of catecholamine receptors, the α-receptors and the β-receptors. The naturally occurring catecholamines are the most potent stimulating hormones of α-receptors, while the synthetic hormone isoproterenol is the most potent stimulator of β-receptors. The α- and β-receptors can also be distinguished by a series of drugs, the

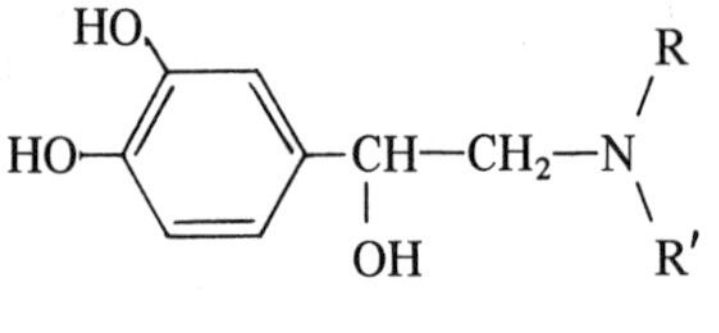

Catecholamines

L−Epinephrine, R $=$ H and R$'$ $=$ CH$_3$

L−Norepinephrine, R $=$ H and R$'$ $=$ H

L−Isoproterenol, R $=$ CH$_3$ and R$'$ $=$ CH$_3$

α- and β-adrenergic agents, which prevent the responses to catecholamines. Tissues containing β-receptors (such as liver and skeletal muscle) respond to catecholamines by accumulating cyclic AMP, while cells having α-receptors (e.g. some smooth muscles) seem to exhibit an entirely different response—an increase in the concentration of another cyclic nucleotide, guanosine-3′,5′-monophosphate (cyclic GMP). Epinephrine is released from the adrenal medulla during exercise, mental stress and other conditions which require increased amounts of glucose for energy. Its major effect is to promote glycogenolysis in both muscle and liver and thence raise the level of glucose in blood. This effect is mediated through an elevation in the concentration of cyclic AMP in liver and muscle. Glucagon, a polypeptide hormone synthesized in the pancreas, also increases blood glucose concentrations by accelerating glycogenolysis in liver by causing an accumulation of cyclic AMP. Glucagon, however, has no effect in muscle, and unlike the catecholamines can only elevate *cyclic AMP* levels. The protein insulin, the other principal hormone of pancreatic origin, has effects on carbohydrate, lipid and protein metabolism. In general, its actions are opposed to those of glucagon, epinephrine or other hormones that elevate cyclic AMP concentrations (cf. Pastan and Perlman, 1971). Some of its effects have been accounted for by a decrease in the concentration of cyclic AMP in the target tissue.

Hormones which cause a transient accumulation of cyclic AMP in cells do so through the reactions shown in Fig. 1. The hormone

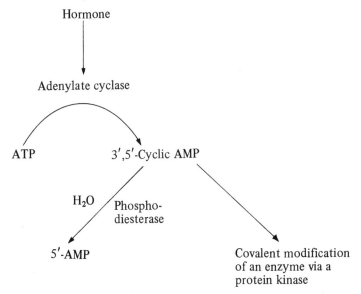

Fig. 1. Scheme showing the steps in the hormone stimulation of enzymatic activity in a mammalian target tissue.

stimulates the activity of adenylate cyclase. The cyclic AMP produced in this reaction exerts its effect by activation of a protein
kinase, and after the required effect has been achieved it is degraded
to adenosine-5'-monophosphate (AMP) by a phosphodiesterase-
catalyzed hydrolysis. The phosphorylation of an inactive enzyme
(or the covalent inactivation of an undesired enzymic activity) is
catalyzed by the cyclic AMP-stimulated protein kinase. The control
of glycogen biosynthesis and degradation in skeletal muscle is the
only cellular phenomenon affected by cyclic AMP characterized in
sufficient detail to provide a mechanism for cyclic AMP action. For
this reason, the covalent modification of the glycogen phosphorylase
system is described below. I shall now discuss each enzyme step in
cyclic AMP metabolism—first the adenylate cyclases, then the
phosphodiesterases and protein kinases, and finally the glycogen
phosphorylase system.

Adenylate Cyclase

Adenylate cyclase (originally called adenyl cyclase) is an enzyme
which is bound to the plasma membrane in animals and bacteria
(reviewed by Perkins, 1973). It catalyzes the following reaction:

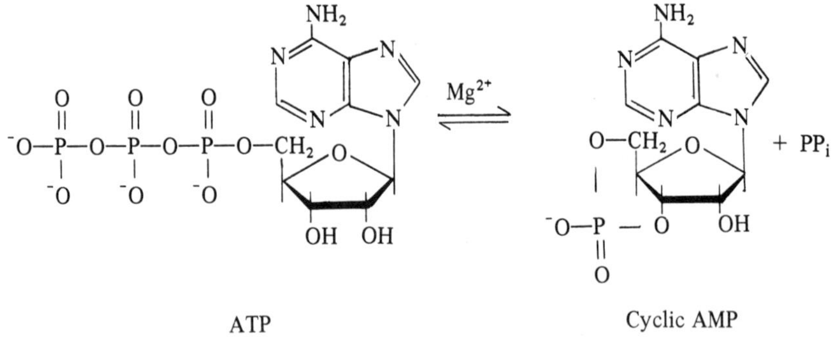

ATP   Cyclic AMP

Although the reaction is written as a reversible reaction, it is not
significantly reversible under physiological conditions. A distinguishing feature of cyclase activity in mammalian tissues is stimulation by
hormones in a tissue-specific fashion. The response to hormones of
the cyclase of a given tissue parallels the physiological effects of the
same hormones on the intact tissue. The manner in which a hormone
stimulates adenylate cyclase in its target tissue is far from completely

understood. Because not all aspects of hormone activation have been demonstrated with a single cyclase, a number of adenylate cyclase preparations are discussed. Hopefully the conclusions and generalizations drawn from these examples will be valid for all the hormonal stimulations of cyclic AMP synthesis.

Research on adenylate cyclase has been hampered by experimental difficulties. Basic problems include obtaining pure cell types from tissues containing mixed cell types, the low content of adenylate cyclase and the hormone receptors coupled to it, and the slow binding and relatively high concentration of hormone needed in most *in vitro* experiments. The latter difficulty has lent support to the theory that there may be many spare or extra hormone receptors in tissues, with activation of only a few of the many receptors necessary for the accomplishment of hormone stimulation. On simple chemical grounds, this theory is questionable; probably exhaustive enzymological study of the cyclase system will explain this anomalous observation. The firm binding of the components of adenylate cyclase to mammalian cell membranes is itself a major barrier, preventing the determination of the physical characteristics of the enzyme. Some tissues contain adenylate cyclase activity which is stimulated by only one hormone. In contrast, other tissues exhibit multi-valent regulation, where as many as seven hormones can cause an increase in cyclase activity. Also, there is always a basal cyclase activity; that is, there is activity in the absence of added hormone.

Assays for adenylate cyclase are difficult to perform accurately. Most assay methods are based upon use of either α-^{32}P-ATP or ^{14}C-ATP as substrate and chromatographic isolation of radioactive cyclic AMP. Assay problems include the use of membrane suspensions rather than soluble proteins, presence of contaminating enzymes, low activities, instability and limited stimulations by added hormones. ATPases and cyclic AMP phosphodiesterase often decompose the substrate and product, respectively, during assays. Both these enzyme activities can be present greatly in excess of the cyclase itself. Removal of the substrate has been prevented by using ATP regenerating systems, very high levels of ATP or the ATP analogue 5′-adenylylimidodiphosphate. This analogue is a substrate for adenylate cyclase but is only slightly hydrolyzed (about 15%) by the membrane ATPase during an assay (Rodbell *et al.*, 1971). In assay systems, the loss of cyclic AMP catalyzed by phosphodiesterase is usually controlled by

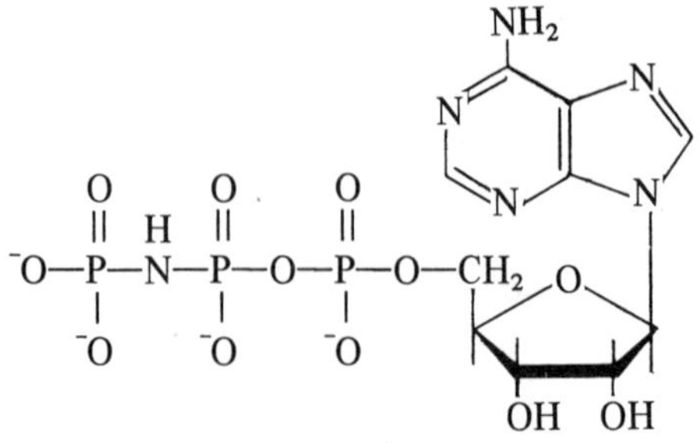

$$\text{NH}_2$$

5′−Adenylylimidodiphosphate

inhibition of the latter enzyme by theophylline, and by addition of large amounts of unlabelled cyclic AMP. An examination of the work done by Drummond and Duncan (1970) with the cyclase of cardiac tissue gives insight into the difficulties associated with the determination of the simpler kinetic properties of the adenylate cyclases. These workers have carefully verified both the identity of the product and the validity of the assays they used. The K_m for ATP with their enzyme is 0.08 mM, and the true substrate seems to be the Mg–ATP complex. The maximal activation by epinephrine was only 1.6-fold, insufficient for examination of the kinetics of hormone activation. From the technical difficulties that all workers have noted, we can see why the characterization of adenylate cyclases is a slow task.

At present, experiments on adenylate cyclase are performed using broken cell preparations. Activity, especially the stimulation by hormones, is not stable and attempts at solubilization usually lead to loss of all activity. The localization of most of the cyclase in cell membranes has been demonstrated by sedimentation experiments with homogenates. In these experiments, the cyclase activity copurifies with plasma membrane marker enzymes. Hormones covalently bound to large supporting matrices such as agarose can produce normal physiological responses with intact cells. Experiments using agarose-insulin have been cited as proof that the receptors are on the cell's exterior surface because the agarose bead cannot enter the intact cell. However, Kolb et al. (1975) have shown that this conclusion is invalid because the hormonal effect is due to free insulin that has leaked from the matrix (p. 58). Vauquelin et al. (1975) have demonstrated that both the hormone and the spacer arm are released by hydrolysis of cyanogen bromide-activated agarose with

Table I. Degree of stimulation of adenylate cyclase of frog erythrocyte membrane fragments. Fluoride gave optimal stimulation at 0.01 M. The concentration of hormone added was 4×10^{-5} M. From the data published by Rosen and Rosen (1969).

Addition	Approximate increase over basal activity
Fluoride	20
Isoproterenol	10
Epinephrine	7
Norepinephrine	4

isoproterenol bound. They suggest that other activation procedures must be used for immobilizing hormones for direct proof that the hormone does not enter the cell. Tissues commonly used for cyclase studies include rat liver (Rodbell, 1973) and fat cells (Birnbaumer and Rodbell, 1969), and frog erythrocyte membrane fragments (Rosen and Rosen, 1969). These erythrocyte preparations are particularly useful because of the low basal activity ($<5\%$ of maximal) and the absence of cyclic AMP phosphodiesterase. With this cyclase preparation, the degree of stimulation by hormones is parallel to their potency as β-adrenergic compounds (Table I).

Adenylate cyclases are less tightly bound to cellular membranes in bacteria than in animals. Several microbial cyclases have been solubilized, and that from *Brevibacterium liquefaciens* (Takai *et al.*, 1974) has been purified and crystallized. It is a dimer of molecular weight 92 000, composed of two seemingly identical monomers of 46 000 daltons. In bacteria, adenylate cyclases do not respond to hormone addition, nor does the cyclic AMP exert its effect through stimulation of a protein kinase reaction. Instead, there is a lowering of the intracellular concentrations of cyclic AMP when cells are grown on glucose, preventing the transcription of some operons, called catabolite repressible operons. Cyclic AMP stimulates the synthesis of the repressed proteins by binding to an activator protein. This phenomenon is mentioned below, in the section on the *lac* operon.

The stimulation of adenylate cyclase of broken cell preparations by sodium fluoride was observed early in the research on the enzyme. Fluoride usually gives a maximal stimulation of the enzyme, so that little or no response to a stimulating hormone can be seen after activation by fluoride. Fluoride stimulates neither bacterial adenylate cyclase nor the cyclase activity of intact mammalian cells. Drummond and Duncan (1970) reported that the only anion that will stimulate the cyclase is fluoride. Kalish *et al.* (1974) have found that very high concentrations of other halide ions can produce partial stimulations. Fluoride stimulation is characteristic of adenylate cyclase, but is of variable magnitude depending on the source of the enzyme preparation. The extent of stimulation is about two- to 20-fold in most cases. As seen below, solubilized preparations of mammalian adenylate cyclase which no longer respond to hormones can retain the ability to be stimulated by F^-. A tentative but reasonable explanation of the stimulation of adenylate cyclase by F^- has been proposed by Najjar, on the basis of the reaction of F^- with the phosphoserine form of phosphoglucomutase (Layne and Najjar, 1975). With phosphoglucomutase, the F^- nucleophile attacks the phosphoryl group on the enzyme to release phosphorofluoridate. Constantopoulos and Najjar (1973) and Layne *et al.* (1973) performed similar experiments on adenylate cyclase using membrane enzyme preparations from leucocytes and platelets. Najjar has postulated that adenylate cyclase can exist in two forms, an inhibited phospho-form and an activated dephospho-form. Interconversion of these forms would be catalyzed by a protein kinase and a fluoride-stimulated phosphoprotein phosphatase. In support of this postulate, these workers found that fluoride-activated and then washed cyclase could be inhibited (about 50–75%) by incubation with ATP and crude preparations of protein kinase. Upon treatment of the deactivated cyclase with either fluoride or the hormone prostaglandin E_1, full adenylate cyclase activity is regained. Further, they demonstrated that both the granulocytes and the platelets contain a membrane-bound and fluoride-stimulated phosphoprotein phosphatase, using ^{32}P-phosphoglucomutase as substrate. Possibly the phosphatase is itself phosphorylated, so that it can catalyze its own activation. Results to date are consistent with the scheme for regulation of adenylate cyclase activity shown in Fig. 2.

Intact membrane fragments seem to be required for retention of hormone-stimulated but not F^--stimulated or basal cyclase activity. Exposure of adenylate cyclase preparations to detergent or to phospholipase can reduce the hormone-stimulated activity. For example,

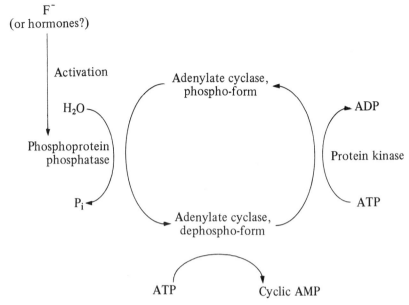

Fig. 2. Mechanism of regulation of adenylate cyclase postulated by Najjar and his colleagues. Nucleophilic attack by F^- on a protein-phosphate group may release phosphofluoridate from the phosphatase, activating it so that it can catalyze the dephosphorylation of the less active or fully inhibited adenylate cyclase phospho-form. Because both fluoride and hormones stimulate to the same maximal extent, it is believed that both activators function by the dephosphorylation of phospho-adenylate cyclase.

Pohl *et al.* (1971) showed that treatment of the adenylate cyclase system of rat liver plasma membranes with phospholipase A at suitable concentrations reduces the glucagon-stimulated activity, but not the fluoride-stimulated or basal activities. By exposing the treated membranes to aqueous dispersions of lipids (either purified lipids or those extracted from membranes), these workers observed a partial (approximately 25%) restoration of the activation of cyclase by glucagon. Treatment of the hepatic adenylate cyclase with phospholipase C (Rubalcava and Rodbell, 1973) does not diminish the binding of glucagon to any great extent, but does block glucagon activation. Levey (1971) has solubilized adenylate cyclase from cat heart by homogenizing the tissue in the non-ionic detergent Lubrol-PX. The solubilized preparation is stimulated by F^-, but not by hormones. When the solubilized cyclase is freed of detergent by chromatography on DEAE-cellulose, it shows no response upon addition of hormones that activate the particulate cyclase. However, addition of phosphatidyl serine restores responsiveness to glucagon, and monophosphatidyl inositol restores responsiveness to norepinephrine. (These hormonal stimulations of cyclase activity are about 50% and 200% of the basal activity, respectively.) Recent studies (Levey *et al.*, 1975) show that ^{125}I-glucagon can bind to solubilized adenylate cyclase in the absence of added phospholipids. The hormone appears to be bound to a dissociable glucagon binding site. This receptor is seen as a polypeptide gel band of about 25 000 molecular weight, separable from the 100 000–200 000 molecular weight band containing adenylate cyclase activity. Although binding of hormone is slow and not quantitatively related to activation, experiments of this type suggest that the phospholipids of the membrane do not participate in hormone binding but that an intact membrane is needed for the coupling of the hormone receptor to the catalytic adenylate cyclase unit. Ryan and Storm (1974) have reported the solubilization of a hormonally sensitive adenylate cyclase from rat liver plasma membranes. They have tested a variety of non-ionic detergents and found that treatment of the membrane particles with Triton X-305 gives an enzyme preparation which is not sedimented at 100 000 X g in one hour, and is stimulated five-fold by fluoride, seven-fold by glucagon and 20-fold by epinephrine. No analysis of this preparation, neither for lipid nor for the number of protein molecules, has been published.

Another complication which must be resolved is the role of GTP in the coupling of receptor(s) to the catalytic site of adenylate cyclase. In 1971, Rodbell and his colleagues noticed that low concentrations (maximal at 1 μM) of GTP enhance activation of hepatic adenylate cyclase by glucagon. This enhancement was only observable at concentrations of ATP below 10^{-4} M, possibly because of contamination of the solutions of ATP substrate with traces of GTP or weak binding of ATP at the GTP site. The GTP requirement for hormone stimulation of cyclase has been confirmed in experiments with many other tissues. The effect of GTP was envisaged (Rodbell, 1973) as a binding to an allosteric site of adenylate cyclase, essential for hormone stimulation of the catalysis of cyclic AMP formation. The synthetic GTP analogue 5'-guanylylimidodiphosphate—a compound not readily hydrolyzed by membranes—can fulfil the role of GTP, causing complete activation of the cyclase (Londos *et al.*, 1974). In fact, guanylylimidodiphosphate is more active as an activator than is GTP. It exerts its full effect only after a lag period, apparently by an irreversible process (Lefkowitz and Caron, 1975; Cuatrecasas *et al.*, 1975). Cyclase which is activated maximally by the GTP analogue is not further stimulated by hormones or F⁻. Irreversible covalent modification of the cyclase appears to be the most logical explanation for stimulation by guanylylimidodiphosphate.* It will be interesting to see how its mechanism of action relates to physiological activation at the same site by GTP.

The structure of the hormone-stimulated adenylate cyclase system is still unknown. Early postulates favoured a structure with two directly interacting subunits, with the receptor being a regulatory subunit (*R*) having its hormone-binding site on the exterior surface of the cell membrane, and the cyclase being a catalytic subunit (*C*) on the inner surface facing the cytoplasm. A hormone-induced conformational change in *R* would cause an increase in the activity of *C* in a manner analogous to the activation of cytoplasmic allosteric enzymes. Most current descriptions of the cyclase system include the membrane lipids as a third essential component, with the membrane physically separating but coupling *R* and *C* (Fig. 3). Several functions, relying on the fluidity of the membrane proteins and lipids, have been suggested for this third component (Perkins, 1973; Cuatrecasas, 1974). The phosphorylation–dephosphorylation system may be considered a fourth essential component of adenylate cyclase. Upon stimulation by binding of the hormone, the receptor protein *R* transmits the stimulus to the enzymic site either by migrating to and showing increased affinity for *C*, or activating *C* by altering the structure of its lipid environment. *A priori,* the former suggestion seems more likely, but there is no experimental data to favour either theory. The site of GTP stimulation is not yet known, nor has the role of the hormone- or fluoride-stimulated phosphatase been confirmed. These effects both must be included in the "coupling" area between receptor and catalyst, though. The adenylate cyclases of cells exhibiting uni-valent stimulation probably have only one type of receptor which can interact with *C*. There is no adequate information on whether cells showing multi-valent regulation have one adenylate cyclase coupled to each type of receptor, or a separate cyclase unit

---

* If there is a cyclase kinase, then its inactivation by covalent modification could also explain these observations.

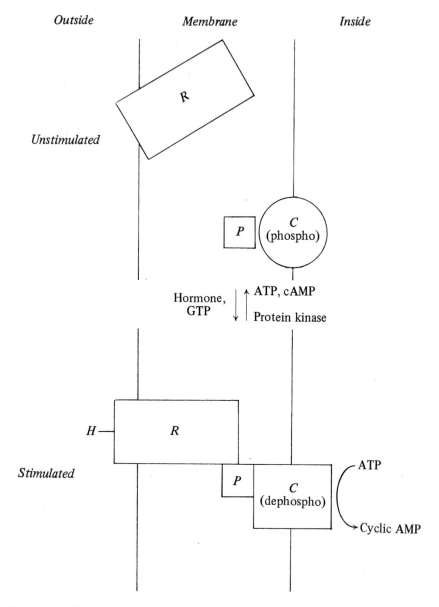

Fig. 3. Possible structure of the hormone-stimulated adenylate cyclase system. H is the hormone, R the receptor protein (or glycoprotein), C the catalytic protein and P the phosphoprotein phosphatase. The unstimulated cyclase is shown in the upper diagram, and the cyclase system after fruitful binding of the hormone molecule is shown below. The site of GTP action is probably either at P or C.

for each receptor type. For example, the adenylate cyclase of fat cell particles, which is stimulated by seven different hormones, gives less than additive stimulation when combinations of hormones are added (Birnhaumer and Rodbell, 1969). These and other results (e.g. Schorr *et al.*, 1971) indicate that a single cyclase is activated by all the hormones acting at discrete receptors. However, this crucial conclusion is being re-examined using newer assay methods, and it is a distinct possibility that each hormone stimulates a different cyclase. Similarly, basal activity may be due either to the lowered activity of the inactive phosphocyclase, to a separate uncontrolled cyclase or to denaturation during isolation. Resolution of these possibilities should solve the riddle of how cells produce the minimal amount of cyclic AMP needed for cellular maintenance.

The reactions by which hormones lower the cellular concentration of cyclic AMP have not yet been clarified. Some hormones, when administered to isolated tissues, can cause the cyclic GMP level to increase and the cyclic AMP level to decrease. As noted above, insulin acts in a manner antagonistic to epinephrine and glucagon, lowering the level of cyclic AMP in some cells. In addition, insulin can cause an increase in the concentration of cyclic GMP, for example in isolated fat cells (Illiano *et al.*, 1973). Goldberg has speculated (Goldberg *et al.*, 1973, 1975) that cyclic GMP promotes many cell events that are opposed to those mediated by increases in cyclic AMP. This dualism between the two cyclic nucleotides has some experimental support but is as yet unproven. Present evidence does indicate that cyclic GMP is also a second messenger, and that it differs from cyclic AMP in its effects.

Cyclic GMP (reviewed by Goldberg *et al.*, 1973) occurs at concentrations even lower than those of cyclic AMP, usually at about 3–10% the cyclic AMP concentration. Its formation from GTP is catalyzed by guanylate cyclase. Guanylate cyclase activity can be detected in the soluble fraction of many tissue homogenates, but some activity remains in the particulate fractions. No one has demonstrated stimulation of cell-free preparations of guanylate cyclase by hormones or fluoride. A cyclic GMP-dependent protein kinase activity was detected in all mammalian tissues examined, after a suitable assay system was devised (Kuo, 1974). It has been separated from cyclic AMP-dependent kinase activities. Gill and Kanstein (1975) have provided evidence for the existence in adrenal cortex cytoplasm of a specific cyclic

GMP receptor protein. This receptor has a high affinity for cyclic GMP (a dissociation constant of 1.4×10^{-8} M), and possibly functions as a regulatory subunit of the protein kinase (cf. the R subunit of cyclic AMP-dependent kinase, p. 555).

In contrast to the controversy over the detection and partial purification of catecholamine (i.e. β-adrenergic) receptors (Lefkowitz, 1974, 1975; Cuatrecasas, 1974), the insulin receptor has been extracted and purified from the plasma membranes of several tissues (Cuatrecasas, 1974). The insulin-binding components can be isolated only by vigorous extraction with non-ionic detergents. Affinity chromatography played an important part in the extensive (almost 500 000-fold) purification required. The receptors appear to be high molecular weight proteins or glycoproteins. Cuatrecasas has theorized (Illiano et al., 1973; Cuatrecasas, 1974) that because of the reciprocal relationship between actions of cyclic AMP and cyclic GMP their synthesis may be catalyzed by a common enzyme system. By this hypothesis, the substrate specificity of the membrane-bound nucleotide triphosphate cyclase (i.e. whether ATP- or GTP-specific) would be determined by the type of hormone bound. If true, this would satisfactorily explain the inhibition of cyclic AMP formation when cyclic GMP is being synthesized. It does not fit with the existence of cytoplasmic guanylate cyclase and the apparently different properties of the particulate adenylate and guanylate cyclases.

Major questions which remain to be answered are:

1. What are the components of mammalian adenylate and guanylate cyclases and how does each function?
2. Is the dualism theory of cyclic AMP and cyclic GMP action valid?
3. What is the protein substrate for the GMP-dependent protein kinase?
4. Do the cytoplasmic and particulate guanylate cyclases have different functions?
5. Is the insulin receptor coupled to adenylate cyclase?
6. Does cyclic GMP mediate any steroid hormone effects?
7. How are cyclic AMP and cyclic GMP related to mammalian cell proliferation?

Finally, a further complication has arisen with the isolation of cyclic CMP from leukemia cells (Bloch, 1974) and cyclic UMP from rat liver (Bloch, 1975). The distribution and function of these newly discovered nucleotides is currently under investigation. No protein kinases specific for activation by either cyclic CMP or cyclic UMP have been discovered.

Cyclic Nucleotide Phosphodiesterases

Regulation of the cellular levels of cyclic AMP depends not only on control of the rate of its synthesis but also on the rate of its degradation. A group of enzymes, the cyclic nucleotide phosphodiesterases (reviewed by Appleman et al., 1973) are responsible for the hydrolysis of both cyclic AMP and cyclic GMP. Sutherland and Rall (1958) found enzymes in heart, brain and liver which could catalyze the hydrolysis of cyclic AMP to 5'-AMP. Because the two former tissues contain the highest activities of the many tissues since tested, they have been used as purification sources. The cyclic AMP phosphodiesterase was activated by Mg^{2+} and inhibited by caffeine and other methylxanthines. It was partially purified from the supernatant fraction of beef heart homogenates by Butcher and Sutherland (1962), but these authors noted that about one-quarter of the total activity remained in the particulate fraction. During the period 1968 to 1971, more evidence appeared suggesting that multiple forms of the phosphodiesterase exist, culminating in the separation of multiple phosphodiesterases from several tissues. These enzymes differ in their substrate specificity and in their K_m values for cyclic nucleotide substrates. Typical results have been reported by Russell et al. (1973) for the activities of rat liver. Three cyclic nucleotide phosphodiesterase fractions have been separated by chromatography of sonicated extracts of liver, with another minor peak observed in some experiments. Of the three major peaks, the first eluted catalyzes the hydrolysis of only cyclic GMP. The second, and major, peak is a soluble phosphodiesterase capable of hydrolyzing both cyclic nucleotides. The third peak, accounting for about one-third of the cyclic AMP phosphodiesterase activity of the extracts, originates in particulate matter. It has a lower K_m for cyclic AMP than does the second fraction. In tests to date, the first peak shows linear kinetics, the second positive co-operativity under some conditions and the third negative co-operativity for cyclic AMP. Understanding of the interrelationships of these (and possible other forms of) phosphodiesterases will require a great deal of work.

One apparent mechanism of control of cyclic AMP phosphodiesterase is through its interaction with a protein activator. In 1971, Cheung noticed that there was a partial loss of activity of the phosphodiesterase from beef brain during purification. Activity could be restored by addition of a heat-stable protein which trailed the enzyme activity on columns of DEAE-cellulose. Other laboratories confirmed the existence of this specific protein activator. By interacting with activator-deficient preparations of the soluble cyclic AMP phosphodiesterase (the second peak described by Russell et al.), it gives a six- to ten-fold stimulation of esterase activity. Pure activator has been obtained from both heart (Teo et al., 1973) and brain (Lin et al., 1974). The activator has a fairly low molecular weight (15 000–19 000). It stimulates the phosphodiesterase only when it is in the form of a Ca^{2+}-protein complex (Teo and Wang, 1973; Lin et al., 1974). Direct stimulatory interaction of insulin with the membrane-bound phosphodiesterase has been postulated as a control mechanism, but there is no firm data to support this suggestion for decreasing concentrations of cyclic AMP.

Cyclic AMP-dependent Protein Kinases

In animal cells, the only known mechanism for the expression of the second messenger role of cyclic AMP is via its stimulation of protein kinase activity (cf. Fig. 1). These cyclic nucleotide-dependent protein kinases have recently been reviewed by Krebs (1972), Langan (1973) and Walsh and Krebs (1973). Because it was the first such enzyme to be discovered, and because its enzymic function can be related to a physiological role, I shall describe only the protein kinase isolated from rabbit skeletal muscle.

The first report of a cyclic AMP stimulation of protein phosphorylation came from the laboratory of Edwin Krebs (Walsh et al., 1968). A cyclic AMP-dependent protein kinase was purified about 300-fold from rabbit muscle. It catalyzes the transfer of the terminal phosphate group of γ-^{32}P-ATP to either casein or protamine, and increases the rate of cyclic AMP-dependent phosphorylation (and thus activation) of phosphorylase kinase by ATP. The protein kinase is stimulated about five- to 20-fold by the cyclic nucleotide and shows a K_m of about 10^{-7} M for cyclic AMP. The protein kinase exhibits a broad specificity toward protein substrates, with glycogen synthetase and the cyclic AMP-independent phosphorylase kinase showing greatest reactivity. In all proteins, the hydroxyl groups of specific serine (or to a lesser extent threonine) residues are the phosphoryl acceptors. Experiments using a synthetic octapeptide substrate (Daile *et al.*, 1975) and genetic variants of casein (Kemp *et al.*, 1975), combined with knowledge of the partial sequences of the phosphorylation sites of other substrates, indicate that local primary structure of a substrate contributes to the specificity of the protein kinase. In all cases, an arginine residue occurs on the amino-terminal side, two to five residues from the phosphorylated hydroxymethyl group. Presumably susceptible Arg – – Ser sequences must be accessible to the enzyme, not buried in the substrate. (Pyruvate kinase from pig liver can be phosphorylated by ATP in the presence of protein kinase. The serine modified is in the sequence Leu-Arg-Arg-Ala-Ser-Leu (Hjelmquist *et al.*, 1974).) Similar cyclic AMP-stimulated protein kinases were found in a number of tissues. In 1969, Kuo and Greengard reported that the enzyme activity was present in every mammalian tissue that they examined, and in both vertebrates and invertebrates. On this basis, they postulated a unifying theory for the mechanism of action of cyclic AMP by activation of tissue-specific protein kinases. The presence of these kinases necessitates the presence of phosphoprotein phosphatases, keeping the phosphoprotein-phosphorus turning over rapidly.

The mechanism by which cyclic AMP activates protein kinases was deduced from experiments in several laboratories using kinases from other tissues. For example, Gill and Garren (1970) found that when a cyclic AMP-activated protein kinase was purified from adrenal cortex, the kinase and a cyclic AMP binding protein were enriched in parallel. The binding protein could be partially separated from the kinase by incubation with cyclic AMP, followed by chromatography on DEAE-cellulose. A model in which the kinase is made up of catalytic (C) and regulatory (R) subunits was proposed, assuming that the binding of cyclic AMP to R would favour dissociation of the inactive $C–R$ complex:

$$C\text{-}R \quad + \quad \text{Cyclic AMP} \; \rightleftharpoons \quad C \quad + \quad \text{Cyclic AMP-}R.$$

$\qquad$ (inactive) $\qquad\qquad\qquad\qquad\quad$ (active)

The protein kinase holoenzyme from rabbit muscle has been resolved into C and R fractions by affinity chromatography on casein-Sepharose in the presence of cyclic AMP (Reimann et al., 1971). The cyclic AMP-independent kinase fraction (C) recovered its requirement for cyclic AMP when the cyclic AMP binding fraction (R) was mixed with it. Physicochemical studies have shown that the rabbit muscle protein kinase has a molecular weight of about 170 000 daltons and is composed of four polypeptide chains—two each of C and R (Beavo et al., 1975). One mole of enzyme binds two moles of cyclic AMP, with the regulatory subunit remaining in a dimeric structure:

$$R_2C_2 + 2 \text{ Cyclic AMP} \rightleftharpoons R_2 (\text{Cyclic AMP})_2 + 2 C.$$

Because cyclic AMP-dependent protein kinases demonstrate anomalously high molecular weights on testing by gel filtration, it has been assumed that they possess a long, thin quaternary structure, C–$(R$–$R)$–C.

At least two other factors—a protein inhibitor and covalent modification—are known to regulate the activity of the cyclic AMP-dependent protein kinases in vitro. A protein inhibitor (or modulator) has been partially purified from rabbit muscle (Walsh et al., 1971). It is a low molecular weight and heat-stable protein which blocks the cyclic AMP-stimulation of protein kinase. The inhibitor does not bind cyclic AMP nor does it prevent binding of cyclic AMP to the kinase. Ashby and Walsh (1973) have observed that there is no interaction of the inhibitor (I) with the protein kinase in the absence of cyclic AMP. The kinetics and stoichiometry of the inhibition favour binding of I to C after its cyclic AMP-stimulated dissociation from the holoenzyme:

$$C\text{-}R + \text{Cyclic AMP} \rightleftharpoons \text{Cyclic AMP-}R + C$$

$$C + I \rightleftharpoons CI$$

The CI complex is stable, but its formation is reversible. Since the inhibitor can only inhibit the active kinase (C), it could turn off the cyclic AMP-stimulated kinase but not prevent the cyclic AMP-promoted dissociation of holoenzyme. Under conditions of cyclic AMP depletion, the holoenzyme could be reformed from R and CI. The inhibitor protein is required for activity in assays for cyclic

GMP-dependent protein kinase activity (Kuo, 1974). Because the low molecular weight protein has a dual role, inhibition of the cyclic AMP-dependent form of the kinase and stimulation of the cyclic GMP-dependent kinase, Kuo has suggested that it should be called "protein kinase modulator," not inhibitor protein.

Recently, Erlichman et al. (1974) have reported that the cyclic AMP-dependent protein kinase of beef heart exists in two forms, a dephospho-form and a phospho-form. This kinase catalyzes its own phosphorylation, with incorporation of two moles of ^{32}P from γ-^{32}P-ATP into the R–R dimer. The dephosphorylated kinase is resistant to dissociation in the presence of cyclic AMP, but the phospho-form readily dissociates when incubated with cyclic AMP. It is not known yet if dephosphorylation of the kinase is catalyzed by a phosphoprotein phosphatase or if it is a reversal of autophosphorylation (Rosen and Erlichman, 1975). The following scheme summarizes the conversion of the dephospho-form of the heart protein kinase to the modified form which is more sensitive to cyclic AMP-induced dissociation:

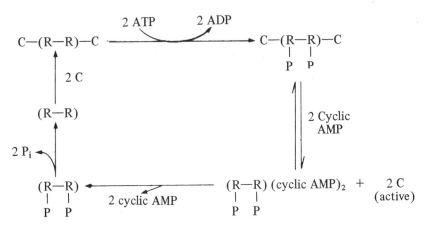

The muscle kinase is not subject to control by phosphorylation and dephosphorylation (Hofmann et al., 1975). Comparison of the properties of the purified beef heart and rabbit muscle enzymes shows that they are representative of two classes of isoenzymes recognized previously on the basis of the extent of their dissociation in 0.5 M NaCl (Corbin et al., 1975). Type I protein kinases (found as the major species in heart and brain) dissociate more readily and elute from DEAE-cellulose at <0.1 M NaCl. Type II kinases (the major isoenzymes in adipose tissue and muscle) are only slightly dissociated in 0.5 M NaCl and elute from DEAE-cellulose at >0.1 M NaCl. The muscle and heart isoenzymes have many common properties (e.g. quaternary structure and dissociation into $2\,C + R-R$) but differ in several ways. The differences appear to be mainly differences in their regulatory subunits, and are exemplified by the distinguishing features, dissociation by salt and phosphorylation, discussed above. The skeletal muscle enzyme requires a lower concentration of cyclic AMP (0.15 μM) for dissociation in the absence of MgATP. In the presence of MgATP, both isoenzymes require about 1.0 μM cyclic AMP for dissociation.

Witt and Roskoski (1975) have examined the effect of ligands on the chemical modification of protein kinases I and II by ethoxyformic anhydride, a relatively nonspecific reagent. The holoenzymes are inhibited in the presence, but not the absence, of cyclic AMP. The substrate, MgATP, protects kinase II against inactivation, so the inactivation probably is occurring at the active site. Kinase I holoenzyme is the less resistant to inactivation in the absence of cyclic AMP, in keeping with its greater spontaneous dissociation. From these data, Witt and Roskoski have concluded that the kinase holoenzymes are inactive because their regulatory subunits shield the active sites of the catalytic subunits.

Other mechanisms for control of the cyclic nucleotide-dependent protein kinases may exist. The importance of these enzymes in the expression of hormonal effects, through cyclic AMP and cyclic GMP, warrants the great deal of research work that will be needed to correlate *in vitro* observations with *in vivo* function.

Glycogen Phosphorylase

Regulation of glycogenolysis by the glycogen phosphorylase system is the prime example of cyclic AMP-mediated hormonal control of enzyme activity. The enzymes involved in the metabolism of glycogen in rabbit muscle have been isolated and characterized to a greater extent than those of any other cyclic AMP-modulated enzyme activity. The roles of epinephrine as the first messenger and cyclic AMP as the second messenger have now been observed both *in vivo* and at the molecular level with these reactions. Four enzymes have their activities altered by covalent modification as a result of the increase in the cyclic AMP level in muscle. Although each step of the metabolic pathway has been examined in depth, additional control factors are still being discovered. The integration of all the regulatory processes in glycogen mobilization is still to be completed.

Glycogen phosphorylase catalyzes the sequential phosphorolysis of the outermost branches of the storage polysaccharide, glycogen:

$$(\text{Glucose})_n + \text{P}_i \; \rightleftharpoons \; \text{Glucose-1-phosphate} + (\text{Glucose})_{n-1} \; .$$

Phosphorylase from skeletal muscle is a multimeric enzyme which can be isolated in two forms, phosphorylase *b*, a dimer, and phosphorlyase *a*, a tetramer (reviewed by Graves and Wang, 1972). Both forms were obtained as crystals and their fundamental properties described in the 1950s (cf. Krebs and Fischer, 1962). The phosphorylase protomer has a molecular weight of 92 500 daltons. It is a single polypeptide chain, with each monomer having one mole of bound pyridoxal phosphate. The role of this prosthetic group is unknown, but it does not function in an aldehyde type reaction (p. 229) because reduction of its internal imine does not inactivate the phosphorylase. Phosphorylase *a* is presumed to be the active form of the enzyme *in vivo*. The interconversion (i.e. covalent modification) of phosphorylases *b* and *a* is catalyzed by two enzymes, phosphorylase kinase and phosphorylase phosphatase (Fig. 4). Although phosphorylase *a* is isolated as a stable tetramer, its dimer— formed by dissociation in the presence of the polysaccharide substrate—appears to have the higher specific activity.

A glycogen–protein complex containing phosphorylase, phosphorylase kinase, and phosphorylase phosphatase can be isolated from muscle (Meyer *et al.*, 1970). This complex, composed of glycogen granules and muscle fragments, is believed to be both a structural and functional entity of the muscle cells. In the complex, phosphorylases *a* and *b* have different kinetic properties than in the purified states, presumably behaving more like the enzymes of intact tissue. For example, phosphorylase *b* has much lower activity *in vivo* and in the protein–glycogen complex than *in vitro* under relatively similar conditions. Perhaps further investigation of the complex will resolve some of the discrepancies between *in vivo* and *in vitro* work. Nonetheless, most findings with the whole animal have been explained by experiments with isolated enzymes.

T

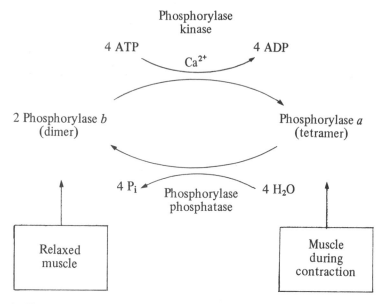

Fig. 4. The activation of phosphorylase *b* by phosphorylation is catalyzed by phosphorylase kinase, and requires the presence of calcium ions. The inactivation of phosphorylase *a* by dephosphorylation is catalyzed by phosphorylase phosphatase.

Phosphorylase kinase also exists in both non-activated and activated forms. The non-activated form has appreciable activity only at pH values above 6.8, while the activated form is active over a much wider pH range. Two factors control the activity of phosphorylase kinase, the presence of Ca^{2+} and the presence of cyclic AMP. Both stimuli are essential for rapid formation of the enzyme form having maximal activity. It is possible that the Ca^{2+} released upon neural stimulation of muscle fibres is sufficient to activate the phosphorylated kinase.

Phosphorylase kinase is a large enzyme. It has a molecular weight of 1.3×10^6 daltons, and seems to be composed of four molecules each of three types of polypeptide chains: the α-chain, the β-chain, and the γ-chain (Cohen, 1973). The intact enzyme, therefore, can be described as $(\alpha\beta\gamma)_4$. When phosphorylase kinase is incubated with cyclic AMP, cyclic AMP-dependent protein kinase and γ-^{32}P-ATP, only the two larger polypeptide chains (α and β) incorporate radioactive phosphate. Phosphorylation of the β-chain parallels the increase in enzyme activity (ten- to 20-fold at pH 6.8), but the α-chain is

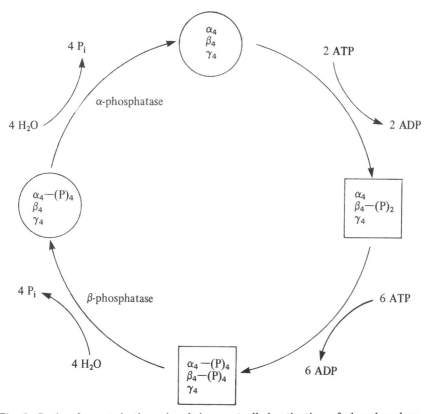

Fig. 5. Cycle of events in the epinephrine-controlled activation of phosphorylase kinase, based on *in vitro* experiments. A number of individual steps have been omitted. The circular form of the enzyme indicates the basal or non-activated state. The square representation of the phosphorylase kinase indicates the activated form, in which phosphorylation of the β-chains has occurred. First, the kinase is activated by modification of the β-chains. Then modification of the α-chains initiates the inactivation process. Experiments in which epinephrine was injected into rabbits, and the site and level of phosphorylation of phosphoryl kinase was measured, indicate that a similar control pathway operates *in vivo* (Yeaman and Cohen, 1975).

phosphorylated more slowly (Hayakawa *et al.*, 1973; Cohen, 1973). Cohen *et al.* (1975b) have isolated and sequenced the phosphopeptides of activated phosphorylase kinase. One seryl residue in the α-chain and one in the β-chain are phosphorylated. Only these two serines, out of 200 serines and 150 threonines in the kinase, are covalently modified; this demonstrates the rather strict specificity of the cyclic AMP-dependent protein kinase. Cohen and Antoniw (1973) noted that following the phosphorylation of the α-chains, there is an enhancement of the rate of phosphatase-catalyzed dephosphorylation of the β-chains. They have called this stimulation of dephosphorylation by modification of a second polypeptide chain *second site phosphorylation*. Phosphorylation of the second site does not start until two of the β-chains have been phosphorylated. Similarly, the phosphorylation of at least two α-chains is necessary for conversion of phosphorylase kinase to a form which is a good substrate for its specific phosphatases. Figure 5 summarizes the current knowledge of how the phosphorylase kinase activity of rabbit muscle is regulated by the two phosphorylation steps. Now that it has been shown that the β-chains are specific activation sites and the α-chains are deactivation sites, work is underway to see if the γ-chain contains the catalytic centre. Purification and characterization of two apparently specific kinase phosphatases, and further elucidation of the role of Ca^{2+} in the phosphorylations is also in progress.

Although it was first thought that glycogen was synthesized by a reversal of the phosphorylase-catalyzed reaction, Leloir and Cardini (1962) described an enzyme, glycogen synthetase, that catalyzes the biosynthesis of glycogen in liver. This enzyme activity has since been detected in and purified from other tissues, including skeletal muscle (reviewed by Stalmans and Hers, 1973). Glycogen synthetase catalyzes the transfer of a glucosyl residue from uridine diphosphate glucose (UDPG) to the non-reducing terminal (C^4 hydroxyl) of a glycogen primer molecule:

$$UDPG + (Glucose)_n \rightleftharpoons UDP + Glucosyl\text{-}(\alpha\text{-}1,4)\text{-}(Glucose)_n .$$

Glycogen synthetase, like phosphorylase and phosphorylase kinase, occurs in two forms interconvertible by covalent modification (Friedman and Larner, 1963). The fully active (I) form of the synthetase, when phosphorylated at a specific serine residue, is converted to the less active (D) form. This phosphorylation is catalyzed by the same cyclic AMP-dependent protein kinase (Schlender *et al.*, 1969; Soderling *et al.*, 1970). In response to this dual control, glycogen is either degraded to form glucose-1-phosphate for glycolysis, with the glycogen synthetase turned off, or synthesized from glucose-1-phosphate via UDPG, with the glycogen phosphorylase turned off (Fig. 6). Glycogen synthetase is an oligomer. The phosphorylation of the two sites per monomeric unit catalyzed by cyclic AMP-dependent protein kinase causes complete conversion of the I

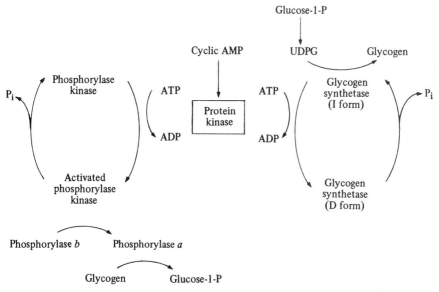

Fig. 6. Regulation of glycogen metabolism by cyclic AMP-dependent protein kinase. The kinase catalyzes the activation of phosphorylase kinase and the inactivation of glycogen synthetase when cyclic AMP is formed at higher levels as a result of hormone stimulation of the tissue (cf. Fig. 1, this chapter). The cyclic AMP-dependent kinase also catalyzes the inactivation of phosphorylase phosphatase.

form to the D form (Soderling, 1975). Incorporation of one mole of phosphate per monomer in this reaction decreases the activity of the synthetase by 75%. A second protein kinase, not stimulated by cyclic AMP, is present in rabbit muscle and catalyzes the phosphorylation of the I form to the extent of one mole per subunit (Huang *et al.*, 1975). The two types of protein kinases appear to act together to regulate the activity of glycogen synthetase, with the synthetase activity varying with the relative activities of the kinases.

The fourth protein involved in glycogen metabolism that is modified by a phosphorylation catalyzed by cyclic AMP-dependent protein is phosphorylase phosphatase (Huang and Glinsmann, 1975). Phosphorylase phosphatase is inactivated by this covalent modification. Much less is known about the reversal of kinase-induced effects than is known about the phosphorylations. There is an Mn^{2+}-stimulated phosphatase partially purified from rabbit muscle, which catalyzes the dephosphorylation of phosphorylase phosphatase (Huang and Glinsmann, 1975) as well as the inactive D form of glycogen synthetase and phosphorylase kinase (Zieve and Glinsmann, 1973). Phosphorylase phosphatase and the β-phosphorylase kinase phosphatase (Fig. 5) copurify through an extensive purification method (Cohen *et al.*, 1975a), and so they may constitute a single protein. Highly purified phosphorylase phosphatase contains glycogen synthetase dephosphorylation activity (Brandt *et al.*, 1975). In short, it is slowly becoming apparent that a single protein phosphatase catalyzes all the reactions which inhibit glycogenolysis and activate glycogen synthesis! The α-phosphorylase kinase phosphatase, which inhibits the reversal of hormone activation, appears to be a distinct enzyme.

By these complex control procedures, epinephrine exerts activation of one main enzyme-catalyzed reaction (glycogen phosphorolysis) and inactivation of an opposing reaction (synthesis of glycogen). The cascade of enzymic controls amplifies the signal given by the relatively low concentration of the hormone. Central to this amplification is the role of cyclic AMP as a stimulator of protein kinase. The elegant research performed on the glycogen phosphorylase control system indicates the careful and complete work that must be done to resolve the other functions of cyclic nucleotides as second messengers of hormones, and to elucidate the mechanism of the first enzyme in the pathway of hormone action, adenylate cyclase.

REPRESSORS

Control of the rate of protein biosynthesis in bacteria is imposed primarily at the level of transcription, the synthesis of the messenger RNA (mRNA) directed by the complementary DNA or gene template (Watson, 1976). The protein biosynthetic apparatus of cells then translates the mRNA into protein molecules. A molecular description of transcriptional regulation was not attainable until there was sufficient knowledge both of bacterial genetic techniques and of the pathway of protein biosynthesis. Early experiments culminated in the proposal of the operon theory of control by Jacob and Monod (1961). This theory suggested the existence of messenger RNA and of specific repressors, substances which inhibit synthesis of certain protein molecules. Subsequent research on gene control in bacteria has amplified and slightly modified this theory. Repressor proteins have some of the properties of allosteric enzymes (e.g. oligomeric structure and several binding sites). Several allosteric enzymes seem to have repressor-like properties. There also is a protein modifier which, in the presence of a specific ligand, can accelerate protein biosynthesis. Because most of the operon–repressor model was based on experiments pertaining to the metabolism of lactose in *E. coli*, I will first describe the lactose operon and its control. I will then describe the biosynthesis of histidine in *Salmonella typhimurium,* a controlled system in which an enzyme participates as a feedback-regulated catalyst and has been postulated to be a regulatory element in protein biosynthesis.

Lactose Operon

The best characterized example of gene control is the biosynthesis of three proteins by *E. coli* when this bacterium is grown using lactose as the sole source of carbon atoms (cf. the review by Zubay and Chambers, 1971). This control of gene expression provides synthesis of enzymes when they are needed for growth, but prevents their synthesis when they would serve no useful function (cf. p. 43). When glucose is the carbon source, there are very few molecules of each of these enzymes per cell, but the presence of lactose induces synthesis of three protein molecules required for its own metabolism. For example, there may be only one or two β-galactosidase molecules in an *E. coli* cell growing in the presence of glucose, but several thousand β-galactosidase molecules in a cell growing in a lactose-containing medium. Although a metabolite of lactose (allolactose) is the natural inducer of these enzymes, the synthetic non-metabolizable compound isopropyl-β-D-thiogalactoside (IPTG) is a more potent inducer and has been used for some of the experiments that I will mention. A set of contiguous genes exists which codes for the three

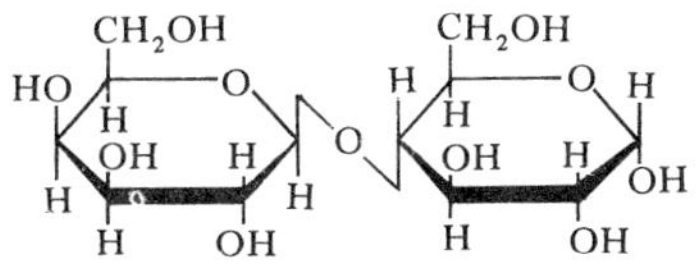

Lactose 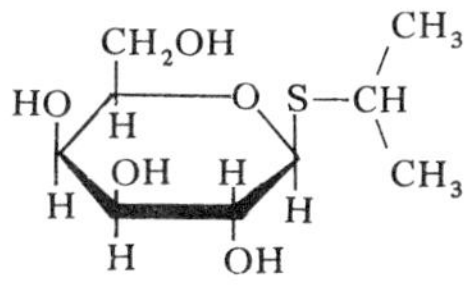   IPTG

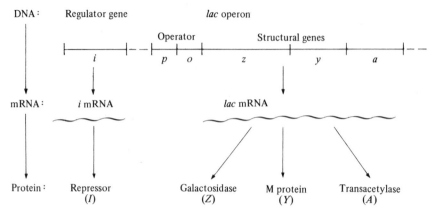

DNA: Regulator gene   *lac* operon

Operator   Structural genes

i   *p*   *o*   *z*   *y*   *a*

mRNA: *i* mRNA   *lac* mRNA

Protein: Repressor   Galactosidase   M protein   Transacetylase
         (*I*)   (*Z*)   (*Y*)   (*A*)

Fig. 7. Summary of the transcription and translation steps of the *lac* operon. The operon consists of the operator gene and the structural genes z, y and a. The operator region has a locus for binding of the RNA polymerase (the promoter or p site) and a binding site for the *lac* repressor (the o site). The associated regulatory or i gene, which codes for the *lac* repressor, is located near the promoter region. The direction of transcription and translation is from the left to the right in the diagram. In the presence of sufficient inducer to release the repressor molecules from the o region, RNA polymerase can bind to p and enzyme synthesis can proceed. RNA polymerase catalyzes the synthesis of polycystronic mRNA which then is translated into the three structural gene products, the proteins Z, Y and A.

proteins. These units of DNA, the *structural genes* of the three proteins, along with the adjacent DNA of the control or *operator gene*, make up the lactose (*lac*) operon (Fig. 7). The structural genes of the *lac* operon are the z gene, which codes for β-galactosidase, the catalyst of the hydrolysis of lactose to galactose and glucose; the y gene, which codes for the M protein (galactoside permease), a component of the sugar transport system; and the a gene, which codes for thiogalactoside transacetylase, an enzyme of unknown function. Jacob and Monod proposed that a *regulator gene* (i) is the template for the synthesis of some sort of cytoplasmic *repressor* molecule which can interact with the operator gene. The reversible binding of the repressor to the operator prevents transcription of the adjacent structural genes, and thus the biosynthesis of large amounts of the three proteins of lactose metabolism. For example, Watson (1976) has estimated that there are 35–50 β-galactosidase mRNA molecules in a cell during maximal β-galactosidase biosynthesis, but either none or a few molecules when enzyme biosynthesis is repressed. The low molecular weight ligands that induce the synthesis of the *lac* gene products act by forming non-covalent repressor–inducer complexes which cannot bind to the operator region of the gene (Fig. 7). With inducible enzymes, inducer–repressor complex formation results in activation of the operon. In repressible systems, the repressor itself is inactive and cannot bind to the operator gene until it forms a repressor–corepressor complex (cf. p. 44). Jacob and Monod originally suggested that the repressor material was an RNA fraction, but later work has shown that all isolated repressors are oligomeric proteins.

Experiments with many classes of mutants of *E. coli* support the operon theory. In Monod's laboratory, constitutive mutants were isolated. These mutants synthesize large quantities of the *lac* proteins, with or without inducer present. Jacob and Monod deduced that these mutants were defective in the *i* gene, so could not synthesize the regulatory element, the repressor. The constitutive mutants invariably affected the galactosidase, transacetylase and the permease, and always increased their amounts in a constant ratio. Genetic experiments have demonstrated that the operator gene contains several distinct sections, so this DNA is often called the *lac* control region. A promoter (*p*) locus has been found, and appears to be the genetic material to which RNA polymerase binds to initiate transcription. There is also a nucleotide length of the promoter DNA which can bind the protein–cyclic AMP complex described below. The order of the genes of the DNA related to lactose catabolism obtained from genetic experiments is *i, p, o, z, y* and *a*. The *lac* operon is composed of the DNA from *p* to *a*.

Although the original theory suggested that the *lac* repressor is RNA, evidence soon accumulated that it is a protein. In 1966, Gilbert and Müller-Hill succeeded in isolating the repressor. They used equilibrium binding of ^{14}C-IPTG inducer (i.e. the capacity to bind radioactivity per weight of protein) as an assay for the purification of the repressor from extracts of *E. coli*. They were able to purify the repressor about 100-fold. Although their repressor fractions were far from pure, the active component was shown to be protein by its pattern of reactivity with hydrolases. The activity of the partially purified repressor was not destroyed by treatment with ribonuclease or deoxyribonuclease, but was lost upon pronase treatment or upon heating above 50° C. Riggs and Bourgeois (1968) further purified the *lac* repressor using column chromatography on phosphocellulose. On the basis of sedimentation properties and SDS gel electrophoresis, they concluded that the repressor contains four subunits of about 40 000–50 000 daltons each. The repressor binds *in vitro* to *lac* operator DNA, and it represses the cell-free synthesis of β-galactosidase. By suitable genetic manipulation, a mutant of *E. coli* was prepared that could produce enough *lac* repressor for protein structural work. The primary structure of the repressor has been established and its molecular weight calculated from the sequence as 37 160 for the protomer and 148 640 for the tetramer (Beyreuther *et al.*, 1975).

Ohshima *et al.* (1974) have performed a quantitative examination of the binding of ^{14}C-IPTG to purified *lac* repressor. The number of inducer molecules bound per tetramer varies from 2.3 to 4.0, depending on the assay conditions. Their results indicate that under conditions found in the bacterial cells the binding of inducer to free repressor is not co-operative, but suggest that there may be positive co-operativity of inducer binding in the presence of the operator DNA. Barkley *et al.* (1975), using similar techniques, have not found any co-operative interactions in the binding of effectors, neither to repressor nor to repressor–operator complex. They reported that inducers reduce the affinity of the repressor for *lac* operator DNA by a factor of about one thousand. The allosteric nature of the repressor is supported by the demonstration that treatment with trypsin inactivates its ability to bind to operator DNA without a loss in ability to bind IPTG (Platt *et al.*, 1973). The terminal regions of the repressor are required for binding to *lac* operator, and the tetrameric core which contains the four inducer-binding sites is resistant to mild proteolysis. This proteolysis experiment is a type of desensitization in which the biological activity is lost but the ability to bind the allosteric ligand (the inducer) is retained.

Control of protein biosynthesis by repressors is a negative process; i.e. it is an inhibition of protein biosynthesis by prevention of expression of the gene. There is also a positive or stimulatory control mechanism in *E. coli* affecting the biosynthesis of the gene products of the lactose, galactose and arabinose operons and of some other enzymes, most of which catalyze catabolic reactions. The formation of these proteins is controlled by catabolite repression, with cyclic AMP playing a regulatory role.

Catabolite repression is the inhibition by glucose of enzyme induction. Physiologically, catabolite repression allows bacterial cells

presented with a mixture of substrates to grow on glucose in preference to the other carbon source, until most of the glucose is consumed.
Then—and only then—are the catabolic enzymes required for growth
on the second substrate synthesized. Makman and Sutherland (1965)
discovered that the cyclic AMP content of *E. coli* cells varies inversely
with the concentration of glucose in the medium. They found that
when cells are grown on glucose, there is an abrupt increase in the
concentration of cyclic AMP which coincides with the complete
utilization of the glucose. Addition of glucose to cells containing the
high levels of cyclic AMP causes a decrease in the level of the cyclic
nucleotide. Although it has not been possible to demonstrate inhibition of *E. coli* adenylate cyclase by glucose *in vitro*, Peterkofsky and
Gazdar (1974) have found that glucose inhibits the cyclase in intact
cells and that this inhibition can account for the decreased cellular
cyclic AMP levels observed. They have ascribed the absence of an
inhibitory effect of glucose *in vitro* to the need for an additional
undiscovered factor. Phosphoenol pyruvate (PEP) confers a higher
activity on adenylate cyclase in an *E. coli* mutant that has a low
activity of Enzyme I (p. 303) in the bacterial carbohydrate transport
system (Peterkofsky and Gazdar, 1975). The adenylate cyclase-
phospho–Enzyme I complex is postulated to have activity, and the
cyclase–dephospho–Enzyme I complex, no cyclase activity. Glucose
might inhibit cyclase by dephosphorylation of Enzyme I by phosphate transfer through the transfer system. This postulate still must
be tested to see if elevated concentrations of glucose lower the
amount of phospho–Enzyme I and if this enzyme complexes with
adenylate cyclase.

More is known about the mechanism of cyclic AMP control of
protein biosynthesis than about the control of cyclic AMP formation
by glucose in *E. coli*. The most definitive research has been performed
on the expression of the *lac* operon, although biosynthesis of most
inducible enzymes seems to be controlled by the same mechanism.
Catabolite repression of β-galactosidase biosynthesis can be overcome by addition of cyclic AMP to repressed wild type *E. coli* cells,
but cyclic AMP is ineffective with mutants defective in the promoter
region. Therefore, cyclic AMP should exert its stimulatory effect at
this locus, where RNA polymerase binds to the gene. It has since been
shown that cyclic AMP forms a complex with a new protein, called
the catabolite gene activator protein (CAP), before binding to the *p*

region. CAP was detected and isolated first in Zubay's laboratory, and soon after in Pastan's laboratory, using different experimental approaches. Zubay's group isolated CAP by classical fractionation of the components of a cell-free protein synthesis system. Their assay required *lac* DNA, RNA polymerase, the substrates and cofactors of RNA synthesis, and the elements of protein synthesis necessary to form β-galactosidase as a measure of the amount of *lac* mRNA formed. Using this couple assay, a partially purified CAP was obtained by Zubay *et al.* in 1970. Pastan's group had found that there were two types of *E. coli* mutants unable to use lactose for growth. One type, defective in adenylate cyclase, could grow normally in the presence of cyclic AMP. The other type did not respond to cyclic AMP supplementation, so required some other factor, presumably a gene product which bound cyclic AMP. These observations were the basis of a search in wild type *E. coli* for a protein which could bind tritiated cyclic AMP. By this method, Anderson *et al.* (1971) obtained a homogeneous preparation of CAP. The activator protein is a dimer of 45 000 daltons, with both 22 500-dalton protomers appearing identical in structure. The dimer only binds one mole of cyclic AMP, as measured by equilibrium dialysis, suggesting a negatively co-operative binding process. CAP binds to the DNA of *E. coli,* and this binding requires the presence of cyclic AMP. It does not bind to RNA polymerase, so the suggestion that CAP–cyclic AMP complex activates protein biosynthesis by binding to the protomer and assisting the binding of RNA polymerase to a polymerase interaction site on the DNA seems plausible. In support of this suggestion, Mitra *et al.* (1975) and Majors (1975) have demonstrated that a ternary cyclic AMP–CAP–DNA complex is formed and that CAP binds most strongly to the promoter region of the *lac* operon. The binding of repressor to the operator and the binding of polymerase to the promoter have been shown to be mutually exclusive events.

The nucleotide sequence of the *lac* control region and the tentatively assigned binding sites of the three interacting proteins—RNA polymerase, *lac* repressor and CAP—have recently been published. Gilbert and Maxam (1973) isolated a double-stranded DNA fragment that was protected by the *lac* repressor against digestion with deoxyribonuclease. They sequenced the RNA transcript of this fragment, and from their results predicted the repressor binding sequence of the *lac* operator. Maizels (1973), who has reported the sequence of

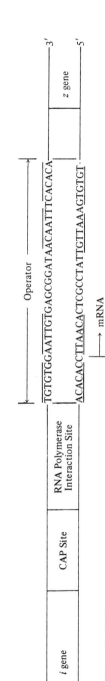

Fig. 8. Arrangement of the protein binding sites of the *lac* operon control region as reported by Dickson *et al.* (1975). The sequence is only shown for the 35 nucleotides of the operator region. Note that there is a pattern of twofold rotational symmetry (marked by the horizontal lines) about the central G–C pair. A smaller symmetrical region occurs in the CAP binding site.

the first 63 bases of the *lac* mRNA, found that the AUG initiation codon for the *z* gene product is at position 39 of the messenger RNA. Dickson *et al.* (1975) have completed the sequence determination of the 122 base pairs of the control region. They have tentatively identified the recognition sites of all three proteins, as shown in Fig. 8. One interesting feature of the base sequence is the symmetrical structure of the operator. This structure suggests that the repressor molecule can recognize the operator region from either direction of the DNA. The 21-nucleotide duplex fragment surrounding the centre of symmetry in the operator has been chemically synthesized (Bahl *et al.*, 1976). This DNA fragment binds *lac* repressor, and the binding is inhibited by IPTG. Attempts now are being made to determine which amino acids of the repressor bind to which nucleotides of the operator, using covalent linkage of the repressor–operator complex and enzymatic digestion. Steitz *et al.* (1974) have obtained data from electron microscopy and powder X-ray diffraction of the *lac* repressor which show the repressor to be a symmetric molecule, with a dumb-bell or waisted shape in cross-section. They have suggested that the groove down the long axis of the tetramer binds to the operator. The symmetries of the repressor and the operator DNA probably match. Other electron microscopy experiments (Ohshima *et al.*, 1975) provide different dimensions for the repressor but also suggest that the operator DNA fits into a cleft in the repressor oligomer. Three-dimensional structures of CAP and repressor are essential for a knowledge of their functions.

Histidine Operon

The biosynthesis of L-histidine by *Salmonella typhimurium*, like the biosynthesis of many other anabolites by bacteria, is controlled in two ways: by feedback control of the first enzyme of the pathway, and by repression of synthesis of the enzymes which catalyze formation of the ultimate product (reviewed by Brenner and Ames, 1971; Goldberger and Kovach, 1972). The repression control of histidine synthesis is more complicated than the *lac* operon control system. From an enzymological viewpoint, its most interesting feature is the *suggestion* that both the allosteric inhibition regulation and the repression control are inter-related, with the feedback-controlled

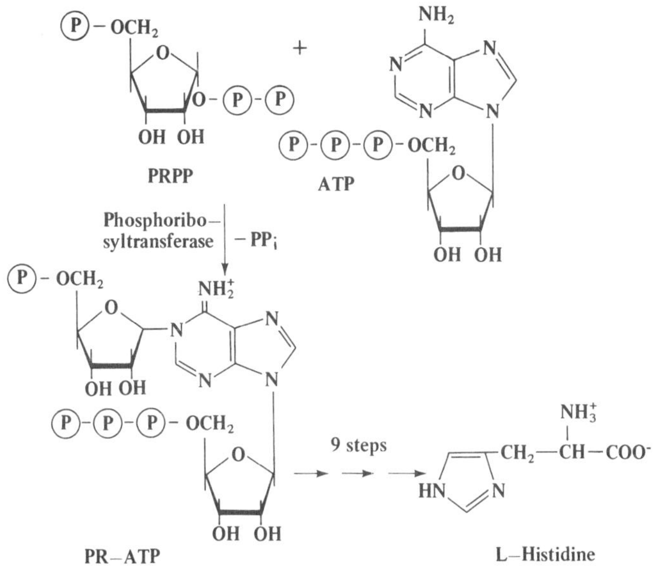

Fig. 9. Outline of the pathway of biosynthesis of histidine in bacteria. The product of the *his* G gene, adenosine triphosphate phosphoribosyltransferase, catalyzes the first step of this pathway. In this step, a phosphoribosyl group is transferred from 5-phosphoribosyl-1-pyrophosphate (PRPP) to ATP to form phosphoribosyl ATP (PR-ATP). This product is converted to L-histidine in nine steps, catalyzed by the other eight enzymes coded for by the *his* operon. Histidine exerts a negative feedback control on ATP phosphoribosyltransferase.

enzyme being a repressor. Histidine is formed from ATP and 5-phosphoribosyl-1-pyrophosphate (PRPP) by a series of ten enzyme-catalyzed reactions, in which the seventh and ninth steps are catalyzed by the same enzyme (Fig. 9). The synthesis of the nine enzymes of this metabolic pathway is directed by the histidine (*his*) operon—a cluster of about 10 000 nucleotide pairs making up nine contiguous structural genes with an associated operator region. The product of the first structural gene (*G*) is the first enzyme of the pathway, but the order of the remaining structural genes does not correspond to the order of the reactions in the pathway (e.g. the assigned gene order is *G D C B H A F I E* and the reaction order is *G E I A H F B C B D*).

ATP phosphoribosyltransferase (formerly called PR-ATP synthetase), as the first enzyme of the histidine biosynthetic pathway, is the site of the negative feedback control by histidine. It has a molecular weight of about 215 000, and is composed of six subunits of 36 000 daltons each (Voll *et al.*, 1967). However, this enzyme is quite unstable, and further studies were hampered until Parsons and Koshland(1974) devised a rapid purification method and ascertained conditions under which the enzyme is stable for storage and examination. The purification scheme utilizes an acid pH treatment and high ammonium sulphate concentrations—both of which stabilize the enzyme. The success of this work demonstrates the value of time spent in determining the fundamental properties of an enzyme, and making use of these properties for purification. Using the phosphoribosyltransferase purified by this technique, Bell *et al.* (1974) have re-examined the kinetic and ligand-binding properties of the enzyme. With kinetic experiments, they detected three different conformation states of the synthetase—a low activity form, a high activity form and a form resulting from incubation with histidine which binds histidine more co-operatively and therefore can be more completely inhibited. Under appropriate conditions, these three forms are slowly and reversibly interconvertible. The K_i they reported for histidine, in agreement with previous estimates, was about 4×10^{-5} M, so that histidine (present at about 1.5×10^{-5} M in wild-type cells) can be an effective feedback inhibitor *in vivo*.

The *his* operon is transcribed into a polycistronic mRNA. If the rate of growth of *Salmonella* is limited by the concentration of histidine, then intracellular activity of all of the enzymes of histidine biosynthesis increases to the same extent. If sufficient histidine is supplied to the *Salmonella*, then it is rapidly transported into the cells and the activity of all the biosynthetic enzymes repressed, also in a co-ordinated fashion. Therefore, the observed repression and derepression of histidine biosynthesis seems typical of a repression control in which a repressor–corepressor complex binds to the operator gene. If the Jacob–Monod repressor theory applied, then only two types of constitutive mutants of *S. typhimurium* would be expected—one defective at a site adjacent to the operon (at the operator locus), and one defective at a chromosomal position unlinked to the *his* operon (at the regulator gene). Constitutive mutants with defects located at the operator (*his O*) have been isolated, as anticipated, but a large number of regulatory mutants mapping at five distinct loci (*his R, S, T, U* and *W*) not linked to the operon have been prepared. These five types of mutants are defective in either the synthesis, the maturation or the aminoacylation of tRNA^{His}. None of these five mutations causes full derepression. Characterization of these latter constitutive mutants supports the conclusion that the effect of histidine in repression is not expressed by the amino acid itself, but by histidyl tRNA^{His}. Although the aminoacylated tRNA might be the controlling factor of the *his* operon, many workers consider that it is a corepressor and that a protein is the repressor. All other known repressors are proteins. However, no mutants having a genetic locus for a simple repressor protein have been identified. To explain this observation, Goldberger and his colleagues have suggested that the repressor protein has more than one function: repression control and enzyme activity. The best candidate for a repressor protein in the *his* operon system was considered to be ATP phosphoribosyltransferase, but there are conflicting data which strongly indicate that this enzyme is a component but not an essential one of the repression system. Research on this topic is far from complete. Even if ATP phosphoribosyltransferase is not its own repressor, the concept of repressor enzymes originating from this research may be applicable with some other repressible enzymes.

Two enzymes have been considered as repressors of the *his* operon. De Lorenzo and Ames (1970) purified histidyl tRNA synthetase (coded by the *his S* gene) from *S. typhimurium*, and determined the

$$\text{Histidine } + \text{ ATP } + \text{ tRNA}^{His} \xrightleftharpoons{\text{Histidyl tRNA synthetase}} \text{Aminoacyl(histidyl) tRNA}^{His} + \text{AMP } + \text{PP}_i$$

K_m value of the purified enzyme for tRNA^{His} to be about 1×10^{-7} M. They calculated that most of the tRNA^{His} and the synthetase in the cell would be in the form of an *ES* complex, and suggested that histidyl tRNA synthetase might be the repressor protein. Derepression occurs in mutations of the structural gene for this synthetase, but the constitutive behaviour can be explained as defective synthesis of the histidyl tRNA^{His} corepressor. Indirect evidence implicating the first enzyme of the pathway, ATP phosphoribosyltransferase, as a regulator of the *his* operon has been accumulating in the last decade. What is not known is how this enzyme is involved in the control of protein biosynthesis. Goldberger (1974), stimulated by his own work on histidine biosynthesis, has searched the literature and compiled a list of examples of proteins that appear to directly control expression of their own structural genes. He has coined the term

U

autogenous regulation for the control of gene expression by a protein molecule specified by the nucleotide sequence of the operon it regulates. Some of these examples have since been discounted, and others have little data to support their autogenous regulation. The largest body of experimental evidence on autogenous regulation describes experiments on the role of ATP phosphoribosyltransferase.

Mutants of *S. typhimurium* have now been isolated in which mutations of the *his G* gene render the ATP phosphoribosyltransferase insensitive to negative feedback inhibition by histidine and also make the operon constitutive. Genetic experiments (Meyers *et al.*, 1975) provide evidence that any regulatory effect of the *his G* gene product is exerted as a freely diffusible protein. Studies with these mutants suggested that histidyl tRNA[His] might bind to the phosphoribosyltransferase to form a corepressor–repressor complex. This thesis has been tested, with some positive results. Histidyl tRNA does bind to the phosphoribosyltransferase with a high affinity. This binding has been demonstrated both by chromatography on Sephadex G-100 and by a filter assay binding technique. Both binding experiments were performed with mixed tRNAs isolated from *Salmonella* to which ^{3}H-histidine and 19 unlabelled amino acids were bound covalently. A series of *in vitro* experiments indicated that the histidyl tRNA is bound to the phosphoribosyltransferase in preference to either unacylated tRNA[His] or other acyl tRNAs. The binding of the histidyl tRNA appears to be at a site distinct from the catalytic site of the ATP phosphoribosyltransferase because selective desensitization of RNA binding (by heat treatment of the enzyme) and of catalysis (by treatment of the enzyme with a sulphydryl inhibitor) has been achieved (Blasi *et al.*, 1971). The most convincing evidence favouring control of the *his* operon by the *his G* gene product is the demonstration that transcription of the operon *in vitro* can be inhibited by the first enzyme of the pathway (Blasi *et al.*, 1973). This transcription experiment was carried out using a transducing phage carrying the DNA of the *his* operon of *E. coli* and measuring the formation of labelled mRNA that could hydridize with the *his* operon DNA. Addition of phosphoribosyltransferase to the transcription system blocked the transcription of the *his* operon specifically, but no effect of histidyl tRNA[His] on the system was reported.

As noted above, regulation of the expression of the *his* operon is much more complicated than a simple repressor control. Autogenous negative control by the *G* enzyme is not the only, and possibly not even the major, regulatory mode. Kasai (1974), using DNA from mutants that lack a small portion of the control region of the operon, has found that RNA polymerase transcribes the *his* operon at a greatly enhanced frequency. He has suggested that there is a transcriptional barrier or *attenuator* present in wild-type DNA but absent in the mutant. He also reported that some positive control factor present in normal cells reverses the transcriptional barrier of the attenuator. Artz and Broach (1975) have proposed an activator–attenuator model based on *in vitro* biosynthesis of the product of the second or *D* gene, histidinol dehydrogenase, in the presence of DNA from several strains of *Salmonella*. Addition of tRNA^{His} inhibits the activation of the *his* DNA expression. The *G* enzyme is unnecessary for repression or depression of the *his* operon, because mutants lacking all or part of the *G* region display normal regulation (Scott *et al.*, 1975). Artz and Broach consider the major controlling element to be the positive factor. The positive control factor has not yet been identified, although it may be histidyl tRNA synthetase. Figure 10 presents the basic ideas of the activator–attenuator control model. The proposal is quite different from a classical repressor–operator mechanism. The roles of ATP phosphoribosyltransferase and tRNA^{His} still must be elucidated and the positive factor identified. This con-

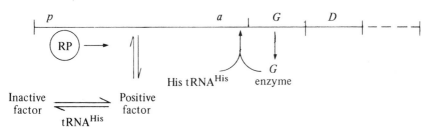

his operon control region

Fig. 10. Activator–attenuator model for the regulation of the *his* operon. During derepression, a positive factor allows RNA polymerase (RP) to bypass the attenuator locus (*a*) and transcribe the *his* structural genes. Repression during histidine excess is effected by inactivation of the positive factor. This is envisaged as binding of tRNA^{His} to the positive factor. The *G* enzyme may have an accessory role in regulation, by interacting at the attenuator to reinforce the repression caused by tRNA^{His}-positive factor formation. Modified from Artz and Broach (1975).

trol system is complicated and unresolved, but much too interesting to omit from a discussion of enzyme biosynthesis.

A second positive control factor in operon expression has been identified recently. This compound is guanosine-5'-diphosphate-3'-diphosphate (ppGpp), an unusual nucleotide produced in large amounts by bacteria during times of amino acid starvation. The ppGpp molecules play an inhibitory role in RNA synthesis. Ribosomal RNA accumulation is selectively inhibited by ppGpp, but the transcription of several operons is also stimulated by it (Reiness et al., 1975). The present concept is that ppGpp is a general transcriptional effector, possibly interacting directly with RNA polymerase to stimulate transcription of the lac and other operons. The his operon of S. typhimurium is one of those affected positively by ppGpp (Stephens et al., 1975). This effect is independent of the activator-attenuator mechanism described above. The two control systems work in conjunction to adjust the rate of histidine biosynthesis to that needed for optimal cell growth. Involvement of cyclic AMP and ppGpp in many control systems differentiates these systems from the simpler controls of negative feedback inhibition, for example. Both of these effectors are metabolically labile, having enzymes that catalyze both their rapid formation and degradation. Both compounds may be involved not only in intracellular controls in bacteria but also in intercellular regulation in higher organisms (Tomkins, 1975).

Another bacterial enzyme that is a candidate for repressor activity is glutamine synthetase (p. 452). The adenylylated synthetase may be part of the repression system for the biosynthesis of nitrogenase (p. 370). Unadenylylated glutamine synthetase activates transcription of the DNA coding for the biosynthesis of histidine-degrading enzymes. In Klebsiella aerogenes (Foor et al., 1975), adenylylated glutamine synthetase has been proposed to be responsible for the repression of glutamine synthetase. Mutants in which P_{II}, UTase and ATase are altered—and thus deadenylation prevented—show greatly decreased ability to synthesize glutamine synthetase. Alterations causing failure to adenylylate produce high levels of synthetase. Perhaps this control system will be confirmed to be one of autogenous regulation.

Protein synthesis in mammals probably involves similar types of controls to the bacterial systems, but neither repressors nor operons have been observed in mammalian tissue. Several examples suggestive of autogenous regulation in mammalian cells have been described. One potential case of autogenous control of the biosynthesis of an enzyme is the human glucose-6-phosphate dehydrogenase. Yoshida (1970) has characterized a mutant variety of this dehydrogenase which differs from the normal enzyme by the single amino acid substitution of a tyrosine residue for a histidine residue. This single alteration appears to be the cause of increased production of the variant dehydrogenase. One possible explanation is that the dehydrogenase affects the rate of its own synthesis.

TASTE AND SMELL

At present, one of the biochemical research areas of greatest mystery is neurochemistry. There are few major neurochemical problems which have yielded positive results, but one topic which may soon be explained is the mechanism of mammalian *chemoreception,* i.e. the chemical basis of both taste (gustation) and smell (olfaction). These two chemical senses are complicated physiological and psychological phenomena. For both senses, it has been postulated that there are specific membrane-bound protein receptor molecules that reversibly bind the stimulant molecules to form a stimulant–receptor complex. Occupancy of the stimulant-binding site of the receptor triggers a nerve impulse to the appropriate taste or smell centres of the brain. Although the receptors are not enzymes, there is no reason to think that an allosteric transition cannot be responsible for the triggering of a nerve impulse. For argument's sake, then, let us assume that the taste and smell receptors are proteins (or lipoproteins or glycoproteins) that undergo a conformational change as a direct result of interaction with a specific type of stimulant molecule. Proteins are the receptor molecules in a similar type of sensory reception in bacteria–chemotaxis, the movement of the bacteria towards or away from chemical stimulants (reviewed by Adler, 1975). Small molecules can bind to the bacterial receptors, and by an unknown mechanism cause the flagella to rotate. We shall discuss some experiments which support the premise that chemosensory receptors in animals are stimulant-specific protein molecules. If the postulate is true, chemical explanations may be achieved for the specificity of recognition and response in chemoreception, just as they have been for enzymic catalysis.

Data about gustation and olfaction have been collected for centuries, from observations made by workers in food and perfume industries. Additional information is being obtained from current studies on the pheromones or odour messages of animals, from chemical analyses of air pollution, and from food research. Nontoxic insect control through attraction to traps by pheromones is an example of the potential value of odour research. Species-specific sex attractant molecules, normally emitted by the female insect, can be used to lure male insects to traps. The extremely low concentrations of the attractants in the air should not be toxic to or attract other species.

Taste is the detection of dissolved molecules by receptors in the taste buds of the tongue, and *odour* is the detection of volatile substances by nasal receptors. *Flavour* is a complex sensation, determined mainly by the combined taste and odour qualities of a compound. There are four dominant basic tastes (so probably there are four types of receptors): sweet, sour, bitter and salty. Perfume experts, though, have recognized and discriminated the individual odours of thousands of pure chemicals, indicating that there must be a larger number of odour receptors. Sour-tasting compounds are acids, and with mineral acids sourness is directly proportional to the $[H^+]$. Bitter compounds are usually basic. Some compounds can stimulate two (or more) types of taste receptors. For example, citric acid is both sweet and sour. Small chemical differences between compounds can alter their taste. Perhaps the most striking example of a minor structural difference altering the taste quality is the sweet taste of α-D-mannose and the bitter taste of β-D-mannose (Stewart *et al.*, 1971). An equilibrium mixture of these anomers is an ambiguous stimulator.

In 1966, Dastoli and Price partially purified from the taste buds of beef tongues a protein which they called "sweet-sensitive protein." As an assay for this protein, which they believe to be the sweet receptor, they used changes of refractive index or ultraviolet absorption spectrum in the presence of sugars or saccharin. The purified sweet-sensitive protein (Dastoli *et al.*, 1968) has a molecular weight of 150 000 daltons, and appears to be cationic at neutral pH values. Dastoli claims that the constants for binding of sweet substances to the protein parallel the intensity of their sweetness to humans. Preliminary reports of physicochemical studies of sweet-sensitivity protein : sucrose interactions (Dastoli, 1974) indicate that the protein undergoes conformational change upon binding of the sugar. Other experiments using ^{14}C-saccharin suggest that the sweet-sensitivity protein activity is membrane-bound. Dastoli's work certainly supports the postulate of specific proteins as taste receptors, but it must be more fully documented. Proof of a specific role will be difficult to obtain. Other workers (e.g. Lum and Henkin, 1976) are attempting to detect and purify the sweet receptor by systematic fractionation of membrane fragments obtained from taste buds. Sucrose binding is inhibited by phospholipase C treatment, suggesting that intact membrane is essential. Considering our knowledge of other receptors, I find it difficult to believe that the sweet taste receptor could be purified easily and in high yield from crude homogenates. It probably will turn out to be a membrane-bound integral protein or glycoprotein and present in relatively low amounts. Research on isolation of the other three taste receptors is even less well developed.

Other experimental observations on taste also support an allosteric protein–receptor mechanism. Several low molecular weight compounds cause enhancement of the taste of foods. Monosodium glutamate and disodium inosinate (IMP) are the best known of these. Perhaps they are allosteric effectors which stabilize a stimulant–receptor complex. Artichokes contain several compounds that cause water to taste sweet (Bartoshuk *et al.*, 1972). Gymnemic acid is a taste inhibitor; it causes table sugar crystals to have no sweet flavour but be only gritty. Three taste-active proteins have recently been isolated (Cagan, 1973). Miraculin is a glycoprotein isolated from the miracle fruit. It has no taste itself, but is a taste-modifying protein. When it is present on the tongue, it changes the taste of sour compounds to sweet. It has two separable effects—blocking of the sourness of compounds like HCl or citric acid, and sweetness provoked by the acidic compounds. Miraculin also attenuates the bitter taste of urea or phenylthiocarbamide. Henkin and Shallenberger (1970) have described two patients who are unable to recognize the sweet taste (a condition called aglycogeusia). To these patients, sucrose and D-fructose taste sour, D-glucose and D-galactose bitter, and D-xylose salty. Apparently they have a genetic deficiency of the sweet receptor. Miraculin does not provoke any sweet taste with the aglycogeusic patients, but does reduce the recognition of both sour and bitter compounds. Two other proteins from tropical plants, monellin and thaumatin, are intensely sweet (up to 10^5 times as sweet as sucrose on a molar basis). Neither is a glycoprotein. Because these macromolecules can elicit taste sensation, the specificity site of the receptor must be on the surface of the taste bud membrane. These two sweet-tasting proteins have a persistent effect, so are probably firmly but non-covalently bound to receptors. Highly radioactive monellin or thaumatin could possibly be used for detection and purification of sweet taste receptors. Studies of the binding of monellin or thaumatin to Dastoli's protein preparations could indicate whether or not the sweet-sensitive protein is the sweet receptor. Other techniques that have been applied to enzymes (e.g. affinity-labelling) could also be used for assay, purification and mechanism studies with the taste receptors and to aid in the determination of the functions of taste inhibitors and enhancing agents.

Let us turn our attention to the even more difficult task, elucidation of the mechanism of olfaction. In principle, the problem should be similar to the detection and isolation of hormone receptors, but it will be much more complicated. First, there must be many primary odours, possibly as many as 30 or more. Only a few of these primary odours have been identified. Because there are thousands of distinct odours, most compounds must stimulate more than one specific receptor. These compounds, which each possess several primary odours, are described as possessing *complex odours*. Recall that some substances have more than one primary taste (e.g. the principal taste of sugars is sweet, but the secondary tastes like the bitterness of D-glucose are discerned in aglycogeusia). Second, the odour stimulants are all volatile compounds, and the nasal receptors are relatively few. Stimulants activate the olfactory receptors at very low concentrations, and probably are weakly bound to the receptors. Because of these properties, the isolation of odour receptors will be extremely difficult. Third, the odour receptors are in nerve cells, not in separate receptor cells as are the taste receptors. Due to the complexity of the olfactory process, some workers still do not believe that there are *specific* odour receptors, and most experiments into odour reception are indirect. The technical difficulties are not insurmountable, but they have forced researchers into using ingenious approaches. Some of these experiments, described below, strongly suggest that olfactory receptors are compounds, probably proteins, having relatively high ligand specificity.

Amoore (1970) has presented the relevant background of the problem of odour quality, and offered a biochemical approach to its solution. In his book, there is a description of some of the anatomy and physiology of olfactory nerve responses. Briefly, after being drawn in through the nostrils, air reaches the smelling organs in the upper nasal cavity. Odour stimulants in the air must dissolve in a thin layer of mucus covering the olfactory epithelial cells. The sensing portion on the membrane of these cells seems to be a group of fine "hairs" with which the odour molecules must come into direct contact. A nerve fibre leads directly from each olfactory cell to the olfactory bulb of the brain, which in turn is connected to the higher centres of the brain where the conscious sensation of odour is perceived. Using specialized electrical equipment, impulses due to stimulation of receptor cells by odours can be recorded with single frog nerve units. This electrophysiological approach has not yet revealed any pattern of specificity to odour stimulants. However, Getchell and Gesteland (1972) have shown that N-ethylmaleimide can irreversibly block the single olfactory nerve fibre response of frogs to odour stimuli. They have found that normal electrical responses to ethyl n-butyrate are obtained if the receptor is protected from the sulphydryl reagent by prior exposure of the frog's nose to ethyl butyrate. Responses to other types of odours are still abolished under these conditions, though, indicating that the receptor for ethyl n-butyrate is highly selective. They have suggested that experiments using treatment with radioactive N-ethylmaleimide after protection might allow isolation and characterization of the substance reacting with the protecting stimulus (i.e. by our assumption, receptor protein).

Stereochemical Theory of Odour

Currently, there are only two prominent theories of odour quality that assume the existence of specific receptors. These are the stereochemical theory, elaborated by Amoore (1964), and the vibrational motion theory suggested by Wright (1966). Both theories attempt to relate chemical properties of compounds to their odours. In support of the vibrational theory, Wright has reported that compounds with similar smells have somewhat similar low energy vibrational modes, as measured by their absorption spectra in the far infrared region. These correlations have not always given clear support to the theory. In addition, compounds having identical spectra have been found to have different odours. Although inadequate in its original formulation, the stereochemical theory of odour quality has been more successful in explaining available data. I will explain this theory, to show how the specificity of each primary receptor may be determined—a project analogous to detecting the existence of a series of enzymes and determining their substrate specificity.

The stereochemical theory of odour explains what the primary odours are and classifies the odours in terms of the shape, size and reactive groups of the odorous molecules. The original expression of this theory listed only seven primary odours, with all other smells being complex odours. Amoore postulated that each primary odour is associated with a particular molecular configuration and size, and therefore odorants are bound to receptor sites having complementary spatial properties. The strongest evidence for the initial theory came

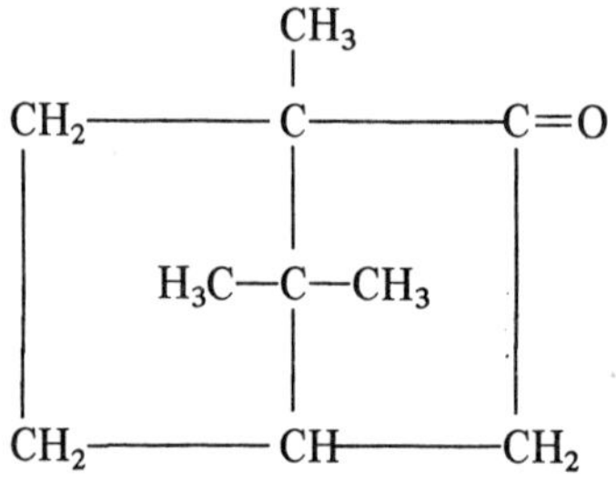

Camphor

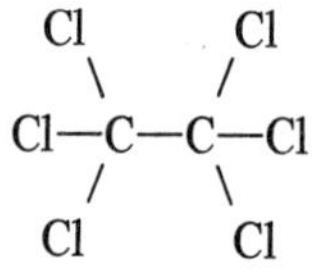

Hexachloroethane

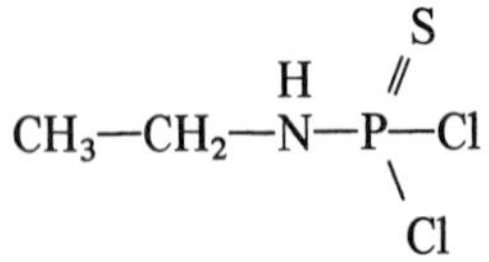

Thiophosphoric acid
dichloride ethylamide

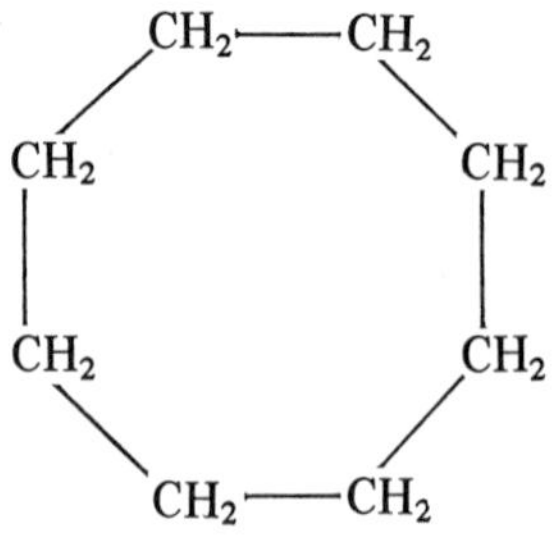

Cyclooctane

Table II. The seven primary odours proposed by Amoore (1964).

Primary Odour	Example	Familiar Description
Ether	1,2-Dichloroethane	Dry-cleaning fluid
Camphor	Camphor	Moth balls
Musk	15-Hydroxy-pentadecanoic acid lactone	Angelica root
Floral	Phenylethylmethylethyl-carbinol	Roses
Mint	Menthone	Peppermint
Pungent	Formic acid	Vinegar
Putrid	Dimethyl disulphide	Bad eggs

from examples of compounds having similar shape and odour but very different chemical reactivity. For example, camphor, hexachloro-ethane, thiophosphoric acid dichloride ethylamide and cyclooctane are chemically unrelated but all have a camphor-like odour. After inspecting space-filling molecular models of these four compounds, Amoore suggested that each could fit into an oval basin-shaped receptor site about 7.5 Å wide, 9 Å long and 4 Å deep. Amoore's seven primary odours (Table II) were ethereal, camphoraceous, musky, floral, minty, pungent and putrid. The latter two scents seem related to the electrophilic and nucleophilic properties of the odour molecules, respectively. Their receptors should have complementary charges. For the other five types of compounds, Amoore envisaged receptors of different sizes and shapes. He proposed that ether, camphor and musk compounds differed in molecular diameter or length, that floral compounds were kite-shaped, and minty compounds were wedge-shaped. The estimate of only seven primary odours has turned out to be too low, because two additional primary odour classes, sweaty and spermous, have since been found. Amoore (1974) now estimates that there are at least 20 to 30 primary odours.

Amoore uses a psychological assay method called the "sniff test" for determination of the odour qualities and intensities of compounds. This test utilizes two flasks of buffered aqueous solutions of purified odorants and three control flasks of buffer. All tests must be performed in an odour-free and temperature-controlled room. Judges are asked to describe the odours of test compounds by matching them to standard odours. The odour threshold of a compound, a quantitative measurement, is determined by asking a subject to correctly identify which two of the five flasks presented contain the odorous compound. Each judge is given a series of dilutions of the compound. The odour threshold is the lowest concentration of solute at which the subject properly identifies both experimental flasks in the tray of five flasks. Using this assay to determine a person's threshold for a particular odour, Amoore is testing for mutant individuals as an approach to the elucidation of primary odour classes.

A number of factors affect the odour sensitivity of individuals. Nasal sensitivity to odours is lowered temporarily by smoking prior to testing, by allergies or by a head cold. This type of effect must be controlled in the psychological experiments. A more dramatic and more useful lowering of sensitivity occurs in *specific anosmia* or "odour blindness." Specific anosmia is the loss of the ability to detect one family of compounds belonging to a primary odour class. This, like aglycogeusia, appears to be an inborn deficiency of one primary receptor. A person who has a specific anosmia can tolerate smell levels which would be extremely unpleasant to a normal person. Because many (about 20-30) types of specific anosmia are known, Amoore has estimated that there are 20 to 30 primary odours. By exploiting the existence of these anosmias, Amoore (1974) already has mapped two primary odours, sweaty and spermous. The specific anosmia approach, although slow, should make the identification of other primary odours possible.

About 2% of people cannot smell the disagreeable sweaty (dirty socks) odour of isovaleric acid. Their odour threshold for isovaleric

$$CH_3 \atop \diagdown \overset{O}{\underset{\parallel}{}}$$
$$CH-CH_2-C-OH$$
$$\diagup$$
$$CH_3$$
Isovaleric acid

$$CH_3 \atop \diagdown \overset{O}{\underset{\parallel}{}}$$
$$CH-C-OH$$
$$\diagup$$
$$CH_3$$
Isobutyric acid

acid is about one hundred times higher than that of a normal person. Some selected data are listed in Table III. Isobutyric acid also smells sweaty, but the degree of anosmia for this acid is not quite as great as for isovaleric acid, the best example of the sweaty odour. Isobutyr-

Table III. Difference in sensitivity to representative compounds in sweat-specific anosmia. Adapted from Amoore (1970, 1974).

Compound	Steps deficiency[a]	Threshold ratio[b]
Isovaleric acid	6.6	95
Isobutyric acid	5.6	50
sec-Valeric acid	4.9	30
tert-Valeric acid	4.7	26
n-Valeric acid	4.4	22
Isobutyraldehyde	2.2	4.6
Isobutyl alcohol	1.7	0.77

[a] Steps deficiency is the difference between the mean odour thresholds for specific anosmic subjects and normal subjects expressed in steps of binary dilution of the odorant solution.
[b] The threshold ratio is the same difference in odour sensitivity expressed as the ratio of elevation of the mean molar threshold concentration in the anosmia.

aldehyde and isobutyl alcohol are both similar in size and shape to isobutyric acid, but have unrelated odours. Therefore, the carboxylic acid group is essential for the smell of sweat. The sweaty receptor responds to many acids, but demonstrates some specificity for branched-chain acids.

The other primary odour to be worked out by the specific anosmia method is the spermous or semen-like odour. 1-Pyrroline has the most pronounced spermous odour. Its next higher homologue,

1–Pyrroline   1–Piperideine

1-piperideine, is the only other compound tested which has the spermous smell, so the spermous receptor must be quite specific.

Enantiomeric isomers can have distinctly different odours. Workers in two laboratories (Russell and Hills, 1971; Friedman and Miller, 1971) have proved that the difference in odour between S-(+)-carvone and R-(−)-carvone is not due to impurities. These isomers

S-(+)-Carvone   R-(−)-Carvone

were interconverted chemically, and prepared by independent synthesis. S-(+)-Carvone has the caraway smell of the caraway oil from which it can be prepared. R-(−)-Carvone has a strong spearmint odour. In a similar fashion, R-(+)-limonene smells like oranges and its S-(−)-isomer smells like lemons. The R(−) and S-(+)-isomers of amphetamine differ by smelling musty and faecal, respectively. There are many examples of enantiomers that have the same odour, but the demonstration that some enantiomeric isomers have different smells supports the stereochemical theory, and for these odours rules out the vibrational theory of odour.

The production of a series of compounds that can "almost instantaneously eliminate from a room perception of any malodours" of at least the pungent and putrid types has been announced (*Chemical and Engineering News,* 1975). These compounds do not mask or overpower bad odours, but counteract them by what is proposed as allosteric inhibition of the malodour receptors. These counteractants have been described as a "fresh air" smell. But our noses provide us with a defence, so that we can detect and avoid odorous materials! If a person who is allergic to tobacco smoke or musty books cannot smell the irritants, the odours of which can be suppressed by a trace of a counteractant, how can he avoid the irritants? Market use of this type of compound must be considered as carefully as that of any drug.

The three examples of complex controls—cyclic AMP-mediated hormonal regulation of metabolism, repression of protein biosynthesis by control of RNA polymerase and allosteric conformational changes in chemoreception—have many differences, but one similarity. All three processes are so complicated that they cannot yet be easily explained. We have reasonable concepts of their mechanisms, but the molecular proofs of these mechanisms will be obtained only after many technical difficulties have been surmounted.

REFERENCES

Adler, J. (1975). *A. Rev. Biochem.* **44**, 341-356.

Amoore, J. E. (1964). *Ann. N.Y. Acad. Sci.* **116**, 457-476.

Amoore, J. E. (1970). "Molecular Basis of Odor". C. C. Thomas, Springfield, Illinois, U.S.A.

Amoore, J. E. (1974). *Ann. N.Y. Acad. Sci.* **237**, 137-143.

Anderson, W. B., Schneider, A. B., Emmer, M., Perlman, R. L. and Pastan, I. (1971). *J. biol. Chem.* **246**, 5929-5937.

Appleman, M. M., Thompson, W. J. and Russell, T. R. (1973). *Adv. Cyclic Nucleotide Res.* **3**, 65-98.

Artz, S. W. and Broach, J. R. (1975). *Proc. natn. Acad. Sci. U.S.A.* **72**, 3453-3457.

Ashby, C. D. and Walsh, D. A. (1973). *J. biol. Chem.* **248**, 1255-1261.

Bahl, C. P., Wu, R., Itakura, K., Katagiri, N. and Narang, S. A. (1976). *Proc. natn. Acad. Sci. U.S.A.* **73**, 91-94.

Barkley, M. D., Riggs, A. D., Jobe, A. and Bourgeois, S. (1975). *Biochemistry* **14**, 1700-1712.

Bartoshuk, L. M., Lee, C.-H. and Scarpellino, R. (1972). *Science, N.Y.* **178**, 988-990.

Beavo, J. A., Bechtel, P. J. and Krebs, E. G. (1975). *Adv. Cyclic Nucleotide Res.* **5**, 241-251.

Bell, R. M., Parsons, S. M., Dubravac, S. A., Redfield, A. G. and Koshland, D. E., Jr. (1974). *J. biol. Chem.* **249**, 4110-4118.

Beyreuther, K., Adler, K., Fanning, E., Murray, C., Klemm, A. and Geisler, N. (1975). *Eur. J. Biochem.* **59**, 491-509.

Birnbaumer, L. and Rodbell, M. (1969). *J. biol. Chem.* **244**, 3477-3482.

Blasi, F., Barton, R. W., Kovach, J. S. and Goldberger, R. F. (1971). *J. Bact.* **106**, 508-513.

Blasi, F., Bruni, C. B., Avitabile, A., Deeley, R. G., Goldberger, R. F. and Meyers, M. (1973). *Proc. natn. Acad. Sci. U.S.A.* **70**, 2692-2696.

Bloch, A. (1974). *Biochem. biophys. Res. Commun.* **58**, 652-659.

Bloch, A. (1975). *Biochem. biophys. Res. Commun.* **64**, 210-218.

Brandt, H., Capulong, Z. L. and Lee, E. Y. C. (1975). *J. biol. Chem.* **250**, 8038-8044.

Brenner, M. and Ames, B. N. (1971). *In* "Metabolic Pathways" (H. J. Vogel, ed.), Third edition, Vol. 5, pp. 349-387. Academic Press, New York.

Buller, R. E. and O'Malley, B. W. (1976). *Biochem. Pharmac.* **25**, 1-12.

Butcher, R. W. and Sutherland, E. W. (1962). *J. biol. Chem.* **237**, 1244-1250.

Cagan, R. H. (1973). *Science, N.Y.* **181**, 32-35.

Chemical and Engineering News (1975). Oct. 13, pp. 24-25.

Cheung, W. Y. (1971). *J. biol. Chem.* **246**, 2859-2869.

Cohen, P. (1973). *Eur. J. Biochem.* **34**, 1-14.

Cohen, P. and Antoniw, J. F. (1973). *FEBS Letters* **34**, 43-47.

Cohen, P., Antoniw, J. F., Nimmo, H. G. and Proud, C. G. (1975a). *Biochem. Soc. Trans.* **3**, 849-854.

Cohen, P., Watson, D. C. and Dixon, G. H. (1975b). *Eur. J. Biochem.* **51**, 79-92.
Constantopoulos, A. and Najjar, V. A. (1973). *Biochem. biophys. Res. Commun.* **53**, 794-799.
Corbin, J. D., Keely, S. L., Soderling, T. R. and Park, C. R. (1975). *Adv. Cyclic Nucleotide Res.* **5**, 265-279.
Cuatrecasas, P. (1974). *A. Rev. Biochem.* **43**, 169-214.
Cuatrecasas, P., Jacobs, S. and Bennett, V. (1975). *Proc. natn. Acad. Sci. U.S.A.* **72**, 1739-1743.
Daile, P., Carnegie, P. R. and Young, J. D. (1975). *Nature, Lond.* **257**, 416-418.
Dastoli, F. R. (1974). *Life Sciences* **14**, 1417-1426.
Dastoli, F. R. and Price, S. (1966). *Science, N.Y.* **154**, 905-907.
Dastoli, F. R., Lopiekes, D. V. and Price, S. (1968). *Biochemistry* **7**, 1160-1164.
De Lorenzo, F. and Ames, B. N. (1970). *J. biol. Chem.* **245**, 1710-1716.
Dickson, R. C., Abelson, J., Barnes, W. M. and Reznikoff, W. S. (1975). *Science, N.Y.* **187**, 27-35.
Drummond, G. I. and Duncan, L. (1970). *J. biol. Chem.* **245**, 976-983.
Erlichman, J., Rosenfeld, R. and Rosen, O. M. (1974). *J. biol. Chem.* **249**, 5000-5003.
Foor, F., Janssen, K. A. and Magasanik, B. (1975). *Proc. natn. Acad. Sci. U.S.A.* **72**, 4844-4848.
Friedman, D. L. and Larner, J. (1963). *Biochemistry* **2**, 669-675.
Friedman, L. and Miller, J. G. (1971). *Science, N.Y.* **172**, 1044-1046.
Getchell, M. L. and Gesteland, R. C. (1972). *Proc. natn. Acad. Sci. U.S.A.* **69**, 1494-1498.
Gilbert, W. and Maxam, A. (1973). *Proc. natn. Acad. Sci. U.S.A.* **70**, 3581-3584.
Gilbert, W. and Müller-Hill, B. (1966). *Proc. natn. Acad. Sci. U.S.A.* **56**, 1891-1898.
Gill, G. N. and Garren, L. D. (1970). *Biochem. biophys. Res. Commun.* **39**, 335-343.
Gill, G. N. and Kanstein, C. B. (1975). *Biochem. biophys. Res. Commun.* **63**, 1113-1122.
Goldberg, N. D., O'Dea, R. F. and Haddox, M. K. (1973). *Adv. Cyclic Nucleotide Res.* **3**, 155-223.
Goldberg, N. D., Haddox, M. K., Nicol, S. E., Glass, D. B., Sanford, C. H., Kuehl, F. A., Jr. and Estensen, R. (1975). *Adv. Cyclic Nucleotide Res.* **5**, 307-330.
Goldberger, R. F. (1974). *Science, N.Y.* **183**, 810-816.
Goldberger, R. F. and Kovach, J. S. (1972). *Curr. Topics cell. Reguln* **5**, 285-308.
Graves, D. J. and Wang, J. H. (1972). *In* "The Enzymes" (P. D. Boyer, ed.), Third edition, Vol. 7, pp. 435-482. Academic Press, New York.
Hayakawa, T., Perkins, J. and Krebs, E. G. (1973). *Biochemistry* **12**, 574-580.
Henkin, R. I. and Shallenberger, R. S. (1970). *Nature, Lond.* **227**, 965-966.
Hjelmquist, G., Andersson, J., Edlund, B. and Engström, L. (1974). *Biochem. biophys. Res. Commun.* **61**, 509-513.
Hofmann, F., Beavo, J. A., Bechtel, P. J. and Krebs, E. G. (1975). *J. biol. Chem.* **250**, 7795-7801.

Huang, F. L. and Glinsmann, W. H. (1975). *Proc. natn. Acad. Sci. U.S.A.* **72**, 3004-3008.

Huang, K.-P., Huang, F. L., Glinsmann, W. H. and Robinson, J. C. (1975). *Biochem. biophys. Res. Commun.* **65**, 1163-1169.

Illiano, G., Tell, G. P. E., Siegel, M. I. and Cuatrecasas, P. (1973). *Proc. natn. Acad. Sci. U.S.A.* **70**, 2443-2447.

Jacob, F. and Monod, J. (1961). *J. molec. Biol.* **3**, 318-356.

Kalish, M. I., Pineyro, M. A., Cooper, B. and Gregerman, R. I. (1974). *Biochem. biophys. Res. Commun.* **61**, 731-737.

Kasai, T. (1974). *Nature, Lond.* **249**, 523-527.

Kemp, B. E., Bylund, D. B., Huang, T.-S. and Krebs, E. G. (1975). *Proc. natn. Acad. Sci. U.S.A.* **72**, 3448-3452.

Kolb, H. J., Renner, R., Hepp, K. D., Weiss, L. and Wieland, O. H. (1975). *Proc. natn. Acad. Sci. U.S.A.* **72**, 248-252.

Krebs, E. G. (1972). *Curr. Topics cell. Reguln* **5**, 99-133.

Krebs, E. G. and Fischer, E. H. (1962). *Adv. Enzymol.* **24**, 263-290.

Kuo, J. F. (1974). *Proc. natn. Acad. Sci. U.S.A.* **71**, 4037-4041.

Kuo, J. F. and Greengard, P. (1969). *Proc. natn. Acad. Sci. U.S.A.* **64**, 1349-1355.

Langan, T. A. (1973). *Adv. Cyclic Nucleotide Res.* **3**, 99-153.

Layne, P. P. and Najjar, V. (1975). *J. biol. Chem.* **250**, 966-972.

Layne, P., Constantopoulos, A., Judge, J. F. X., Rauner, R. and Najjar, V. A. (1973). *Biochem. biophys. Res. Commun.* **53**, 800-805.

Lefkowitz, R. J. (1974). *Biochem. biophys. Res. Commun.* **58**, 1110-1118.

Lefkowitz, R. J. (1975). *Biochem. Pharmacol.* **24**, 1651-1658.

Lefkowitz, R. J. and Caron, M. G. (1975). *J. biol. Chem.* **250**, 4418-4422.

Leloir, L. F. and Cardini, C. E. (1962). *In* "The Enzymes" (Boyer, P. D., Lardy, H. and Myrbäck, K., eds), Second edition, Vol. 6, pp. 317-326. Academic Press, New York.

Levey, G. S. (1971). *J. biol. Chem.* **246**, 7405-7407.

Levey, G. S., Fletcher, M. A. and Klein, I. (1975). *Adv. Cyclic Nucleotide Res.* **5**, 53-65.

Lin, Y. M., Liu, Y. P. and Cheung, W. Y. (1974). *J. biol. Chem.* **249**, 4943-4954.

Londos, C., Salomon, Y., Lin, M. C., Harwood, J. P., Schramm, M., Wolff, J. and Rodbell, M. (1974). *Proc. natn. Acad. Sci. U.S.A.* **71**, 3087-3090.

Lum, C. K. L. and Henkin, R. I. (1976). *Biochim. biophys. Acta* **421**, 380-394.

Maizels, N. M. (1973). *Proc. natn. Acad. Sci. U.S.A.* **70**, 3585-3589.

Majors, J. (1975). *Proc. natn. Acad. Sci. U.S.A.* **72**, 4394-4398.

Makman, R. S. and Sutherland, E. W. (1965). *J. biol. Chem.* **240**, 1309-1314.

Meyer, F., Heilmeyer, L. M. G., Jr., Haschke, R. H. and Fischer, E. H. (1970). *J. biol. Chem.* **245**, 6642-6648.

Meyers, M., Levinthal, M. and Goldberger, R. F. (1975). *J. Bact.* **124**, 1227-1235.

Mitra, S., Zubay, G. and Landy, A. (1975). *Biochem. biophys. Res. Commun.* **67**, 857-863.

Ohshima, Y., Mizokoshi, T. and Horiuchi, T. (1974). *J. molec. Biol.* **89**, 127-136.

Ohshima, Y., Horiuchi, T. and Yanagida, M. (1975). *J. molec. Biol.* **91**, 515-519.
Parsons, S. M. and Koshland, D. E., Jr. (1974). *J. biol. Chem.* **249**, 4104-4109.
Pastan, I. and Perlman, R. L. (1971). *Nature, New Biology* **229**, 5-7 and 32.
Perkins, J. P. (1973). *Adv. Cyclic Nucleotide Res.* **3**, 1-64.
Peterkofsky, A. and Gazdar, C. (1974). *Proc. natn. Acad. Sci. U.S.A.* **71**, 2324-2328.
Peterkofsky, A. and Gazdar, C. (1975). *Proc. natn. Acad. Sci. U.S.A.* **72**, 2920-2924.
Platt, T., Files, J. G. and Weber, K. (1973). *J. biol. Chem.* **248**, 110-121.
Pohl, S. L., Krans, H. M. J., Kozyreff, V., Birnbaumer, L. and Rodbell, M. (1971). *J. biol. Chem.* **246**, 4447-4454.
Reimann, E. M., Brostrom, C. O., Corbin, J. D., King, C. A. and Krebs, E. G. (1971). *Biochem. biophys. Res. Commun.* **42**, 187-194.
Reiness, G., Yang, H.-L., Zubay, G. and Cashel, M. (1975). *Proc. natn. Acad. Sci. U.S.A.* **72**, 2881-2885.
Riggs, A. D. and Bourgeois, S. (1968). *J. molec. Biol.* **34**, 361-364.
Robison, G. A., Butcher, R. W. and Sutherland, E. W. (1971). "Cyclic AMP". Academic Press, New York.
Rodbell, M. (1973). *Fedn Proc.* **32**, 1854-1858.
Rodbell, M., Birnbaumer, L., Pohl, S. L. and Krans, H. M. J. (1971). *J. biol. Chem.* **246**, 1877-1882.
Rosen, O. M. and Erlichman, J. (1975). *J. biol. Chem.* **250**, 7788-7794.
Rosen, O. M. and Rosen, S. M. (1969). *Archs Biochem. Biophys.* **131**, 449-456.
Rubalcava, B. and Rodbell, M. (1973). *J. biol. Chem.* **248**, 3831-3837.
Russell, G. F. and Hills, J. I. (1971). *Science, N.Y.* **172**, 1043-1044.
Russell, T. R., Terasaki, W. L. and Appleman, M. M. (1973). *J. biol. Chem.* **248**, 1334-1340.
Ryan, J. and Storm, D. R. (1974). *Biochem. biophys. Res. Commun.* **60**, 304-311.
Schlender, K. K., Wei, S. H. and Villar-Palasi, C. (1969). *Biochim. biophys. Acta* **191**, 272-278.
Schorr, I., Rathnam, P., Saxena, B. B. and Ney, R. L. (1971). *J. biol. Chem.* **246**, 5806-5811.
Scott, J. F., Roth, J. R. and Artz, S. W. (1975). *Proc. natn. Acad. Sci. U.S.A.* **72**, 5021-5025.
Soderling, T. R. (1975). *J. biol. Chem.* **250**, 5407-5412.
Soderling, T. R., Hickenbottom, J. P., Reimann, E. M., Hunkeler, F. L., Walsh, D. A. and Krebs, E. G. (1970). *J. biol. Chem.* **245**, 6317-6328.
Stalmans, W. and Hers, H. G. (1973). *In* "The Enzymes" (P. D. Boyer, ed.), Third edition, Vol. 9, pp. 309-361. Academic Press, New York.
Steitz, T. A., Richmond, T. J., Wise, D. and Engelman, D. (1974). *Proc. natn. Acad. Sci. U.S.A.* **71**, 593-597.
Stephens, J. C., Artz, S. W. and Ames, B. N. (1975). *Proc. natn. Acad. Sci. U.S.A.* **72**, 4389-4393.
Stewart, R. A., Carrico, C. K., Webster, R. L. and Steinhardt, R. G., Jr. (1971). *Nature, Lond.* **234**, 220.
Sutherland, E. W. and Rall, T. W. (1958). *J. biol. Chem.* **232**, 1077-1091.

Takai, K., Kurashina, Y., Suzuki-Hori, C., Okamoto, H. and Hayaishi, O. (1974). *J. biol. Chem.* **249**, 1965-1972.

Teo, T. S. and Wang, J. H. (1973). *J. biol. Chem.* **248**, 5950-5955.

Teo, T. S., Wang, T. H. and Wang, J. H. (1973). *J. biol. Chem.* **248**, 588-595.

Tomkins, G. M. (1975). *Science, N.Y.* **189**, 760-763.

Vauquelin, G., Lacombe, M.-L., Hanoune, J. and Strosberg, A. D. (1975). *Biochem. biophys. Res. Commun.* **64**, 1076-1082.

Voll, M. J., Appella, E. and Martin, R. G. (1967). *J. biol. Chem.* **242**, 1760-1767.

Walsh, D. A. and Krebs, E. G. (1973). *In* "The Enzymes" (P. D. Boyer, ed.), Third edition, Vol. 9, pp. 555-581. Academic Press, New York.

Walsh, D. A., Perkins, J. P. and Krebs, E. G. (1968). *J. biol. Chem.* **243**, 3763-3765.

Walsh, D. A., Ashby, C. D., Gonzalez, C., Calkins, D., Fischer, E. H. and Krebs, E. G. (1971). *J. biol. Chem.* **246**, 1977-1985.

Watson, J. D. (1976). "Molecular Biology of the Gene", Third edition, Chapter 14. W. A. Benjamin, Menlo Park.

Witt, J. J. and Roskoski, R., Jr. (1975). *Biochemistry* **14**, 4503-4507.

Wright, R. H. (1966). *Nature, Lond.* **209**, 551-554 and 571-573.

Yeaman, S. J. and Cohen, P. (1975). *Eur. J. Biochem.* **51**, 93-104.

Yoshida, A. (1970). *J. molec. Biol.* **52**, 483-490.

Zieve, F. J. and Glinsmann, W. H. (1973). *Biochem. biophys. Res. Commun.* **50**, 872-878.

Zubay, G. and Chambers, D. A. (1971). *In* "Metabolic Pathways" (H. J. Vogel, ed.), Third edition, Vol. 5, pp. 297-347. Academic Press, New York.

Zubay, G., Schwartz, D. and Beckwith, J. (1970). *Proc. natn. Acad. Sci. U.S.A.* **66**, 104-110.

17 Concluding Remarks

Inquisitiveness is expressed to varying degrees in all people. I have always been curious about the operation of machines and have been able to spend the past 20 years learning about the exquisite machines that support life—enzymes. In order to understand how a device operates, we must find out its composition, learn how the components interact and see how it is controlled. This statement describes not only the approach of this book but also the evolution of research on enzymes. The material in this monograph and in the references cited demonstrates the goals, achievements and some of the future directions of enzymology. The examples of enzymes discussed in the text also show how enzyme chemistry impinges on all areas of biology.

We have seen that enzymes are proteins, often working in association with cofactors. The cofactors are either organic molecules or metals. At one extreme, there are the simple enzymes that do not require cofactors, and at the other extreme are oligomeric enzymes that require several cofactors including protein molecules as coenzymes.

How do enzyme proteins cause reactions to proceed so rapidly? Their catalytic mechanisms involve two essential and inter-related phenomena, the suitable binding of substrates and the chemical catalysis. Knowles and Gutfreund (1974) have described these two phases of enzyme catalysis as *complexation* and *reaction*. They correctly assert that the formation of an enzyme–substrate complex (approaching an enzyme-transition state complex) provides a major kinetic advantage, probably as great in magnitude as the actual catalytic event in which bond rearrangement occurs. The reaction chemistry now has been elucidated for many catalytic centres. For example, side chains of amino acid residues and prosthetic groups that participate in catalysis have been identified for many enzymes, and reasonable mechanisms described for their functions. The onus is now on biochemists to describe more fully the complexation effects or structural environments that protein molecules provide for their substrates.

One obvious role for proteins in catalysis is the specificity they confer on chemical reactions. In providing specificity, protein catalysts differ from simple chemical catalysts. A second role is the regulation of catalysis, controlled by the conformational changes of certain regulated enzyme molecules. These changes in conformation are caused by binding, whether by non-covalent or covalent bonds, of allosteric ligands. Effectors range in size from low molecular weight metabolites or phosphoryl groups to protein modifier molecules. Metabolism seldom is regulated by control of just one enzyme in each pathway. Controls probably are exerted at several points and in many ways along most pathways (Srere, 1969). Aside from learning which enzymes can exert controls, and what conditions alter the activities of these enzymes, we must also determine the amounts of all enzymes and ligands in cells and in the appropriate intracellular compartments. Integration of metabolism is managed by variability in the concentrations of substrates and effectors, and by control of the rates of enzyme synthesis and degradation. Quantitative analysis of the processes that govern the maintenance and reproduction of cells, especially animal cells, will be a formidable task.

The future for enzymology is auspicious. Applications of enzyme chemistry, based on present knowledge, are being made to many endeavours. Answers will be available soon to many questions, extending the usefulness of present accomplishments. The major problems are delineated, and their answers are appearing. However, research must be carefully executed. As a prime example, the mechanism of energy conservation during substrate oxidation in mitochondria is being elucidated slowly and surely (p. 507). For decades, we have suspected that ATP formation is catalyzed by a latent ATPase, but the time between suspicion and proof can be long. I have found that the phenomenon of oxidative phosphorylation is the most difficult biochemical concept to explain to students of introductory biochemistry. The reasons for their difficulties in comprehension of it are obvious; it is a very complex process, and they cannot be told a simple, exact mechanism for ATP formation. Until the latter reaction is explained fully, I only can present them with the known facts and the conclusions based upon them, and try to avoid unwarranted simplicity. As in this book, I do not hide unsolved problems or controversies. Racker (1975) has summarized the technical difficulties of mitochondrial energy conservation studies and the personal approaches to these difficulties taken by different investigators. He recalls that in 1963 "anyone who was not thoroughly confused simply did not understand the situation." There had been "an unfortunate medley of honest errors, wishful thinking and even alleged fraud" associated with claims of several high energy intermediates in oxidative phosphorylation. As I noted in Chapter 14, the high energy intermediate hypothesis has all but been abandoned. Then came Mitchell's original suggestion of a pH difference and a membrane potential. Good scientists are not like Stephen Leacock's hero Lord Ronald who "flung himself from the room, flung himself upon his horse and rode madly off in all directions." They follow logical, though difficult and often differing, directions. Racker considers that the basic principle of coupling of oxidation and phosphorylation has been solved (a proton pump acting in reverse to form ATP, using the latent ATPase), and that research is well on the way to showing *how* the ion pump reverses (ion binding producing energy). This would be a welding of the Mitchell and Boyer theories, and would be typical of the resolution of many difficult biochemical questions. A geographic analogy is applicable to many scientific endeavours. Consider two people leaving the North Pole and travelling south by different but direct southerly routes. They are farthest from each other at the equator (about ten years ago in oxidative phosphorylation), but they continuously get closer as they approach the South Pole, or the scientific truth. We do not know yet what they will find there, but fortunately many projects, like oxidative phosphorylation, are approaching their south polar regions.

Support for basic research in many countries has always been relatively modest, especially in relation to the value of the results. I started this book in a period of decreasing research resources, and am finishing it at a time of even more stringent support for basic research. The accomplishment of professional scientific research costs money and time. The rewards of research, however, greatly outweigh the original costs! I hope that all my colleagues in science will receive the assistance that they deserve. My greatest disappointment would be to see the young scientists, with their well-honed enthusiasm and excellent educations, deprived of a chance to test their skills against the problems the world has supplied them. The talents of the world's scientific community must not be wasted.

REFERENCES

Knowles, J. R. and Gutfreund, H. (1974). *In* "Chemistry of Macromolecules" (H. Gutfreund, ed.), pp. 375-397. Butterworths, London.
Racker, E. (1975). *Biochem. Soc. Trans.* **3**, 785-802.
Srere, P. A. (1969). *Biochem. Med.* **3**, 61-72.

Author Index

A number in italics indicates the page on which the full reference appears for an author whose name does not appear at the point of citation in the text. This will enable the reader to locate most rapidly all references to scientists.

A

Abbott, E. H., 230
Abbott, S. J., 312, *346*
Abdulaev, N. G., 236, *271*
Abeles, R. H., 140, 222, 223, 224, *267, 270, 272,* 339, 340, 343, 344, *346, 347,* 381, 382, 383, *395, 396.*
Abell, C. W., 534
Abelson, J., 571, 572, *590*
Abernethy, J. L., 364, *397*
Ackers, G. K., 417
Adair, G. S., 389
Adair, L. B., 481
Adams, M. J., 214, 215, *271*
Adler, J., 579
Adler, K., 568, *589*
Adler, S. P., 59, *79,* 459
Adman, E. T., 277, 278, 279, *306*
Agrawal, K. C., 531, *537*
Agro, A. F., 363, *397*
Ahmad, F., 296, *306*
Aitken, D. M., 482
Åkeson, Å., 140, *141,* 214, *268*
Akhtar, M., 251, 253
Albert, A., 194, *200*
Albert, Adrien, 513
Alberts, A. W., 294, 299, *307*
Alberty, R. A., 129, 131, 135, 136, 137

Aldanova, N. A., 236, *271*
Alden, R. A., 279, *306,* 323, 324, 325, *347*
Alderton, G., 172
Allen, J. C., 497, *511*
Allewell, N. M., 190, *202*
Allison, W. S., 214, 215, *267, 272*
Amano, T., 183, *201*
Ambler, R. P., 19
Ames, B. N., 67, 572, 575, 578, *592*
Amoore, J. E., 583, 585, 586, 587
Anderson, B., 303
Anderson, B. M., 211
Anderson, M. L., 280
Anderson, W. B., 457, 570
Andersson, J., 555, *590*
Andreesen, J. R., 292
Andrews, P., 67
Anfinsen, C. B., 9, 24, 58, 190, *200*
Ansfield, F. J., 524
Antoniw, J. F., 561, 564, *589*
Anwar, R. A., 405
Aparicio, P. J., 282, *307*
Apella, E., 574, *593*
Appleman, M. M., 553, *592*
Archer, M. C., 262, 263
Arnon, D. I., 282, *307, 308*
Arnone, A., 390
Artz, S. W., 577, 578, *592*

Subject Index

A

Acetoacetate decarboxylase, 313, 337-339

Acetylcholinesterase, 101, 314

Acetyl CoA carboxylase, 293-296, 472

Aconitase, 29, 30

Acid-base catalysis, 92-100, 137, 145, 172, 180, 185, 193, 196-199, 212, 230, 242, 263, 312, 368, 374

Activator-attenuator model, 577

Active site, 4, 27-35, 152, 176-180, 191, 203, 289

Active site-directed irreversible inhibitors, 34, 197, 222, 316-317, 320, 521-523, 526

Acyl carrier protein, 293, 298-302, 473-475

Adenosyl methionine decarboxylase, 339

Adenyl kinase, 24

Adenylate cyclase, 539-553, 569

Adenylosuccinase, 30

Adenylosuccinate synthetase, 529

Adenylyltransferase, 455-461

Affinity chromatography, 57-63, 77, 165, 240-241, 447, 465, 552, 556

Agarose, *see* Gel filtration and Affinity chromatography

Alcohol dehydrogenase, 118, 137-141, 208, 214, 353, 365

Aldehyde oxidase, 374

Aldolase, 410, 413, 471

Alkaline phosphatase, 44, 46, 313, 335-337, 414, 531

Allopurinol, 532

Allosteric control, 401-431, 432-469, 538-593

Allosteric effector, 386, 394, 404

Allosteric site, 402, 404

Allosteric transition, 404, 421

Alumina C_γ gel, *see* Gel adsorption

Amethopterin, 518-521

Amino acid
analysis, 16-17, 173-174
oxidase, 28, 223

Ammonium sulphate fractionation, 40, 41, 51-52, 53, 62, 63, 65, 73-76, 187, 276, 283

Amphiphile, 484

α-Amylase, 356

Antimetabolite, 514-516, 518, 528

Apoenzyme, 3, 203, 231, 362, 363, 426

Arylsulphatase, 162-163

Asparaginase, 533-534

Aspartate aminotransferase, 233-237

Aspartate transcarbamylase, 13, 44, 402, 409, 435-445, 479-483

Aspartokinase, 433-435

Association, 13, 248, 356, 407, 409, 476

ATP phosphoribosyltransferase, 572-578

ATPase
membrane, 543
mitochondrial, 502, 506-509

ATPase–$(Na^+ + K^+)$–ATPase, 332, 496-501, 506, 508

Attenuator locus, 577

Autogenous regulation, 576-579

X

Xanthine oxidase, 363, 374, 375, 532
X-ray crystallography, 13, 21-24, 27,
140, 176-179, 190-191, 194-195,
208, 213-216, 277-279, 283-285,
287, 289, 319-326, 330-331, 364,
366-368, 381, 387, 388-393, 412,
426, 441-442

Z

Zero-order reaction, 110, 114
Zinc (Zn^{2+}), 138-141, 213, 336, 362,
363, 365-368, 442
Zymogen, 3, 62, 314-316